ECONOMIC RISKS OF CLIMATE CHANGE

ECONOMIC RISKS OF CLIMATE CHANGE

AN AMERICAN PROSPECTUS

TREVOR HOUSER ▪ SOLOMON HSIANG ▪ ROBERT KOPP
KATE LARSEN ▪ MICHAEL DELGADO ▪ AMIR JINA
MICHAEL MASTRANDREA ▪ SHASHANK MOHAN
ROBERT MUIR-WOOD ▪ D. J. RASMUSSEN
JAMES RISING ▪ PAUL WILSON

*With contributions from Karen Fisher-Vanden,
Michael Greenstone, Geoffrey Heal, Michael Oppenheimer,
Nicholas Stern, and Bob Ward*

AND A FOREWORD BY MICHAEL R. BLOOMBERG,
HENRY M. PAULSON JR., AND THOMAS F. STEYER

Columbia University Press New York

Columbia University Press
Publishers Since 1893
New York Chichester, West Sussex
cup.columbia.edu

Library of Congress Cataloging-in-Publication Data

Houser, Trevor.
Economic risks of climate change : an American prospectus /
Trevor Houser, Solomon Hsiang, Kate Larsen, Robert Kopp,
D. J. Rasmussen, Michael Mastrandrea, Robert Muir-Wood,
Paul Wilson, Amir Jina, James Rising, Michael Delgado,
Shashank Mohan With contributions from Karen Fisher-Vanden,
Michael Greenstone, Geoffrey Heal, Michael Oppenheimer,
Nicholas Stern, and Bob Ward And a foreword by Michael R. Bloomberg,
Henry Paulson, and Tom Steyer.
pages cm
Includes bibliographical references and index.
ISBN 978-0-231-17456-5 (cloth : alk. paper) — ISBN 978-0-231-53955-5 (e-book)
1. Climatic changes—Economic aspects—United States.
2. Climatic changes—Risk management—United States. I. Title.

QC903.2.U6 H68 2015
363.738'74—dc23
2014045703

Columbia University Press books are printed on permanent
and durable acid-free paper.
This book is printed on paper with recycled content.
Printed in the United States of America
c 10 9 8 7 6 5 4 3 2
Cover Design: Noah Arlow

CONTENTS

TECHNICAL APPENDIXES

FOREWORD

MICHAEL R. BLOOMBERG, HENRY M. PAULSON JR., AND THOMAS F. STEYER

COCHAIRS, RISKY BUSINESS PROJECT

HOW much economic risk does the United States face from climate change? The answer has profound implications for the future of our economy and the American way of life. But until recently there was no systematic, analytically rigorous effort to identify, measure, and communicate these risks.

It was the looming, unknown scale of these risks that led us to launch the Risky Business Project in summer 2013 and to commission the research that became the *American Climate Prospectus* report, published here in its entirety as *Economic Risks of Climate Change: An American Prospectus*. Our aim is to quantify the economic risks of climate change to the U.S. economy and then communicate these risks to the business sector.

In applying a standard risk-assessment approach to future climate impacts, this research provides specific, local, and actionable data for businesses and investors in both the public and private sectors. We hope its findings help spur an active, rigorous conversation among economists, business executives, investors, and public-policy makers about how best to manage these risks, including taking prudent action to prevent them from spiraling out of control.

Over the years, the scientific data have made it increasingly clear that a changing climate, driven by carbon pollution from human activities, will lead to overall global warming. These rising temperatures in turn lead to specific and measurable impacts such as sea-level rise, melting ice and glaciers, and more observable weather events such as droughts, wildfires, coastal and inland floods, and storms. But, until recently, scant analytical work has been done to connect these broad climate changes to the daily workings of our economy.

In our view, the significant and persistent gap between the fields of climate science and economics makes businesses, investors, and public-sector decision makers dangerously vulnerable to long-term and unmanageable risks. How can we make wise financial decisions without understanding our exposure to such risks as severe floods or prolonged drought or storm surge? How can we plan for and build new, more resilient infrastructure and manage our limited public resources responsibly without taking into account the probable changes to our coastlines, our agricultural lands, and our major population centers?

These were the questions that led to the formation of the Risky Business Project. We knew from the outset that, to be effective, the project must be grounded in the same sort of rigorous analytical framework typically used by investors and business leaders in other areas of risk management. The American business community has been slow to assess and address climate risk in part because of a lack of actionable data. Without these data, businesses cannot create risk-assessment models that effectively capture the potential impact of climate change. So it's no surprise that most corporate risk committees, even in industries and sectors at significant risk of climate-driven disruption, do not consistently include climate risk in their disclosures to investors or overall management priorities.

The success of our efforts was dependent on our ability to point business leaders toward exactly the kind of pathbreaking analysis contained in this book. To be credible, the research had to be methodologically unassailable and strictly independent. To be useful, the data it produced had to be detailed, relevant, and highly localized—what climate modelers call "downscaled"—in a way that would allow businesses to incorporate it into their existing risk-management protocols and strategies.

The Risky Business Project and this book are critical first steps toward this goal. The study does not tackle the entire U.S. economy but instead focuses on a few important sectors (agriculture, energy demand, coastal property, health, and labor). In examining how climate change will introduce new risks to these sectors, this research builds on the best available climate science and econometric research, reviewed by a panel of world-class scholars.

This work is also unusual—and unusually relevant to the business sector—in its level of detail and specificity to particular geographic regions. In the following chapters, readers will find a nearly unprecedented level of geographic granularity. Probable climate impacts have been modeled down to the county level, which is the scale at which many business decisions—such as crop planting and harvesting and real estate development—are actually made. This level of geographic detail also underscores the broad regional disparities we can expect from climate change. In a country as large and diverse as the United States, not all states or even counties will face the same type or level of risk. Economy-wide studies, focused on Gross Domestic Product impact or national productivity, completely mask these disparities.

When we undertook this project, it was clear that simply quantifying the economic risks of climate change would not be enough. The data needed to take a form that was meaningful within companies' existing risk-assessment frameworks. Thus, while this report is in many ways novel and groundbreaking, it's also notable in that it makes use of the same risk-assessment approach that businesses and investors use on a daily basis.

In the wake of Hurricane Sandy, New York City created a comprehensive resilience blueprint that measures climate risk across all major vulnerable areas, from the power grid to hospitals to the coastline. We should not wait for a national disaster to create the same blueprint for the U.S. economy as a whole. We hope that this analysis is useful not only for the data it provides but also as a framework for a more effective dialogue among scientists, economists, and the business community—one that will provide decision makers with the information they need to decide how much climate risk they are comfortable taking on.

As we said in the October 2013 *Washington Post* op-ed that launched this entire effort: We believe the Risky Business Project and this book bring a critical missing piece to the national dialogue about climate change while helping business leaders and investors make smart, well-informed, financially responsible decisions. Ignoring the potential costs could be catastrophic—and that's a risk we cannot afford to take.

PREFACE

ROBERT KOPP, SOLOMON HSIANG, KATE LARSEN, AND TREVOR HOUSER

HUMAN civilization is reshaping Earth's surface, atmosphere, oceans—and climate. In May 2013, at the peak of its seasonal cycle, the concentration of carbon dioxide (CO_2) in the atmosphere spiked above 400 parts per million (ppm) for the first time in more than 800,000 years; within the next couple of years, it will exceed 400 ppm year-round. This elevated CO_2 concentration is the result of human activities—primarily the combustion of coal, oil, and natural gas and, secondarily, deforestation. The physics linking increased concentrations of greenhouse gases like CO_2 to higher global average temperatures has been known since the work of Joseph Fourier and Svante Arrhenius in the nineteenth century. And as early as 1938, Guy Stewart Callendar provided evidence that an elevated CO_2 concentration was, in fact, warming the planet. By the early twenty-first century, the scientific evidence of human-caused warming (briefly summarized in chapter 2) was unequivocal.

It is equally certain that climate change will affect the economy and human well-being. Quantifying these impacts and the value of avoiding them has, however, been a major challenge, because the climate, the economy, and their interface are all highly complex. Modern economic analyses of climate change date to the pioneering works of William Nordhaus, William Cline, Samuel Fankhauser, and others in the early 1990s. One central insight from this early work was that investing in heavy-emissions mitigation too early can carry substantial opportunity costs because investments elsewhere in the economy may yield larger returns. However, subsequent work showed that accounting for uncertainty in climate damage could, when combined with risk aversion, motivate more rapid mitigation.

In 2007, Lord Nicholas Stern (co-commentator for part 4) led a groundbreaking analysis of the macroeconomic costs and benefits of climate-change policies. The Stern Review and the dialogue it triggered clarified the critical role of social discount rates in economic evaluations of climate-change policies. In 2010, the U.S. government attempted to quantify the economic cost of climate change and benefits of mitigation. In that year, a working group cochaired by Michael Greenstone (commentator for part 2) issued the U.S. government's first estimates of the social cost of carbon, which are used to integrate climate change into the benefit-cost analyses that guide regulatory decision making.

These contributions have played a central role in both building our understanding of the economics of climate change and elucidating critical gaps in our existing knowledge. One such gap was the weak understanding of the way in which economies are affected by the climate. In previous global analyses, it was often simply assumed that total economic costs grew as a theorized function of global average temperature. This assumption originally arose out of necessity, as there was little empirical research to constrain these "economic damage functions," and evaluating localized impacts *en masse* would have been too computationally challenging.

Early in 2012, two of us (Solomon Hsiang and Bob Kopp) met for the first time and realized that we could fill this knowledge gap by leveraging a recent explosion in econometric analyses of climate impacts, decades of research in climate modeling, and advances in modern computing. Together with Michael Oppenheimer (commentator for part 1), we designed a new framework for assessing the economic costs of climate change that took advantage of these three recent advances. We proposed the development of an assessment system that would automate the calculations needed to stitch together results from econometricians and climate modelers to calibrate the mathematical machinery used in integrated policy models (Kopp, Hsiang, & Oppenheimer 2013). Using modern computing, we could provide the necessary "translation" needed for the physical science, econometric, and integrated assessment communities to share results with one another efficiently and effectively. Furthermore, we wanted to achieve this goal in a risk-based framework: one that took into account uncertainty in projections of future changes, uncertainty in statistical analyses of the past, and the natural uncertainty of the weather, and which could be used by decision makers accustomed to managing other forms of risk. Presenting this ambitious vision at a national conference of academics in December 2012, we were told by a grinning colleague, "good luck with that!"

Luck we had. In 2013, shortly after we ironed out these ideas, the opportunity to implement them arose through the Risky Business Project. The Risky Business Project—led by New York City mayor Michael Bloomberg, former Bush administration treasury secretary Hank Paulson, and former hedge-fund manager Tom Steyer—aimed to move the discourse and U.S. response to climate change beyond its partisan stalemate. Their primary objective was to engage risk managers in the investment and business communities and provide them the basis for incorporating climate risk into their decision making. Bloomberg, Paulson, and Steyer convened and chaired a nonpartisan "Climate Risk Committee" that also included former treasury secretaries Robert Rubin and George Shultz, former Housing and Urban Development secretary Henry Cisneros, former Health and Human Services secretary Donna Shalala, former U.S. Senator Olympia Snowe, former Cargill CEO Greg Page, and Al Sommer, dean emeritus of the Bloomberg School of Public Health at Johns Hopkins University. The Risky Business Project commissioned Rhodium Group, the economic research company where two of us (Trevor Houser and Kate Larsen) are employed, to conduct an independent climate-risk assessment to inform its deliberations. Trevor invited Bob and Solomon to implement a U.S.-focused version of their proposed assessment system, integrating Rhodium's energy sector and macroeconomic analysis and the coastal storm modeling capabilities of Risk Management Solutions (RMS), another project partner. The *American Climate Prospectus*, which forms the core of this volume, was thus born.

The primary goal of the *American Climate Prospectus* is to provide decision makers, the public, and researchers with spatially resolved estimates of economic risks in major sectors using real-world data and reliable, replicable analyses. Achieving this goal requires the careful evaluation of uncertainty in climate projections and economic impacts at a local level, as well as the harmonization and integration of findings and methods from multiple disciplines. In practice, these tasks are difficult; in many cases, the underlying research needed to implement the assessment for specific sectors or effects does not yet exist. The *American Climate Prospectus* platform is therefore designed to grow and expand with the frontier of scientific and economic knowledge, as we learn more about the linkages between the planet's climate and the global economy. The analysis in this volume is novel, and we hope its substance is useful, but we are acutely aware that our findings will not be the last word on these questions. We are building on the work of our predecessors, and we hope that others will build on this contribution. Because of this, we intentionally designed our analysis system to be *adaptive* to new discoveries and better models that will be achieved in the future. As we learn more about our world and ourselves, the assessment system we have built will incorporate this new information, allowing our risk analysis to reflect this new understanding. This may be the first *American Climate Prospectus*, but we do not expect it to be the last.

To help place our findings in context and to point the way forward for researchers to build on this work, we have invited six distinguished researchers—Michael Oppenheimer, Michael Greenstone, Karen Fisher-Vanden, Nicholas Stern, Bob Ward, and Geoffrey Heal—to provide commentaries on each of the five sections of this analysis. We have asked them, as experts on these topics, to be critical of our analysis, to help readers digest both the benefits and the weaknesses of our work, and to highlight future avenues of investigation that will improve our collective understanding.

While we fully recognize that future analyses will revise the numbers we present here, we believe our analysis makes several methodological innovations. Some highlights are:

- We provide new, probabilistic projections of climate changes that are localized to the county level while also being consistent with the estimated probability distribution of global mean temperature change. These projections include information on the distribution of daily temperatures and rainfall, wet-bulb temperature, and sea-level rise.
- We developed a Distributed Meta-Analysis System (DMAS) that continually and dynamically integrates new empirical findings, which can be crowd-sourced from researchers around the world, using a Bayesian framework. DMAS allows our assessment to be easily updated with new results in the future.
- We used econometrically derived empirical findings to develop fully probabilistic impact projections that account for climate-model uncertainty, natural climate variability, and statistical uncertainty in empirical econometric estimates.
- We modeled the energy-market consequences of empirically validated climate-driven changes in heating and cooling demand.
- We conducted the first nationwide assessment of the impact of sea-level rise on expected losses from hurricanes and other coastal storms that combines probabilistic local sea-level rise projections with both historical and projected rates of hurricane activity.
- We developed spatially explicit impact projections at the county level, allowing us to characterize the distribution of winners and losers in different sectors. These projections allowed us to compute the first estimate of the equity premium arising from the distributional impact of climate change within the United States.
- We developed a framework for integrating empirically based dose-response functions into computable general equilibrium models so that damage functions no longer need to be based on theoretical assumptions.

Taking advantage of these innovations, we are able to characterize how climate change will increasingly affect certain dimensions of the U.S. economy. The novel quantitative risk assessment of the *American Climate Prospectus* focuses on six particular impacts that we felt we could reliably estimate given the state of both scientific and economic research in early 2013. These six impacts are:

- the impact of daily temperature, seasonal rainfall, and CO_2 concentration changes on major commodity crops—wheat, maize, soy, and cotton (chapter 6);
- the impact of daily temperature on the number of hours people work, especially in "high risk" outdoor and manufacturing sectors (chapter 7);
- the impact of daily heat and cold on mortality rates across different age groups (chapter 8);
- the impact of temperature on violent and property crime rates (chapter 9);
- the impact of daily temperature on energy demand and expenditures (chapter 10); and
- the impact of sea-level rise and potential changes in hurricane activity on expected future coastal storm–related property damage and business-interruption costs (chapter 11).

For the first four impacts, we implemented the statistical framework we sketched out with Michael Oppenheimer in 2013. For changes in energy demand and expenditures, we used Rhodium's version of the National Energy Modeling System—the tool developed by the U.S. Energy Information Administration for projecting the future of the U.S. energy system. For coastal impacts, we used RMS's North Atlantic Hurricane Model, which is used by RMS to advise its insurance and finance industry clients.

The *American Climate Prospectus* does not attempt to *predict* the costs the future United States *will* experience from climate change. Rather, it is an estimate of the risks the country faces *if it maintains its current economic and demographic structure and if businesses and individuals continue to respond to changes in temperature, precipitation, and coastal storms as they have in the past.* It is not a projection of likely damage given the socioeconomic changes that necessarily will take place; in this, it differs from integrated assessment models such as those developed by Nordhaus and others and used in the Stern Review and by the U.S. government in estimating the social cost of carbon. Rather, we use the structure of the modern economy as a benchmark to inform decision makers as they evaluate how to manage climate risk.

In a risk assessment, it is important to be aware of the different sources of uncertainty (chapter 3). Our assessment focuses on five key sources of uncertainty: (1) emissions, (2) the global temperature response to changes in the atmosphere, (3) the regional temperature and precipitation response to global change, (4) natural variability on timescales ranging from daily weather to multidecadal variations, and (5) statistical uncertainty in our estimation of historical economic impacts.

Future greenhouse-gas emissions are controlled by economics, technology, demographics, and policy—all inherently uncertain, and some a matter of explicit choice. The climate-modeling community has settled upon four Representative Concentration Pathways (or RCPs) to represent a range of plausible emissions trajectories. They are named RCP 8.5, RCP 6.0, RCP 4.5, and RCP 2.6, based on the climate forcing from greenhouse gases that the planet would experience from each pathway at the end of this century (respectively, 8.5, 6.0, 4.5, and 2.6 watts per square meter). RCP 8.5 is the closest to a business-as-usual trajectory, with continued fossil-fuel–intensive growth; RCP 4.5 represents a moderate emissions mitigation trajectory, while RCP 2.6 represents strong emissions control. (RCP 6.0, for idiosyncratic reasons having to do with the construction of the pathways, is of limited use in impact analyses comparing different pathways.) Throughout the *American Climate Prospectus*, we present results for RCP 8.5, 4.5, and 2.6; we focus on RCP 8.5 as the pathway closest to a future without concerted action to reduce future warming.

To generate the projections of temperature and precipitation underlying the risk assessment, we combined projections of the probability of different levels of global average temperature under different RCPs with spatially detailed projections from advanced global climate models (chapter 3 and appendix A). In addition to regional spatial patterns, the resulting projections also incorporate weather and climate variability on timescales ranging from days to decades. To assess impacts on coastal property, we developed new, localized estimates of the probability of different levels of sea-level change that are consistent with the expert assessment of the Intergovernmental Panel on Climate Change. Our approach provides full probability distributions and takes into account all the major processes that cause sea-level change to differ from place to place.

The projections paint a stark picture of the world in the last two decades of the twenty-first century under the business-as-usual RCP 8.5 pathway (chapter 4). In the median projection, with average temperatures in the continental United States 7°F warmer than those in the period 1980–2010, the average summer in New Jersey will be hotter than summers in Texas today. Most of the eastern United States is expected to experience more dangerously hot and humid days in a typical summer than Louisiana does today. By the end of the century under RCP 8.5, global mean sea level is *likely* to be 2.0 to 3.3 feet higher than it was in the year 2000, and there is an approximately 1-in-200 chance it could be more than 5.8 feet higher. Regional factors in some parts of the country—most especially the western Gulf of Mexico and the mid-Atlantic states—could add an additional foot or more of sea-level rise. On top of these higher seas, higher sea-surface temperature may drive stronger Atlantic hurricanes.

Combining these probabilistic physical projections with statistical and sectoral models yields quantitative risk estimates for the six impact categories identified earlier. Were the current U.S. economy to face the climate projected for late in the century in the median RCP 8.5 case, the costs of these six impacts would total 1.4 to 2.9 percent of national GDP; there is a 1-in-20 chance that they would exceed 3.4 to 8.8 percent of GDP. (The low ends of the ranges assume no increase in hurricane intensity and value mortality based on lost labor income; the high ends include hurricane intensification and use the $7.9 million value of a statistical life discussed later to account for mortality.) For a sense of scale, other researchers estimate that, on average, civil wars and currency crises in other countries cause their GDPs to fall by roughly 3 and 4 percent, respectively (Cerra & Saxena 2008). These potential costs are distributed unevenly across the country. The projected risk in the Southeast is about twice the national average, while that in the Northeast is about half the national average; the Pacific Northwest may even benefit from the impacts that we have assessed.

Of the six impacts we quantified, the risk of increased mortality poses the greatest economic threat (chapter 8). The statistical studies underlying this projection account for all causes of death. The most important causes of heat-related deaths are cardiovascular and respiratory disease; low-temperature deaths are dominated by respiratory disease, with significant contribution from infections and cardiovascular disease.

In the median projection for RCP 8.5 toward the end of the century, the United States is projected to experience about 10 additional deaths per 100,000 people each year—roughly comparable to the current national death rate from traffic accidents. There is a 1-in-20 chance the hotter climate could cause more than three times as many deaths. The additional deaths are not spread evenly across the United States but are instead concentrated in southeastern states, along with Texas and Arizona. Florida, Louisiana, and Mississippi are all projected to experience more than 30 additional deaths per 100,000 people annually by late century in the median case, with a 1-in-20 chance of more than 75 additional deaths. The colder regions of the country are *likely* to see reduced mortality from warmer winters, with the greatest reductions in Alaska, Maine, New Hampshire, and Vermont.

Climate-change mitigation significantly reduces the mortality risk, both nationally and regionally. In RCP 4.5, the nation is projected to experience about 1 additional death per 100,000 each year by the end of the century in the median case, with a 1-in-20 chance of 12 additional deaths—a threefold to ninefold reduction in risk. Even Florida, the hardest-hit state, sees a twofold to fourfold reduction in risk under RCP 4.5. Further mitigation to RCP 2.6 has only a modest effect at the national level but in Florida gives rise to a sixfold to sevenfold reduction in mortality risk relative to RCP 8.5.

When the U.S. Environmental Protection Agency quantifies the benefits and costs of regulations, it uses a value of a statistical life—an estimate of the amount a typical American is willing to pay to reduce societal mortality risk—equal to about $7.9 million per avoided death. Using such a value to translate lives lost into dollar terms, the cost of increased mortality under RCP 8.5 amounts to about 1.5 percent of GDP in the median case, with a 1-in-20 chance of a loss of more than 5.4 percent of GDP.

Increased mortality has a smaller economic price if we consider only the labor income lost, although this is an admittedly limited way to value human lives. The expected income lost under RCP 8.5 by late century amounts to about 0.1 percent of GDP, with a 1-in-20 chance of a loss exceeding 0.4 percent of GDP. The economic consequences of these losses are amplified because reduced labor supply in a particular year affects economic growth rates in subsequent years; we assess this amplification when combining impacts in a computable general equilibrium model.

The second greatest economic risk comes from the reduction in the number of hours people work (chapter 7). This effect is most pronounced for those who engage in "high-risk," physically intensive work, especially outdoors. The high-risk sectors identified by statistical studies include agriculture, construction, utilities, and manufacturing. The labor-supply risk is spread more evenly across the country than mortality risk but is highest in states such as North Dakota and Texas, where a large fraction of the workforce works outdoors. It yields a late-century reduction of about 0.5 percent in GDP in RCP 8.5 in the median case, with a 1-in-20 chance of a loss exceeding 1.4 percent of GDP. The labor-supply risk can be moderately reduced through mitigation—by about a factor of 2 by switching to RCP 4.5 and by another factor of 2 by further reducing emissions to RCP 2.6.

The next two largest risks come from impacts on energy demand (chapter 10) and coastal communities (chapter 11).

Nationally, energy expenditures are expected to increase by about 12 percent by late century under RCP 8.5 (with a 1-in-20 chance that they will increase by more than 30 percent) as a result of climate-driven changes in energy demand. These increased energy expenditures amount to about 0.3 percent of GDP (with a 1-in-20 chance of exceeding 0.8 percent of GDP). They are concentrated in the southern half of the country, with the Pacific Northwest even seeing a reduction in energy expenditures in the median projection. RCP 4.5 reduces energy demand risk by a factor of about 2 to 3; further reducing emissions to RCP 2.6 reduces the risk by another factor of 2 to 3. These estimates do not include temperature-related reductions in the efficiency of power generation and transmission, which will likely further increase energy costs.

Both sea-level rise and potential changes in hurricane activity will be costly for the United States, with geographically disparate impacts. Considering only the effects of sea-level rise on coastal flooding, the percentage increase in average annual storm losses is likely to be largest in the mid-Atlantic region, with New Jersey and New York experiencing a median increase of about 250 percent by 2100 under RCP 8.5 (with a 1-in-20 chance of an increase greater than 400 percent). The absolute increases in coastal storm risk are largest in Florida, with losses increasing by about $11 billion per year (relative to current property values) in the median RCP 8.5 case by 2100. If hurricanes intensify with climate change, as many researchers expect, losses may increase nationally by a further factor of 2 to 3. The effects of greenhouse-gas mitigation on sea-level rise are more muted than for many other impact categories, as the oceans and ice sheets respond to warming relatively slowly; switching from RCP 8.5 to RCP 2.6 yields about a 25 percent reduction in coastal storm risk.

The national economic risk from both agriculture (chapter 6) and crime (chapter 9) is relatively small as a fraction of output (about 0.1 percent of GDP in the median late-century RCP 8.5 case for agriculture, with a 1-in-20 chance of about 0.4 percent of GDP; and a 19-in-20 chance of less than 0.1 percent for crime). That is not to say they are not significant—agriculture accounts for a small fraction of U.S. economic activity but is nonetheless of great importance to the nation's well-being, and increases in crime also affect human well-being in ways that do not show up in simple measures of economic output.

Agricultural risk is highly uneven across the country. Provided they have a sufficient water supply—a key uncertainty that remains a topic of investigation—irrigated crops, as are common in the western half of the United States, are less sensitive to temperature than the rain-fed farms that dominate in the eastern half. In addition, higher CO_2 concentrations are expected to increase crop yields. Accordingly, major commodity crops in the Northwest and upper Great Plains may benefit from projected climate changes, while in the eastern half of the country they are likely to suffer if farmers continue current practices. Differences between emissions scenarios are considerable, with median projected losses in RCP 8.5 three times those in RCP 4.5 by mid-century (a 3 percent reduction in crop yield vs. a 1 percent reduction in crop yield) and more than four times by late century (15 percent vs. 3 percent). It is important to bear in mind that the treatment of agriculture in the *American Climate Prospectus* omits some potential key factors; these include risks arising from sustained drought, inland flooding, and pests.

The relationship between crime and climate is well known in law, sociology, and popular culture—even figuring in an episode of the HBO show *The Wire*. Only recently, however, have statistical analyses clearly quantified this relationship in ways that are useful for climate-risk analysis. Applying the observed relationship to the *American Climate Prospectus* temperature projections indicates that violent crime is likely to increase by about 2 to 5 percent across the country under RCP 8.5 by late century, with smaller changes for property crimes. Mitigation moderately reduces these risks; the projected increase in violent crime is lower by about a factor of 2 in RCP 4.5 relative to RCP 8.5 and by another factor of 2 in RCP 2.6 relative to RCP 4.5.

The six economic risks quantified here are—as already noted—far from a complete picture (chapter 16). In the agricultural sector alone, the *American Climate Prospectus* does not cover impacts on fruits, nuts, vegetables, or livestock (chapter 6). Reductions in water supplies and increases in inland flooding from heavy rainfall (chapter 17), weeds and pests (chapter 6), wildfires (chapter 18), changes in the desirability of different regions as tourist destinations (chapter 19), and ocean acidification all pose economic risks. Impacts may interact to amplify each other in unexpected ways. Changes in international trade, migration, and conflict will have consequences for the United States (chapter 20). The Earth may pass tipping points that amplify warming, devastate ecosystems, or accelerate sea-level rise (chapter 3). In the twenty-second century under RCP 8.5, the combination of heat and humidity may make parts of the country uninhabitable during the hottest days of the summer (chapter 4).

To cope with climate risk, decision makers have two main strategies: to work toward global greenhouse-gas emissions mitigation (chapter 21) and to adapt to projected impacts (chapter 22). The comparison between the different RCPs highlights both the power of and limits to mitigation as a risk-management tool. However, decision makers should utilize these insights in conjunction with information on the costs of mitigation policies and technologies. The *American Climate Prospectus* does not address these costs, estimates of which are abundantly covered elsewhere. The Intergovernmental Panel on

Climate Change Working Group 3 report, the publications of the Energy Modeling Forum 27 exercise, and the International Energy Agency's World Energy Outlook and Energy Technology Perspective reports are useful starting points for interested readers.

Many of the impacts we assess can be moderated through adaptation, although most adaptations will come with their own costs (chapter 22). Expanded air-conditioning may reduce mortality impacts, although projections for the Southeast—where air-conditioning is already ubiquitous—suggests that benefits may be limited, concentrated in areas where adoption is not already saturated. Labor-productivity risks can be managed by shifting outdoor work to cooler parts of the day or through automation, but there are other constraints that may prevent a complete shift away from all outdoor exposure. Crop production may become more resilient to temperature extremes, perhaps by use of more irrigation or by migrating toward cooler locations, both of which come at substantial cost. Coastal impacts can be managed through protective structures, building codes, and abandonment of coastlines, all of which will be critical to our future economic well-being, but which will not come for free. We point to the importance of adaptation in limiting the economic cost of future climate changes by demonstrating how our empirically based techniques can be leveraged to estimate the potential size of these gains. This exercise, however, makes it clear that we know very little about the potential scope, effectiveness, and economic cost of potential adaptations—so much so that uncertainty over these values easily dominates all other uncertainty in projections. This result indicates the importance of future research and analysis into the drivers and constraints of adaptation.

In 2013, we set out with both a research goal (i.e., to pilot an innovative framework for fusing detailed physical climate modeling with modern economic studies of the historical effects of climate variability) and a practical goal (i.e., to provide private- and public-sector decision makers with a prospectus surveying key economic risks the United States faces as a result of our planet's changing climate). The success of this seemingly overwhelming endeavor depended on many factors—most critically the members of our team, all of whom made key contributions and shaped the *American Climate Prospectus* into the volume in your hands. D. J. Rasmussen transformed the products of large-scale global climate models into probabilistic climate projections useful for risk analysis. Amir Jina constructed our econometric analysis and designed most of the figures in this book. James Rising built DMAS and integrated climate and economic data into projections. Robert Muir-Wood and Paul Wilson led RMS's work developing high-resolution forecasts of the impact of sea-level rise and potential changes in hurricane activity on expected coastal storm damage. Michael Mastrandrea provided invaluable support in qualitatively describing climate impacts we were unable to quantify in the *American Climate Prospectus*. Shashank Mohan and Michael Delgado modeled energy-sector impacts and integrated all the impact estimates in a consistent economic framework. Without this eclectic team of mavericks, who have been a joy to work with, the *American Climate Prospectus* would not exist.

Trying to peer into the future, one always sees a fuzzy picture. However, thoughtful consideration of the blurry image provides us with far more information than shutting our eyes tight. As a nation, we are making difficult decisions that will determine the structure of the economy in which we, our children, and our grandchildren will compete and make our livings. In navigating these decisions, we need the best possible map—and if it is blurry, we need to know how blurry. The last thing we want is to drive off a cliff that is nearer to the road than we expect. Rational risk management is about identifying when it is safe to drive fast around a turn and when we should slow down. In your hands is the best map we could assemble for navigating America's economic future in a changing climate. Like any map, it has blank regions and will improve in the future . . . but ignoring the information we have now is just as dangerous as driving with our eyes closed.

ACKNOWLEDGMENTS

MEMBERS of our Expert Review Panel—Kerry Emanuel, Karen Fisher-Vanden, Michael Greenstone, Katharine Hayhoe, Geoffrey Heal, Douglas Holt-Eakin, Michael Spence, Larry Linden, Linda Mearns, Michael Oppenheimer, Sean Ringstead, Tom Rutherford, Jonathan Samet, and Gary Yohe—provided invaluable critiques during the development of this report. We also thank Lord Nicholas Stern, who provided excellent input and guidance, and William Nordhaus for his pioneering work in climate economics and for providing suggestions early in the project.

The authors thank Malte Meinshausen for providing MAGICC global mean temperature projections. The sea-level rise projections were developed in collaboration with Radley Horton, Christopher Little, Jerry Mitrovica, Michael Oppenheimer, Benjamin Strauss, and Claudia Tebaldi. We thank Tony Broccoli, Matthew Huber, and Jonathan Buzan for helpful discussion on the physical climate projections.

We acknowledge the World Climate Research Programme's Working Group on Coupled Modeling, which is responsible for the Coupled Model Intercomparison Project (CMIP), and we thank the participating climate-modeling groups (listed in appendix A) for producing and making available their model output. We also thank the Bureau of Reclamation and its collaborators for their downscaled CMIP5 projections. For CMIP, the U.S. Department of Energy's Program for Climate Model Diagnosis and Intercomparison provides coordinating support and led development of software infrastructure in partnership with the Global Organization for Earth System Science Portals.

For their contributions to the impact assessment, the authors thank Max Auffhammer, Joshua Graff Zivin, Olivier Deschênes, Justin McGrath, Lars Lefgren, Matthew Neidell, Matthew Ranson, Michael Roberts, and Wolfram Schlenker for providing data and additional analysis; Marshall Burke, William Fisk, David Lobell, and Michael Greenstone for important discussions and advice; and Sergey Shevtchenko for excellent technology support. We acknowledge the Department of Energy Office of Policy and International Affairs and the U.S. Climate Change Technology Program for providing seed funding for the Distributed Meta-Analysis System.

We thank Kerry Emanuel for supplying hurricane activity rate projections for RMS's coastal-flood modeling, as well as for invaluable discussions along the way. We also thank the RMS consulting group that facilitated the analytical work, specifically Alastair Norris and Karandeep Chadha, and all the members of the RMS development team that have contributed to RMS's models over the years, especially Alison Dobbin and Alexandra Guerrero for their expert contribution in modifying the RMS North Atlantic Hurricane Model to account for climate change.

We partnered with Industrial Economics, Inc. (IEc), the developer of the National Coastal Property Model, to assess the extent to which investments in seawalls, beach nourishment, and building enhancements can protect coastal property and infrastructure. We are grateful to Jim Neumann and Lindsey Ludwig of IEc for their excellent work on this project.

Our assessment of energy-sector effects was made possible by the hard work of the U.S. Energy Information Administration in developing, maintaining, and making publicly available the National Energy Modeling System (NEMS).

We thank Tom Rutherford, Karen Fisher-Vanden, Miles Light, and Andrew Schreiber for their advice and guidance in developing our economic model, RHG-MUSE. We acknowledge Andrew Schreiber and Linda Schick for providing customized support and economic data. Joseph Delgado provided invaluable technical assistance.

We thank Michael Oppenheimer for his help in envisioning our overall approach and for his role in shaping the career paths of two of the lead authors in a way that made this collaboration possible.

This assessment was made possible through the financial support of the Risky Business Project, a partnership of Bloomberg Philanthropies, the Paulson Institute, and TomKat Charitable Trust. Additional support for this research was provided by the Skoll Global Threats Fund and the Rockefeller Family Fund. We thank Kate Gordon and colleagues at Next Generation for providing us with the opportunity to perform this assessment and their adept management of the Risky Business Project as a whole. We are grateful to our colleagues at the Rhodium Group, Rutgers University, the University of California at Berkeley, and Columbia University for their assistance in this assessment. Most important, we thank our friends and families for their seemingly endless patience and support over the past two years.

ECONOMIC RISKS OF CLIMATE CHANGE

CHAPTER 1

INTRODUCTION

WEATHER and climate—the overall distribution of weather over time—shape our economy. Temperature affects everything from the amount of energy we consume to heat and cool our homes and offices to our ability to work outside. Precipitation levels determine not only how much water we have to drink but also the performance of entire economic sectors, from agriculture to recreation and tourism. Extreme weather events, such as hurricanes, droughts, and inland flooding, can be particularly damaging, costing Americans more than $110 billion in 2012 (NOAA 2013a).

Economic and technological development has made us less vulnerable to the elements. Lighting allows us to work and play after the Sun goes down. Buildings protect us from wind and water. Heating and air-conditioning allow us to enjoy temperate conditions at all times of the day and year. That economic growth, however, has begun to change the climate. Scientists are increasingly certain that carbon dioxide (CO_2) emissions from fossil-fuel combustion and deforestation, along with other greenhouse gases (GHGs), are raising average temperatures, changing precipitation patterns, and elevating sea levels. Weather is inherently variable, and no single hot day, drought, winter storm, or hurricane can be exclusively attributed to climate change. A warmer climate, however, increases the frequency or severity of many extreme weather events.

ASSESSING CLIMATE RISK

The best available scientific evidence suggests that changes in the climate observed over the past few decades are likely to accelerate. The U.S. National Academy of Sciences and the UK's Royal Society (National Academy of Sciences & The Royal Society 2014) recently concluded that continued GHG emissions "will cause further climate change, including substantial increases in global average surface temperatures and important changes in regional climate." Given the importance of climate conditions to U.S. economic performance, this presents meaningful risks to the financial security of American businesses and households alike.

Risk assessment is the first step in effective risk management, and there is a broad need for better information on the nature and magnitude of the climate-related risks we face. National policy makers must weigh the potential economic and social impact of climate change against the costs of policies to reduce GHG emissions (mitigation) or make our economy more resilient (adaptation). State and city officials need to identify local vulnerabilities in order to make sound infrastructure investments. Utilities are already grappling with climate-driven changes in energy demand and water supply. Farmers and ranchers are concerned about the commercial risks of shifts in temperature and rainfall, and American families confront climate-related threats—whether storm surges or wildfires—to the safety and security of their homes.

While our understanding of climate change has improved dramatically in recent years, predicting the severity and timing of future impacts remains a challenge. Uncertainty surrounding the level of GHG emissions going forward and the sensitivity of the climate system to those emissions makes it difficult to know exactly how much warming will occur and when. Tipping points, beyond which abrupt and irreversible changes to the climate occur, could exist. Because of the complexity of Earth's climate system, we do not know exactly how changes in global average temperatures will manifest at a regional level. There is considerable uncertainty about how a given change in temperature, precipitation, or sea level will affect different sectors of the economy and how these impacts will interact.

Uncertainty, of course, is not unique to climate change. The military plans for a wide range of possible conflict scenarios, and public health officials prepare for pandemics of low or unknown probability. Households buy insurance to guard against myriad potential perils, and effective risk management is critical to business success and investment performance. In all these areas, decision makers consider a range of possible futures in choosing a course of action. They work off the best information at hand and take advantage of new information as it becomes available. They cannot afford to make decisions based on conditions that were the norm ten or twenty years ago; they look ahead to what the world could be like tomorrow and in coming decades.

OUR APPROACH

A financial prospectus provides potential investors with the facts about material risks and opportunities, and they need these facts in order to make a sound investment decision. In this report, we aim to provide decision makers in business and in government with the facts about the economic risks and opportunities climate change poses in the United States. We use recent advances in climate modeling, econometric research, private-sector risk assessment, and scalable cloud computing (a system we call the Spatial Empirical Adaptive Global-to-Local Assessment System, or SEAGLAS) to assess the impact of potential changes in temperature, precipitation, sea level, and extreme weather events on different sectors of the economy and regions of the country (figure 1.1).

TIPPING POINTS

Even the best available climate models do not predict climate change that may result from reaching critical thresholds (often referred to as tipping points) beyond which abrupt and irreversible changes to the climate system may occur. The existence of several such mechanisms is known, but they are not understood well enough to simulate accurately at the global scale. Evidence for threshold behavior in certain aspects of the climate system has been identified from observations of climate change in the distant past, including ocean circulation and ice sheets. Regional tipping points are also a possibility. In the Arctic, destabilization of methane trapped in ocean sediments and permafrost could potentially trigger a massive release, further destabilizing global climate. Dieback of tropical forests in the Amazon and northern boreal forests (which results in additional CO_2 emissions) may also exhibit critical thresholds, but there is significant uncertainty about where thresholds may be and the likelihood of their occurrence. Such high-risk tipping points are considered unlikely in this century but are by definition hard to predict, and as warming increases, the possibilities of major abrupt change cannot be ruled out. Such tipping points could make our most extreme projections more likely than we estimate, though unexpected stabilizing feedbacks could also act in the opposite direction.

Physical Climate Projections

The scientific community has recently released two major assessments of the risks to human and natural systems from

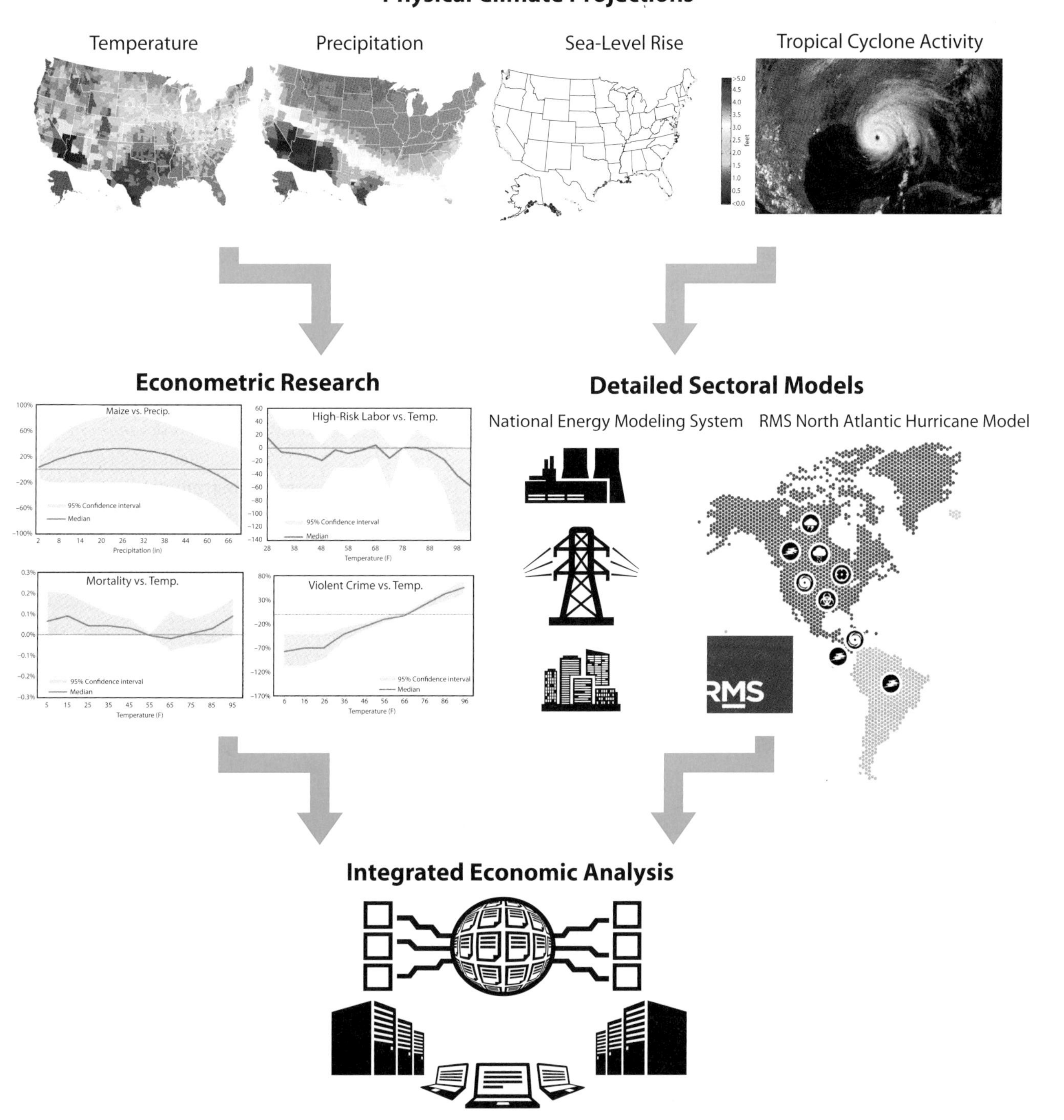

FIGURE 1.1. Spatial Empirical Adaptive Global-to-Local Assessment System (SEAGLAS)

climate change. The Fifth Assessment Report (AR5) of the United Nations' Intergovernmental Panel on Climate Change (IPCC) provides a global outlook, while the U.S. government's third National Climate Assessment (NCA) focuses on regional impacts within the United States. These assessments consolidate the best information that science can provide about the effects of climate change to date and how the climate may change going forward.

Building on records of past weather patterns, probabilistic projections of future global temperature change, and the same suite of detailed global climate models (GCMs) that informed AR5 and the NCA, we explore a full range of potential changes in temperature and precipitation at a daily, local level in the United States as a result of both past and future GHG emissions. Because variability matters as much in shaping economic outcomes as averages, we assess potential changes in the number of hot and cold days each year in addition to changes in annual means. Using the observed, local relationships between temperature and humidity, we also project changes in the number of hot, humid summer days. Synthesizing model projections, formal expert elicitation, and expert assessment, we provide a complete probability distribution of potential sea-level rise at a local level in the United States. While there is still considerable uncertainty surrounding the impact of climate change on hurricane and other storm activity, we explore potential changes, drawing on the work of leading tropical-cyclone modelers at NOAA's Geophysical Fluid Dynamics Laboratory and at the Massachusetts Institute of Technology (MIT).

Econometric Research

Economists have studied the impact of climate change on macroeconomic activity for nearly a quarter century, starting with the pioneering work of the Yale professor William Nordhaus and the Peterson Institute for International Economics fellow William Cline in the early 1990s (Nordhaus 1991; Cline 1992). Just as our scientific understanding of climate change has improved considerably, so has our ability to assess the impacts of climate change on particular sectors of the economy and, in particular, regions of the country. Such finer-scale assessments are necessary to provide useful information to individual decision makers. For example, coastal-property developers need to assess whether, when, and to what extent climate change increases the risk of flooding where they are looking to build. Farmers will want to understand the commercial risks of shifts in temperature and rainfall in their regions rather than the country as a whole. Electric utilities need to prepare for changing heating and cooling demand in their service territories, and the impact of climate change on labor productivity will vary by industry as well as geography. Natural variability in temperature and precipitation provides a rich data set from which to derive insights about the potential economic impact of future climate changes. A wealth of new findings from micro-econometric research has become available in recent years, enabling us to evaluate the effects of climatic changes on certain segments of the economy using historically observed responses.

Detailed Sectoral Models

Complementing our meta-analysis of micro-econometric research, we use detailed, empirically based public- and private-sector models to assess the risk of climate change to key economic sectors or asset classes. These models are not traditionally used for climate-change impact analysis but offer powerful, business- and policy-relevant insights. For example, to assess the impact of greater storm surges during hurricanes and nor'easters on coastal property as a result of climate-driven increases in local sea levels, we use the North Atlantic Hurricane Model and the building-level exposure data set of Risk Management Solutions (RMS). More than 400 insurers, reinsurers, trading companies, and other financial institutions trust RMS models to better understand and manage the risks of natural and human-made catastrophes, including hurricanes, earthquakes, floods, terrorism, and pandemics. To model the impact of changes in temperature on energy demand, power generation, and electricity costs, we use RHG-NEMS, a version of the U.S. Energy Information Administration's National Energy Modeling System (NEMS) maintained by the Rhodium Group. NEMS is used to produce the Annual Energy Outlook, the most detailed and widely used projection of U.S.-energy-market dynamics.

Integrated Economic Analysis

We use geographically granular U.S. economic data to put projected climate impacts in a local economic context. This is critical given how widely climate-risk exposure varies across the country. We also integrate sectoral impact estimates into a state-level model of the U.S. economy to measure the knock-on effects of climate-related changes in one sector or region to other parts of the economy and to assess their combined effect on long-term economic growth.

Cloud Computing

Both the individual components of the analysis and their integration to produce probabilistic, location-specific climate-risk assessments are possible only because of the advent of scalable cloud computing. All told, producing this report required more than 200,000 CPU-hours processing more than 20 terabytes of data, a task that would have taken months, or even years, to complete not long ago. Cloud computing also enables us to make our methodology, models, and data available to the research community, which is critical given the iterative nature of climate-risk assessment and the limited number of impacts we were able to quantify for this report.

USING THIS ASSESSMENT

In part 1, we provide projections of the physical changes facing the United States. In part 2, we assess the direct effects of these changes on six impact categories amenable to quantification: commodity agriculture, labor productivity, heat-related mortality, crime, energy demand, and storm-related coastal damage. In part 3, we assess the economic costs of these impacts. Part 4 provides an overview of the many types of additional impacts that we have not attempted to quantify. Part 5 concludes by presenting principles for climate-risk management.

This assessment does not attempt to provide a definitive answer to the question of what climate change will cost the United States. Nor does it attempt to predict what *will* happen or to identify a single "best estimate" of climate-change impacts and cost. While great for making headlines, best-guess economic cost estimates at a nationwide level are not helpful for effective risk management. Instead, we attempt to provide American policy makers, investors, businesses, and households with as much information as possible about the probability, timing, and scope of a set of economically important climate effects. We also identify areas of potential concern, where the state of knowledge does not permit us to make quantitative estimates at this time. How decisions makers choose to act upon this information will depend on where they live and work, their planning time horizon, and their appetite for risk.

Probability

For many decision makers, low-probability, high-impact climate events matter as much, if not more, than those futures most likely to occur. Nuclear safety officials, for example, must consider worst-case scenarios and design reactors to prevent catastrophic impacts. National security planners, public health officials, and financial regulators are likewise concerned with "tail risks." Most decision makers will not make day-to-day decisions with these catastrophic risks in mind, but for those with little appetite for risk and high potential for damage, the potential for catastrophic outcomes is a data point they cannot afford to ignore. Thus, in addition to presenting the most likely outcomes, we discuss those at each end of the probability distribution.

Throughout the report, we employ the same formal probability language as the IPCC did in AR5. We use the term "*more likely than not*" to indicate probabilities greater than 50 percent, the term "*likely*" for probabilities greater than 67 percent, and the term "*very likely*" for probabilities greater than 90 percent. The formal use of these terms is indicated by italics. For example, where we present "*likely* ranges," that means there is a 67 percent probability that the outcome will be in the specified range.

In some contexts, we also discuss "tail risks," which our probability estimates place at less than 1 percent probability. While we judge these outcomes as exceptionally unlikely to occur within the current century (though perhaps more likely thereafter with continued warming), we could plausibly be underestimating their probability. For example, carbon-cycle feedbacks of the sort discussed in chapter 3 could increase the temperature response of the planet or the destabilization of West Antarctica might amplify sea-level rise. Though our formal probability calculation places low likelihood on these possibilities, the true probability of these scenarios is difficult to quantify.

As described in chapter 4, our analyses include the four global concentration pathways generally used by the scientific community in climate-change modeling.

Timing

Most of our analysis looks out over the next eighty-five years to 2100, extending just three years beyond the expected lifetime of a baby girl born in the United States the day this book was first published. While climate change is already affecting the United States, the most significant risks await us in the decades ahead. How much a decision maker worries about these future impacts depends on his or her age, planning or investment time horizon, and level of concern about long-term economic or financial liabilities. Individuals often care less about costs borne by future generations than those incurred in their own lifetimes. A small start-up does less long-term planning than a multigenerational family-owned company. Property and infrastructure developers have longer investment horizons than commodities or currency traders, and while some politicians are focused purely on the next election cycle, others are focused on the economy's health long after they leave office. We present results in three periods—2020–2039, 2040–2059, and 2080–2099—to allow individual decision makers to focus on the time horizon most relevant to their risk-management needs. For a few physical changes, we also discuss effects beyond 2100 to highlight the potential challenges facing the future children of today's newborns.

Scope

Nationwide estimates of the economic cost of climate change average out important location- or industry-specific information. Climate risk is not evenly spread across regions, economic sectors, or demographic groups. Risks that appear manageable on an economy-wide basis can be catastrophic for the communities or businesses hardest hit. To ensure this risk assessment is useful to a wide range of decision makers, we report and discuss sector-specific impacts as well as nationwide results. We also analyze economic risk by state and region.

A FRAMEWORK TO BUILD ON

Given the complexity of Earth's climate system, uncertainty in how climatic changes affect the economy, and ongoing scientific and economic advances, no single report can provide a definitive assessment of the risks we face. Our work has a number of limitations, which are important to keep in mind when considering the findings presented in this report (see figure 1.2).

First, the universe of potential impacts Americans may face from climate change is large and complex. No study to date has adequately captured them all, and this assessment is no different. We have necessarily been selective in choosing which economic risks to quantify—focusing on those where there is a solid basis for assessment and where sector-level impacts are of macroeconomic significance. This excludes well-known impacts that could be catastrophic for particular communities or industries, as well as poorly understood impacts that pose risks for the economy as a whole. We describe these impacts to the extent possible, drawing on recent academic, government, and private-sector research, but they are not included in our economic cost estimates.

Second, this analysis is limited to the direct impact of climate change within the United States. Of course, climate change is a global phenomenon, and climate impacts elsewhere in the world will have consequences for the United States as well, whether through changes in international trade and investment patterns or new national security concerns. While we discuss some of these dynamics, we have not attempted to quantify their economic impact.

Third, individual climate impacts could very well interact in ways not captured in our analysis. For example, we assess the impact of changes in temperature on electricity demand and the impact of changes in precipitation on water supply, but not changes in water supply on the cost of electric-power generation. These types of interactions can be limited in scope or pose systemic risks.

Finally, economic risk is a narrow measure of human welfare. Climate change could result in a significant decline in biodiversity, lead to the extinction of entire species of plants and animals, and permanently alter the appearance and utility of national parks and other natural treasures. Very little of this is captured in standard

Science

 Temperature: averages and extremes

 Precipitation: averages and extremes

 Local sea-level rise

 Humidity: wet-bulb temperature

 Strong positive carbon-cycle feedbacks

 Ice-sheet collapse

 Ocean temperature and acidification

 Ecosystem collapse

 Unknown unknowns

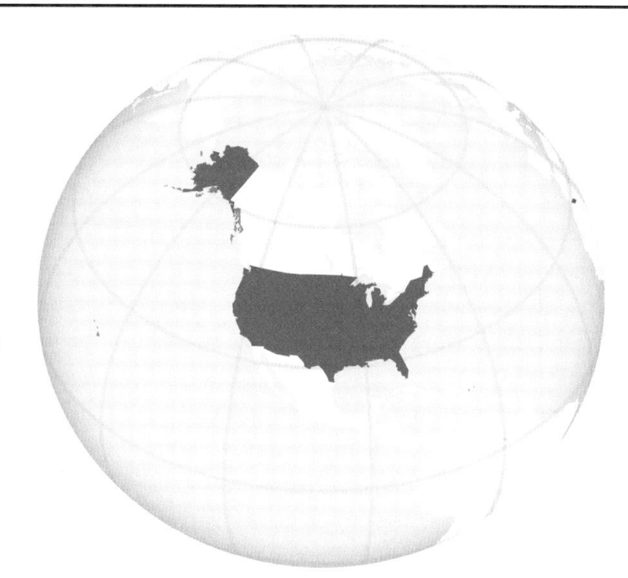

Methodology

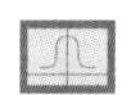 Full probability distribution, tail risks

 Market impact

 Quality of life

 Biodiversity, ecosystem loss

 Ecosystem services

 International trade

Other Impacts

 Water supply and demand

 National security

 International civil conflict

 Aid and disaster relief

 Tourism, outdoor recreation

 Fisheries

 Forests

 Wildfire

Impacts

Coastal Damages

 Inundation from sea-level rise

 Hurricanes and nor'easters

 Changes in hurricane activity

 Transportation

 Infrastructure

Health

 Heat/cold-related mortality

Respiratory effects

 Extreme weather

 Vector and water-borne disease

Agriculture

 Grains, soy, cotton yields

 Other crops: fruit, vegetables, nuts

 Livestock

Energy

Energy demand

Energy supply

Labor Productivity

Hours worked

Labor quality, health impacts

Crime

 Property crime

 Violent crime

 Included

 Limited

 Excluded

FIGURE 1.2. **Scope of this Assessment**

economic indicators like GDP. While understanding the economic risk of climate change is important, it is only one facet of the climate-related risks we face. A number of the economic risks we quantify have noneconomic impacts as well, which we describe alongside the economic findings.

Figure 1.2 highlights the impacts we have included in our quantitative analysis of risks of climate change to the United States, those we include in a limited or purely qualitative way, and those that are excluded from our assessment altogether.

Given these limitations, our goal is to provide a research framework rather than a definitive answer. Our climate is complex, and our understanding of how it is changing and what that means for our economy is constantly evolving. The U.S. National Academy of Sciences has suggested that this kind of "iterative risk management" is also the right way to approach climate change (National Research Council 2010), and we believe the approach we took in preparing this report provides a useful model for future climate-risk assessments. Our team included climate scientists, econometricians, economic modelers, risk analysts, and issue experts from both academia and the private sector. We found this interdisciplinary, intersectoral collaboration unique, enjoyable, and extremely helpful in better understanding such a complicated issue. While taking an integrated approach, our research is modular so that individual components can be updated, expanded, and improved as the science and economics evolve, whether the global climate models we use for local temperature and precipitation projections, our sectoral impact estimates, or the U.S. macroeconomic model we employ. We provide a complete description of our methods and information sources in the technical appendices of this report and will be making our data and tools available online at www.climateprospectus.org. We hope others build on and improve upon our work in the months and years ahead.

PART 1

AMERICA'S CLIMATE FUTURE

OPENING COMMENTARY

MICHAEL OPPENHEIMER

ALBERT G. MILBANK PROFESSOR OF GEOSCIENCES AND INTERNATIONAL AFFAIRS, PRINCETON UNIVERSITY

SCIENTISTS and interested laypeople alike carry a vision of the hazards posed by climate change—for instance, increasing temperatures, higher sea levels, and intensification of precipitation leading to higher risks of heat-related deaths and coastal and inland flooding. These outcomes follow from nearly fifty years of refinement of computer modeling of the global climate system. However, where the rubber meets the road, in the world of policy, planning, and political action aimed at stemming greenhouse-gas emissions to reduce the risk, these models have proved to be of limited value. The same is true of attempts to plan and implement measures to increase resilience in the face of a changing climate. Parochial as it may seem, our representatives in Congress are moved to action not by global threats but by risk and damage to their districts and constituents. It is this gap, between modeling at the large scale and risk management at the geographic scale where political traction resides, that this report attempts to fill.

Climate models can reproduce observed long-term (multidecade) trends in temperature at the global or continental scale, and, as a consequence, scientists have substantial confidence in projection of temperature extremes (like the future frequency of very hot days). Such projections begin to fall short when held against trends on the scale of a cluster of several states, much less a metropolitan area. Models successfully simulate the history of sea-level rise. But one factor that is of growing importance, the behavior of ice sheets, is simulated poorly. Precipitation and storm trends, even at the largest scales, are modeled with much lower confidence than temperature trends. This shortfall in confidence creates enormous difficulties for attempts to project future risk at an actionable scale.

While this book hasn't circumvented these difficulties, it has further developed the existing approaches for doing so and, even more importantly, points the way toward new methods that promise to revolutionize the field of climate-risk analysis. Here, I'll elaborate briefly on only one aspect of the method. Unlike most of the climate-change literature, this work presents outcomes (for example, projected temperature or precipitation changes, as well as the resulting effects on humans and society) in fully probabilistic form, a necessity for analyzing risk. Among the most daunting problems that must be surmounted in order to do so is representing the so-called tail of the probability distributions of temperature, precipitation, or resulting effects. The tail describes the likelihood of the most infrequent but generally most damaging outcomes, like the advent of Hurricane Sandy.

Observations of past climate do not provide sufficient information about the tails because critical events have in the past been so rare that few of them occurred in recorded experience (roughly two centuries, compared to the return time of roughly ten centuries for a Sandy-like storm). But with the climate changing, risk must be viewed as a dynamic feature, with probabilities of rare events increasing over time as the unusual becomes the quotidian. Elaborating the tail, especially for events at a small scale, would require, as a first step, multiple simulations of complex computer models, which is unaffordable. So the authors developed an innovative method that combines output from models of different complexity with expert judgment to flesh out the details of the tail. While the method is new and surely will be improved, it points the way toward a new era in climate-risk analysis.

Using this and related approaches, this book elaborates outcomes that were well known but previously not well quantified, like the future impact of hurricanes. But it also uncovers risks that previously received little attention, specifically that of heat so extreme that people attempting normal outdoor activities would be placing their lives at high risk. In doing so, the authors not only provide a basis for rational judgments by policy makers but also open a new avenue toward progressive improvement in our understanding of risk.

CHAPTER 2

WHAT WE KNOW

OVER the nearly eight decades since the groundbreaking work of Guy Stewart Callendar (Callendar 1938), scientists have become increasingly confident that humans are reshaping Earth's climate. The combustion of fossil fuels, deforestation, and other human activities are increasing the concentration of carbon dioxide (CO_2) and other greenhouse gases in the planet's atmosphere. These gases create a greenhouse effect, trapping some of the Sun's energy and warming Earth's surface. The rise in their concentration is changing the planet's energy balance, leading to higher temperatures and sea levels and to shifts in global weather patterns. In this chapter, we provide an overview of what scientists currently know about climate change and what remains uncertain. In the following two chapters, we discuss the factors that will shape our climate in the years ahead and the approach we take to modeling future climate outcomes in the United States. We present projections of changes in temperature, precipitation, humidity, and sea level between now and the end of the twenty-first century.

SEPARATING THE SIGNAL FROM THE NOISE

The climate is naturally variable. Temperature and precipitation change dramatically from day to day, month to month, and year to year. Ocean circulation patterns result in climate variations on decadal and even multidecadal timescales. Scientists have identified changes in Earth's climate, however, that cannot be explained by these natural variations and are increasingly certain they are caused by human activities (National Research Council 2010; Molina et al. 2014).

Since the late nineteenth century, Earth's average surface air temperature has increased by about 1.4°F (Hartmann et al. 2013). At the global scale, each of the past three decades has been successively warmer than the decade before (figure 2.1). Comparing thermometer records to indirect estimates of temperature, such as the isotopic composition of ice core samples, suggests that, at least in the Northern Hemisphere, the period between 1983 and 2012 was *very likely* the warmest 30-year period of the

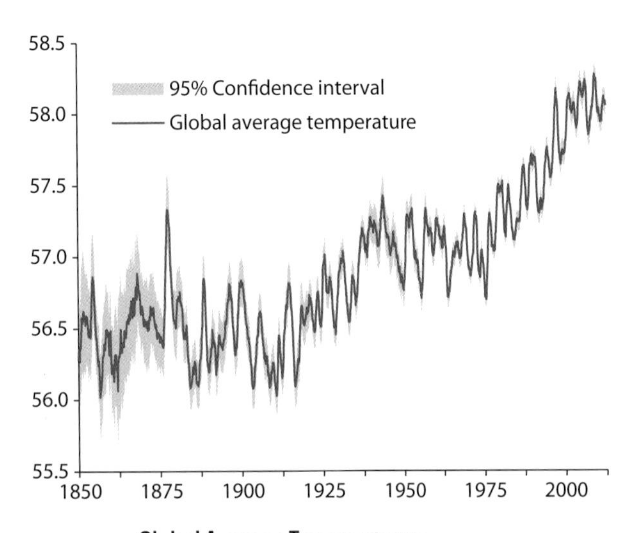

FIGURE 2.1. **Global Average Temperatures**

1850–2013, degrees Fahrenheit

Source: Berkeley Earth (www.berkeleyearth.org)

past 800 years and *likely* the warmest of the past 1,400 years (Masson-Delmotte et al. 2013). Other evidence supports these surface-temperature measurements, including observed decreases in snow and ice cover (from glaciers to sea ice to the Greenland ice sheet), ocean warming, and rising sea levels.

Over the contiguous United States, the average temperature has risen about 1.5°F over the past century, with more than 80 percent of the increase occurring in the past 30 years (Menne, Williams, & Palecki 2010; Walsh et al. 2014). Glaciers are retreating, snowpack is melting earlier, and the growing season is lengthening. There have also been observed changes in some extreme weather events consistent with a warmer United States, including increases in heavy precipitation and heat waves (Walsh et al. 2014).

The increase in both U.S. and global temperatures over the past century transcends the regular annual, decadal, or even multidecadal climate variability. It is a disruption far beyond normal changes in the weather.

A HISTORY OF CLIMATE DISRUPTION

This is not the first time Earth has experienced a climate disruption lasting more than a century. Indeed, over the past 800,000 years, variations in Earth's orbit around the Sun have triggered glacial cycles spanning roughly 100,000 years during which Antarctic temperatures (estimated using ice core samples) have fluctuated by 10°F to more than 20°F (figure 2.2).

The amount of heat a body radiates increases as its temperature rises. For a planet to have a stable global average temperature, the heat it absorbs from the Sun must equal the heat it radiates to space. If it is absorbing more than it is radiating, its surface and atmosphere will warm until energy balance is achieved. CO_2 and other gases in the atmosphere hinder the escape of heat from Earth's surface to space. As the atmospheric concentrations of these gases rise, so, too, do average surface temperatures. This is known as the greenhouse effect, and its fundamental physics have been well understood by scientists since the late nineteenth century (Arrhenius 1896).

Variations in Earth's orbit alter the way the heat that Earth receives from the Sun is distributed over the planet's surface and over the course of the year. These variations cause changes in surface temperatures that can increase or decrease natural emissions of CO_2 and methane (another greenhouse gas), amplifying the temperature impact of the orbital changes (figure 2.2). As the great ice sheets of the last ice age began to retreat about 18,000 years ago, atmospheric concentrations of CO_2 rose from a low of 188 parts per million (ppm), reaching 260 ppm over the following 7,000 years. Concentrations stayed in the 260 to 285 ppm range until the 1860s, when they started rising again. Today's CO_2 levels seasonally exceed 400 ppm, and, within a couple of years, they will do so year-round. This level is far above the range experienced over the past 800,000 years (Luthi et al. 2008). Indeed, the last time CO_2 concentrations exceeded 400 ppm was likely more than 3 million years ago (Seki et al. 2010), a period when global average temperature was about 5°F warmer than today (Lunt et al. 2010) and global average sea level may have been as much as 70 feet higher than today (Miller et al. 2012; Rovere et al. 2014).

The pace of the recent rise in atmospheric concentrations of CO_2 has also been far faster than what occurs under normal glacial cycles—rising more over the past 60 years than during the 7,000 years after the last ice age (figure 2.2).

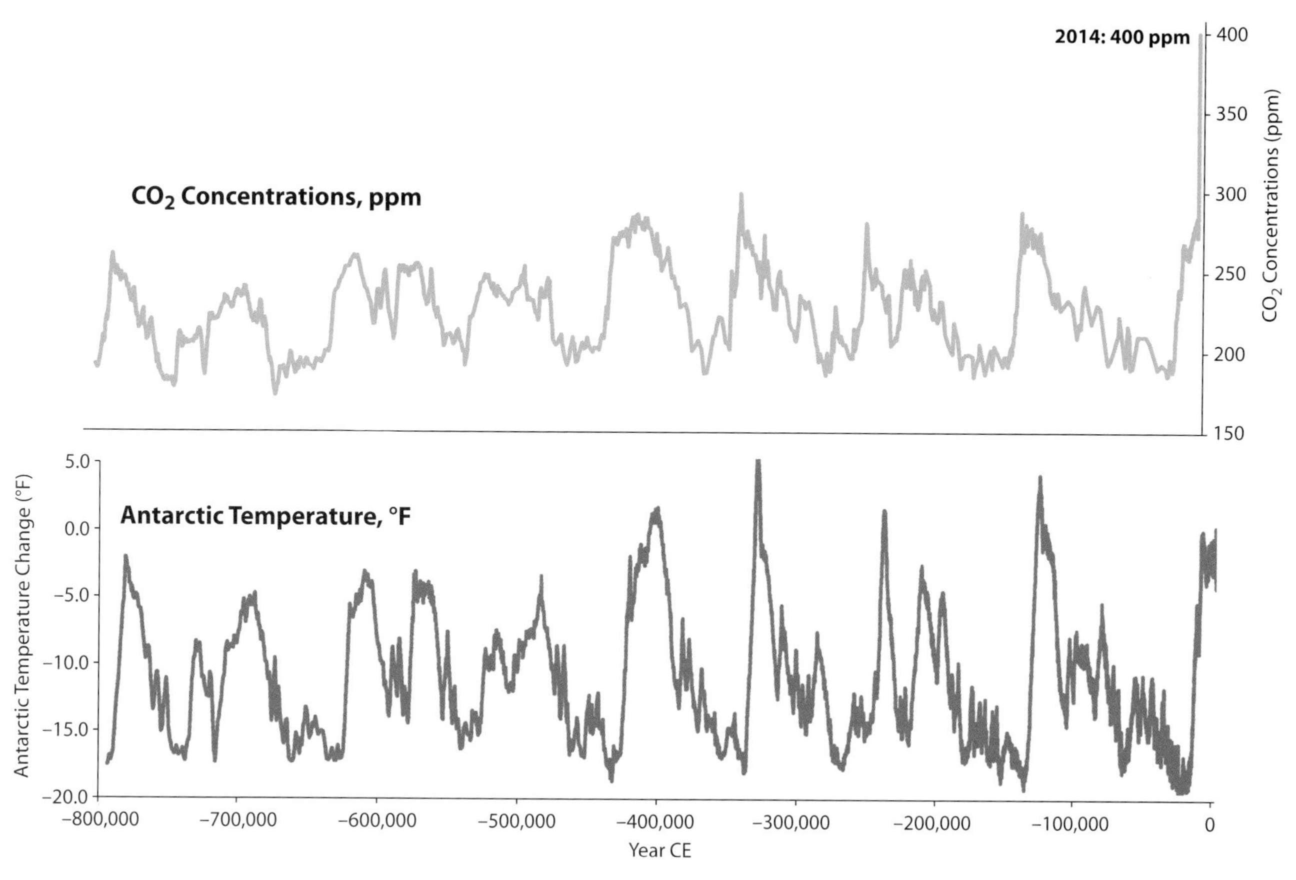

FIGURE 2.2. **Temperature and CO_2**

Historical record

CHAPTER 3

WHAT COMES NEXT

If past greenhouse-gas emissions from fossil-fuel combustion and other human activities have already changed our climate, what risks do we run if we continue on our current course? As discussed in chapter 1, this report attempts to help answer that question. While our focus is the economic risks of climate change, the analysis necessarily starts with an assessment of ways in which the climate may change in the years ahead.

A growing body of evidence shows conclusively that continued emission of CO_2 and other greenhouse gases will cause further warming and affect all components of Earth's climate system. While there have been significant advances in climate science in recent years, Earth's climate system is complex, and predicting exactly how global or regional temperatures and other climate variables will change in the coming decades remains a challenge. It's important to be honest about the uncertainty involved in forecasting our climate future if we are to provide policy makers, businesses, and households with the information they need to manage climate-related risks effectively (Heal & Millner 2014). Scientists face five major sources of uncertainty in predicting climate outcomes: (1) socioeconomic uncertainty, (2) global physical uncertainty, (3) regional physical uncertainty, (4) natural climate variability, and (5) tipping points. In this chapter we discuss each and provide an overview of how they are addressed in our analysis.

SOCIOECONOMIC UNCERTAINTY

Future levels of greenhouse-gas emissions will depend on the pace of global economic and population growth, technological developments, and policy decisions—all of which are challenging to predict over the course of a decade, let alone a century or more. As a consequence, the climate-science community has generally preferred to explore a range of plausible, long-run socioeconomic scenarios rather than rely on a single best guess (Bradfield et al. 2005; Moss et al. 2010). Each scenario includes assumptions about economic development, energy-sector evolution, and policy action—capturing potential futures that range from slow economic growth, to rapid economic growth powered primarily by fossil fuels, to vibrant economic development in a world transitioning to low-carbon energy sources. Each scenario results in an illustrative greenhouse-gas emission and atmospheric concentration pathway.

A broadly accepted set of global concentration pathways was recently developed by the Integrated Assessment Modeling Consortium (IAMC) and used in the Fifth Assessment Report (AR5) of the Intergovernmental Panel on Climate Change (IPCC). These four pathways, termed "Representative Concentration Pathways" (RCPs), span a plausible range of future atmospheric greenhouse-gas concentrations. They are labeled based on their *radiative forcing* (in watts per square meter, a measure of greenhouse-gas concentrations in terms of the amount of additional solar energy the gases retain) in the year 2100 (van Vuuren et al. 2011). The pathways also include different assumptions about future changes in emissions of particulate pollution, which reflects some of the Sun's energy to space and thus dampens regional warming. The RCPs are the basis for most global climate modeling undertaken over the past few years.

At the high end of the range, RCP 8.5 represents a modest increase in recent global emissions growth rates, with atmospheric concentrations of CO_2 reaching 940 ppm by 2100 (figure 3.1) and 2,000 ppm by 2200. These are not the highest possible emissions: Rapid conventional economic growth could lead to a radiative forcing 10 percent higher than RCP 8.5 (Riahi 2013). But RCP 8.5 is a reasonable representation of a world where fossil fuels continue to power relatively robust global economic growth and is often considered closest to the most likely "business-as-usual" scenario absent new climate policy by major emitting countries.

At the low end of the range, RCP 2.6 reflects a future achievable only by aggressively reducing global emissions (even achieving net negative emissions by the end of the twenty-first century) through a rapid transition to low-carbon energy sources. Atmospheric CO_2 concentrations remain below 450 ppm in RCP 2.6, declining to 384 ppm by 2200. Two intermediate pathways (RCP 6.0 and RCP 4.5) are consistent with a modest slowdown in global economic growth and/or a shift away from fossil fuels more gradual than that in RCP 2.6 (Riahi 2013). In RCP 6.0, CO_2 concentrations stabilize around 750 ppm in the middle of the twenty-second century. In RCP 4.5, CO_2 concentrations stabilize around 550 ppm by the end of the twenty-first century.

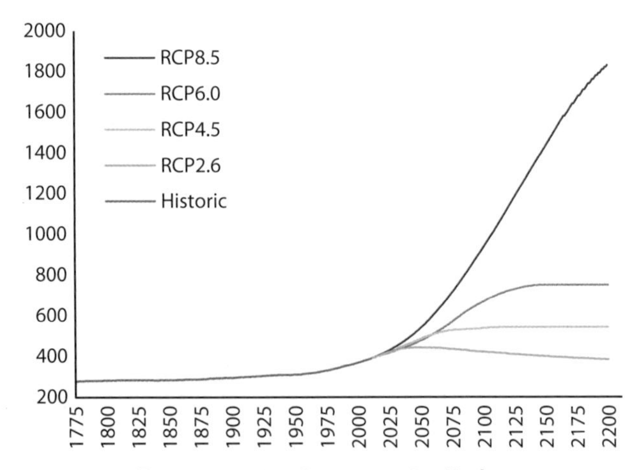

FIGURE 3.1. **Representative Concentration Pathways**

Atmospheric concentration of CO_2 in parts per million

Source: van Vuuren et al. (2011)

We include all four RCPs in our analysis for two reasons. First, an individual RCP is not uniquely associated with any particular set of population, economic, technological, or policy assumptions; each could be attained through a variety of plausible combinations of assumptions. For example, a rapid emissions decline in the United States combined with continued emissions growth in the rest of the world could result in a concentration pathway similar to RCP 8.5. Likewise, if the current decline in U.S. emissions reverses course but the rest of the world makes a rapid transition to a low-carbon economy, a concentration pathway similar to RCP 4.5 is still potentially possible. Given the uncertainty surrounding emissions pathways in other countries, American policy makers must assess the risks associated with a full range of possible concentration futures. This is especially true for local officials and American businesses and households, as these local stakeholders have little control over America's overall emission trajectory, let alone global concentration pathways.

The second reason is to identify the extent to which global efforts to reduce greenhouse-gas emissions can reduce climate-related risks associated with the absence of deliberate mitigation policy (i.e., RCP 8.5 or, under a slower global economic growth scenario, RCP 6.0). This is not to recommend a particular emission-reduction pathway, but to identify climate outcomes that are potentially avoidable versus those that are already locked in.

GLOBAL PHYSICAL UNCERTAINTY

Even if we knew future emissions growth rates with absolute certainty, we would still not be able to predict their impact precisely because of the complexity of Earth's climate system. At a global level, the largest source of physical uncertainty resides in the magnitude and timescale of the planet's response to a given change in radiative forcing, commonly represented by *equilibrium climate sensitivity* and *transient climate response*. The former, typically reported as the response to a doubling of CO_2 concentrations, reflects the long-term response of global mean temperature to a change in forcing; the latter reflects how that response plays out over time.

The effect on global temperature of the heat absorbed and emitted by CO_2 alone is fairly well understood. If CO_2 concentrations doubled but nothing else in the Earth system changed, global average temperature would rise by about 2°F (Hansen et al. 1981; Flato et al. 2013). Across the entire climate system, however, there are several feedback mechanisms that either amplify or diminish this effect and respond on different timescales, complicating precise estimates of the overall sensitivity of the climate system. These feedbacks include an increase in atmospheric water vapor concentrations; a decrease in the planet's reflectivity because of reduced ice and snow coverage; changes in the rate at which land, plants, and the ocean absorb carbon dioxide; and changes in cloud characteristics. Significant uncertainties remain regarding the magnitude of the relatively fast cloud feedbacks and of longer-term or abrupt feedbacks, such as high-latitude permafrost melt or release of methane hydrates, which would amplify projected warming (see the discussion in the section "Tipping Points" later in this chapter). Such longer-term feedbacks are not included in the equilibrium climate sensitivity as conventionally defined.

Uncertainty in the equilibrium climate sensitivity is a major contributor to overall uncertainty in projections of future climate change and its potential effects. Based on observed climate change, climate models, feedback analysis, and paleoclimate evidence, scientists have high confidence that the long-term climate sensitivity (over hundreds to thousands of years) is *likely* 3°F to 8°F of warming per CO_2 doubling, *extremely likely* (95 percent probability) greater than 2°F of warming per CO_2 doubling, and *very likely* (90 percent probability) less than 11°F of warming per CO_2 doubling (Collins et al. 2013). This warming is not realized instantaneously because the ocean serves as a heat sink, slowing temperature rise. A more immediate measure, the transient climate response, indicates that a doubling of CO_2 over 70 years is *likely* to cause a warming of 2°F to 5°F over that period of time (Collins et al. 2013).

These ranges of climate sensitivity values are associated with significantly different projections of future climate change. Many past climate-impact assessments have focused only on the "best estimates" of climate sensitivity. To capture a broader range of potential outcomes, we use MAGICC, a commonly employed simple climate model (Meinshausen, Raper, & Wigley 2011) that can emulate the results of more complex models and can be run hundreds of times to capture the spread in estimates of climate sensitivity and other key climate parameters. MAGICC's model parameters are calibrated against historical observations (Meinshausen et al. 2009; Rogelj, Meinshausen, & Knutti 2012) and the IPCC's estimated distribution of climate sensitivity (Collins et al. 2013). A more detailed description of our approach is provided in appendix A.

REGIONAL PHYSICAL UNCERTAINTY

Because deliberate planetary-scale climate experiments are largely infeasible and would raise profound ethical questions, scientists must rely on computer models to conduct experiments on Earth's complex climate system, including projecting how climate will change at a regional scale in response to changes in greenhouse gases. Global climate models are descended from the first numerical weather-prediction models developed after World War II (Phillips 1956; Manabe & Wetherald 1967; Edwards 2011). Over time, they have been expanded to include the dynamic effects of oceans and sea ice, atmospheric particulates, atmospheric-ocean carbon cycling, atmospheric chemistry, vegetation, and most recently land ice. Model projections of the central components of long-term, human-induced climate change have grown increasingly robust, and recent generations of increasingly complex models provide greater detail and spatial resolution than ever before.

There are dozens of global climate models, with a range of different model structures and parameter assumptions. Since the 1990s, the global climate modeling research community has engaged in structured model comparison exercises, allowing them to compare experiments run in different models to one another and to the observational record. The differences identified among the models allow estimates to be made of the uncertainties in projections of future climate change and highlight which aspects are robust and where to focus future research efforts to improve results over time. By comparing and synthesizing many models, clear trends emerge.

Analysis of the range of potential climate effects on the United States for this report is based on climate projections developed as part of the Coupled Model Intercomparison Project Phase 5 (CMIP5) with a suite of 35 different global climate models (Taylor, Stouffer, & Meehl 2012). This suite of complex models has become the gold standard for use in global climate assessments (including by the IPCC in AR5) and for regional assessments (including the third U.S. National Climate Assessment, released in 2014). Major U.S.-based models participating in CMIP5 have been developed by teams led by the NASA Goddard Institute for Space Studies, the NOAA Geophysical Fluid Dynamics Laboratory, and the National Center for Atmospheric Research.

The global climate models that participated in CMIP5 typically have spatial resolutions of ~1° to 2° (about 70 to 150 miles at midlatitudes). To produce projections at a finer spatial resolution, researchers have used a variety of downscaling approaches. The projections in this report build upon one particular downscaling technique: bias-corrected spatial disaggregation (BCSD; Wood et al. 2002; Brekke et al. 2013). We use a BCSD data set generated by the Bureau of Reclamation (Brekke et al. 2013) from the CMIP5 archive. In addition to the uncertainty in the global climate models themselves, further uncertainty is introduced by the downscaling step. Alternative downscaling approaches can give rise to different localized projections, particularly of extremes (Bürger et al. 2012).

It is important to recognize that the CMIP5 model projections are not a probability distribution, but instead an "ensemble of opportunity" (Tebaldi & Knutti 2007). The models are not fully independent of one another, instead sharing overlapping lineages and a common intellectual milieu (Edwards 2011). Moreover, every modeling team that participates in the Coupled Model Intercomparison Project has striven to develop a model that captures a suite of important physical processes in the oceans and atmosphere and has tuned some of the parameters of its model to reproduce historical behavior reasonably. Attempts to interpret the CMIP5 ensemble as a probability distribution will accordingly undersample the distribution tails and oversample the best estimates.

For this reason, we use estimates from MAGICC of the probability of different temperature outcomes at the end of the century to weight the projections of more complex global climate models. For those parts of the probability distribution for global temperature not covered by the CMIP5 models, primarily in the tails, we create "model surrogates" by scaling spatial patterns of temperature and precipitation change from the CMIP5 models using temperature projections from MAGICC. In appendix A, we compare our key results to those we would estimate if we treated the CMIP5 projections as though they formed a probability distribution.

NATURAL CLIMATE VARIABILITY

As discussed earlier, natural climate variability can range in timescale from day-to-day temperature variations, to interannual patterns such as El Niño, to longer-term patterns such as the Pacific Decadal Oscillation. In addition to the trends in climate associated with climate change, global climate models simultaneously simulate natural climate variability. The magnitude of such variability renders the differences in climatic response among plausible emissions pathways essentially undetectable at a global scale until about 2025. The relative magnitude of climate variability, physical uncertainty, and scenario uncertainty differs from place to place and for different variables. For example, in the British Isles, internal variability in decadal mean surface air temperature dominates scenario uncertainty through the middle of the century (Hawkins & Sutton 2009). Internal variability, not fully captured by climate models, probably accounts for a significant fraction of the slowdown in global warming over the past decade (Trenberth & Fasullo 2013) and for the absence of net warming in parts of the southeastern United States over the past century (Kumar et al. 2013). While unprecedentedly warm years will occur with increasing frequency, climate variability means that the annual mean temperatures of cooler years in most of the United States will be in the range of historical experience until at least the middle of the century (Mora et al. 2013).

Extreme weather events like heat waves, hurricanes, and droughts are examples of natural climate variability experienced on more compressed timescales. By nature, the probability of these events occurring is low, putting them at the far "tails" of statistical weather distributions. There is increasing evidence, however, that climate change is altering the frequency or severity of many types of these events (Cubasch et al. 2013). Although most individual extreme events cannot be directly attributed to human-induced warming, there is relatively high confidence that heat waves and heavy rainfall events are generally becoming more frequent (Hartmann et al. 2013). As the climate continues to warm, certain types of storms such as hurricanes are expected to become more intense (though not necessarily more frequent), although less is known about how other types of storms (such as severe thunderstorms, hailstorms, and tornadoes) may respond (National Academy of Sciences & The Royal Society 2014).

Because of the tremendous damage caused by hurricanes in recent decades, there has been significant interest in understanding global and regional trends in cyclone activity and the causes of any observed changes. At a global level, the evidence for long-term changes and the influence of human-induced climate changes on hurricane activity over the past century is unclear (Knutson et al. 2010). That is not to say that human-induced warming played no role; rather, because of limitations in the quality of historical records, it is possible that such influence is simply not yet detectable or is not yet properly modeled given the uncertainty in quantifying natural variability and the effects of particulate pollution, among other factors (Knutson et al. 2010; Seneviratne et al. 2012; Christensen et al. 2013). Short-term and regional trends vary, however; hurricane activity has increased in the North Atlantic since the 1970s (Christensen et al. 2013).

Our confidence in projecting future changes in extremes (including the direction and magnitude of changes) varies with the type of extreme, based on confidence in observed changes, and is thus more robust for regions where there is sufficient and high-quality observational data (Seneviratne et al. 2012). Temperature extremes, for example, are generally well simulated by current global climate models, though models have more difficulty simulating precipitation extremes (Randall et al. 2007). The ability to project changes in storms, including hurricane activity, is more mixed. There is a growing consensus that, around the world, the strongest hurricanes (categories 4 and 5) and associated rainfall levels are *likely* to increase (Knutson et al. 2010; Seneviratne et al. 2012; Christensen et al. 2013). There is low confidence, however, in climate-induced changes in the origin and track of future North Atlantic hurricanes (Bender et al. 2010).

TIPPING POINTS

Many components of the Earth system exhibit critical thresholds (often referred to as *tipping points*) beyond which abrupt or irreversible changes to the climate or the biosphere may occur (Lenton et al. 2008; Collins et al. 2013; National Academy of Sciences 2013). Many of these tipping points are poorly represented in the current generation of climate models. Some may have direct societal or economic effects, while others may affect the global carbon cycle and amplify climate change. Such feedbacks could increase the probability of our most extreme projections (although unexpected stabilizing feedbacks could also act in the opposite direction).

Summer Arctic sea-ice cover has fallen faster than most of the previous generation of climate models had projected (Stroeve et al. 2007), although the current generation of models appears to perform better (Stroeve et al. 2012). The Arctic appears on track for nearly ice-free Septembers in the coming decades. Reduced sea-ice coverage amplifies warming in the Arctic and may also lead to slower-moving weather patterns at lower latitudes (Francis & Vavrus 2012). Slow-moving weather patterns supported the long-lived cold winter experienced by much of North America in 2013–2014, which had a significant economic effect. However, the linkage with low summer Arctic sea ice remains highly controversial (Barnes 2013).

Past mass extinctions have been tied to global climate change (Blois et al. 2013). Human activities, primarily land-use changes, have increased the global species-extinction rate by about two orders of magnitude above the background rate (Barnosky et al. 2011), and climate change is beginning to exacerbate extinction further (Barnosky et al. 2012). The economic effects of mass extinction and the associated loss of ecosystem services are difficult to estimate, but they are likely to be substantial.

Past climate change has also driven rapid ecosystem shifts (Blois et al. 2013). Some research suggests that the Amazon rain forest and northern boreal forests may be vulnerable to a climatically driven die-off, which would increase global CO_2 emissions, but there is significant uncertainty about the climatic threshold for such a die-off and its likelihood (Collins et al. 2013).

Destabilization of methane trapped in ocean sediments and permafrost may have played a major role in the geologically rapid 10°F global warming of the Paleocene-Eocene Thermal Maximum, which occurred about 56 million years ago (McInerney & Wing 2011). Global warming today may trigger a similar destabilization of methane reservoirs, amplifying projected warming significantly, although such a methane release would be expected to play out over centuries (Collins et al. 2013).

Reconstructions of past sea level and physical models of ice-sheet dynamics suggest that the West Antarctic Ice Sheet can collapse and raise sea level by many feet over the course of a few centuries (Kopp et al. 2009; Pollard & DeConto 2009). Indeed, recent evidence suggests that such a collapse may be under way (Joughin, Smith, & Medley 2014; Rignot et al. 2014). The possibility of a rapid collapse is included in the projections of sea-level rise described in chapter 4, which indicate a 1-in-1,000 probability of 8 feet of global mean sea-level rise by 2100 and 31 feet of global mean sea-level rise by 2200, but its likelihood may be underestimated.

Other potential tipping points include drops in ocean oxygen content, changes to monsoons, and changes to patterns of climatic variability such as El Niño (National Academy of Sciences 2013). There may be other critical thresholds not yet considered by science. High-impact tipping points with consequences realized primarily in this century are considered unlikely, but confidence in many of these projections is low (Collins et al. 2013). As warming increases, the possibility of major abrupt changes cannot be ruled out.

CHAPTER 4

U.S. CLIMATE PROJECTIONS

THIS report seeks to assess how potential climate futures may differ from the conditions we know today. Results are provided for three future time periods: 2020–2039, midcentury (2040–2059), and late century (2080–2099). We also report results for a historical reference period, in most cases the period 1981–2010 used by the National Climate Data Center in defining the latest release of Climate Normals. Using multidecadal averages rather than a single year ensures that results are not excessively influenced by natural interannual variability.

Under all scenarios, average global and U.S. temperatures rise over the course of the century. By midcentury, global average temperature will *likely* (67 percent probability) be between 2.2°F and 3.7°F warmer under the continued high global emissions pathway (RCP 8.5). The increase will be somewhat less under RCP 6.0 and RCP 4.5, with *likely* warming of 1.4°F to 2.5°F and 1.5°F to 2.8°F, respectively. Even under RCP 2.6, average temperatures continue to increase to a *likely* range of 1.1°F to 2.2°F by midcentury. By the end of the century, the differences between future pathways are larger: the *very likely* (90 percent probability) warming is 4.7°F to 8.8°F for RCP 8.5, 2.8°F to 5.4°F for RCP 6.0, 2.1°F to 4.5°F for RCP 4.5, and 0.9°F to 2.6°F for RCP 2.6 (figure 4.1).

The land warms faster than the oceans, and as a consequence the mean temperature increase in the United States over the twenty-first century will, *more likely than not*, be greater than the global average. (Note that these are average temperatures; just as they do today, individual years will vary by about 1°F to 2°F around the average; see figure 4.2.) Across the continental United States, by midcentury the average temperature will *likely* be between 2.6°F and 5.8°F warmer under RCP 8.5 and between 1.9°F and 3.5°F warmer under RCP 2.6. By the end of the century, the differences between future pathways are larger, with *likely* warming of 6.1°F to 12.5°F for RCP 8.5, 4.1°F to 7.7°F for RCP 6.0, 2.9°F to 6.9°F for RCP 4.5, and 1.1°F to 3.7°F for RCP 2.6. These *likely* ranges, however, do not reflect the small but not insignificant chance that average U.S. temperatures may rise even further. Under RCP 8.5, by the end of the century there is a 1-in-20 chance that average temperatures could rise by more than 14°F and a very small chance (which we estimate at 1 percent or less) of temperature increases above 19°F. In the continental United States, RCP 6.0 is associated with cooler temperatures than those

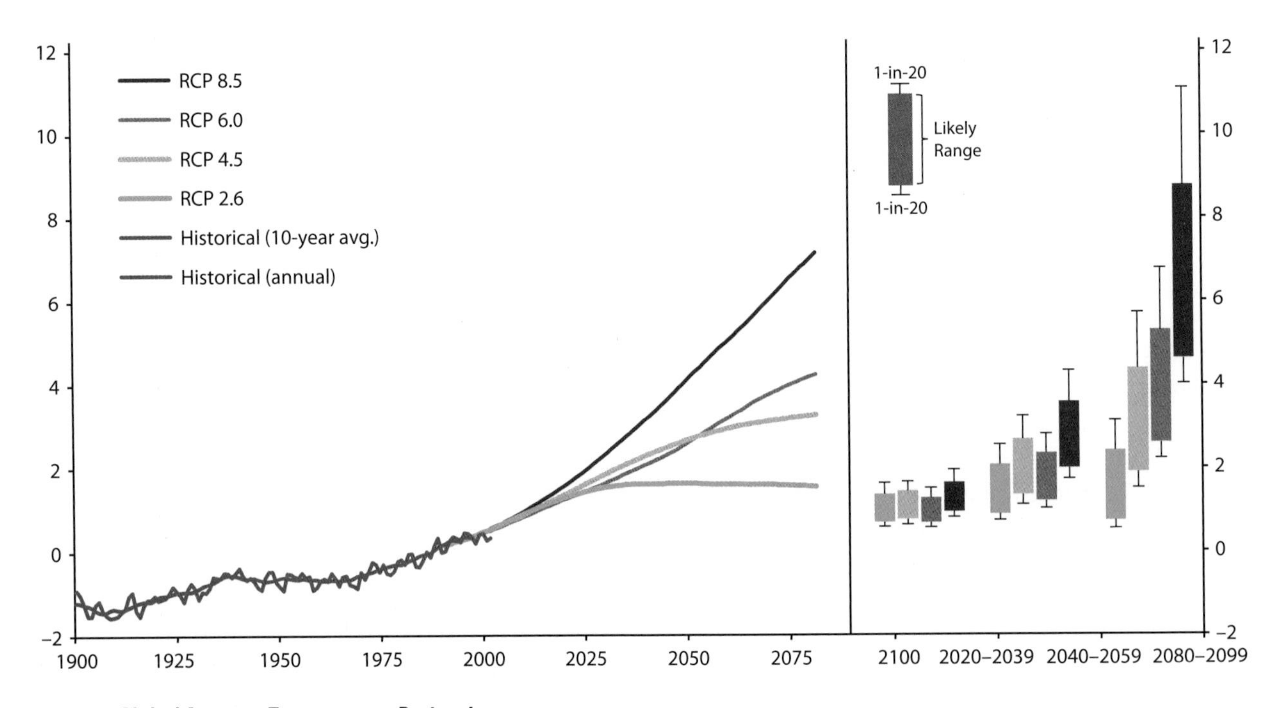

FIGURE 4.1. **Global Average Temperature Projections**

Degrees Fahrenheit relative to 1981–2010 averages, historical median projections (left side) and ranges (right side)

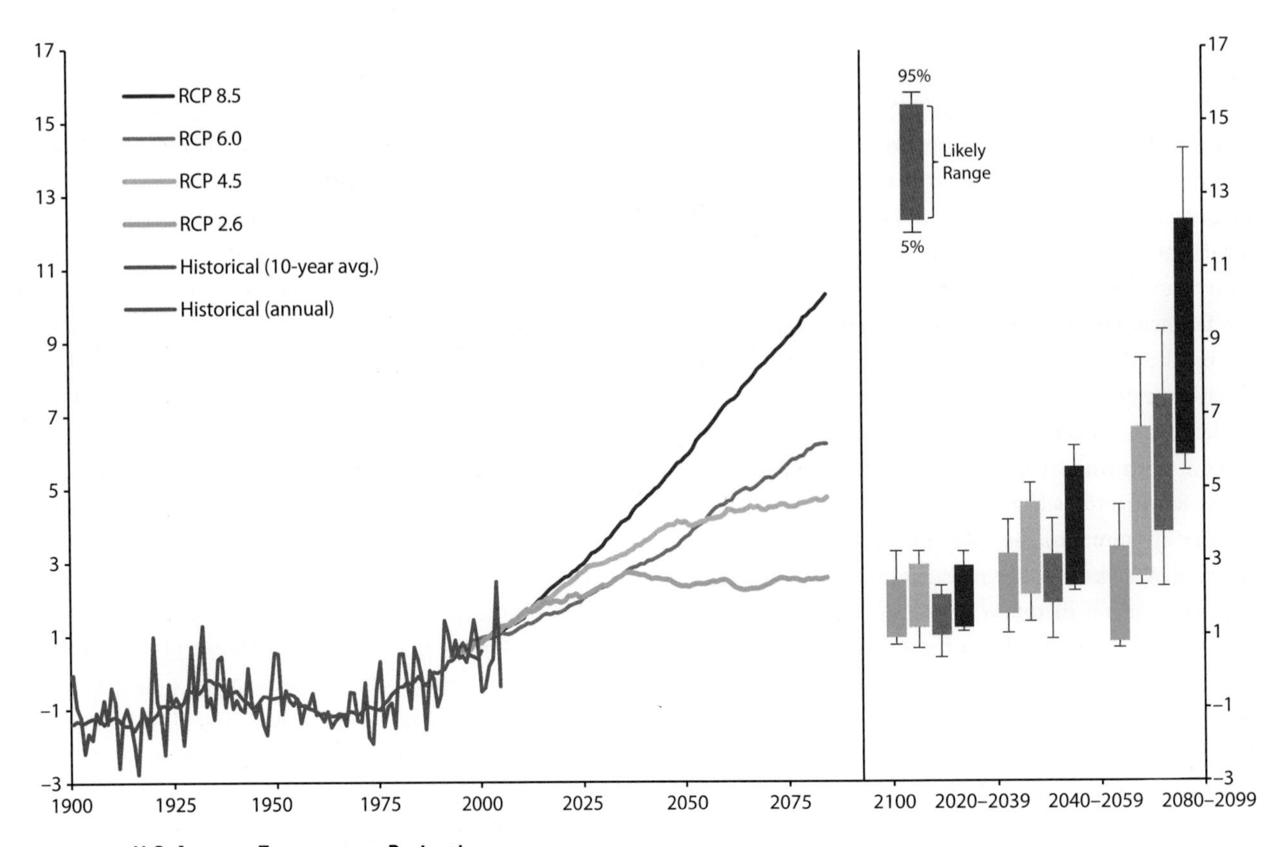

FIGURE 4.2. **U.S. Average Temperature Projections**

Degrees Fahrenheit relative to 1981–2010 averages, historical temperatures and median projections (left side) and ranges (right side)

of the other three RCPs through midcentury because it projects greater emissions of particulate pollution from power plants and industrial sources, offsetting some of the warming that would otherwise have occurred.

The rise in global and U.S. average temperatures will be reflected in increased daily high temperatures. Since 1950, global maximum and minimum air-surface temperatures have increased by more than 1.1°F, about 0.2°F per decade (Hartmann et al. 2013). Over the past 30 to 40 years, the ratio of record daily high temperatures to record daily low temperatures for the continental United States has steadily increased (Walsh et al. 2014). The past decade experienced twice as many record highs as record lows, a larger difference than even the 1930s—a time of record heat and drought in much of the United States (Blunden & Arndt 2013). Extreme summer temperatures have also approached or exceeded those in the 1930s over much of the United States, and the average summer temperature is projected to rise (figure 4.3).

One metric of changes in extreme temperatures is the number of days with temperatures reaching 95°F or more, a measure that is projected to increase dramatically across the contiguous United States as a result of climate change. Under RCP 8.5, by midcentury (assuming the geographic distribution of the population remains unchanged) the average American will *likely* experience an average of 27 to 50 days above 95°F each year. This represents a near doubling to more than tripling of the average 15 days per year above this threshold from 1981 to 2010. By late century, the average American will *likely* see an average of 46 to 96 days per year above 95°F, or around 1.5 to 3 months out of the year. By the end of the century, the average Coloradan will *likely* experience more days above 95°F in a typical year than the average Texan does today (figure 4.4).

There are similarly large projected changes in average winter temperatures (figure 4.5) and number of extremely cold days (figure 4.6). Again, northern states see the largest shift, with average winter temperatures *likely* rising by 2.9°F to 6.5°F in the Northeast by midcentury under RCP 8.5 and by 6.9°F to 13.2°F by the end of the century (figure 4.5). Of the 25 states that currently have subfreezing average winter temperatures, only six (Vermont, Maine, Wisconsin, Minnesota, North Dakota, and Alaska) are still *likely* to do so under RCP 8.5 by the end of the century. In that scenario, the average number of days with temperatures dropping below 32°F that the average resident of New York state experiences will *likely* fall from 93 to less than 51. The number of days dipping below 32°F in Washington, D.C., will *likely* fall from 87 to less than 37.

HUMIDITY

"It's not the heat; it's the humidity," the common saying goes. The combination of high temperatures with high humidity is significantly more uncomfortable and potentially more dangerous than high temperatures under drier conditions. Wet-bulb temperature is an important climatic and meteorological metric that reflects the combined effect of temperature and humidity (Sherwood & Huber 2010; Buzan 2013). Measured with a ventilated thermometer wrapped in a wet cloth, it reflects the ability of mammals to cool by sweating. In order for humans to maintain a stable body temperature around 98°F, skin temperature must be below 95°F, which for a well-ventilated individual at rest in the shade requires a wet-bulb temperature below 95°F. Exposure to sunlight and exertion will increase body temperature. About an hour of vigorous, shaded activity at a wet-bulb temperature of 92°F leads to skin temperatures of 100°F and core body temperatures of 104°F (Nielsen et al. 1997; Liang et al. 2011). Higher core temperatures are associated with heat stroke, which can be fatal (Bouchama & Knochel 2002).

Such high wet-bulb temperatures almost never occur on the planet today. The highest heat-humidity combinations in the United States in the past 30 years occurred in the Midwest in July 1995, during the middle of that summer's heat wave. Wet-bulb temperatures then approached 90°F; one weather station in Appleton, Wisconsin, recorded a temperature and dew point that correspond to a wet-bulb temperature of 92°F (Burt 2011).

We developed the ACP Humid Heat Stroke Index, which divides daily peak wet-bulb temperature into four categories (table 4.1). Category I reflects uncomfortable conditions typical of summer in much of the Southeast, while category II reflects dangerous conditions typical of the most humid days of summer in the Southeast, as far north as Chicago and Washington, D.C. Category III conditions are rare and extremely dangerous, occurring only a few times in the United States between 1981 and 2010, including during the 1995 Midwest heat wave. The extraordinarily dangerous category IV conditions exceed U.S. historical experience.

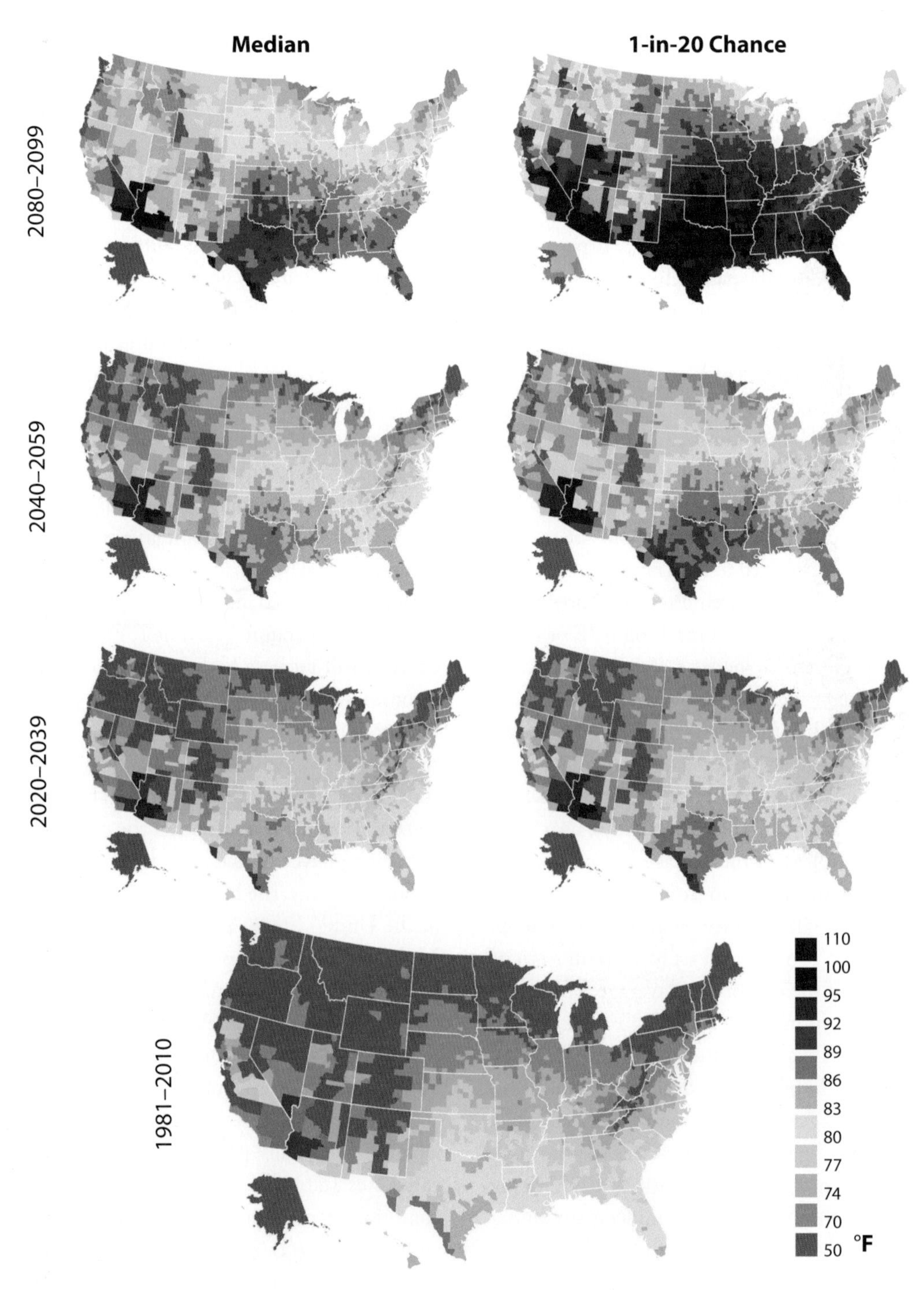

FIGURE 4.3. **Change in Average Summer Temperatures**

Daily average summer (June-July-August) temperature (degrees Fahrenheit) under RCP 8.5

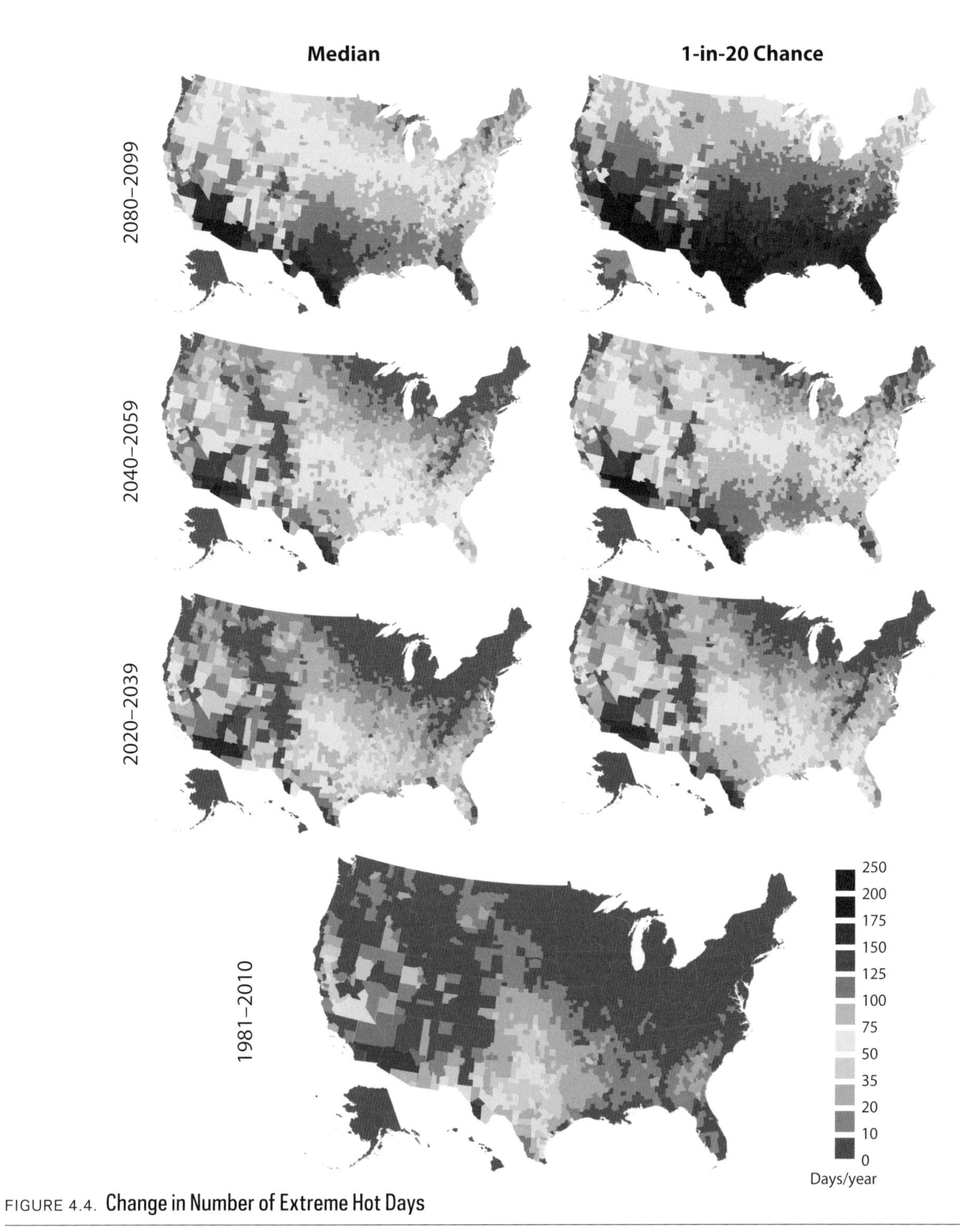

FIGURE 4.4. **Change in Number of Extreme Hot Days**

Number of days with maximum temperatures above 95°F under RCP 8.5

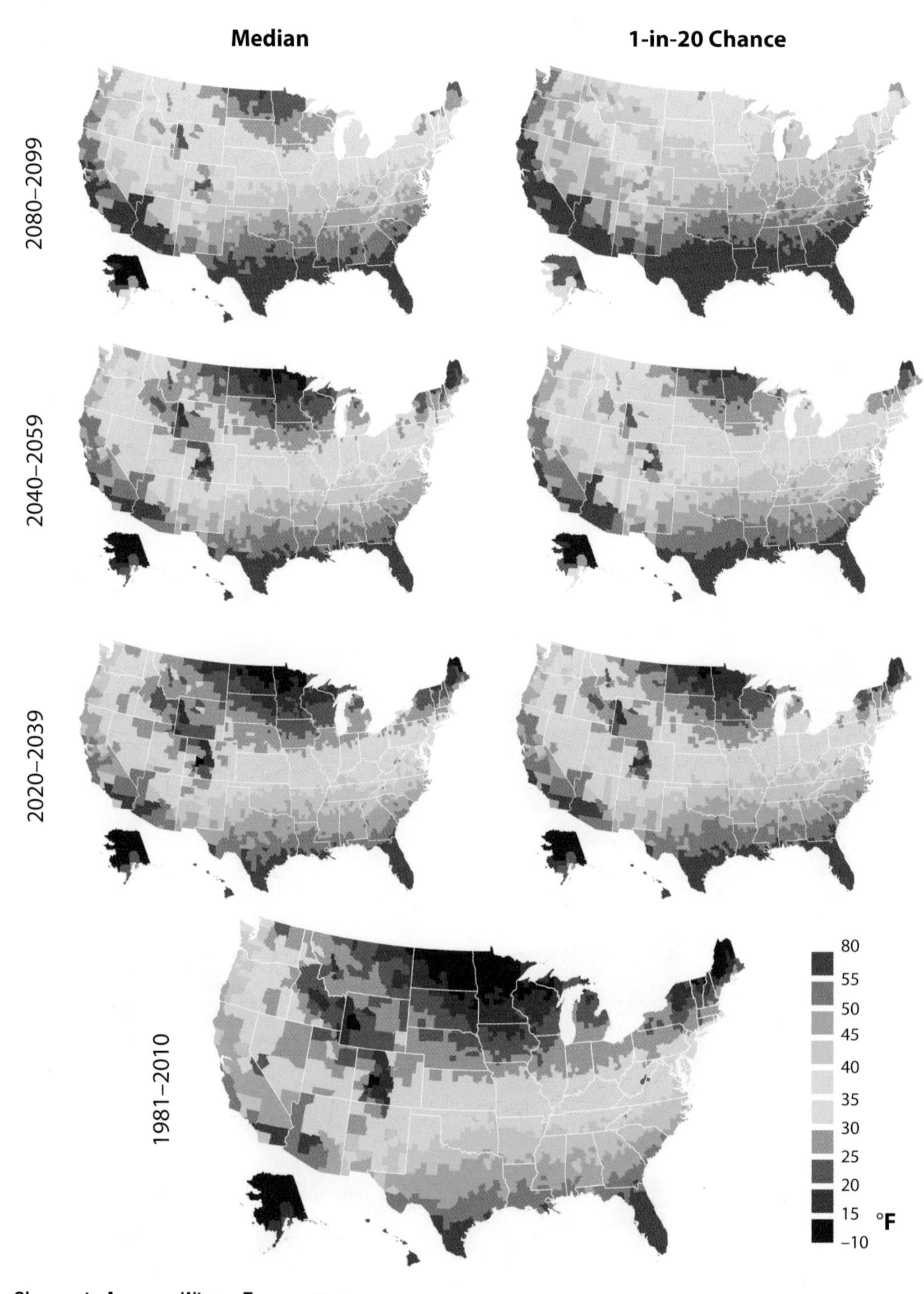

FIGURE 4.5. **Change in Average Winter Temperatures**

Daily average winter (December-January-February) temperature (degrees Fahrenheit) under RCP 8.5

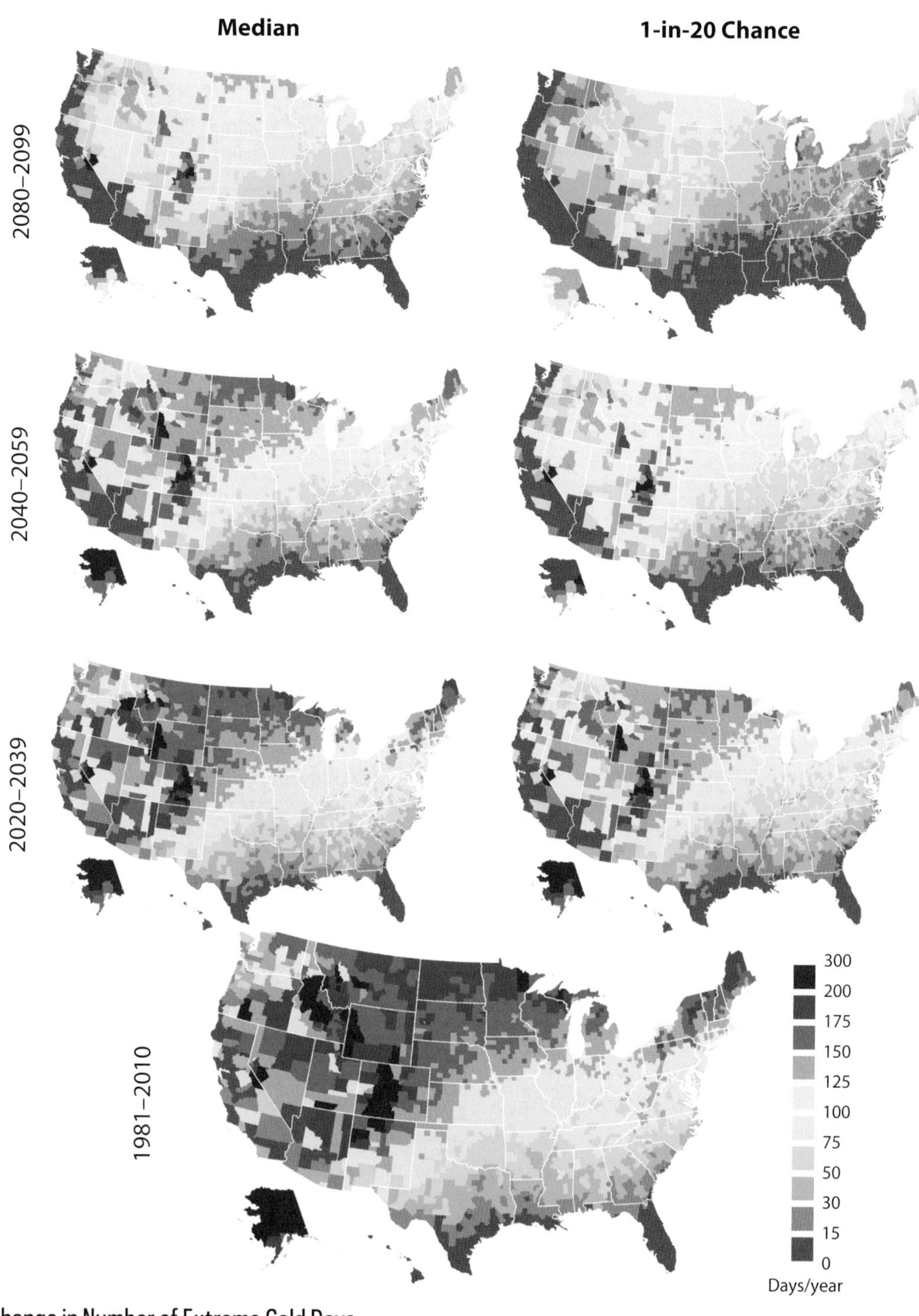

FIGURE 4.6. **Change in Number of Extreme Cold Days**

Number of days with minimum temperatures below 32°F under RCP 8.5

TABLE 4.1 The ACP Humid Heat Stroke Index

ACP Humid Heat Stroke Index	*Peak Wet-Bulb Temperature*	*Characteristics of the Hottest Part of Day*
I	74°F to 80°F	Uncomfortable. Typical of much of summer in the Southeast.
II	80°F to 86°F	Dangerous. Typical of the most humid parts of Texas and Louisiana in the hottest summer month and of the most humid summer days in Washington, D.C., and Chicago.
III	86°F to 92°F	Extremely dangerous. Comparable to conditions in the Midwest during the peak days of the 1995 heat wave.
IV	>92°F	Extraordinarily dangerous. Exceeds all U.S. historical records. Heat stroke likely for fit individuals undertaking less than one hour of moderate activity in the shade.

Projecting future increases in wet-bulb temperatures at the same resolution as the other analyses in this report is challenging. Indeed, assessing past wet-bulb temperatures precisely is tricky as well; differences between analytical methods and variations in humidity near weather stations can produce differences in historical estimates of up to about 4°F. Nonetheless, we can make some projections of future changes in wet-bulb temperature based on the observed relationships between dry-bulb (conventional) temperature and wet-bulb temperature. Note that in the Midwest, humidity is enhanced by transpiration from crops (Changnon, Sandstrom, & Schaffer 2003); changing agricultural practices and suitability, as well as the response of crops to higher CO_2 concentrations, may affect the likelihood of future extreme wet-bulb temperatures in this region in a way that we cannot account for in our projections.

Currently, the area expected to experience more than a month of dangerous category II+ conditions in a typical year is confined to coastal Texas and Louisiana. Under RCP 8.5, by midcentury it will expand to cover most of the Southeast up to Washington, D.C., and much of the Midwest as far north as Chicago (figure 4.7). A day or more of extremely dangerous category III conditions is expected in a typical summer in counties currently home to a quarter of the U.S. population. By the end of the century, dangerous category II+ conditions are expected to characterize most of a typical summer in most of the eastern half of the country. A week or more of extremely dangerous category III+ conditions is expected in counties currently home to half the U.S. population, with a third of the population expected to experience a day or more of record-breaking, extraordinarily dangerous category IV conditions in a typical year.

While projections for midcentury are similar across emissions scenarios, projections for the end of the century diverge significantly (figure 4.8). Under RCP 4.5, only a third of the population would be expected to experience at least one extremely dangerous category III+ day in a typical year; under RCP 2.6, only 1 in 25 Americans would thus be exposed. The expected number of extraordinarily dangerous category IV days drops dramatically relative to RCP 8.5 under lower emission scenarios. Under RCP 4.5, only one eighth of the population is expected to have at least a 1-in-10 chance of experiencing a category IV day in a typical year. Under RCP 2.6, category IV days remain rare for the entire population, with 97 percent of Americans having less than a 1 percent chance of experiencing such a day in a typical year.

Climate change does not stop in 2100, and under RCP 8.5 increasing numbers of category III and IV days could transform the face of the eastern half of the country in the twenty-second century. By the end of the next century, extremely dangerous category III+ days are expected to characterize most of the summer in most of the eastern United States, with extraordinarily dangerous category IV days expected for about a month of a typical year. RCP 4.5 limits the expected number of category III+ days to less than two weeks and the number of category IV days to less than half a week for almost the entire country. Under RCP 2.6, conditions at the end of the next century resemble those at the end of this one.

The combination of projected summer temperatures and projected ACP Humid Heat Stroke Index days

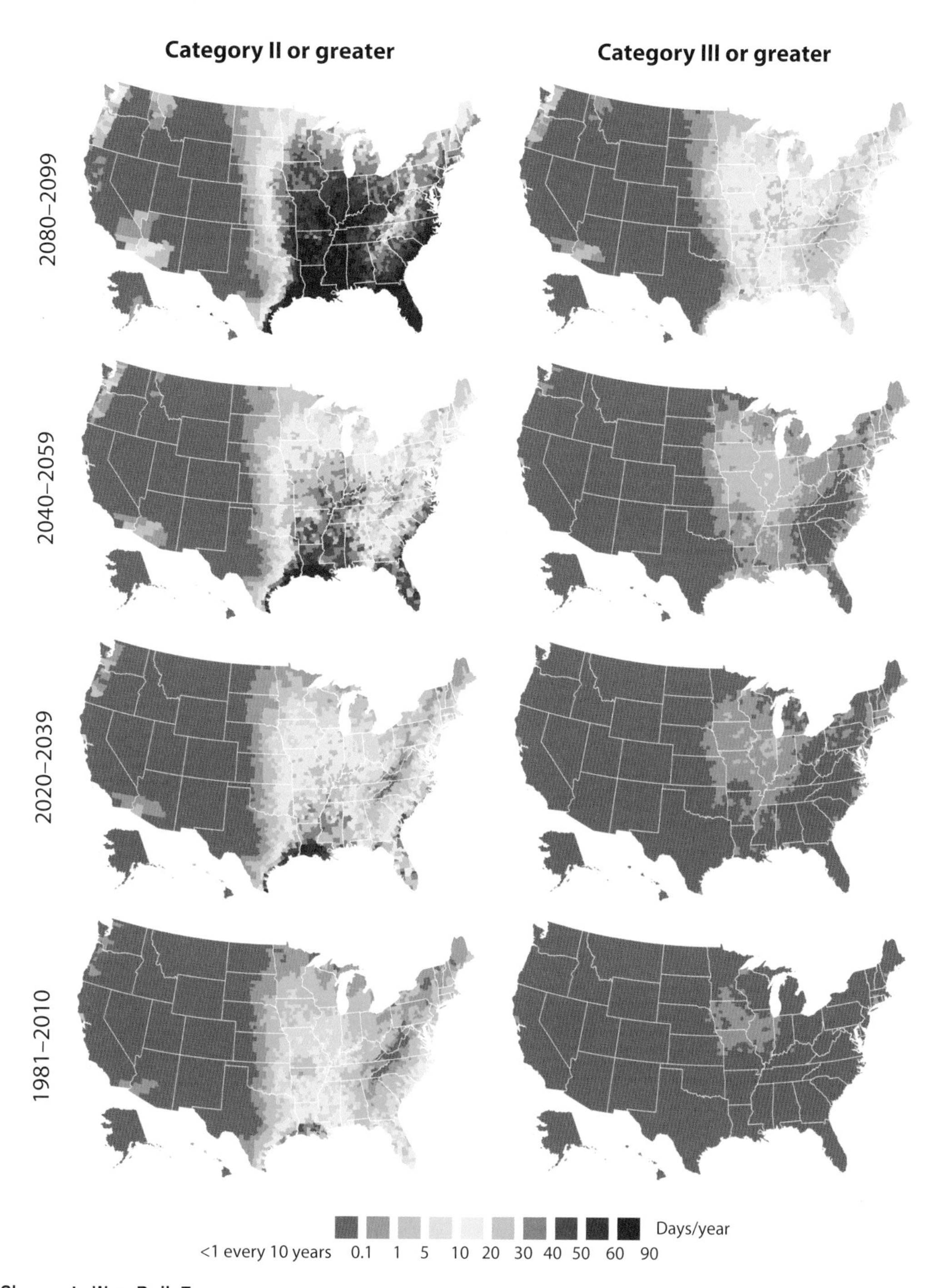

FIGURE 4.7. **Change in Wet-Bulb Temperatures**

Expected number of category II+ and category III+ ACP Humid Heat Stroke Index days in a typical summer under RCP 8.5

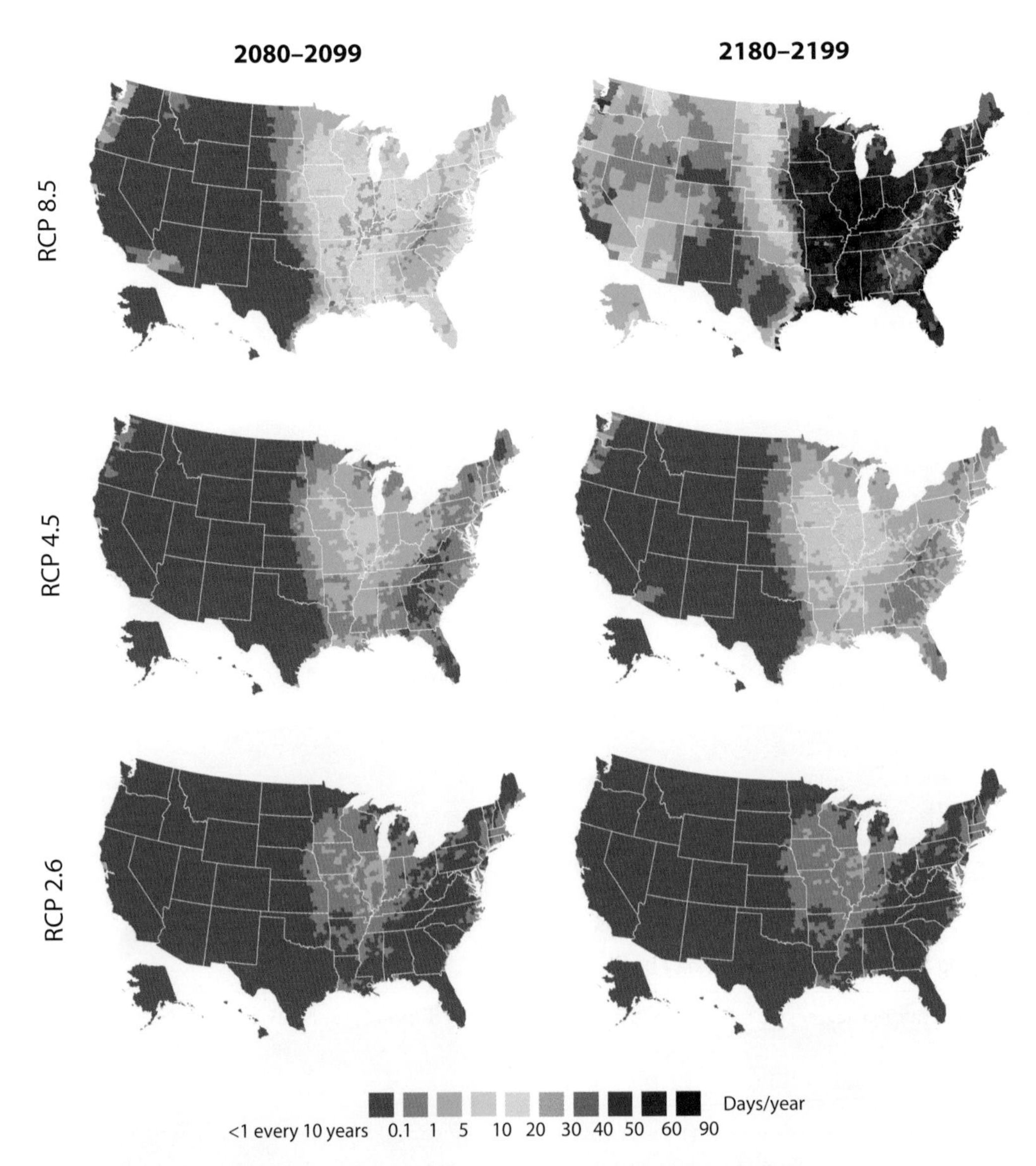

FIGURE 4.8. **Increasing Wet-Bulb Temperatures Under Different Long-term Emissions Pathways**

Expected number of category III+ ACP Humid Heat Stroke Index days in a typical summer

provides a sense of how the experience of future summers will change. By the period 2020–2039, for example, median projected average summer temperatures in Washington, D.C., match, and the expected number of hot, humid category II+ days experienced by Washingtonians exceeds, those in Mississippi today. By midcentury, median projected summer temperatures in Missouri under RCP 8.5 approach those in Florida today, while the expected number of hot, humid category II+ days experienced by the average Missourian exceeds those of Louisiana today. By the end of the century under RCP 8.5, the Northeastern, Southeastern, and Midwestern states south of the Mason-Dixon line have higher median projected summer temperatures than Louisiana today, and the residents of almost the entirety of those three regions—including the states north of the Mason-Dixon line—have more expected hot, humid category II+ days than does Louisiana today (figure 4.9).

While air-conditioning can allow humans to cope with extreme wet-bulb temperatures, habitability in the face of sustained extreme wet-bulb temperatures would require fail-safe technology and time-shifting of outdoor work to

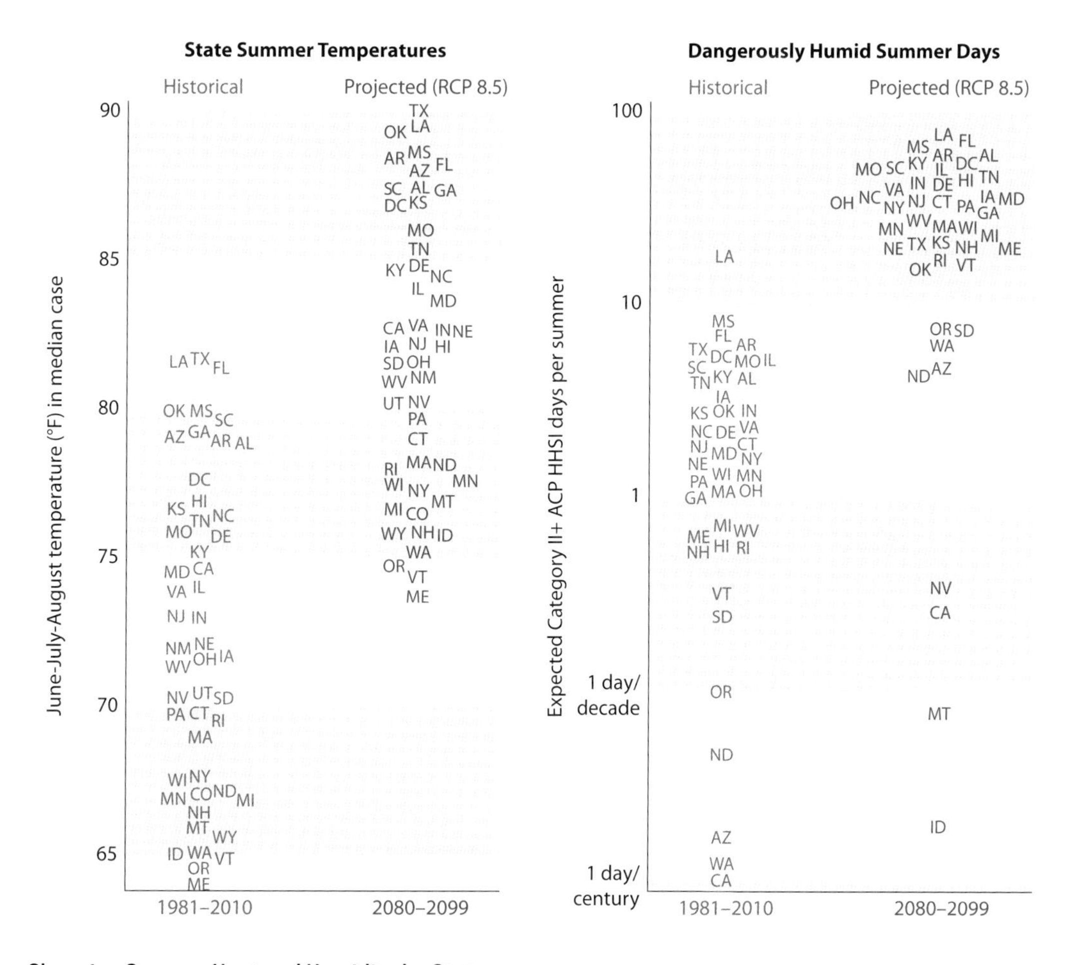

FIGURE 4.9. **Changing Summer Heat and Humidity by State**

Median summer temperature (degrees Fahrenheit) and category II+ ACP Humid Heat Stroke Index days per summer, 2080–2099

cooler (but likely still extremely unpleasant) parts of the day. Other species may not be as fortunate.

PRECIPITATION

Precipitation changes are more challenging to predict than temperature changes. Higher atmospheric temperatures will in general increase the absolute humidity of the atmosphere, making extreme precipitation events more likely. Higher temperatures will also increase evaporation, however, making extreme drought more likely. In general, wetter areas are expected to get wetter and drier areas drier, but much will depend upon changes in atmospheric circulation patterns, which could shift the dry subtropics poleward. High latitudes and wet midlatitude regions are *likely* to experience an increase in annual mean precipitation by the end of this century under RCP 8.5, while many midlatitude and subtropical dry regions will *likely* see decreases. Extreme precipitation events over most of the midlatitude land masses and over wet tropical regions will *very likely* become more intense and more frequent by the end of this century, as global mean surface temperature increases (Collins et al. 2013).

Across the contiguous United States, average annual precipitation will *likely* increase over the course of the twenty-first century. The spatial distribution of median projected changes in seasonal precipitation under RCP 8.5 is shown in figure 4.10. The Northeast, Midwest, and Upper Great Plains are *likely* to experience more winter precipitation. Wetter springs are *very likely* in the

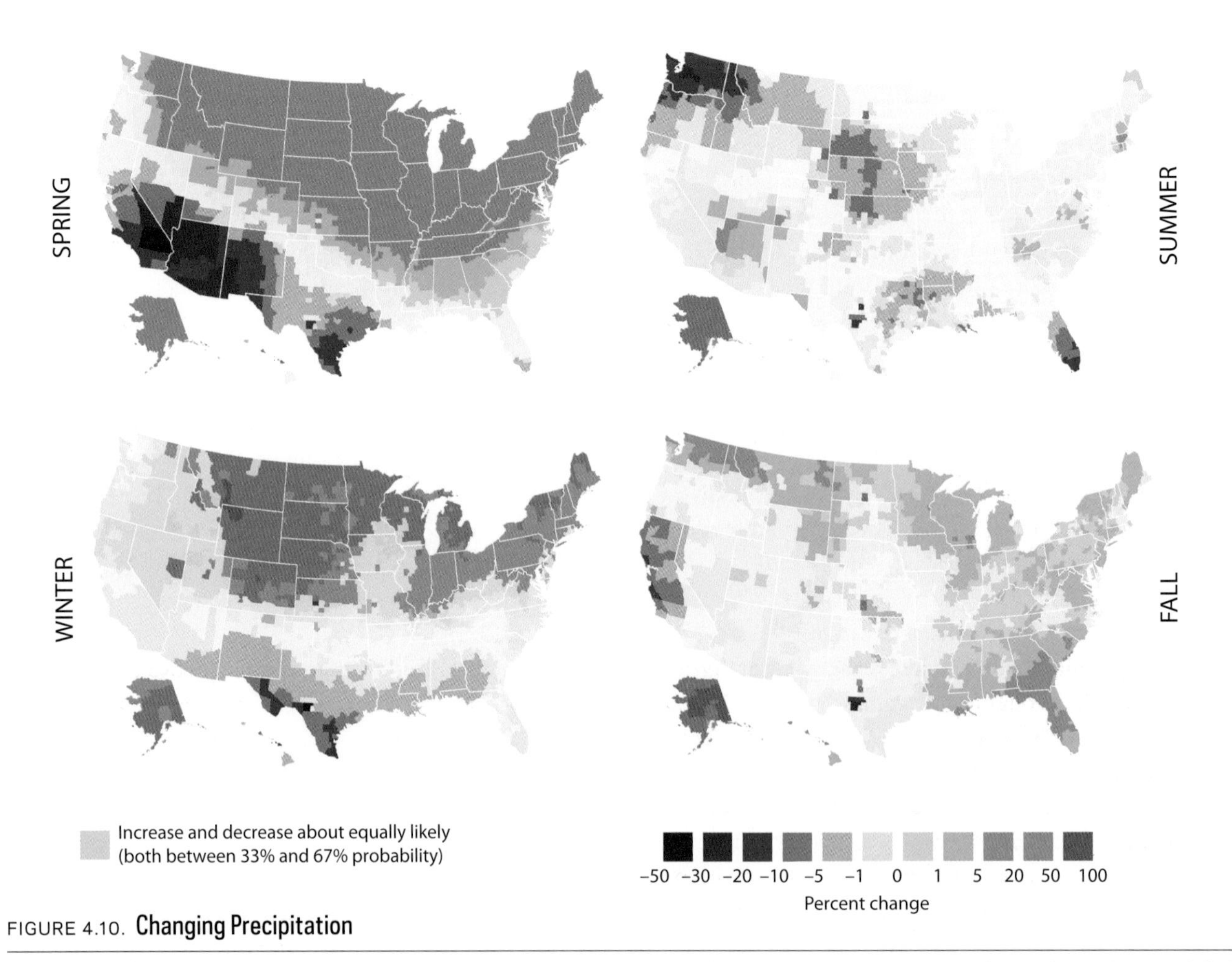

FIGURE 4.10. **Changing Precipitation**

Percentage change in average seasonal precipitation in 2080–2099 under RCP 8.5, relative to 1981–2010; for median values, in faded areas an increase and decrease are about equally likely to occur (both between 33 percent and 67 percent probability)

Northeast, Midwest, and Upper Great Plains and *likely* in the Northwest and Southeast. An increase in fall precipitation is *likely* in the Northeast, Midwest, Upper Great Plains, and Southeast. The Southwest is *likely* to experience drier springs, while drier summers are *likely* in the Great Plains and the Northwest (figure 4.10).

DROUGHT

Drought has multiple definitions. Meteorological droughts are defined by abnormally low precipitation, agricultural drought by abnormally low soil moisture, and hydrological drought by reductions in water supply through groundwater, reservoirs, or streams (Heim 2002). Projected decreases in precipitation in some regions and seasons—for example, the *likely* spring decrease in the Southwest and the *likely* summer decrease in the Great Plains and Northwest—make meteorological drought increasingly likely over the course of the century.

Projecting agricultural droughts is more challenging than simply projecting precipitation. Soil moisture is also affected by temperature—which increases evaporation—and by transpiration of plants, which may decrease in response to carbon fertilization (conserving soil moisture). Global climate models can explicitly model changes in soil moisture: Of the models participating in the Coupled Model Intercomparison Project Phase 5 (CMIP5) and reporting soil moisture, more than 90 percent projected a decrease in annual mean soil moisture in the Southwest by late century in RCPs 4.5, 6.0, and 8.5. More than 90 percent also projected a decrease in soil moisture in the Northwest and Great Plains under RCP 8.5

(Collins et al. 2013). In the western half of the country as a whole, averaging model results together, drought extent by area in a typical year, defined as soil moisture below the 20th percentile, is projected to increase from about 25 percent to about 40 percent (Wuebbles et al. 2013). Summer droughts are projected to become more intense in most of the continental United States because of longer dry periods and more extreme heat that increases moisture loss from plants and soils (Georgakakos et al. 2014; Walsh et al. 2014).

Our agricultural projections (see chapter 6), which incorporate historical relationships between temperature, precipitation, and crop yield, as well as future responses to changing carbon dioxide concentrations, implicitly estimate agricultural droughts. The 2012 drought provides a benchmark for assessment. That year saw drops in corn yield of 13 percent nationally, 16 percent in Nebraska, 20 percent in Iowa, and 34 percent in Illinois (U.S. Department of Agriculture 2013a). The Department of Agriculture dubbed the drought the nation's "worst agricultural calamity since 1988." Chapter 6 projects a 1-in-3 chance that, nationwide, grain yields will decrease by more than 33 percent under RCP 8.5 by late century, with drops in production concentrated in the generally unirrigated eastern half of the country—an indicator of the need to adapt to recurring agricultural shocks of the scale of the 2012 drought.

Hydrological droughts are even more challenging to project, as this involves tracing changes in precipitation and evaporation through to runoff and stream flow, and ultimately to implications for surface and groundwater supply. With continued high emissions of greenhouse gases, surface and groundwater supplies in the Southwest and parts of the Southeast and Southern Rockies are expected to be affected by runoff reductions and declines in groundwater recharge, increasing the risk of water shortages (Seager et al. 2013; Georgakakos et al. 2014). See chapter 17 for more on changes in water resources.

SEA-LEVEL CHANGE

Sea-level change is an important physical consequence of global warming, one that will increase flood risk along American coastlines severalfold over the course of the century (Kopp et al. 2014). At a global level, sea-level rise is driven primarily by thermal expansion (the increase in the volume of the ocean that occurs as it absorbs heat) and land-ice melt, with an additional contribution from groundwater withdrawal that over the twentieth century has been largely counterbalanced by dam construction. Both thermal expansion and land-ice melt are expected to increase significantly over the course of the twenty-first century; indeed, observations indicate an ongoing acceleration of ice-sheet melt in both Antarctica and Greenland (Shepherd et al. 2012; Church et al. 2013). However, there is considerable uncertainty regarding the future behavior of the ice sheets. This is particularly true for the West Antarctic Ice Sheet, much of which sits below sea level and may therefore be vulnerable to positive feedbacks that could lead to more than 3 feet of global mean sea-level rise over the century from this one source alone.

Under RCP 8.5, global mean sea level will *likely* rise by about 0.8 to 1.1 feet between 2000 and 2050 and by 2.0 to 3.3 feet between 2000 and 2100 (figure 4.11; Kopp et al. 2014). There is a 1-in-200 chance that sea level could rise by 5.8 feet, and in a worst-case projection reflecting the maximum physically plausible sea-level rise, global mean sea level could rise by as much as 8 feet. It is important to note that the estimates of tail probabilities involve a particular set of assumptions about likely ice-sheet behavior; feedbacks could render these extreme outcomes more likely than we project.

The uncertainty in ice-sheet physics plays a larger role in sea-level projections than scenario uncertainty, but lower greenhouse-gas emissions will lower projected sea-level rise, particularly in the second half of the century. Under RCP 2.6, global mean sea level will *likely* rise by about 0.7 to 0.9 feet by 2050 and by 1.2 to 2.1 feet by 2100. Under RCP 2.6, there is a 1-in-200 chance of a sea-level rise of 4.6 feet, and the worst-case projection is reduced to 7 feet.

Sea-level rise will not occur evenly across all regions of the globe. Understanding what global mean sea-level rise will mean for U.S. coasts requires consideration of several specific local factors (Kopp et al. 2014). First, ocean dynamics and the uneven distribution of ocean heat and salinity can cause unevenness in the height of the sea surface. For example, the height of the sea surface off the coast of New York is about 2 feet lower than off the coast of Bermuda (Yin & Goddard 2013). Climate change can affect these factors, with some models suggesting that changes in them could cause more than a foot of sea-level rise off New

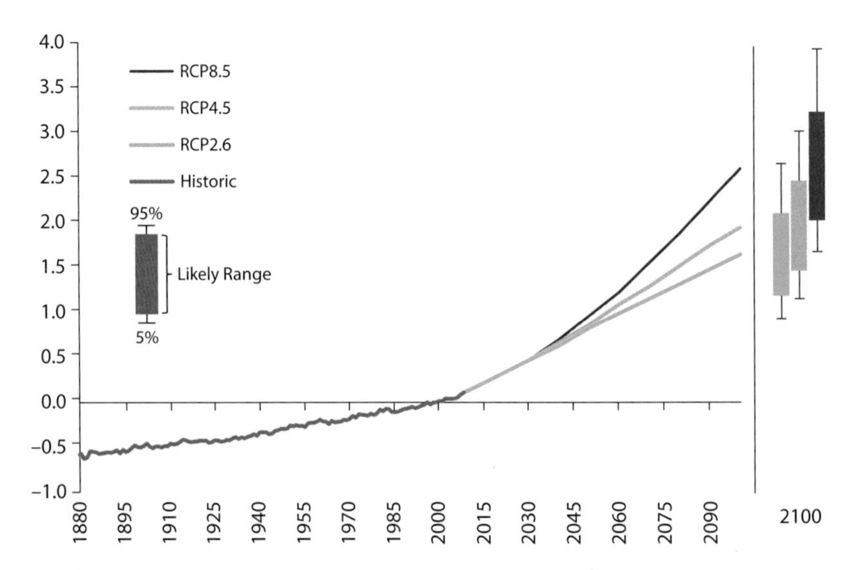

FIGURE 4.11. **Projected Global Mean Sea-Level Rise**

Measured in feet

York during the twenty-first century (Yin, Schlesinger, & Stouffer 2009). Second, redistributing mass—including land-ice mass—on the surface of Earth affects Earth's gravitational field, its rotation, and the way Earth's crust bends underneath loads (Mitrovica et al. 2011). Because of changes in Earth's gravitational field, sea level actually falls near a melting ice sheet: If the Greenland ice sheet melts, sea level will fall in Scotland, and the northeastern U.S. will experience less than half the associated rise in global mean sea level. Third, in areas that were near the margins of the great North American ice sheet of the last ice age, the solid Earth's ongoing adjustment to the loss of that ice sheet is continuing to drive sea-level rise. Fourth, in tectonically active regions such as the western United States, sea-level change can occur as a result of uplift or subsidence of the land driven by plate tectonics. Finally, in regions, such as the coastal plains of the mid-Atlantic and southeastern U.S., that rest on sand and other sediments rather than bedrock, regional sea-level rise can be driven by the compaction of these sediments. Such compaction can occur naturally, due to the weight of additional sediment deposited on the coastal plain, or artificially, due to the withdrawal of water or hydrocarbons from the sediments (Miller et al. 2013).

As discussed in greater detail in chapter 11, sea-level change will vary around the country (figure 4.12). Both the Atlantic and Pacific coasts of the continental United States will experience greater-than-global sea-level rise in response to West Antarctic melt. The highest rates of projected sea-level rise occur in the western Gulf of Mexico, due to the effects of hydrocarbon withdrawal, groundwater withdrawal, and sediment compaction. In the mid-Atlantic region, sea-level rise is heightened by the ongoing response to the end of the last ice age, potential changes in ocean dynamics, and—on the coastal plain sediments of the Jersey Shore and Delaware, Maryland, and Virginia—groundwater withdrawal and sediment compaction. In Alaska and, to a lesser extent, in the Pacific Northwest, sea-level rise is reduced by the changes in Earth's gravitational field associated with melting Alaskan glaciers. In Hawaii, far from all glaciers and ice sheets, sea-level rise associated with melting land ice will be greater than the global average.

EXTREME EVENTS

Projecting changes in the future occurrence of storms across the United States is subject to much greater uncertainties than temperature or sea-level rise. Relatively little is known about the influence of climate on the frequency and severity of winter storms and convective storms like tornadoes and severe thunderstorms. Since 1950, there has

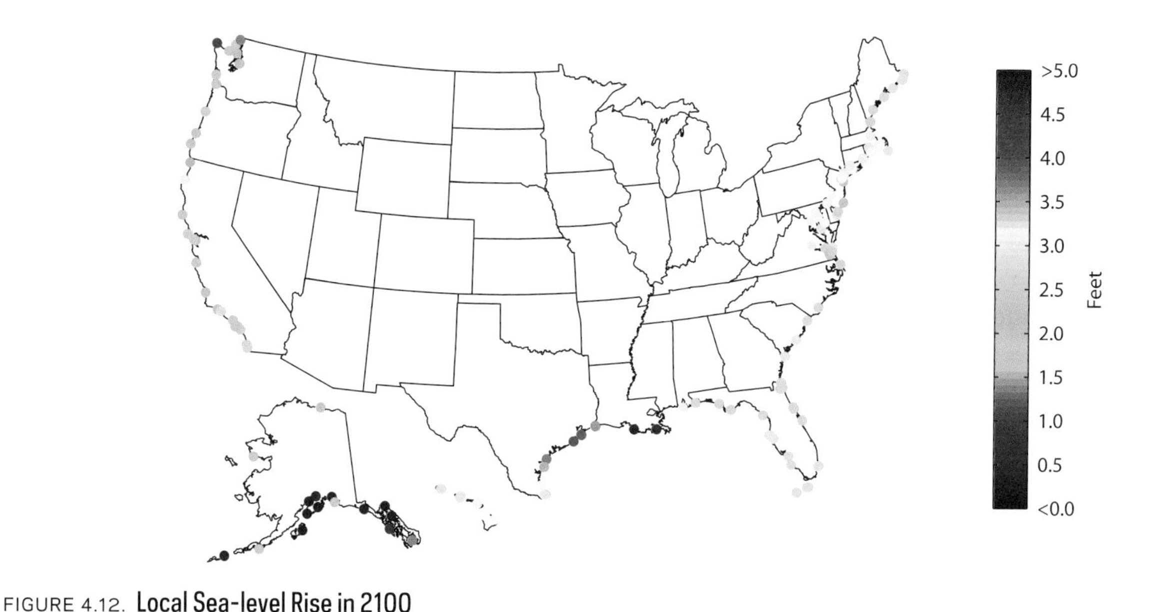

FIGURE 4.12. **Local Sea-level Rise in 2100**

Median projected change (feet) from 2000 levels under RCP 8.5

Source: After Kopp et al. (2014)

been no significant change in winter storm frequency and intensity across the United States, though the northeastern and northwestern coasts have experienced an increase in winter storm activity in the period since 1979 (Vose et al. 2012). The focus of most studies to date has been on understanding the relationship between changes in climate and Atlantic hurricane activity including changes in frequency, intensity, and duration—an area that is only just beginning to be understood.

Observational data show a marked increase in hurricane activity in the Atlantic since the 1970s. There is a robust correlation between increases in hurricane activity, as measured by the Power Dissipation Index (PDI), and rising sea-surface temperatures over that time (Emanuel 2007). The statistical correlation suggests the possibility of an anthropogenic influence on Atlantic hurricanes over the past few decades; however, numerous factors influence local sea-surface temperatures, and hurricanes respond to more than just sea-surface temperature. It remains difficult to attribute past changes in Atlantic hurricane activity to anthropogenic factors with any certainty. Although data going back to 1880 indicate a pronounced upward trend in the number of tropical storms in the Atlantic, much of this increase may be due to improved monitoring, making it difficult to determine long-term trends (Vecchi & Knutson 2008, 2011; Villarini et al. 2011).

Sea-surface temperatures in the North Atlantic basin are expected to continue to rise over the next century, along with changes in wind shear and other climate variables that influence hurricane formation (Vecchi & Soden 2007). Incorporating the best understanding of the complex interaction among these factors, several studies project further increases in the frequency and intensity of the strongest Atlantic hurricanes (Emanuel 2013; Knutson et al. 2013). One study, based on the RCP 4.5 pathway, found that although anthropogenic warming in the Atlantic basin over the twenty-first century will lead to a moderate reduction in tropical storms and hurricanes overall (of approximately 20 percent), the frequency of very intense hurricanes (categories 4 and 5) and the overall intensity of Atlantic hurricanes across the basin will *likely* increase (Knutson et al. 2013). While very intense hurricanes are relatively rare (accounting for only 15 percent of cyclones that make landfall in the United States), the damage they inflict on coastal communities is considerable, contributing more than half of historical U.S. hurricane damage over the past century (Pielke et al. 2008). According to Knutson et al. (2013), tropical storms and hurricanes are also expected to have higher rainfall rates under future warming, with increases by late century of approximately 10 percent and even larger increases (approximately 20 to 30 percent) near a hurricane's core.

Given the large uncertainties in changes in hurricanes over the course of the twenty-first century, we do not assume any changes in storm distribution in our base case. We run two side cases, one using the six climate models downscaled for RCP 8.5 by Emanuel (2013) and the other using the ensemble mean of the models downscaled for RCP 4.5 by Knutson et al. (2013).

Regardless of changes in storm activity, expected sea-level rise under all future concentration pathways will enhance flooding in coastal communities when storms do strike (Strauss et al. 2012). In the coming decades, even small changes in sea level will affect low-lying coastal areas as they push water levels associated with storm surges to progressively greater heights. This will increase the risk of extreme flood events along the U.S. coast, with once-rare floods occurring at a substantially higher frequency. In many areas, sea-level rise will by midcentury transform "1-in-100 year" coastal floods (which currently have a 1 percent chance of happening in any given year) into "1-in-10 year" events (a 10 percent chance in any given year) (Tebaldi, Strauss, & Zervas 2012).

PART 2

ASSESSING THE IMPACT OF AMERICA'S CHANGING CLIMATE

OPENING COMMENTARY

MICHAEL GREENSTONE

MILTON FRIEDMAN PROFESSOR IN ECONOMICS AND THE COLLEGE, DIRECTOR OF THE ENERGY POLICY INSTITUTE AT CHICAGO, UNIVERSITY OF CHICAGO

THIS report has changed my understanding of what climate change means for the United States, providing incredibly detailed visibility into regional changes in climate and their economic consequences to specific sectors. Given its unique ability to touch home for so many, I suspect the study will ultimately be considered a landmark contribution in our understanding of this complex and vital subject. Since its release, it has already influenced political discourse around climate change—no small feat in a world where people's views on this topic can seem immutable. Further, it has set a new bar for climate change research, shaping our exploration of the topic for decades to come.

In this short piece, I will highlight a few of the report's most striking findings. I will then survey the most urgent directions for research to build on this work and further our understanding of climate change and the best ways to confront and adapt to it moving forward.

One of the most important and illuminating aspects of the report is its recognition that climate change will affect different regions in different ways. Up until this point, researchers have emphasized climate change's impact on global temperatures. But this work recognizes that families, businesses, and governments need to know how climate change will affect them at a local level. Armed with this knowledge, such decision makers will be better equipped to adapt to a changing climate.

Two of many figures within the report (figures 4.3 and 4.4) that illustrate this approach focus on the changes states will encounter in their daily average summer temperatures. How many more ninety-five-degree days will your state experience? Right now, only a small region of the country—roughly, parts of Arizona and a good chunk of the Texas-Mexico border—experience this very hot weather. But by the end of the century, the report projects that very hot weather experienced by someone in Arizona today will envelope the entire Southeast and stretch as far north as the California Bay Area and far up the Atlantic Coast. Even for people who like hot weather (I am not one of them!), this dramatic shift is difficult to comprehend—I know it is for me.

The changes projected in sea level (see figure 4.12), are similarly daunting. Major cities that are the lifeblood of entire industries, that millions of people call home, and that serve as historical and cultural American hotbeds—New York, Miami, New Orleans, and many others—are threatened by rising seas. At a time when the government is having trouble maintaining core infrastructure, such as highways and bridges, how will it protect entire metropolitan areas from harm? My guess is: it can't. So how will the political system choose which places to save? Will it decide that Miami is of greater significance than New Orleans? I cannot even begin to guess how these decisions will be made.

A second major breakthrough in this work is its focus on the deep and expansive economic losses we can expect from these changes in climate. Its efforts in this respect are impressive for several reasons: they impose high standards for what qualifies as credible evidence; they unpack the effects in very different sectors; and they help shine a light on the sectors where the losses are expected to be the greatest.

Again, the ways in which different regions will experience these losses is particularly striking. My wife, Katherine, is from Arkansas, so I have great affection for the state. It is currently a relatively poor state yet seems poised to be among the hardest-hit areas from climate change. I'm sure that insurance, both private and public, as well as markets, will help to spread the costs more evenly around the country. But land cannot migrate, and today's residents of Arkansas and some other states appear likely to suffer disproportionately.

How does this report change the course of research moving forward? Thanks to the bar it sets, there are three urgent ways the research community can proceed.

First, the approach these researchers take in detailing how climate change would affect different U.S. regions and industries should now be expanded globally. There is a growing literature shedding light in this area, but the world needs something similar to this report to pull together these fragmented studies. An even better approach would be to construct a model—in the spirit of the one developed here—that could continuously update as new studies emerge.

Second, efficient policy depends on a reliable estimate of the social cost of carbon (SCC) or the damages from an extra ton of carbon dioxide (CO_2) emissions. These damages include effects estimated across the global economy, certainly including the sectors outlined in the report, but others too. The SCC can be used by governments around the world to value the benefits gained from various climate change policies, allowing them to evaluate which policies have net benefits and which are not worth undertaking. The most influential SCC estimates to date are derived from models that do not draw from the frontier evidence on the estimates of climate change's economic effects. Marrying a global analysis that follows the style of this report to the estimation of the social cost of carbon would be valuable.

Third, the standard approach used to estimate the social cost of carbon has three weaknesses that future work should confront. All of these weaknesses may cause the estimated social cost of carbon to be too low.

The first weakness in the current approach to estimating the social cost of carbon is that it treats climate damages as a known quantity. However, the estimates of climate damages are uncertain for a variety of reasons, including uncertainty about the climate sensitivity parameter, underlying damage estimates, and a variety of other factors. The problem is that people do not like uncertainty, especially when large losses are possible. This is reflected in economics jargon by "the coefficient of relative risk aversion" but can be seen in the robust markets for insurance around the world across a variety of domains (e.g., home, auto, life, etc.). This report developed a new, integrated approach for combining all of these different sources of uncertainty into its probabilistic damage projections. Future analyses should follow suit.

The second weakness: it models the losses for a "representative agent." At the heart of this assumption is the idea that all people in the world are identical in income, preferences, etc., as well as in exposure to climate damages. So when a model like the one used in this work predicts an average 3% loss in GDP in 2090, standard models would interpret this as *everyone* in the United States losing 3% of their income—ignoring that the losses may be unequal. The result is that it treats the following cases as identical to the 3% loss for everyone: 10% of the population loses 30% of their income and 90% suffer no loss; 3% of the population dies and 97% are unaffected. Obviously, society treats these examples as substantially worse outcomes than the equal 3% loss. This is because of an aversion to a subgroup of individuals bearing a disproportionate share of society's loss—an issue which is also touched on, albeit incompletely, in chapter 15.

An emerging literature, as well as this work, demonstrates that the assumption of equal damages is unlikely to hold. This literature finds that the damages in Miami will be much greater than in Minneapolis. It also suggests that the damages will be larger for certain categories of individuals, e.g., construction workers versus office workers or the poor versus the rich. As discussed above, it strains credibility to assume that insurance markets are developed enough to spread the costs equally around the country or the globe (e.g., what insurance product is available to people living in coastal Bangladesh?).

The degree of regional and sectoral disaggregation in the report makes it clear that the assumption of uniform impact is unnecessary, and future work should vigorously explore the degree of unequal impact. Figures 13.18 and 13.19, for example, show that coastal damages are concentrated in specific parts of specific counties.

One level of disaggregation not present in this report but important for future work is the distinction between rich and poor within a single geographic area. This work shows that projected mortality, for instance, is greatest in our nation's poorest states. But it does not show what is likely also true: that within poor states, the poorest will suffer the most. Internationally, some of my own work finds that the mortality effects of higher temperatures will be larger in rural India than in wealthier India, supporting this broader conclusion

The third weakness in the standard approach to estimating the social cost of carbon assumes that climate damages are proportional to GDP. This assumption would justify using a discount rate that is comparable to the average rate of return in the stock market, say 5–7%. But some research has found that climate-mitigation investments—unlike investments in the overall stock market—may pay off when the economy is doing poorly and additional income is especially valuable, not when it is doing well. If this alternative assumption is correct, it would push society to use a substantially lower discount rate. It is instructive to consider gold, which is an asset class that pays off in tough times and has an annual rate of return of about 2%. The reason that people are willing to hold gold and other assets with low mean annual rates of return is exactly because they pay off when other investments and income opportunities are poor.

The choice of a discount rate in the range of the returns of assets like gold would greatly increase the present value of climate damages from the release of an additional ton of CO_2 today and, in turn, the social cost of carbon. This notion was not addressed in the report, which, as discussed in chapter 13 of this book, calculated risks that today's economy would face under the climate of the future and examines how these risks evolve over time rather than discounting them to estimate a net present value.

As greenhouse-gas emissions continue to accumulate in the atmosphere, it is critical that decision makers have a reliable estimate of the social cost of carbon, as well as a fuller sense of the ways in which climate change will affect them and their constituencies. This landmark study has opened some promising pathways toward improved approaches. With further efforts that build on this report, researchers will be in a position to provide decision makers with the tools and knowledge they need to confront and adapt to climate change.

CHAPTER 5

AN EVIDENCE-BASED APPROACH

How do we assess the impact of the potential changes in temperature, precipitation, sea level, and storm patterns described in the previous chapter on our homes, businesses, and communities? Anticipating climate impacts is in many ways even more analytically challenging than projecting climatic changes, as human systems are not constrained by laws as rigid as those of physics and chemistry that shape the natural world. Yet, by piecing together evidence from the distant and not-so-distant past, including what we have experienced in our own lifetimes, we can begin to identify common patterns in how populations respond to climatic conditions, and then use this information to assess the impact of climate change, both positive and negative, in the United States in the years ahead.

PALEOCLIMATIC EVIDENCE

Hints of the physical effects of climate change and suggestions of their possible impact on humans and ecosystems can be found buried deep in the geologic record. As discussed in chapter 3, the greenhouse-gas concentrations and temperatures projected for the twenty-first century have never before been experienced by human civilization, but they have occurred in our planet's past.

The last time global mean temperature was warmer than that of today was during the Last Interglacial stage, some 125,000 years ago. Temperatures during that period may have been as much as 2.5°F warmer than at present (Turney & Jones 2010), comparable to levels expected by midcentury under all scenarios. The geologic record shows that global mean sea level during this interval was 20 to 30 feet higher than that of today (Kopp et al. 2009; Dutton & Lambeck 2012)—a magnitude of change that will not be realized in this century but could occur over the coming centuries in response to warming. Such dramatic sea-level rise would swamp nearly all of Miami, Norfolk, New Orleans, Savannah, and Charleston.

The rate at which we are putting greenhouse gases into the atmosphere has no known precedent in the geologic record before at least 56 million years ago. At that distant time, within a period that may have been as short as a decade or as long as a few millennia, the Paleocene-Eocene Thermal Maximum (PETM) began with a massive release of carbon dioxide and methane that caused global mean temperatures to rise by 9°F to 14°F, on top of a baseline already several degrees warmer than today (Zachos, Dickens, & Zeebe 2008; McInerney & Wing 2011; Wright & Schaller 2013). While there were no humans around to experience it, other animals did. The warming—comparable to that possible in the twenty-second century under RCP 8.5—lasted tens of thousands of years and led to dramatic ecological shifts, including the dwarfing of land mammals as a result of heat stress (Gingerich 2006; Sherwood & Huber 2010).

In more recent millennia, human populations have been subjected to long-term climatic shifts lasting decades to centuries. By linking paleoclimatic reconstructions to archeological data, researchers have amassed a growing body of evidence that these historical shifts are systematically related to the migration, destabilization, or collapse of these premodern societies (Hsiang, Burke, & Miguel 2013). For example, abrupt drying or cooling events have been linked to the collapse of populations in ancient Mesopotamia, Saharan Africa, Norway, Peru, Iceland, and the United States (Ortloff & Kolata 1993; Cullen et al. 2000; Kuper & Kröpelin 2006; Patterson et al. 2010; D'Anjou et al. 2012; Kelly et al. 2013). The iconic collapse of the Mayan civilization has been linked to extreme droughts superimposed on sustained multicentury regional drying (Haug et al. 2003; Kennett et al. 2012), the fifteenth-century collapse of the Angkor city-state in modern-day Cambodia occurred during sustained megadroughts (Buckley et al. 2010), and the collapse of almost all Chinese dynasties coincided with periods of sustained regional drying (Zhang et al. 2006; Yancheva et al. 2007).

Economic development and technological advances (like air-conditioning) have, of course, made humans of today more resilient to climatic changes than humans in the past, so we do not think it is appropriate to use paleoclimatic evidence in contemporary climate risk assessment. These historical examples demonstrate how ecologically and economically disruptive climatic change has been in the past, even though we have archeological evidence that these past societies attempted to adapt to the climatic changes they faced using innovative technologies. Thus these anecdotes, if nothing else, motivate us to carefully consider low-probability but high-cost outcomes.

EMPIRICAL ESTIMATES

Another strategy for understanding climate's potential impact on human and natural systems—what we'll call the empirical approach—is based on evidence of actual impacts and damages experienced in the not-so-distant past. This approach also uses the historical record to assess future risks, rather than rely on "proxy" data buried in the geologic record, and it draws on data recorded and analyzed during modern times. There has been an explosion of econometric research in recent years examining the relationship between temperature and precipitation and current human and economic activity. When combined with the high-resolution output from global climate models, this research enables a granular assessment of the risks particular regions of the country or sectors of the economy face in the years ahead. The Spatial Empirical Adaptive Global-to-Local Assessment System (SEAGLAS) approach uses these empirical findings for impact categories with a sufficiently robust body of econometric research. This includes

1. Agriculture: The impact of projected changes in temperature and precipitation on maize, wheat, soy, and cotton yields
2. Labor: The change in number of hours employees in high-risk (construction, utilities, mining, and other) and low-risk (indoor services) sectors of the economy work in response to projected temperature change
3. Health: Changes in all-cause mortality for different age groups resulting from projected changes in temperature
4. Crime: The sensitivity of violent and property crime rates to projected temperature and precipitation
5. Energy: The impact of temperature change on U.S. electricity demand

When trying to use data from the real world to understand the influence of climate on society, the key challenge is separating the influence of the climate from other factors. For example, if we tried to study the effect of warmth on mortality by comparing a warm location like Florida to a cooler location like Minnesota, it might look like Florida had a higher mortality rate due to climate alone, but there are many other factors that make Florida different from Minnesota—such as the fact that the population of Florida tends to be older on average.

To get at these questions more reliably, we could imagine an ideal (but impossible) scientific experiment where we take two populations that are identical and assign one to be a "treatment" group that is exposed to climate change and one to be a "control" group that is exposed to a preindustrial climate. If we then observed how outcomes, such as mortality or productivity, changed between these two groups, we could be confident that the change in the climate caused the change in these outcomes.

Because we cannot do this ideal experiment, econometricians have looked for situations where natural conditions approximate this experiment; that is, "natural experiments." In these situations, individuals or populations that are extremely similar to one another are assigned to slightly different climates that result from random circumstances, and we observe how those small changes in climate are then reflected in economically important outcomes. To ensure that "control" and "treatment" populations are extremely similar to one another, the strongest studies compare a single population to itself at different moments in time when it is exposed to different climatic conditions. In this way, we know that most or all other important factors, such as local geography, politics, demographics, and so forth, are the same and that changes we observe are driven by the observed random changes in the climate. This approach allows researchers to construct *dose-response functions* (a term adopted from medicine), which describe a mathematical relationship between the "dose" of a climate variable that a population experiences and the corresponding "response" that the population exhibits in terms of economic outcomes.

In developing empirically based dose-response functions, we rely only on studies that account for temporal patterns that are often important factors in the outcomes we observe and might be correlated with small changes in the climate. For example, there is seasonality in crime rates and mortality, and these seasonal patterns may differ by locality, so it is critical that seasonality is accounted for because it will also be correlated with climatic conditions. Thus, the studies we rely on only compare how an outcome for a specific location, at a specific time of year, compares to that same outcome at that same location and time under slightly different climatic conditions. For example, a study might examine the number of minutes an average individual works on a Tuesday in May in Rockland County on a day that is 80°F compared to that on a day that is 70°F.

The climate of a location is neither the conditions on a specific day nor the average conditions throughout the year, but rather the distribution of conditions throughout the year. Two locations might have the same annual average temperature, but while one location may have very little daily or seasonal variation around that temperature (e.g., San Francisco), the other location may have tremendous daily and seasonal variation (e.g., New York). Therefore, we rely primarily on studies that measure responses to a complete distribution of daily temperature and rainfall measures. By decomposing outcomes as a response to the full distribution of daily temperatures that are experienced, we can more accurately characterize how populations will respond to changes in those distributions. For example, in the future some locations might see higher rainfall variance (more intense storms and more dry periods) but little shift in their average conditions.

Some critics suggest that considering the distribution of daily conditions conflates "weather" or "climate variability" with "climate." Often, these critics argue in favor of more simplistic approaches where outcomes are simply correlated with average conditions, but this alternative ignores the fact that individuals experience their local climate one day at a time, making decisions about their actions based on these daily events that they experience. Few individuals make choices about their daily activities based on what they expect annual mean temperature to be for the coming year, and in fact most individuals do not even know the average climatic conditions of the location they live in. Often, individuals will adapt to their local climate based on what they perceive the distribution of daily conditions to be; for example, Chicagoans buy winter coats because they expect some days in the winter to be cold. It is therefore essential that we consider daily distributions to model these adaptive decisions. While it is true that Chicagoans may have less need of their coats if average Chicago temperatures increase, it is unlikely that winter coats will be discarded entirely as long as there is a reasonable likelihood that some days in a year will be below freezing. Thus, information on the distribution of daily outcomes is more informative for adaptive behavior than averages.

For additional reasons, understanding responses to daily climatic conditions is a particularly powerful approach for economic policy analysis. First, it enables us to carefully identify nonlinear responses that have proved to be critical in these sorts of analyses. In many cases, such as agricultural yields, variations in temperature or rainfall do not have a substantial effect on outcomes until sufficiently extreme conditions are reached, at which point outcomes may respond dramatically. Disentangling these nonlinearities is essential to our analysis, as many of the important changes in the climate will occur at these extremes.

Second, by examining how different populations respond to the same daily conditions, we can begin to understand how populations adapt to climatic conditions in the long run. Reactions on hot days often differ in regions that are usually hot from those in regions that are usually cool. Because populations in hot regions may have adapted to their climate, for example, through infrastructure investments or behavior changes, we will be able to observe the effectiveness of this adaptation by comparing how the two populations respond to physically identical events (e.g., a 90°F day). In some cases, we are even able to study how populations at a single location change their response to the climate over time—allowing us to observe how outcomes might change (or fail to change) as new technology is developed and adopted (or not).

Finally, by identifying the effect of specific daily events on outcomes, we are able to naturally link empirically derived responses to climate models that simulate future environmental conditions on a daily basis. Because we are able to compute the daily average, minimum, and maximum temperature, as well as rainfall, at each location throughout the country on a daily basis for each run of a suite of climate models, it is straightforward to estimate how outcomes at each location will be expected to respond to any of these future scenarios.

For each scenario that we model using empirical dose-response functions, we project changes in future outcomes relative to a future in which climate conditions are unchanged from those of the year 2012. Although it is common practice to compare future temperature changes to a preindustrial baseline, climate change has been under way for many decades. We therefore focus our economic risk assessment on the ways in which future climate change may alter our economic future relative to the economic reality that we know today, which has already been partially influenced by the climate changes that have already occurred. This idea is illustrated in figure 5.1, which shows how the probability distribution of potential future outcomes in an example sector (low-risk labor supply) changes in RCP 8.5 relative to historical impacts estimated using the same method. Throughout this report, we present impacts relative to recent conditions, when recent conditions may already be somewhat different from long-term historical patterns.

For each of the impacts we examine, we carefully scrutinize the existing literature to select those studies that we think most credibly identify the effect of climate on specific outcomes. (We have engaged extensively with the original researchers of these studies to ensure an accurate representation of their results, and in several cases these authors provided additional analysis to help us better integrate their findings.) We restrict our analysis to studies that (1) are nationally comprehensive (or representative)

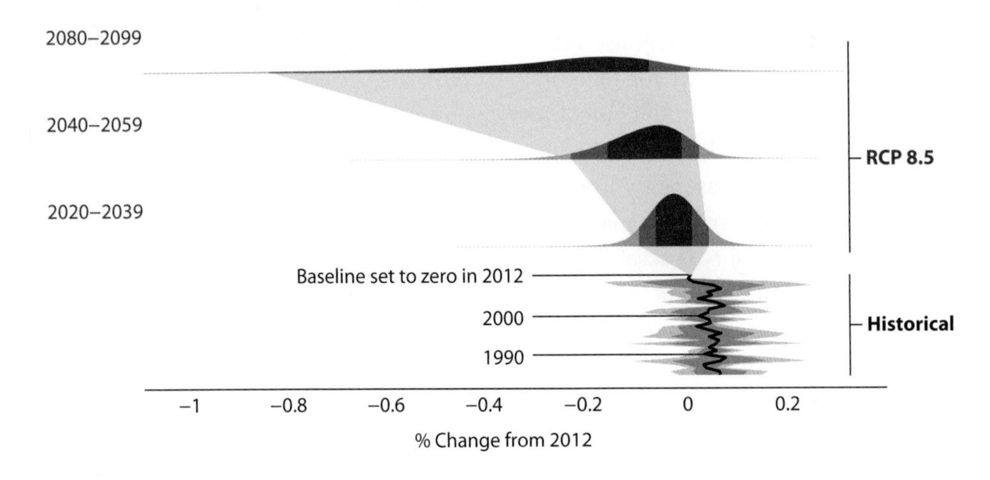

FIGURE 5.1. **Example of Projected RCP 8.5 Impacts Relative to 2012 Baseline Estimates**

Future probability distributions of 20-year average lost low-risk labor supply relative to impact estimates in 2012, with historical annual estimates also shown in black (likely range and 90 percent confidence interval based on statistical uncertainty shown in gray)

in scope, (2) analyze recent history in the United States, (3) account for all unobserved nonclimate factors that differ across locations (usually counties), and (4) break down the outcomes into responses to daily or weekly conditions. For this assessment, we have been especially conservative in the selection of empirical studies we rely on, although we expect that, in the future, additional studies will further improve our understanding of these relationships. To ensure that future assessments can build on our analysis, we have developed the Distributed Meta-Analysis System (accessible at http://dmas.berkeley.edu) (Rising, Hsiang, 2014), which enables future researchers from around the world to introduce new findings seamlessly.

There are limits to what econometric research alone can tell us about the future, however. Some of the projected temperatures and precipitation patterns described in chapter 2 are outside the range of empirical evidence, and we limit our econometrically derived impact functions to historical experience. In a United States that is on average 10°F warmer, most days will have past analogs that can be used for developing empirical models, but many will be record-breaking and have no such precedents. We have not experienced sea-level rise of 3 feet (our median projection for RCP 8.5 in 2100) or seen what that will do to Miami. Empirically based sectoral models can help fill in some of these gaps.

DETAILED SECTORAL MODELS

There are a number of sector-specific models developed by academic researchers, government agencies, and private industry that can be useful in exploring the potential impact of future changes in climate. These models are calibrated using empirical estimates but can also be used to explore temperature, precipitation, sea-level rise, or storm changes outside the range of historical experience or to analyze market interactions not captured in econometric research. We use two such models in this assessment.

Energy

Existing econometric research on the relationship between temperature and energy demand is limited to electricity. To capture a broader range of fuels, as well as the impact of changes in demand on energy prices, we employ RHG-NEMS, a version of the U.S. Energy Information Administration's National Energy Modeling System (NEMS) maintained by the Rhodium Group. NEMS is the most detailed publicly available model of the U.S. energy system and is used to produce the Energy Information Administration's Annual Energy Outlook, the most commonly used forecast of U.S. energy supply and demand. We model changes in regional residential and commercial electricity, natural gas, oil, and coal demand that would likely occur in potential climate futures (comparing modeled results to empirical estimates where possible), and what that implies for energy prices and the composition of energy supply.

Coastal Communities

The insurance and finance industries use sophisticated and extremely detailed models of hurricanes and other storm activity and impacts on coastal property and infrastructure. Risk Management Solutions, Inc. (RMS), is the world's leading developer of hurricane and other catastrophic risk models and a partner in this assessment. RMS's North Atlantic Hurricane Model combines extensive empirical evidence of past hurricane activity with a wind and surge model that simulates the wind and flooding damage likely to result from a given storm. We use RMS's building-level exposure database to identify property at risk from mean local sea-level rise and the North Atlantic Hurricane Model to assess the increase in hurricane and nor'easter flood damage likely to occur as result of that sea-level rise. Using input from the leading cyclogenesis models, we also explore how changes in hurricane activity as a result of climate change could shape wind and flood damage in the future.

OTHER IMPACTS

There are many potential climate impacts beyond those listed above that are of profound importance to the functioning of the U.S. economy and the lives of most Americans, among them impacts on national security, tourism, wildfires, water resources, and ecosystems. To date, there is not yet a sufficient body of U.S.-based econometric research from which to develop an econometrically derived damage function or an empirically based sectoral model capable of robustly analyzing potential climate impacts in these areas. That does not, however, mean these impacts should be ignored. Indeed, the impacts that are hardest to quantify could end up being the most costly. In part 4, we describe the universe of potential climate impacts not captured in this assessment. More important, we provide a framework and a platform for quantifying these impacts in the future as research improves.

ADAPTATION

An important question in any climate impact assessment is the extent to which businesses, households, investors, and policy makers will be able to adapt to potential changes in temperature, precipitation, sea levels, and storm activity. Will coastal communities build walls to guard against rising seas? Will farmers develop and deploy heat- and drought-resistant seeds? As a principal objective of our research is to give decision makers the information they need to make those long-term adaptation investments, we exclude them from our baseline assessment. In part 5, we explore the extent to which both adaptation investments and global greenhouse-gas emission reductions can shield the U.S. economy from future climate risks.

CHAPTER 6
AGRICULTURE

AGRICULTURE has long been an economic and cultural foundation for the United States. Known for historical boom and bust cycles, agricultural productivity and incomes are often influenced by and in turn influence the U.S. economy as a whole (Landon-Lane, Rockoff, & Steckel 2011; Hornbeck & Keskin 2012; Feng, Oppenheimer, & Schlenker 2013). In agriculture-dependent regions, the extreme drought and environmental mismanagement of the Dust Bowl in the early 1930s exacerbated the already dire economic conditions of the Great Depression (Egan 2006; Hornbeck 2012). Climate and weather variability have played roles to varying degrees in the cycles of U.S. agriculture. Extremes in local and regional weather patterns and climate variability have disrupted agricultural production in the past. American farmers have developed production practices and strategies appropriate for their local conditions, taking into account long-term historical trends as well as the risks of short-term variability. Despite the flexibility of the U.S. agricultural system and advances in agricultural practices and technologies, U.S. production and prices remain highly dependent on climate, making the sector particularly vulnerable to both gradual climate change and extreme climate events.

The agricultural sector's central role in rural and local economies and the national economy, as well as its importance for human health and security, make understanding the economic risks posed by climate change important not only for agricultural states but also for farmer livelihoods, rural communities, and the U.S. economy as a whole. The United States produced more than $470 billion in agricultural commodities in 2012. Although agriculture has traditionally contributed less than 2 percent of U.S. GDP, it is a much more significant source of income for many Midwest and Great Plains states such as North Dakota, South Dakota, Nebraska, and Iowa. Although a small share of California's overall economy, the state's agricultural contribution is significant, producing more than 10 percent of the value of all U.S. agricultural commodities in 2014, and nearly half of U.S.-grown fruits, nuts, and vegetables.

American farmers, ranchers, and the agriculture sector as a whole are familiar with making decisions in the face of uncertainty, which arises not just from variability in weather patterns but also from fluctuations in a whole host of other factors including trade dynamics, shifts in market demands and consumer preferences, evolution of agricultural technologies, and ever-changing state and federal policies. Risk-based decision making must take each of these factors into account. Managing the risks associated with climate change will require the integration of the potential risks of climate on agricultural productivity and prices into decision making by those involved in the full value chain of agricultural production.

In assessing the risks that climate change poses to agricultural productivity, there are a whole host of variables to consider, including temperature; precipitation; availability of water resources for irrigation; CO_2 concentrations; ozone and other pollutant concentrations; and climate-driven changes in pests, weeds, and diseases. The relative importance of each of these variables will vary based on the region and the crop or livestock type. In this analysis, we focus on the impact of changing temperatures and precipitation on commercial crop yields (including grains, cotton, and oilseeds) in areas where they are currently grown in the United States. We discuss other effects in more detail in the sections that follow.

BACKGROUND

On the whole, agricultural yields have increased across the United States during the past quarter of a century due primarily to dramatic improvements in agricultural techniques and secondarily to increases in temperature and precipitation. Studies isolating climate-related effects observed to date have shown that, on average, crops were more affected by changes in temperature than by precipitation, though temperature played a greater role in increased yields in central and northern regions, with higher precipitation contributing in the southern United States (Sakurai, Iizumi, & Yokozawa 2011). However, in the past 15 years there has been a marked increase in crop losses attributed to climate events such as drought, extreme heat, and storms, with instability between years creating significant negative economic effects (Hatfield, Cruse, & Tomer 2013). Understanding the potential risks to the highly varied agricultural regions across the United States requires an assessment of both the changes in average climate variables and changes in the intensity and frequency of extremes.

Historical changes in temperature have varied both across regions of the United States, with more significant changes in the Midwest and Southwest, and by season, with greater winter and spring warming. Overall, warming has lengthened the growing season by 4 to 16 days since 1970 (U.S. Environmental Protection Agency 2012). Final spring frost is now occurring earlier than at any point since 1895, and the first fall frosts are arriving later (U.S. Environmental Protection Agency 2012). Changes in the length of the growing season can have both positive and negative effects, as they may allow farmers to have multiple harvests from the same plot. However, they may preclude certain crops, lead to significant changes in water requirements, or disrupt normal ecosystem functions such as the timing of pollination and natural protection against weeds and invasive species.

Rising temperatures are expected to further lengthen the growing season across most of the United States (by as much as a month or two over the course of the century) and reduce the number of frost days, particularly in the West (Walthall et al. 2013). While longer growing seasons may be a boon to agriculture in some regions, the overall impact on yields will also be influenced by associated increases in exposure to warmer temperatures over greater time spans. While warmer average temperatures and increased precipitation over the past few decades have contributed to increased yields, this trend is unlikely to continue as temperatures rise across much of the United States. Crop species display temperature thresholds that define the upper and lower boundaries for growth, and the current distribution of crops across the United States corresponds to temperatures that match their thresholds (Hatfield et al. 2014). The impacts on yield are nonlinear as temperatures reach and then exceed a crop's threshold. When paired with declining precipitation and increased evaporation in areas like the Southwest and southern Great Plains, warmer temperatures result in even greater declines in yield. In most regions of the United States, optimum temperatures have been reached for dominant crops, which means that continued warming would reverse historic gains from warmer temperatures and instead lead to reduced yields over time. As temperatures increase over this century, crop production areas may shift to follow the temperature range for optimal growth.

Rising temperatures and shifting precipitation patterns will also affect productivity through altered water requirements and water-use efficiency of most crops. The differential effect of these various factors will lead to regional production effects that alter regional competitiveness, potentially altering the agricultural landscape significantly by midcentury.

Changes in average conditions will be compounded by changes in extremes on a daily, monthly, and seasonal scale (Schlenker & Roberts 2009) and by changing intensity and frequency of extreme weather events (Seneviratne et al. 2012). Many extreme weather events of the past decade are outside of the realm of experience for recent generations, and, as we've seen, these events can have devastating effects. The drought that plagued nearly two thirds of the country for much of 2012 was the most extensive to affect the United States since the 1930s, resulting in widespread crop failure and other impacts estimated to have had a cost of $30 billion, with states in the U.S. heartland—Nebraska, Iowa, Kansas, South Dakota—experiencing the greatest impacts as maize and soybean yields were severely reduced, dealing a serious blow to the states' economies (NOAA 2013a). Temperature fluctuations need not be long in duration to cause widespread destruction. In 2008, heavy rain and flooding, with up to 16 inches in parts of Iowa, caused significant agricultural losses and property damage in the Midwest totaling more than $16 billion (NOAA 2013b).

Changing frequency, severity, and length of dry spells and sustained drought can significantly reduce crop yields. At their most extreme, crop death and reduced productivity due to drought can result in billions of dollars of damage; the 1988 drought that hit the central and eastern United States resulted in severe losses to agriculture and related industries totaling nearly $80 billion (NOAA 2013a). As the IPCC notes, it is not possible to attribute historic changes in drought frequency to anthropogenic climate change (Romero-Lankao et al. 2014). However, observations of emerging drought trends are consistent with projections of an increase in areas experiencing droughts in several regions of the United States (Walthall et al. 2013). There has been no overall trend in the extent of drought conditions in the continental United States, although more widespread drought conditions in the Southwest have been observed since the beginning of the twentieth century (Hoerling et al. 2012a; Georgakakos et al. 2014). Summer droughts are projected to become more intense in most of the continental United States, with longer-term droughts projected to increase in the Southwest, southern Great Plains, and parts of the Southeast (Cayan et al. 2010; Wehner et al. 2011; Dai 2012; Hoerling et al. 2012b; Georgakakos et al. 2014; Walsh et al. 2014).

Excess precipitation can be as damaging as too little precipitation, as it can contribute to flooding, erosion, and decreased soil quality. Surface runoff can deplete nutrients, degrading critical agricultural soils, and contribute to soil loss, which reduces crop yields and the long-term capacity of agricultural lands to support crops. In some critical producing states such as Iowa, there have been large increases in days with extremely heavy rainfall even though total annual precipitation has remained steady (Hatfield et al. 2013). Greater spring precipitation in the past two decades has decreased the number of days for agricultural field operations by more than three days when compared to the previous two decades, putting pressure on spring planting operations and increasing the risk of planting on soils that are too wet, reducing crop yields and threatening the ability of soils to support crops in the long-term (Hatfield, Cruse, & Tomer 2013). Greater rainfall quantities and intensity across much of the northern United States are expected to contribute to increased soil erosion (Pruski & Nearing 2002).

The projected higher incidence of heat, drought, and storms in some regions will influence agricultural productivity. The degree of vulnerability will vary by region and will depend on both the severity of events and adaptive capacity. Due to projected increases in extreme heat, drought, and storms, parts of the Northeast and Southeast have been identified as "vulnerability hotspots" for corn and wheat production by 2045, based on expected exposure and adaptive capacity, with increased vulnerability past midcentury (Romero-Lankao et al. 2014). Livestock production is also vulnerable to temperature stresses, as animals have limited ability to cope with temperature extremes, and prolonged exposure can lead to reduced productivity and excessive mortality. These effects increase the production cost associated with all animal products, including meat, eggs, and milk.

Extremes that last for only short periods are still often critical to productivity because annual agricultural output may be driven largely by conditions during narrow windows of time when crops and livestock undergo important developments. The impact of variability in precipitation and water resource availability as well as temperature extremes will depend on the timing of such events in relation to these critical periods. Warmer spring temperatures within a specific range may accelerate crop development, but extremely high temperatures during the pollination or critical flowering period can reduce grain or seed production and even increase risk of total crop failure (Walthall et al. 2012). Warmer nighttime temperatures during the critical grain, fiber, or fruit production period will also result in lower productivity and reduced quality. Such effects were already noticeable in 2010 and 2012, as high nighttime temperatures across the Corn Belt were responsible for reduced maize yields. With projected increases in warm nights, yield reductions may become more prevalent (Walthall et al. 2012). Fewer days with cold temperatures can also have significant effects, reducing the frequency of injury from chilling in some cases, while in others yields may be negatively affected as chilling requirements for some crops are not satisfied. Many fruit and nut tree species must be exposed to the winter chill to generate economically sufficient yields. The state of California is home to 1.2 million hectares of chill-dependent orchards, supporting an estimated $8.7 billion industry. With warmer temperatures expected by the middle to the end of this century, one study concludes that conditions will not be sufficient to support some of California's primary fruit and nut tree crops (Luedeling, Zhang, & Girvetz 2009).

Although the effect is less well understood than temperature- and precipitation-related impacts, rising CO_2 concentrations are expected to affect plant growth and therefore agricultural yields. Elevated atmospheric CO_2 concentrations stimulate photosynthesis and plant growth, with some plant species (e.g., C3 crops such as wheat, cotton, soybean) exhibiting a greater response than others (e.g., C4 crops including maize) (Leakey 2009). Increased atmospheric CO_2 since preindustrial times has enhanced water-use efficiency and yields, especially for C3 crops, although these benefits have contributed only minimally to overall yield trends (Amthor 2001; McGrath & Lobell 2013). Experiments and modeling indicate that the impact of CO_2 on yields depends highly on crop species, and even subspecies, as well as on variables such as temperature, water supply, and nutrient supply. The interactions between CO_2 concentrations and these variables are nonlinear and difficult to predict (Porter et al. 2014). Elevated CO_2 concentrations can also increase weed growth rates and alter species distribution, and there is some indication that elevated CO_2 may contribute to a reduction in the effectiveness of some herbicides (Archambault 2007).

An important consideration for determining the impacts of climate change on U.S. agriculture is the degree to which farmers, ranchers, and the industry as a whole can adapt to changes over time. Agriculture is a complex system and has proved to be extremely adept at responding to changes over the past 150 years, though these adaptations were made during a period of relative climatic stability. Producers have continually adapted management practices in response to climate variability and change by using longer-maturing crop varieties, developing new cultivars, planting earlier, introducing irrigation, or changing the type of crop altogether (Olmstead & Rhode 1993, 2011).

However, the effectiveness of strategies used in the past may not be indicative for the types of changes expected in the future. Technological improvements, for example, improve yields under normal conditions but may not protect harvests from extremes expected in the future (Schlenker, Roberts, & Lobell 2013), such as increased drought in the Southwest and southern Great Plains or increased flooding in the Midwest and Northeast. Catastrophic crop or livestock losses are likely to affect the financial viability of production enterprises in a fundamentally different way than moderate losses over longer periods of time. In addition, many adaptive actions may be costly (e.g., requiring increased energy consumption) or constrained by climate change (e.g., increasing groundwater use may not be an option in areas with declining precipitation) (Romero-Lankao et al. 2014). Decisions about future adaptive action will need to take into account the potential risks of climate-related damage and the costs of adaptation, as well as complex changes in domestic and international markets and policies, all of which will determine the cost of doing business.

OUR APPROACH

To quantify the potential impacts of climate change on agricultural production, we rely on statistical studies that isolate the effect of temperature and rainfall on crop yields in the United States. Because there are strong cross-county patterns in crop yields, as well as strong trends over time (that may differ by location), we rely on studies that account for these patterns when measuring the effects of climate variables. Schlenker and Roberts (2009) provide nationally representative estimates that satisfy these criteria, which we use to construct quantitative projections. They examine county-level agricultural production during the period 1950–2005 and identify the incremental influence of temperature and rainfall variability on maize, soy, and cotton yields using data collected by the U.S. Department of Agriculture's National Agricultural Statistical

Service. While they focus their analysis on the eastern United States, they also provide parallel results for the western United States, which we also utilize. To estimate yield impacts on wheat, we apply a similar approach to yield data from the same source (see appendix B). We also consider how projections change when future adaptation is modeled explicitly by linking the results from Schlenker and Roberts to an analysis by Burke and Emerick (2013), who use similar econometric strategies to measure rates of agricultural adaptation in the United States (see part 5).

Figure 6.1 displays the temperature impact function for maize yield. In general, rising daily temperatures increase yields slightly until a breakpoint is reached, after which higher daily temperatures dramatically reduce yields. For maize, soy, and cotton, these breakpoints occur respectively at 84°F, 86°F, and 90°F.

This nonlinear response has been broadly replicated in multiple studies that are more local in character and is consistent with quadratic temperature responses in studies that use seasonal mean temperature. Seasonal precipitation has a nonlinear inverse-U-shaped relationship with yields (figure 6.2), again broadly consistent with local studies.

Schlenker and Roberts assess whether there is evidence that farmers adapt by examining whether there are changes in the sensitivity of crop yields to temperature over time. They find that the relationship between heat and yields has changed slightly since 1950, providing only weak evidence of adaptation. This finding is consistent with a more detailed analysis on the evolution of heat tolerance in maize in Indiana counties during the period 1901–2005 (Roberts & Schlenker 2011) and analysis of how yields in the eastern United States have responded to long-term trends in temperatures during the period 1950–2010 (Burke & Emerick 2013). Thus, while there is evidence that farmers are adapting over time, the evidence indicates that this process is extremely slow.

Schlenker and Roberts also look for evidence of adaptation by examining if counties that are hotter on average (in the Southeast) or drier and/or hotter on average (in the West) have a different sensitivity to climate. They find strong evidence that crop yields in counties in the South or in the West are less sensitive to temperature, suggesting that these locations have adapted somewhat to their local climatic conditions, probably through the adoption of heat-tolerant cultivars and/or irrigation (Butler & Huybers 2013). These adaptations come at a cost, such as lower average yields (Schlenker, Roberts, & Lobell 2013), but they might be more consistently adopted in the future in the Midwest and East if rising temperatures make them cost-effective strategies in these regions.

Schlenker and Roberts are unable to account for the effect that rising CO_2 concentrations have on agricultural yields because gradual trends in CO_2 cannot be statistically distinguished from other trends (e.g., technological

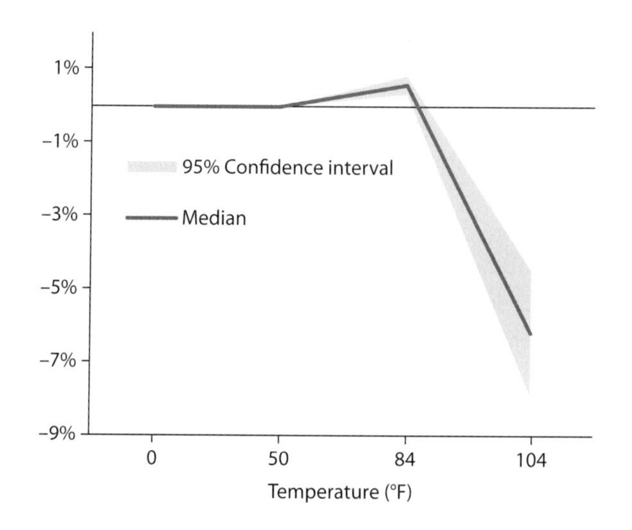

FIGURE 6.1. **Impact Function: Temperature and Maize Yields**

Observed change in maize yields (percent) vs. daily temperature (degrees Fahrenheit)

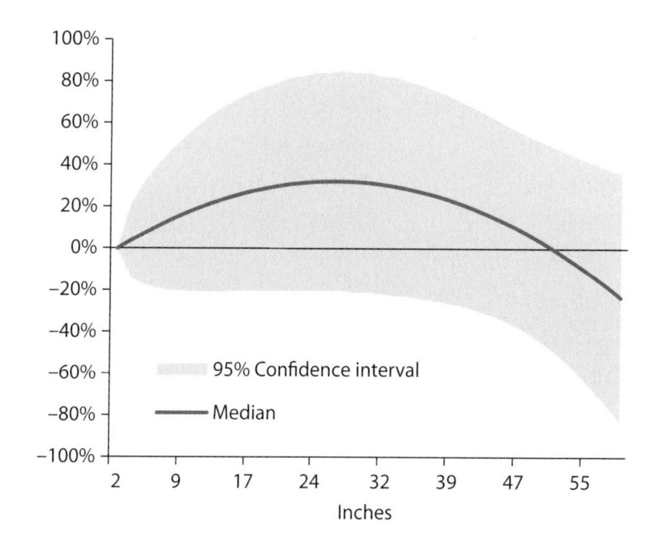

FIGURE 6.2. **Impact Function: Precipitation and Maize Yields**

Observed change in maize yields (percent) vs. seasonal precipitation (inches)

progress). Thus, to account for increasing CO_2, we must draw on a body of literature that combines field experiments in CO_2 enrichment with simple models. We obtain estimates for the incremental effect that CO_2 enrichment has on yields for different crops from McGrath and Lobell (2013), who collect results from multiple field experiments and use these results to construct estimates for the effect of CO_2 fertilization on U.S. crops.

To assess potential future impacts of climate change on national agricultural production, we simulate changes in production of major crop varieties (maize, wheat, soybeans, and cotton) under different climate scenarios relative to a future in which the climate does not drive economic changes after 2012—although other social and economic trends are assumed to continue. Within each scenario, we account for uncertainty in climate models, weather, and statistical results, causing our projection to be a probability distribution of potential outcomes at each moment in time.

When we consider the potential impact of changes in temperature, precipitation, and CO_2 fertilization on national yields, we find that the value of total production generally declines as early as the period 2020–2039—even under RCP 2.6—although the range of *likely* outcomes spans positive values through 2099 under all scenarios (table 6.1). Under RCP 8.5, total production is *likely* to change by −14 percent to +7 percent by midcentury and −42 percent to +12 percent by late century, with a 1-in-20 chance that late-century changes are below −56 percent or exceed +19 percent of current production. Impacts on maize are generally negative throughout all periods because maize is strongly heat sensitive and benefits least from CO_2 fertilization, while impacts on wheat are overwhelmingly positive because wheat benefits more from CO_2 fertilization than it is harmed by heat. Impacts on cotton and soybeans are about as likely to be positive as negative until late century in RCP 8.5, when they become generally negative. The *likely* ranges for all crops are shown in table 6.1.

Projected changes are smaller in magnitude for RCP 4.5 and RCP 2.6, and the distribution of projected changes is more skewed toward negative yield changes relative to RCP 8.5. The *likely* range of late-century production changes for total production spans −25 percent

TABLE 6.1 Impacts of future climate change on U.S. agricultural yields with CO_2 fertilization

Crop Type	*RCP 8.5*			*RCP 4.5*			*RCP 2.6*		
	1 in 20 Less Than (%)	*Likely (%)*	*1 in 20 Greater Than (%)*	*1 in 20 Less Than (%)*	*Likely (%)*	*1 in 20 Greater Than (%)*	*1 in 20 Less Than (%)*	*Likely (%)*	*1 in 20 Greater Than (%)*
Maize									
2080–2099	−84	−73 to −18	−8.1	−64	−44 to −2.8	1.9	−27	−19 to 0.4	2.8
2040–2059	−39	−30 to −2.3	2.8	−34	−25 to 0.1	3.6	−23	−18 to −1.0	1.3
2020–2039	−19	−15 to 4.3	12	−19	−15 to 5.2	9.7	−21	−14 to −3.1	0.4
Wheat									
2080–2099	8.6	19 to 42	50	−1.1	4.7 to 15	17	−2.6	−0.9 to 4.4	5.3
2040–2059	3.0	6.0 to 14	17	1.0	3.7 to 10	12	−0.8	0.6 to 5.1	6.2
2020–2039	0.6	1.8 to 5.6	8.3	−0.3	1.2 to 6.5	7.7	−0.9	0.2 to 4.4	5.3
Oilseeds									
2080–2099	−74	−56 to 18	29	−55	−30 to 8.6	16	−18	−13 to 6.3	8.4
2040–2059	−23	−16 to 11	17	−24	−15 to 7.6	14	−15	−8.8 to 5.8	9.9
2020–2039	−9.7	−6.6 to 9.9	15	−15	−10 to 6.9	13	−16	−7.4 to 3.8	6.8
Cotton									
2080–2099	−74	−52 to 16	31	−38	−18 to 9.8	18	−17	−9 to 3.0	5.7
2040–2059	−20	−12 to 13	18	−15	−7.3 to 8.0	13	−15	−7.3 to 4.9	8.6
2020–2039	−7.7	−3.6 to 5.6	7.8	−8.9	−4.8 to 5.8	9.2	−11	−5.4 to 4.3	6.3

Note: Percentage change from 2012 production levels for maize, wheat, oilseeds, and cotton.

to +6 percent for RCP 4.5 and −11 percent to +3 percent for RCP 2.6. The skewed distribution is most apparent when considering 1-in-20 outcomes: production changes below −43 percent or above 10 percent for RCP 4.5 and below −17 percent or above 5 percent for RCP 2.6. The skewed distribution of total production is mainly driven by maize and soy, which have especially skewed outcomes with a 1-in-20 chance that yields are below −64 percent and −55 percent, respectively, in RCP 4.5 by late century. The skewness for total production in RCP 4.5 is sufficiently large that potential downside losses are similar in magnitude to downside losses in RCP 8.5; however, in RCP 4.5 there is a lower probability of ending up with the largest losses.

Across all RCPs, the distribution of potential yields broadens over time. The rate of spreading increases dramatically with increasing emissions. For total production, the late-century *very likely* range spans 15 percentage points in RCP 2.6 and widens to span 31 and 54 percentage points in RCP 4.5 and 8.5, respectively (figure 6.3). Climate

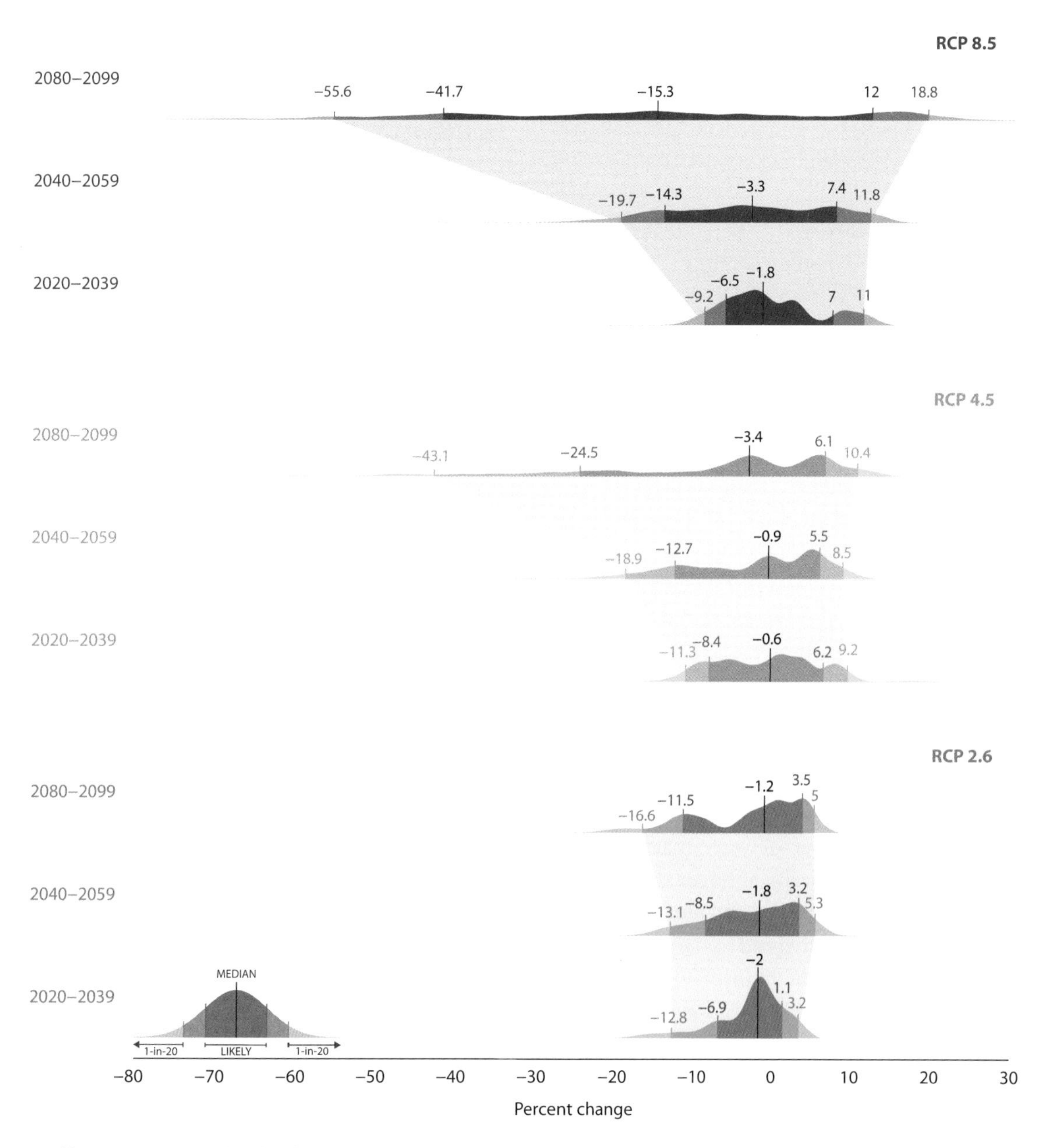

FIGURE 6.3. **Change in National Yield of Grains, Oilseeds, and Cotton**

Percent change, including CO_2 fertilization

change not only decreases expectations for national production but also increases uncertainty regarding future national production in a warming world.

In percentage terms, the spatial distribution of projected impacts is uneven across the country, with the South and East regions suffering the largest projected yield losses while the Rockies, Northwest, and northern Great Plains regions achieve yield gains in the median RCP 8.5 projection (figure 6.4). The eastern United States is hardest hit primarily because the dose-response function is more sensitive to extreme heat in the East, in part because irrigation infrastructure is not as widespread as in the West (Schlenker & Roberts 2009). The Southeast suffers the largest percentage losses because the dose-response function is sensitive to extreme temperatures and because southern counties experience the highest number of additional extreme temperature days in future projections. Projected yields in the Rockies, Northwest, and northern Great Plains benefit from both moderate warming and moderate wetting from a current climate that is both cool and dry. Projected changes in total national output are dominated by production losses in central Midwest states that are not heavily irrigated, that warm substantially, and that currently have large land areas dedicated to high-yield production.

The effects noted above are described in terms of average changes over 20-year intervals. These averages are useful for describing persistent economic changes in future periods, but they mask short-lived events that may only last a year or two but have substantial economic consequences. Within each 20-year window, the likelihood of extreme annual events, such as a very low-yield year, evolves with the climate. One way to describe how the likelihood of extreme events changes is to examine how frequently we expect to experience years that are as damaging as the worst year experienced during two decades of recent history, a so-called 1-in-20 year event. In figure 6.5, we plot the estimated number of years that will have yield losses larger than historically observed 1-in-20 year losses. For each year, we plot the expected number of these extreme years that will be experienced in the 20 years that follow; that is, we plot what the immediate future looks like to an individual in a given year. For a long-term investor with a 20-year time horizon, these are expected risks to take into account. By 2030, in all scenarios, production losses that used to occur only once every 20 years will be expected to occur roughly five times in the following 20 years. By 2080, these events will be occurring roughly eight times every 20 years in RCP 4.5 and 12 times every 20 years in RCP 8.5.

These projections suggest there is a possibility that national yields will be higher in the future, with the benefits of CO_2 fertilization counterbalancing the adverse effects of extreme heat. We advise caution in interpreting these results, as the magnitude of carbon fertilization effects have not been measured empirically with the same level of consistency as temperature and rainfall effects (Long et al. 2006), and they have not been measured empirically in nationally representative samples (McGrath & Lobell 2013). Thus, we also consider the distribution of potential yield changes due only to temperature and precipitation changes—not because the CO_2 fertilization effect is likely to be zero, but because separating the effect of CO_2 fertilization allows evaluation of how large these uncertain effects must be to offset temperature and rainfall effects. When the effect of CO_2 fertilization is removed, agricultural output declines much more dramatically in projections that use only temperature and precipitation changes (table 6.2 and figure 6.6). The likely range of late-century losses in RCP 8.5 are unambiguously negative and large, spanning 20 to 59 percent for total production. The effect of removing CO_2 fertilization has different effects for different crops, although in all cases removing CO_2 fertilization causes projected losses to be larger. It is unlikely that losses this large will occur, as carbon fertilization will offset some of these losses, as it did in our main projections, so these estimates should be considered a worst-case scenario for the situation where the benefits from carbon fertilization have been overestimated.

It is important to note that these estimates assume that the national distribution of crop production remains fixed relative to the period 2000–2005. It is extremely likely that farmers' decisions regarding what they plant will change as they observe their climate changing, but this response could not be evaluated here because systematic analysis of this response is absent from the body of existing research. We hope that future analyses will incorporate this response, and we will update our projections accordingly.

Prior analyses have not examined how planting decisions change in response to the climate, although recent work has examined how farmers who always plant the same crop adapt to changes in their local climate over time. We consider how these results can be incorporated into our analysis in part 5.

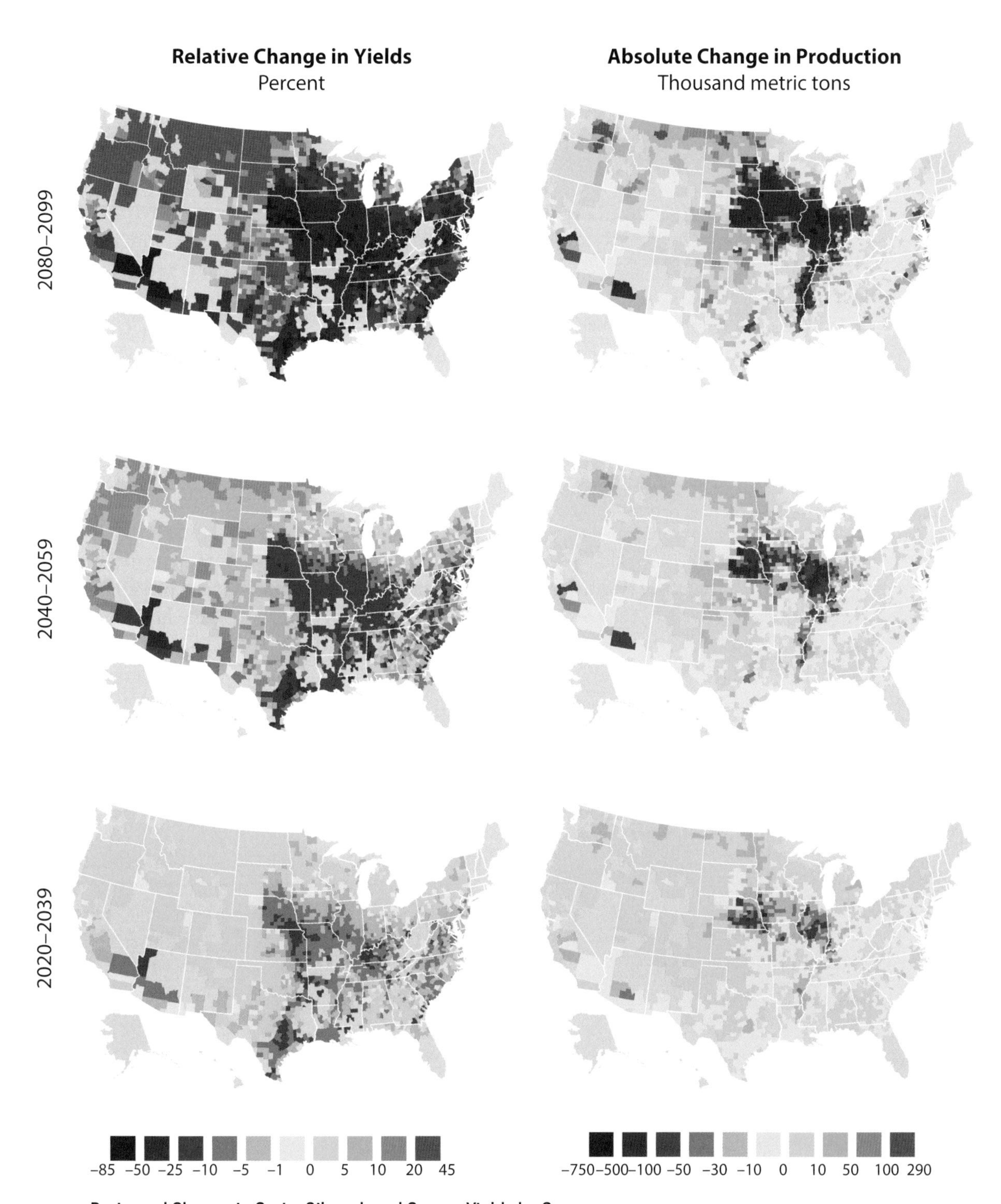

FIGURE 6.4. **Projected Change in Grain, Oilseed, and Cotton Yields by County**

RCP 8.5 median projection; gray counties are those where no grain, oilseed, or cotton production currently occurs

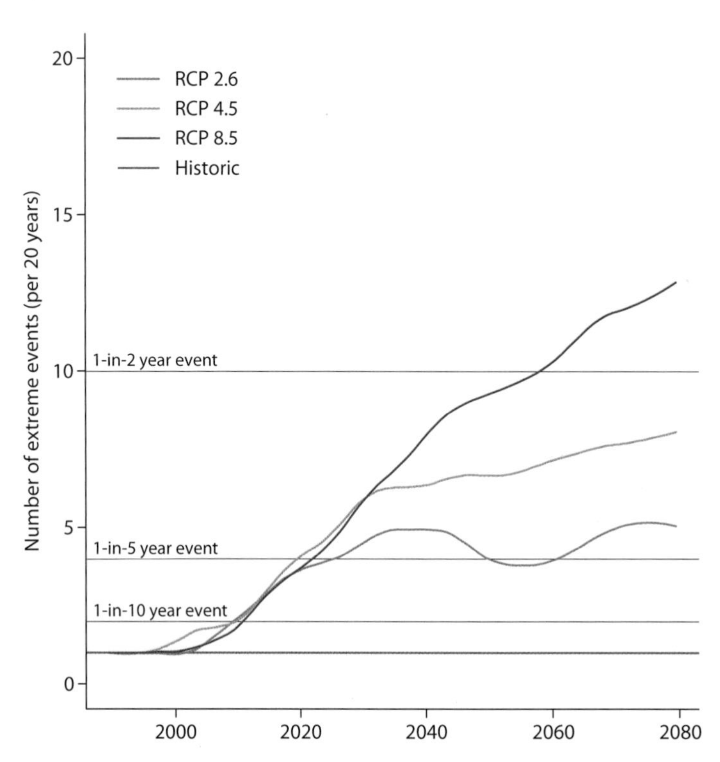

FIGURE 6.5. **Projected Change in the Frequency of National Yield Losses Equal to or Worse than Historical 1-in-20 Year Losses**

TABLE 6.2 Impacts of future climate change on U.S. agricultural yields without CO_2 fertilization

Crop Type	*RCP 8.5*			*RCP 4.5*			*RCP 2.6*		
	1 in 20 Less Than (%)	*Likely (%)*	*1 in 20 Greater Than (%)*	*1 in 20 Less Than (%)*	*Likely (%)*	*1 in 20 Greater Than (%)*	*1 in 20 Less Than (%)*	*Likely (%)*	*1 in 20 Greater Than (%)*
Maize									
2080–2099	−87	−76 to −29	−22	−66	−47 to −7.5	−3.6	−28	−20 to −0.8	1.5
2040–2059	−41	−33 to −7.1	−2.4	−36	−28 to −3.2	0.1	−25	−19 to −2.7	−0.5
2020–2039	−21	−16 to 2.5	9.4	−20	−16 to 3.7	8.0	−22	−15 to −4.3	−0.9
Wheat									
2080–2099	−27	−20 to −7.0	−4.0	−15	−9.8 to −1.5	−0.8	−6.2	−4.7 to 0.4	0.9
2040–2059	−11	−8.6 to −2.9	0.1	−8.2	−6.0 to 0.1	0.7	−6.0	−4.7 to −0.5	−0.2
2020–2039	−4.9	−3.9 to −0.9	2.0	−4.6	−3.3 to 1.9	2.5	−4.5	−3.7 to 0.4	0.9
Oilseeds									
2080–2099	−82	−70 to −20	−14	−61	−40 to −6.6	−0.3	−21	−16 to 2.2	4.2
2040–2059	−33	−27 to −4.0	0.7	−31	−23 to −2.5	3.2	−19	−14 to 0.0	3.9
2020–2039	−15	−12 to 4.0	8.5	−19	−14 to 2.4	8.3	−19	−11 to −0.1	2.6
Cotton									
2080–2099	−83	−68 to −24	−15	−47	−30 to −6.4	0.5	−21	−13 to −1.2	1.5
2040–2059	−31	−24 to −3.7	0.3	−23	−16 to −2.8	1.3	−19	−13 to −1.1	2.5
2020–2039	−13	−9.4 to −0.8	1.2	−13	−9.1 to 0.8	4.0	−14	−9.3 to 0.0	1.9

Note: Percentage change from 2012 production levels for maize, wheat, oilseeds, and cotton.

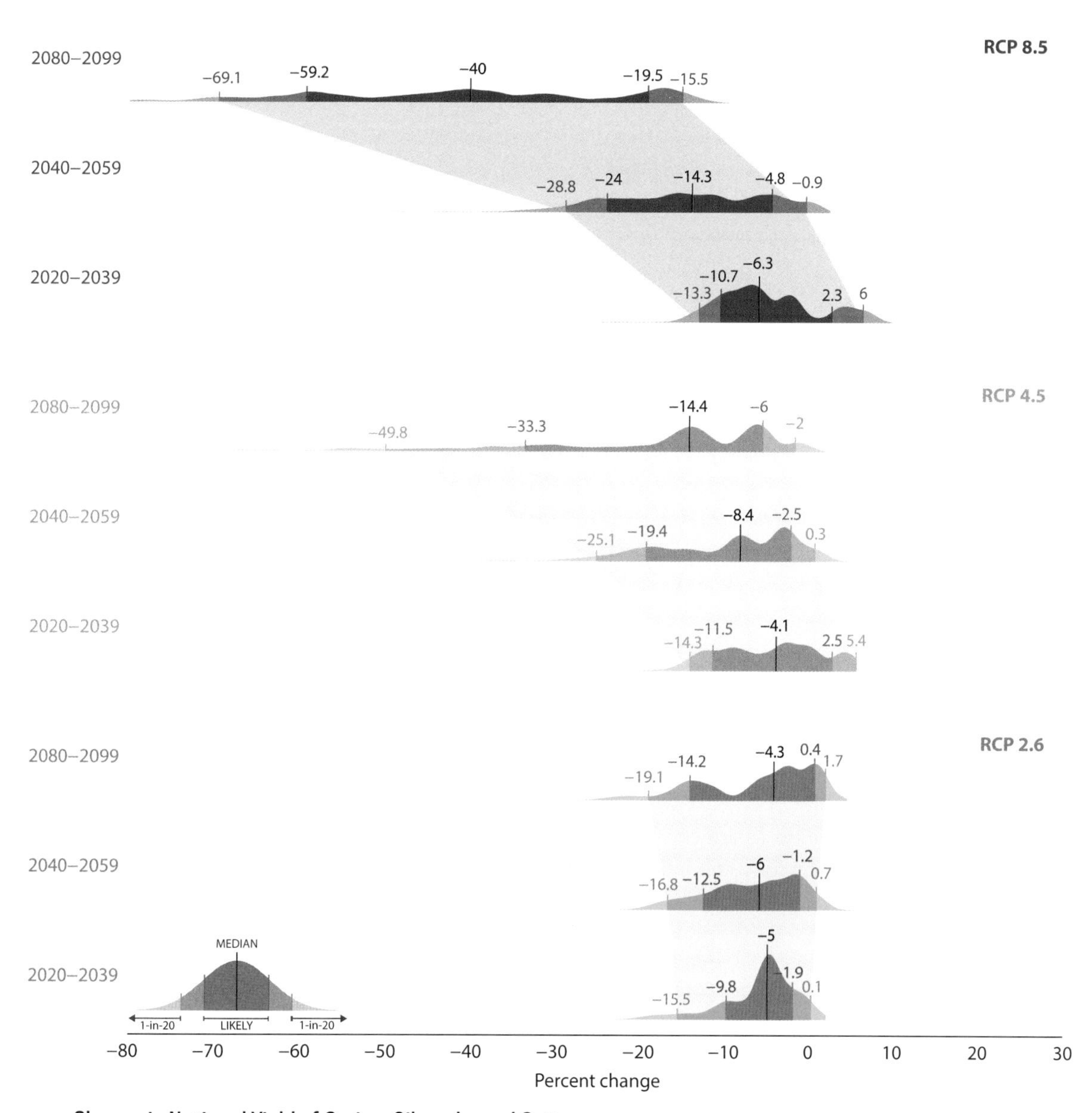

FIGURE 6.6. **Change in National Yield of Grains, Oilseeds, and Cotton**

Percent change, not including CO_2 fertilization

OTHER IMPACTS

There is a whole host of impacts that we were not able to include in this round of our analysis. We discuss some of them in this section.

Water Resources

Changing climate—including shifting precipitation patterns and greater frequency and intensity of precipitation extremes like heavy rainfall and drought in some regions—are likely to affect water resource availability, with wide-ranging implications for the U.S. agricultural sector and crop production in particular.

While irrigation reduces the risk from variable seasonal rainfall, producers that rely on irrigation to maintain yields may be at greater risk from volatility in cost and availability of water supplies. Climate change will have important implications for the extent and distribution of future U.S. irrigated crop production. Although only 7.5 percent of all U.S. cropland and pastureland are irrigated, farms that use irrigation accounted for

55 percent of the total value of crop sales in 2007, the last year for which U.S. Department of Agriculture census data are available (U.S. Department of Agriculture 2010). Irrigated agriculture accounts for more than a third of the nation's freshwater withdrawals and approximately 80 to 90 percent of overall consumptive use (Kenny et al. 2009). Nearly three quarters of irrigated acreage is in the western United States, though in recent decades much of the expansion in irrigated acreage has occurred in the eastern areas.

Reduced water availability for agriculture may lead to contraction in irrigated acreage in some areas, particularly in the western United States (Elliot et al. 2013). Warmer temperatures at the same time will also increase crop water needs and demand for irrigation, although increasing CO_2 concentrations can also increase water-use efficiency of some crops (Hatfield et al. 2013; Elliott et al. 2013; Prudhomme et al. 2013; Wada et al. 2013). Irrigation, which has traditionally been relied on to offset the negative effect of high temperatures, has been particularly effective in areas with intensive cultivation and irrigation such as the Corn Belt (Sakurai, Iizumi, & Yokozawa 2011). Such strategies may not be available or will be much more costly in regions with increased water scarcity where the cost of irrigation is likely to increase, as are energy costs associated with irrigation, including for water pumping.

Ozone Pollution

Carbon dioxide is not the only ambient pollutant that affects plant growth. Emissions of nitrogen oxides (NOx) and volatile organic compounds (VOCs) from farm processes and industrial sources react to form ground-level ozone (O_3), which can damage vegetation by reducing photosynthesis and other important physiologic functions resulting in stunted crops, decreased crop quality, and decreased yields (Mills et al. 2007). High temperatures increase ozone formation, especially during the warm "ozone season" of May to September (Bloomer et al. 2009). The impacts on a range of U.S. agricultural crop yields is an area of emerging study; initial studies indicate that the impacts of elevated ozone concentrations are evident for soybean crops in the Midwest, with annual yield losses in 2002–2006 estimated at 10 percent (Fishman et al. 2010). The interactions between elevated ozone and CO_2 concentrations have been found to dampen these effects, with ozone partially counteracting CO_2 fertilization. More study is necessary to understand the interactions between CO_2, ozone, and temperature on a variety of species.

Weeds, Disease, and Pests

Agriculture is a complex system, and the mechanisms through which climate can affect productivity are many. While changing climatic conditions affect crop yield directly, they also affect a whole array of other competing and complementary organisms that have varying effects on crop yields. Changes in temperature and precipitation patterns, combined with increasing atmospheric CO_2, change weed-infestation intensity, insect population levels, the incidence of pathogens, and potentially the geographic distribution of all three.

The relationship between climate change and agricultural crop yield losses due to increased competition from weeds, for example, is not fully understood because of the complex relationships between temperature, CO_2 concentration, and crop-weed interactions, as well as artificial factors such as herbicide use (Archambault 2007). Weeds are generally hearty species, and several weeds benefit more than crops from higher temperatures and CO_2 levels (Ziska 2010). The geographic distribution of native and invasive weeds will likely be extended northward as temperatures warm, exposing farms in northern latitudes to new or enhanced threats to crop productivity from weeds like privet and kudzu, already present in the South (Bradley, Wilcove, & Oppenheimer 2010; Ziska 2010). Weed control costs the United States more than $11 billion a year, with most of that spent on herbicides. Use of herbicides is expected to increase as several of the most widely used herbicides in the United States, including glyphosate (also known by the brand name Roundup), have been found to lose efficacy on weeds grown at CO_2 levels projected to occur in the coming decades (Ziska, Teasdale, & Bunce 1999).

Climate change is also expected to affect the geographic ranges of specific species of insects and diseases across regions of the United States, potentially altering yield losses as a result. Changes in average temperature can result in gradual shifts in geographic distribution as earlier spring and warmer winters affect species overwintering and survival. In wet years, high humidity can

The Impacts of Climate on Agriculture *Do Not* Stop at U.S. Borders

Although for the purposes of this book we isolate our analysis of climate impacts to those that occur within the United States, the global nature of food production cannot be overlooked (Roberts & Schlenker 2013). The response of global agricultural systems to a changing climate may mean production shifts as some regions become more or less suitable for agriculture. The effects of climate on crop and food production are already evident in several key producing regions of the world, with recent periods of rapid food and cereal price increases following climate extremes (Porter et al. 2014). By the 2030s global average yields will *likely* be negatively affected, with reductions *more likely than not* to be as much as 5 percent beyond 2050 and *likely* by the end of the century (Porter et al. 2014). The reductions will coincide with growing global demand, which is projected to increase by approximately 14 percent per decade until midcentury (Alexandratos & Bruinsma 2012; Porter et al. 2014).

These shifts will be reflected in changing global production and commodity prices, all of which will impact U.S. producers and, in turn, how they choose to respond. Because of the complexity of estimating the impacts of climate change on global agricultural production, price, and trade, we focus on only those impacts that occur within the United States in the absence of any changes to global trade or prices. In addition, we do not model how farmers will change which crops they grow, as we lack robust empirical evidence to quantify these changes. Historical anecdotes—such as the Dust Bowl—suggest this may be an important margin for future adjustments (Hornbeck 2012; Feng, Oppenheimer, & Schlenker 2013).

In an increasingly interconnected global market, the effects of climate change on global food production and prices will affect U.S. farmers and other agricultural producers, as well as American consumers. Regional climatic changes may shift the distribution and costs of production across the globe over time, while extreme events may affect food security and price volatility. As the United States is a significant agricultural exporter, price and production shocks from extreme climate events in the United States can have reverberations globally, though the globalized system can also act as a buffer to reduce the localized effects of events in the United States (Godfray et al. 2010).

The United States imports about a fifth of all food consumed in the United States, making food prices and supply vulnerable to climate variations in other parts of the world. Climate extremes in regions that supply the United States with winter fruits and vegetables, and in particular tropical products such as coffee, tea, and bananas, can cause sharp reductions in production and increases in prices. Volatility in supplies and prices of internationally traded food commodities have a significant effect on decisions made by U.S. agricultural producers and determine prices U.S. consumers pay for such goods. Fluctuations and trends in food production are widely believed to have played a role in recent price spikes for wheat and maize, which followed climate extremes in 2008 and 2011. Between 2007 and 2008, the Food and Agriculture Organization food price index doubled; this was due to a confluence of factors, one of which was extreme weather conditions in major wheat and maize exporters including the United States, Australia, and Russia (Food and Agriculture Organization of the United Nations 2011). Such extreme events have become more likely as a result of recent climate trends and may be more frequent in the future, contributing additional volatility to an already complex global agricultural system.

The IPCC has reported that projected changes in temperature and precipitation by 2050 are expected to increase food prices, with estimates ranging from 3 to 84 percent. Projections of food prices that also account for the CO_2 fertilization effect (but not ozone and pest and disease effects) range from −30 percent to +45 percent by 2050, with price increases about as likely as not. This does not take into account variations in regional effects or the effect of extremes, which can be a major contributor

(*continued*)

to variability in productivity and prices. Compound events where extremes have simultaneous effects in different regions (as was witnessed in 2008 and 2011), driven by common external forcing (e.g., El Niño), climate system feedbacks, or causally unrelated events, may have additional negative impacts on food security and production, though there are very few projections of such compound extreme events, and the interactions between multiple drivers are difficult to predict.

Quantifying these effects, in their agricultural and economic terms, is an extremely difficult task, requiring assumptions about the myriad climate and nonclimate factors that interact to determine food security and prices, both at home and abroad. While all aspects of food security are potentially affected by climate change, including food access, utilization, and price stability, there is limited direct evidence that links climate change to food security impacts (Porter et al. 2014).

help insects and diseases flourish, with negative indirect impacts on animal health and productivity (Garrett et al. 2006, 2011). Climate affects microbial and fungal populations and distribution, the distribution of diseases carried by insects and rodents, animal and plant resistance to infections, food and water shortages, and food-borne diseases (Baylis & Githeko 2006; Gaughan et al. 2009). Regional warming and changes in rainfall distribution may change the distributions of diseases that are sensitive to temperature and moisture, such as anthrax, blackleg, and hemorrhagic septicemia (Baylis & Githeko 2006; Gaughan et al. 2009).

Livestock

Although livestock is a major component of the U.S. agricultural system, with nearly 1 million operations generating nearly half of total U.S. commodity sales, the impact of climate change on livestock production has received less study than impacts on agricultural crops. Climate change will affect the livestock sector both directly, through impacts on productivity and performance due to changes in temperature and water availability, and indirectly, through price and availability of feed grains and pasture and changing patterns and prevalence of pests and diseases (Walthall et al. 2013).

Livestock productivity will be most directly impacted by changes in temperature, which is an important limiting factor for livestock in the United States. High temperatures tend to reduce feeding and growth rates as animals alter their internal temperatures to cope; the resulting increase in animals' metabolism reduces production efficiency (André et al. 2011; Porter et al. 2014). For many livestock species, increased body temperatures 4°F to 5°F above optimum levels disrupts performance, production, and fertility, limiting an animal's ability to produce meat, milk, or eggs. Livestock mortality increases as optimums are exceeded by 5°F to 13°F (Gaughan et al. 2002). Animals managed for high productivity, including most meat and dairy animals in the United States (e.g., cattle, pigs, and chickens), are already operating at a high metabolic rate, decreasing their capacity to tolerate elevated temperatures and increasing the risk of reduced production or death (Zumbach et al. 2007).

Livestock and dairy production will be more affected by changes in the number of days of extreme heat than by changes in average temperature, though the effect of warmer average nighttime temperatures, especially multiple hot nights in a row, can exacerbate animal heat stress (Mader 2003). The negative effects of hotter summer weather will likely outweigh the benefits of warmer winters, with the potential for only about half of the decline in domestic livestock production during hotter summers to be offset by milder winter conditions (Adams et al. 1999).

The majority of American livestock raised in outdoor facilities, and therefore exposed to rising temperatures and increased heat stress, are ruminants (goats, sheep, beef and dairy cattle). Within limits, these animals can adapt to most gradual temperature changes but are much more susceptible to extreme heat events (Mader 2003). Impacts are less acute for confined operations that use temperature regulation, which house mostly poultry and pigs, though management and energy costs associated with increased temperature regulation will increase. Confined operations are not immune to the effect of rising temperatures, which can contribute to livestock heat stress. Despite modern heat-abatement strategies, heat-induced productivity declines during hot summers—including reduced performance and reproduction as well as mortality—cost the American swine industry, for example, nearly $300 million annually (St-Pierre, Cobanov, & Schnitkey 2003).

Current economic losses incurred by the U.S. livestock industry from heat stress, most from effects on dairy and beef cattle, have been valued at $1.7 billion to $2.4 billion annually. Nearly half of the losses are concentrated in a few states (Texas, California, Oklahoma, Nebraska, and North Carolina). Exposure to high-temperature events can be extremely costly to producers, as was the case in 2011, when heat-related production losses exceeded $1 billion (NOAA 2013b). Large-scale commercial dairy and beef cattle farmers are most vulnerable to climate change and the expected rise in high-heat events, particularly as these farmers are less likely to have diversified.

Other, less well-studied impacts on the livestock sector from expected climate change include indirect effects of warmer, more humid conditions on animal health and productivity through promotion of insect growth and spread of diseases. Warming is also expected to lengthen the forage growing season but decrease forage quality, with important variations due to rainfall changes (Craine et al. 2010; Izaurralde et al. 2011; Hatfield et al. 2014). One study identified significant expected declines in forage for ranching in California, even under more modest climate changes (Franco et al. 2011).

Studies of the potential effects of climate change have projected the resulting effects on productivity through factors such as change in days to market and decrease in annual production. One study found that, given expected warming by 2040, days to market for swine and beef may increase 0.9 to 1.2 percent, with a 2.1 to 2.2 percent decrease in dairy milk production (Frank et al. 2001). By 2090, days to market increased 4.3 to 13.1 percent and 3.4 to 6.9 percent for swine and beef, respectively, with a 3.9 to 6.0 percent decrease in dairy production as a result of heat stress.

Relatively few economic-impact studies have estimated the costs of climate-related effects with respect to productivity and management costs of the livestock and dairy sectors, as they involve accounting for the complex and interactive direct and indirect effects, such as lowered feed efficiency, reduced forage productivity, reduced reproduction rates, and assumptions about adaptive actions such as modifying livestock housing to reduce thermal stress. In the absence of such estimates, most system-wide economic-impact assessments do not account for the potential direct costs and productivity effects of climate change on livestock, forage, and rangeland production (Antle & Capalbo 2010; Izaurralde et al. 2011).

CHAPTER 7

LABOR

LABOR is a critical component of our economy. Even slight changes in the productivity of the American workforce have a significant effect on overall economic output. Labor productivity improvements have been an important source of past GDP growth in the United States and, as a result, is an area of extensive study. Of particular interest has been the identification of optimal working conditions—including workplace environment and exposure to a variety of climate-related factors—in various economic sectors (Wyon 2000; Seppanen, Fisk, & Lei 2006). Suboptimal environmental conditions do more than simply make workers uncomfortable; they also affect the ability of workers to perform tasks and can influence work intensity and duration, all of which affect overall labor productivity. Thus, the environmental sensitivity of individual workers represents a pathway through which climate change can influence all economic sectors, even those previously thought to be insensitive to climate (Hsiang 2010).

Climate change will affect workers, workplace environments, and ultimately worker productivity. Rising average and extreme temperatures will likely have the most direct effect on working conditions. Climate change may also affect the U.S. labor force indirectly through increased storm damage, flooding, wildfires, and other climate-related changes, resulting in disruption of business and production in some areas. Health-related effects, both negative and positive, will affect the ability of Americans to work. While we provide an overview of the range of potential climate change impacts on U.S. labor, our analysis focuses specifically on the effect of changing temperatures on labor supply.

BACKGROUND

Rising average temperatures, greater temperature variability, and more frequent and severe temperature extremes will make it harder to sustain optimal working conditions. Higher temperatures can change the amount of time allocated to various types of work as individuals spend more time indoors to beat the heat or as outdoor laborers take more frequent breaks to cool off (Graff Zivin & Neidell 2014). Climate-related factors can also affect worker performance, affecting cognitive capacity and endurance (Mackworth 1948; Ramsey & Morrissey 1978). Increased use of air-conditioning for indoor labor and schedule changes for outdoor labor can mitigate some, but not all, of the effects.

Not all American workers will be equally affected; the impact of climate differs across sectors of the economy. Workers in agriculture, construction, utilities, and manufacturing are among the most exposed (Graff Zivin & Neidell 2014). These high-risk sectors, which account for roughly a quarter of the U.S. labor force, are affected by changes both in average temperatures and temperature extremes. Workers in high-risk sectors are at particular risk of heat stress because of the internal body heat produced during physical labor. Higher temperatures and heat strain, however, can also affect workers in stores and offices (Kjellstrom & Crowe 2011). Thermal conditions inside commercial buildings are often not well-controlled and can vary considerably over time as outdoor conditions change, making it difficult to ensure optimum temperatures for worker comfort and productivity (Seppanen, Fisk, & Lei 2006). The impact of projected temperature changes on these low-risk sectors is considerably lower than that on their high-risk peers.

The first empirical study of the impact of climate on labor productivity observed that performance in labor-intensive sectors declined nonlinearly at high temperature (Hsiang 2010), mirroring the response of subjects in laboratory experiments (Mackworth 1948; Ramsey & Morrissey 1978; Wyon 2000; Seppanen, Fisk, & Lei 2006). Since then, studies have found that labor supply, measured in work hours, declines moderately at higher temperatures. This is true for a range of industrial sectors, though there are substantial differences in climate exposure among them. Temperature affects endurance, fatigue, and cognitive performance, all of which can contribute to diminished "work capacity" and mental task ability, as well as increased accident risk (Kjellstrom & Crowe 2011). To cope with heat, workers often reduce the pace or intensity of their work or take additional breaks, which reduces overall worker output. One study found that at temperatures above 85°F, workers in high-risk industries reduce daily output by as much as one hour, with much of the decline occurring at the end of the day when fatigue from prolonged heat exposure sets in (Graff Zivin & Neidell 2014).

Extreme heat stress, brought on by more intense or extended days of exposure to high temperatures, can induce heat exhaustion or heat stroke and can significantly reduce ability to carry out daily tasks. Estimates of the impact of higher average temperatures and heat stress on work capacity indicate that labor productivity in high-risk sectors is highly vulnerable to temperature extremes, despite our ability in many instances to mitigate these effects. According to Centers for Disease Control and Prevention records, from 1992 to 2006 there were 423 worker deaths attributed to heat exposure in the United States, nearly a quarter from the agriculture, forestry, fishing, and hunting industries (Luginbuhl et al. 2008).

Humidity can exacerbate these effects even further, particularly in midlatitudes during summer months of peak heat stress. Occupational thresholds developed for industrial and U.S. military labor standards provide guidelines for assessing labor capacity, or the ability to perform sustained labor safely under heat stress. Studies using these thresholds have found that the southeastern United States is particularly vulnerable (Dunne, Stouffer, & John 2013). In our analysis, the Southeast is projected to continue to have the country's highest wet-bulb temperatures (the combination of heat and humidity) over the coming century, though the Midwest and Northeast will likely see larger increases.

OUR ANALYSIS

To quantify the potential impact of climate change on labor, we rely on statistical analyses that isolate the effect of temperature and other climatic variables on labor supply by individuals in the United States. Because there are strong cross-county patterns in labor markets, as well as strong trends over time (that may differ by location) and over seasons, we rely on the only analysis that accounts for these patterns when measuring the effect of temperature on labor supply.

Graff Zivin and Neidell (2014) provide nationally representative estimates that satisfy these criteria, which we use to construct quantitative projections. They examine how individuals around the country allocated their time on randomly selected days between 2003 and 2006, identifying the incremental influence of daily maximum temperature on the number of minutes individuals work, using data collected through the American Time Use Survey (Hofferth, Flood, & Sobek 2013). The individuals in the survey are considered nationally representative, and

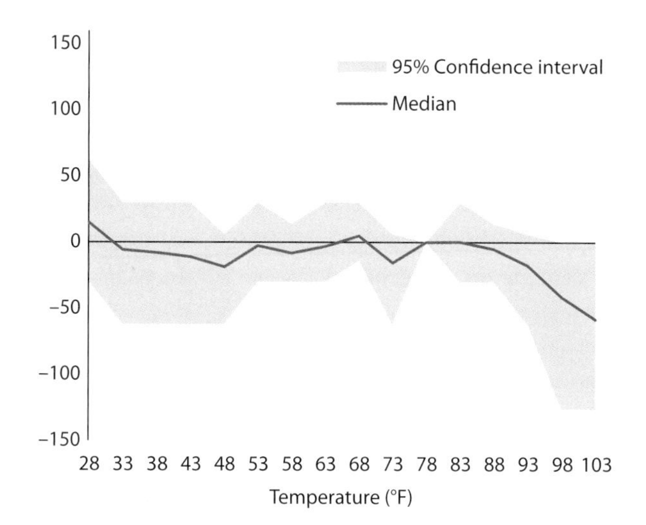

FIGURE 7.1. **Temperature and High-Risk Labor Productivity**

Change in minutes worked for high-risk laborers as a function of daily maximum temperature (degrees Fahrenheit)

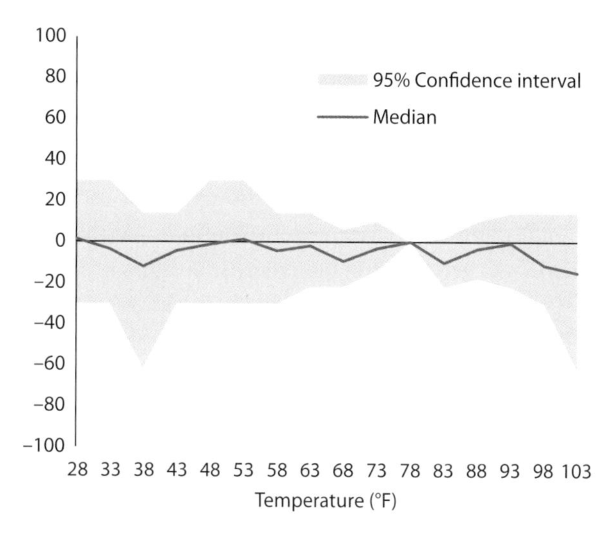

FIGURE 7.2. **Temperature and Low-Risk Labor Productivity**

Change in minutes worked for low-risk laborers as a function of daily maximum temperature (degrees Fahrenheit)

each individual records the allocation of his or her time during a single 24-hour period.

Figure 7.1 and figure 7.2 display the impact functions derived from Graff Zivin and Neidell for individuals working in high-risk and low-risk industries, that is, industries where individuals are likely and unlikely, respectively, to be strongly exposed to unregulated temperatures according to the National Institute for Occupational Safety and Health (Graff Zivin & Neidell 2014).

Temperature has little influence on labor supply in either category until very high daily maximum temperatures are reached, at which point individuals begin to supply less labor approximately linearly—the nonlinear structure of this response is broadly consistent with both laboratory studies (Mackworth 1947) and macroeconomic evidence (Hsiang 2010). As one might expect, the response in high-risk industries is more negative at high temperatures, probably because the working environment of these individuals more closely reflects ambient outdoor temperatures. On average, increasing a county's daily maximum temperature from 76°F to 80°F to greater than 96°F decreases high-risk labor supply by 41 to 58 minutes per day and low-risk labor supply by 11 to 15 minutes per day. As baseline hours worked are 7.66 and 6.92 for these two groups of workers, these reductions represent roughly 9 to 13 percent and 3 to 4 percent of total labor supplied in these two classes of industries. Graff Zivin and Neidell consider whether decreased working hours caused by high temperatures cause workers to supply additional labor at other times (to make up for lost work). They find there is essentially no temporal displacement of labor across days. They also examine whether there is evidence that populations adapt by examining if the sensitivity of labor supply to temperature is lower in counties that are hotter on average. They find ambiguous evidence, due in part to statistical uncertainty. Taken at face value, their estimates suggest warmer counties have a shallower response to temperature (consistent with adaptation), although adverse labor responses emerge at a lower temperature threshold in warmer counties (inconsistent with adaptation). It is possible that these differences are spurious and due to sampling variability, as the statistical uncertainty of these estimates is large and the responses across hot and cool counties are not statistically distinguishable. As with other impacts, we explore possible adaptive responses in part 5.

Using the dose-response functions obtained by Graff Zivin and Neidell, we simulate changes in labor supply under different climate scenarios relative to a future in which the climate does not change after 2012. Importantly, within each scenario we separately account for uncertainty in climate models, weather, and statistical results, causing our future projection to be a probability distribution of potential outcomes at each moment in time.

To assess potential future impacts of climate change on labor productivity, we project changes in national labor supply, using the dose-response functions obtained by Graff Zivin and Neidell, under different climate scenarios relative to a future in which the climate does not drive economic changes after 2012—although other social and economic trends are assumed to continue. Within each scenario we account for uncertainty in climate models, weather, and statistical results, causing our projection to be a probability distribution of potential outcomes at each moment in time.

When we consider the potential impact of changes in temperature on labor supply, we find that high-risk labor supply generally declines by midcentury, and the range of likely changes are unambiguously negative by late century for all scenarios (figure 7.3). In RCP 8.5, we

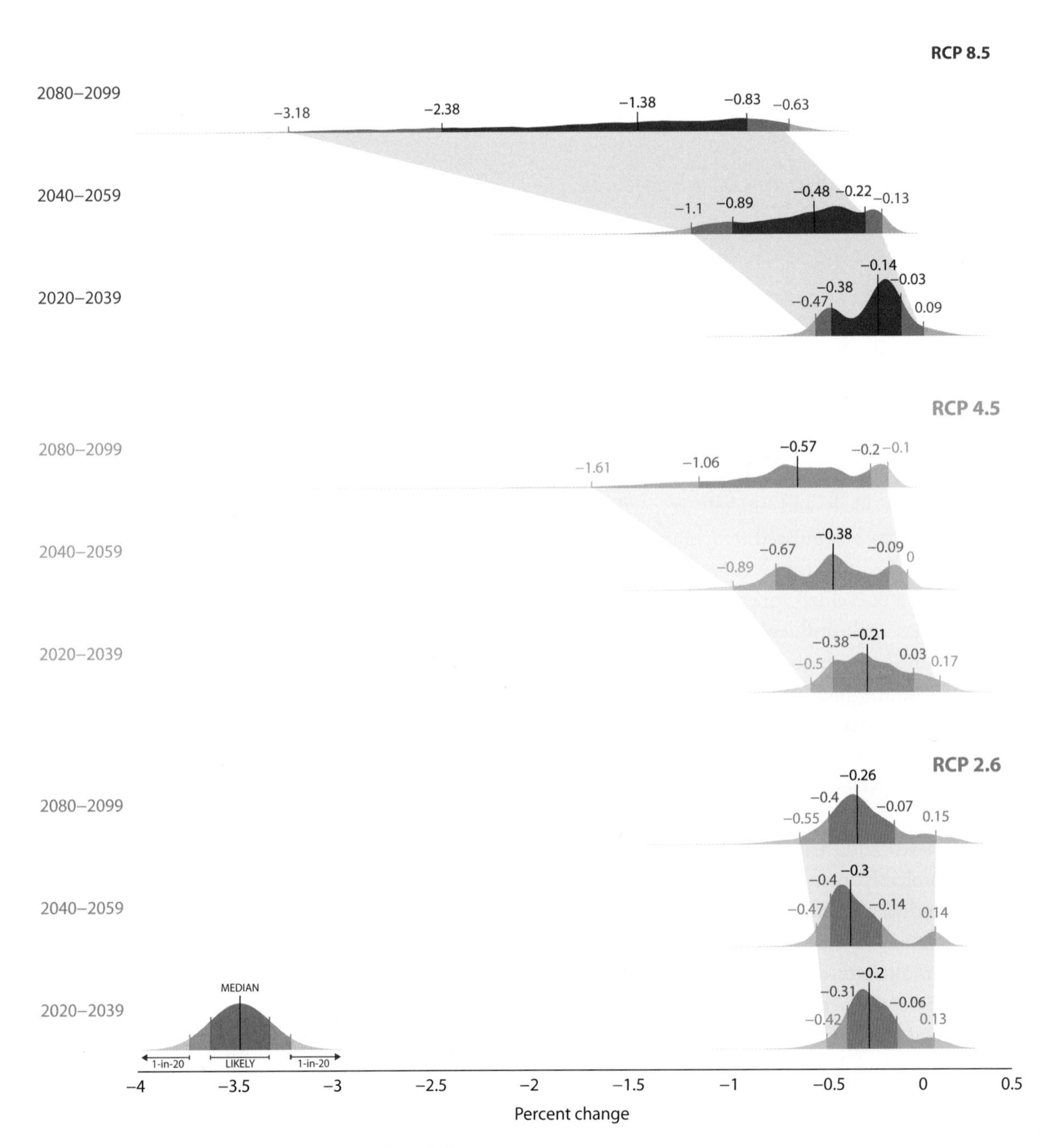

FIGURE 7.3. **Change in Labor Productivity in High-Risk Sectors**

find high-risk labor *likely* declines by 0.2 to 0.9 percent by midcentury and by 0.8 to 2.4 percent by late century, with a 1-in-20 chance that late-century labor supply falls either by more than 3.2 percent or less than 0.6 percent. Examining low-risk labor supply, we find that losses are more modest, with *likely* late-century losses in RCP 8.5 of 0.1 to 0.5 percent, with a 1-in-20 chance that labor supply falls more than 0.8 percent or less than 0.01 percent (figure 7.4).

Projected changes are smaller in magnitude for RCP 4.5 and RCP 2.6, with the *likely* range for high-risk labor supply spanning −0.2 to −1.1 percent for RCP 4.5 and −0.1 to −0.4 percent for RCP 2.6 by late century. The 1-in-20 outcomes span a narrower range than RCP 8.5, with high-risk

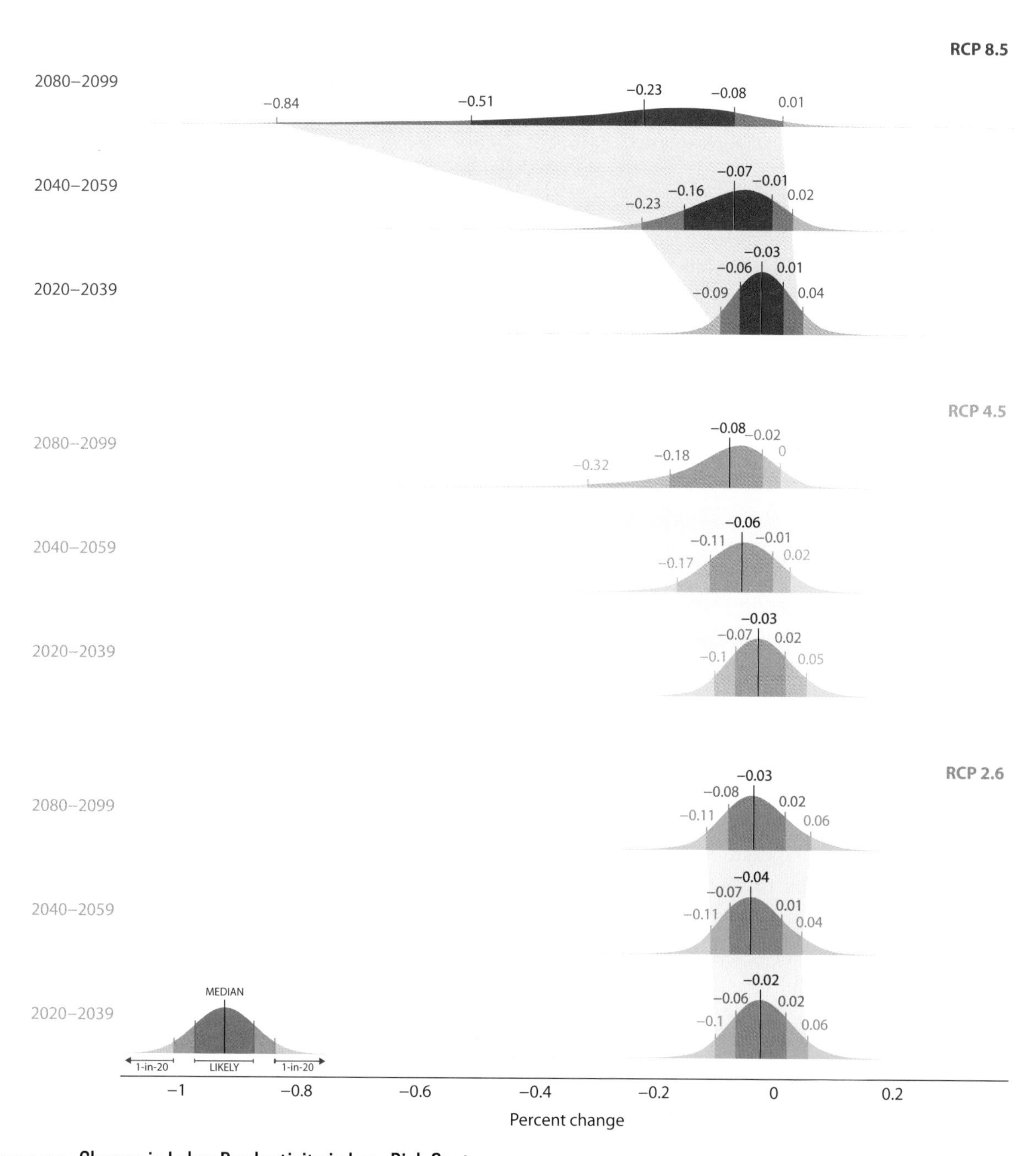

FIGURE 7.4. **Change in Labor Productivity in Low-Risk Sectors**

labor having a 1-in-20 chance of falling by more than 1.6 percent or less than 0.1 percent for RCP 4.5 and more than 0.6 percent or increase by 0.1 percent for RCP 2.6. Projected changes for low-risk labor are very small in RCP 2.6 and are modest in RCP 4.5 where the *likely* range of changes span −0.02 to −0.18 percent.

For both classes of labor, the distribution of potential changes broadens moving forward in time for RCP 8.5 and 4.5, but modestly so for low-risk labor in RCP 4.5 (figures 7.3 and 7.4). By late century in RCP 8.5, changes in labor supply exhibit a 90 percent confidence interval spanning 2.5 percentage points for high-risk labor and 0.8 percentage points for low-risk labor. There is essentially no change in the spread of potential outcomes under RCP 2.6 for either type of labor.

In percentage terms, the spatial distribution of projected effects is similar across the country in the median RCP 8.5 projection (figure 7.5, left column), with only a slight reduction in productivity losses for northern and mountainous regions and small gains in productivity in a few coastal northwestern counties. For a sense of scale, we compute changes in total productivity losses assuming there is no growth in the labor force and then convert this lost productivity to "full-time equivalent workers" (figure 7.5, right column). These losses are more strongly influenced by the locations of urban centers (e.g., Dallas, Atlanta, and Chicago) than regional spatial patterns of climate change.

The effects above are described in terms of average changes over 20-year intervals. These averages are useful for describing persistent economic changes in future periods, but they mask short-lived events that may only last a year or two but have substantial economic consequences. Within each 20-year window, the likelihood of extreme annual events, such as a very hot year, evolves with the climate. One way to describe how the likelihood of extreme events changes is to examine how frequently we expect to experience years that are as damaging as the worst year experienced during two decades of recent history, a so-called 1-in-20 year event. In figure 7.6, we plot the estimated number of years that will have high-risk labor supply losses larger than historically observed 1-in-20 year losses (due to temperature). For each year, we plot the expected number of these extreme years that will be experienced in the 20 years that follow; that is, we plot what the immediate future appears like to an individual in a given year. By 2030, labor supply losses that used to occur only once every 20 years will be expected to occur roughly ten times in the following 20 years under RCP 2.6, with even higher frequencies in RCP 4.5 and 8.5. By 2080, these events will be occurring roughly 16 times every 20 years in RCP 4.5 and every year in RCP 8.5. The relative frequency of analogous events in low-risk labor is essentially the same for all scenarios and time periods.

It is important to note that these estimates assume the national distribution of workers remains fixed relative to the average distribution from 2000 to 2005 and that air-conditioning and other time-allocation behaviors remain fixed. It is likely that some amount of additional adaptation will occur in the presence of climate change, but existing research is currently insufficient to conduct a systematic evaluation of adaptive behaviors that affect labor supply.

We also note that our estimates only account for changes to labor supply, which reflects a change in the total quantity of hours that each individual works (the extensive margin). It is extremely likely that the intensity of each worker's effort (the intensive margin) will also change with warming, as has been observed in numerous laboratory experiments (Mackworth 1948; Ramsey & Morrissey 1978; Wyon 2000), although the magnitude of this effect has not been measured in nationally representative and ecologically valid samples, so it is not included in this analysis. The laboratory-derived dose-response function of labor intensity is similar in magnitude (in percentage terms) to the response of high-risk labor, so accounting for this effect would roughly double the size of the impacts that we present here.

OTHER IMPACTS

The U.S. labor force is composed of people, so any one of the number of climate-related effects on the working population will ultimately affect the supply and quality of the U.S. labor market. In the following chapter, we report the impact on the health of Americans, including increased respiratory illness due to increases in pollution, allergens, and pollens as a result of rising temperatures.

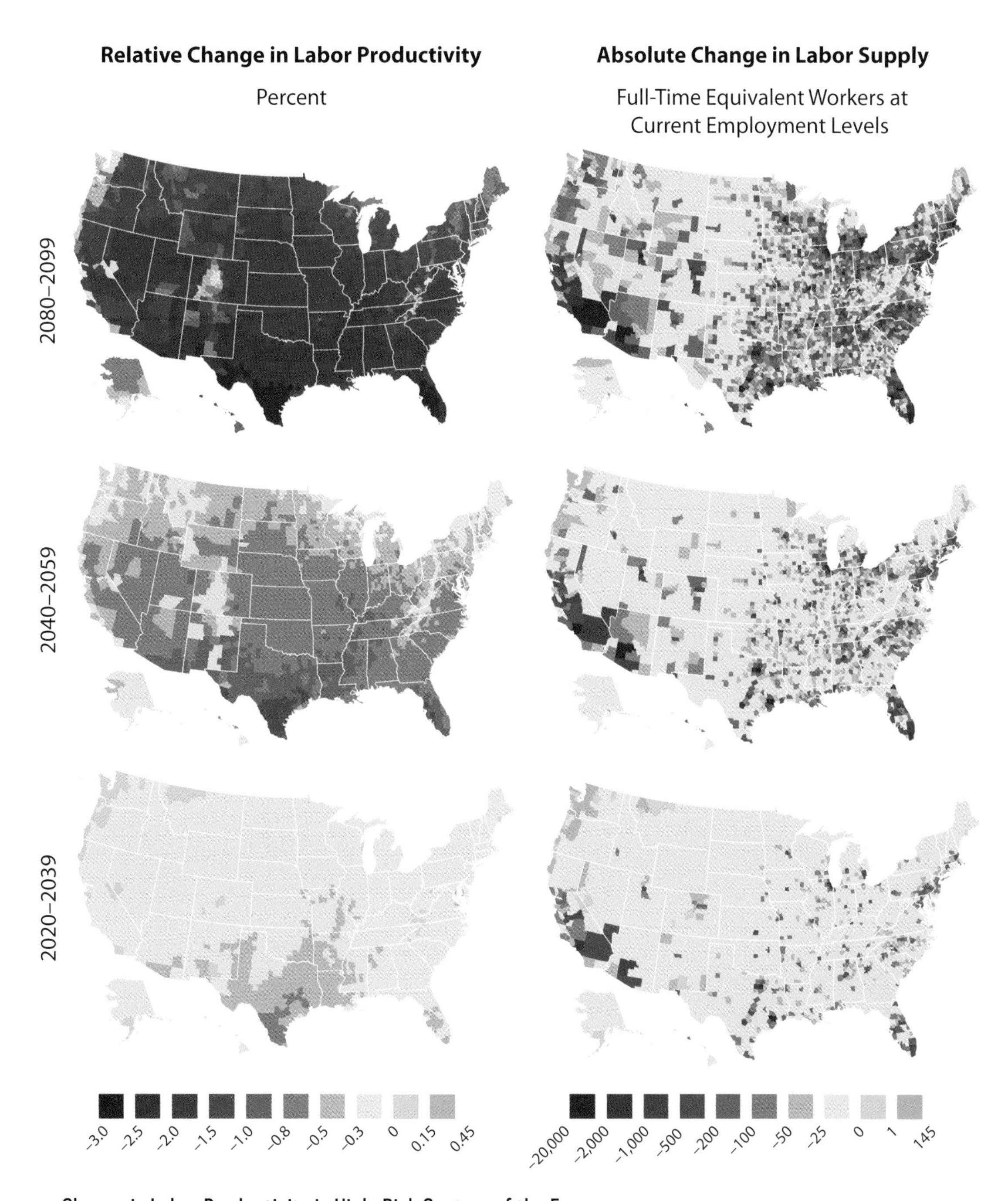

FIGURE 7.5. **Change in Labor Productivity in High-Risk Sectors of the Economy**

RCP 8.5 median

Changes in the geographic ranges and seasonality of vector-borne infectious diseases will also affect health and productivity for some working populations. Outdoor workers are most at risk for vector-borne infections because of their exposure to species that carry disease (Bennett & McMichael 2010). The risk of transmission may increase under climate change as warm, wet conditions contribute to a greater number of vectors, a change in the ranges of their habitats, and as transmission becomes more efficient. Vector-borne diseases are

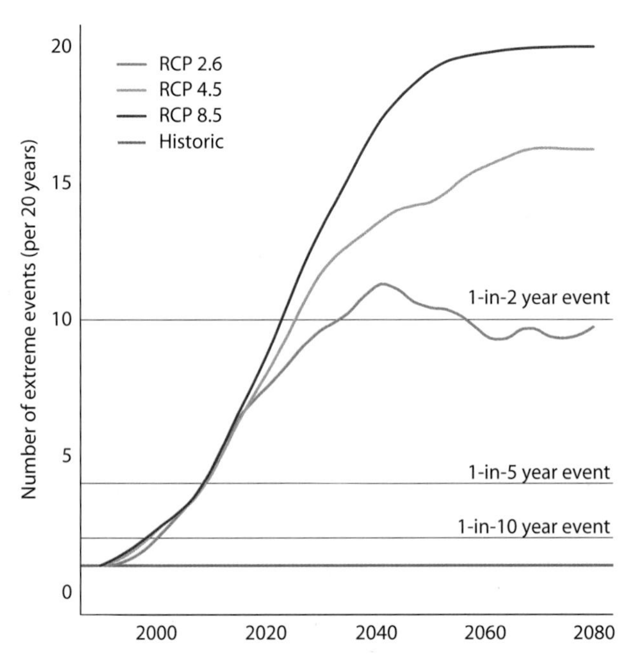

FIGURE 7.6. **Projected Change in the Frequency of National Labor Supply Losses Caused by Temperature Equal to or Worse than Historical 1-in-20 year Losses**

Expected frequencies across climate models and weather; frequencies are plotted in the first year of each 20-year period

already responsible for considerable losses in economic productivity every year, primarily in regions where a vector-borne disease is endemic. We have not captured losses due to illness in our analysis, but these will certainly affect labor supply and productivity.

Illness, injury, and even death from increased damage caused by hurricanes and other storms, flooding, wildfire, and other extreme weather will also affect the labor market, although quantifying their impact is not straightforward. In the months following an extreme storm, depending on the severity of the storm, there can be negative impacts on total employment and earnings, as well as disturbances across labor sectors (Deryugina 2011; Camargo & Hsiang 2012). Labor impacts are not limited to the areas hit by the storm; a study of 19 hurricanes that hit Florida between 1988 and 2005 found that labor markets in counties neighboring an affected county became more competitive, with falling wages, due to the movement of skilled workers out from the affected county (Belasen & Polachek 2009). These effects were found to dissipate over time, though long-run impacts are not generally understood as they are more complex and more difficult to measure.

CHAPTER 8
HEALTH

THE American public-health system has been designed to promote the health of American communities and residents and to prevent disease, injury, and disability. Huge strides in public health have been made over the past century, thanks in part to greater wealth, scientific advances and innovation, increased education, and a more developed public-health infrastructure (e.g., water-treatment plants, sewers, and drinking-water systems). Improvements in disaster planning and emergency response have also improved public health outcomes after responses to disease outbreaks and epidemics, floods, heat waves, storms, and other disasters that can harm public health. These advances have relied on years of assessment of the evolving vulnerabilities and resilience of American communities and populations, the various factors that contribute to health impacts in those communities, and the potential risks, both likely and unlikely, that threaten public health.

Climate change is an emerging factor in the risk landscape for American public health. According to the U.S. National Climate Assessment, there is very high confidence that a wide range of health effects will be exacerbated by climate change in the United States (Joyce et al. 2013). Human health and well-being are likely to be impacted by climate change through gradual changes in average temperature and precipitation and also through changing patterns of extreme events. Incremental effects of increasingly warmer summer temperatures will lead to increased rates of ozone formation and exacerbate respiratory problems. However, milder winters may reduce cold-related deaths. Such changes will also create conditions that can disrupt natural systems that affect public health; for example, altering the length and severity of allergy seasons and changing the patterns and spread of vector-borne diseases, such as Lyme disease. A changing frequency and intensity of extreme weather events, such as heat waves, floods, droughts, storms, and wildfires, creates a less predictable but potentially serious risk of disease, exposure to dangerous pollutants, injury, and even death.

Climate impacts will be wide ranging and highly variable across regions and populations. There will be positive as well as negative impacts, depending on the local circumstances, but on the whole, net impacts are likely to be negative for most regions in the United States. The cause-and-effect chain between climate and health impacts is complex, and climate change is one of many critical factors affecting public-health outcomes. The magnitude and distribution of these effects will depend on the baseline vulnerability of populations over time, which are determined by a whole host of variables including population age, socioeconomic status, and race, as well as regional and local differences in critical public-health infrastructure and investments. Evidence indicates that, absent changes that go beyond current prevention and adaptation activities and with increasing population susceptibilities (aging, limited economic resources, etc.), some existing health threats will intensify and new health threats will emerge (Joyce et al. 2013). While we provide an overview of a number of potential climate-change impacts on health, our analysis focuses specifically on the temperature-related impacts on mortality.

BACKGROUND

One of the most well-studied impacts of climate on public health is the effect of temperature and, in particular, extreme hot and cold days. Impacts will be felt differently across the United States, with some northern regions experiencing milder winters and reduced exposure to extreme cold and snow, while other regions will see longer and more frequent heat waves. The level of vulnerability of populations to these risks will depend on the severity of the extremes and on society's adaptive response.

Across the United States, most regions are already experiencing the types of impacts that could be exacerbated by a changing climate. Many people remember the 1995 Chicago heat wave that brought nine consecutive days of record-setting daytime and nighttime temperatures, unprecedented in the preceding 120 years over which records have been kept. Nearly 800 people died from heat exposure, and thousands of excess emergency room and hospital visits were recorded as a result of the heat wave (Hayhoe et al. 2010). The 2006 heat wave that hit much of California was exceptional both for its intensity and duration, setting records for most consecutive days over 100°F, and resulted in 140 deaths, more than 16,000 excess emergency room visits, and 1,180 excess hospitalizations (Knowlton et al. 2009). Children under the age of 4 years and the elderly tend to be particularly vulnerable.

Heat stress can lead to increased hospitalizations due to heat exhaustion and heat stroke; dehydration and electrolyte disorders; acute renal failure, nephritis, and nephrotic syndrome; and other heat-related illnesses (Kovats & Ebi 2006; Knowlton et al. 2009). Increased mortality during heat waves has been attributed mainly to cardiovascular illness and diseases of the cerebrovascular and respiratory systems, especially among the elderly (Anderson & Bell 2011). Heat stress can rapidly become life threatening among those with limited access to immediate medical attention, and often people with severe heat-stroke symptoms have little time to seek treatment. These impacts have been most severe for people over 65 and those with preexisting conditions (Zanobetti et al. 2012).

Some of the most well-documented effects of heat stress are in big cities, in part because large populations are simultaneously exposed to extreme heat events, generating large numbers of coincident cases. In addition, urban residents may be exposed to higher temperatures than residents of surrounding suburban and rural areas because of the *heat-island effect* resulting from high thermal absorption by dark, paved surfaces and buildings, heat emitted from vehicles and air conditioners, lack of vegetation and trees, and poor ventilation (O'Neill & Ebi 2009).

Heat stress on local populations can also stress the public-health system. In 2009 and 2010, there were an estimated 8,251 emergency department visits for heat stroke in the United States, yielding an annual incidence rate of 1.34 visits per 100,000 population (Wu et al. 2014). In times of excessive heat, these figures can jump considerably. During the 1995 Chicago heat wave, for example, excess hospital admissions totaled 1,072 (up 11 percent) among all age groups for the days during and immediately after the event, including 838 admissions among those 65 years of age and older (an increase of 35 percent), with dehydration, heat stroke, and heat exhaustion as the main causes (Semenza et al. 1999).

Extreme heat is increasing in parts of the United States and is expected to be more frequent and intense (Joyce et al. 2013). Changes in the intensity, duration, and seasonal timing will influence mortality and morbidity effects within communities. Under all future pathways, the number of days with temperatures reaching 95°F or higher across the continental United States is expected to increase from the historic (1981–2010) baseline of 15 days per year. As described in chapter 4, under RCP 8.5, by midcentury the average American will *likely* experience two to three times the number of days over 95°F than on average from 1980 to 2011 (an additional 12 to 35 days). By late century, the average American will *likely* experience

1.5 to 3 months of days over 95°F each year on average (46 to 96 days). National averages, however, say little about regional and local effects, which may be more extreme in some areas. The average resident of the Southeast, for example, will *likely* see 56 to 123 days over 95°F on average by the end of the century, up from only nine per year on average from 1981 to 2010.

At the other end of the spectrum, milder winters may actually have a positive impact on public health. Deaths and injuries related to extreme winter weather, as well as respiratory and infectious disease related to extreme cold, are projected to decline because of climate change (Medina-Ramón & Schwartz 2007). Across the continental United States, winter temperatures will *likely* be 2.5°F to 6.2°F warmer on average by midcentury and 5.4°F to 11.8°F warmer by the end of the century under RCP 8.5. The number of average days with low temperatures below freezing for the average county in the contiguous United States is also expected to decrease dramatically. Under RCP 8.5, the average number of days below freezing is *likely* to drop from the 1981–2010 average of 113 days per year to 81 to 100 days by midcentury, and to 52 to 81 days by late century.

OUR APPROACH

Given what we know about observed climate impacts on health in the United States over the past few decades, what can we predict in terms of likely future impacts? The most systematically documented relationship is the impact of rising temperatures on mortality, which can be estimated nationally because the Centers for Disease Control and Prevention (CDC) compiles and releases national mortality data. Morbidity impacts are more difficult to study because national data are not readily available.

To quantify the potential impact of climate change on mortality, we rely on statistical studies that isolate the effect of temperature on mortality in the United States. Because there are strong cross-county patterns in mortality rates, as well as strong trends over time (that may differ by location) and over seasons, we rely on studies that account for these patterns when measuring the effect of climate variables on mortality rates. Two studies provide nationally representative estimates that satisfy these criteria, which we use to construct quantitative projections. Deschênes and Greenstone (2011) examine county-level annual mortality rates during 1968–2002, and Barreca et al. (2013) examine state-level monthly mortality rates during 1960–2004 (Deschênes & Greenstone 2011; Barreca et al. 2013). Both studies identify the incremental influence on mortality for each additional day at a specified temperature level using data collected by the CDC on all recorded deaths in the United States. Deschênes and Greenstone provide greater spatial resolution, while Barreca et al. provide greater temporal resolution, thus the studies may be complementary.

Figure 8.1 displays the impact function derived from Deschênes and Greenstone and Barreca et al. Both studies largely agree that both low and high daily average temperatures increase overall mortality rates relative to the lowest-risk temperature range of 50°F to 59°F, with annual mortality rates increasing roughly 0.08 percent for an additional single day with mean temperature exceeding 90°F. In both sets of results, low temperatures tend to pose a differentially high risk to middle-age (45 to 64 years) and older (>64 years) individuals. Deschênes and Greenstone estimate that high temperatures pose a differentially high risk to infants (<1 year) and older individuals, while Barreca et al. find that high temperatures pose a differential high risk to infants and younger (1 to 44 years) individuals with modest proportional risks imposed on the elderly. Barreca et al. examine causes of mortality and find that high-temperature deaths are usually attributed

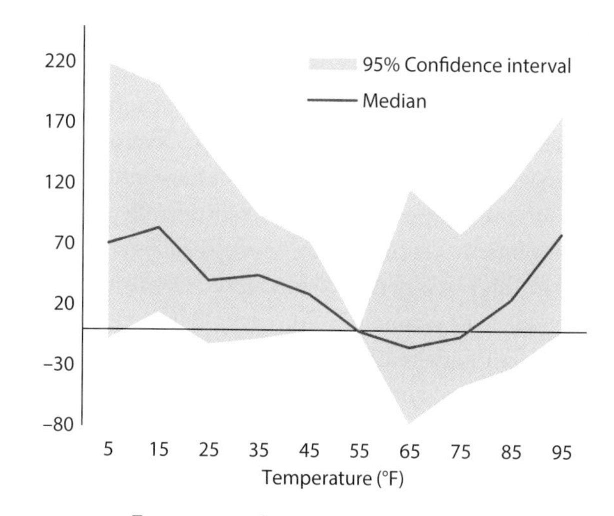

FIGURE 8.1. **Temperature Impact on Mortality**

Change in mortality rate (deaths per 100,000 people) vs. daily maximum temperature (degrees Fahrenheit)

to cardiovascular or respiratory disease, while low-temperature mortality is driven most strongly by respiratory disease as well as infectious and cardiovascular disease. These findings are generally consistent with studies of specific cities, regions, and subpopulations during extreme climatic events (e.g., Curriero et al. 2002; Barnett 2007; Anderson & Bell 2009) and are nationally representative.

It is thought that one effect of high temperatures is to induce an acceleration or *forward displacement* of mortality that would have occurred in the near future anyway, even in the absence of a high-temperature event (a phenomenon sometimes known as "harvesting") (Deschênes & Moretti 2009). Deschênes and Greenstone and Barreca et al. do not extensively consider the extent to which increases in mortality rates caused by high and low temperatures cause forward displacement of mortality, although the authors attempt to account for possible temporal displacement by examining mortality over relatively long windows of time: a year for Deschênes and Greenstone and 2 months for Barreca et al. As long as temporal displacement occurs within these time frames, then these estimates will describe the net effect of a temperature event. Deschênes and Greenstone note that as total mortality is fixed in the very long run, the welfare impact of climatic changes are best measured by changes in total life-years. However, for simplicity and clarity, we focus here on total premature mortality by age group.

Climate variables are not the only factor influencing mortality effects on U.S. populations. Adaptation, primarily through increased use of air-conditioning, mitigates the mortality risk of extreme temperatures. Barecca et al. study whether there is evidence that populations adapt by examining if the sensitivity of mortality to temperature declines over time or is lower in counties that are hotter on average. Barecca et al. find that the response of mortality to temperature has declined substantially since the early twentieth century (1929–1959), with larger reductions in high-temperature mortality. Consistent with this evidence of adaptation over time, Barecca et al. find that modern high-temperature mortality is lowest in hot southern counties and modern low-temperature mortality is highest in these counties. They argue that these patterns of sensitivity inversely reflect patterns of air-conditioning adoption, a likely mechanism through which populations adapt. Here we assume that the sensitivity of mortality to temperature does not change relative to the present, and in part 5 of this book we explore the extent to which increased use of air-conditioning and other adaptations can mitigate these deaths. However, it is important to note that air-conditioning is unlikely to mitigate all temperature-related deaths because only a portion of deaths occur from inadequate air-conditioning, and the impact function we use here is similar to the response recovered when we examine only populations in the American South, where air-conditioner penetration is roughly 100 percent (Barreca et al. 2013).

To assess potential future effects of climate change on mortality, we simulate changes in mortality rates under different climate scenarios relative to a future in which the climate does not drive changes in health after 2012—although other social and economic trends are assumed to continue. Within each scenario, we account for uncertainty in climate models, weather, and statistical results, causing our projection to be a probability distribution of potential outcomes at each moment in time.

When we consider the potential impact of changes in temperature, we find that mortality rates do not generally increase until late-century except for populations aged 1 to 44 years, which exhibit elevated mortality for all time periods (table 8.1) because this age group does not benefit from reductions in cold weather. In RCP 8.5, we find annual all-age mortality rates are *likely* to change by −0.5 to 6.6 deaths per 100,000 by midcentury and 3.7 to 21 deaths per 100,000 by late century, with a 1-in-20 chance that late-century increases are below 0.6 deaths or exceed 36 deaths per 100,000 relative to baseline mortality rates (figure 8.2). Results are roughly similar for all age groups (table 8.1) except the over-65 age group, where mortality rates are more responsive to both warming and cooling climate changes in all periods, with a *likely* range of changes spanning −23 to +18 additional deaths per 100,000 in 2020–2039, −24 to +22 by midcentury, and −21 to +90 additional deaths per 100,000 by late century.

Projected changes are modest in magnitude for RCP 4.5 and RCP 2.6, with the *likely* range of changes for all-age annual mortality spanning a change of −2.5 to 5.9 deaths per 100,000 for RCP 4.5 and −2.3 to 3.2 deaths per 100,000 for RCP 2.6 by late century. The 1-in-20 outcomes span a narrower *likely* range than RCP 8.5, with a 1-in-20 chance mortality rates change by less than −4.5 deaths or more than 12 deaths per 100,000 for RCP 4.5 and less than −3.9 or more than 5.0 deaths per 100,000 for RCP 2.6. Projections for the over-65 age group are universally more extreme in magnitude, with a *likely* range of

TABLE 8.1 Impact of future climate change on U.S. mortality rate

	Mortality Rate								
	RCP 8.5			*RCP 4.5*			*RCP 2.6*		
Age Group	*1 in 20 Less Than*	*Likely*	*1 in 20 Greater Than*	*1 in 20 Less Than*	*Likely*	*1 in 20 Greater Than*	*1 in 20 Less Than*	*Likely*	*1 in 20 Greater Than*
	Deaths per 100,000			*Deaths per 100,000*			*Deaths per 100,000*		
<1 year old									
2080–2099	0.7	3.2 to 17	29	−1.8	−0.4 to 4.9	9.2	−2.1	−1.1 to 2.5	3.5
2040–2059	−1.4	0.1 to 4.5	7.6	−1.9	−0.9 to 2.8	5.1	−2.2	−1.0 to 2.4	−3.6
2020–2039	−1.6	−0.9 to 2.4	3.7	−2.3	−1.4 to 1.7	3.3	−2.2	−1.0 to 2.3	−3.7
1–44 years old									
2080–2099	2.8	3.1 to 7.6	9.6	0.5	1.1 to 3.6	5.1	−0.1	0.2 to 1.5	1.9
2040–2059	0.9	1.0 to 3.0	3.5	0.3	0.5 to 2.3	2.8	−0.1	0.6 to 1.4	1.7
2020–2039	0.0	0.2 to 1.3	1.5	−0.3	0.1 to 1.2	1.5	−0.1	0.3 to 1.0	1.2
45–64 years old									
2080–2099	1.3	2.8 to 14	23	−2.4	−1.2 to 3.6	7.5	−2.2	−1.3 to 2.0	2.6
2040–2059	−1.1	0.2 to 3.6	5.3	−2.6	−1.8 to 2.3	4.3	−2.0	−1.2 to 2.0	3.1
2020–2039	−1.6	−0.9 to 2.1	3.4	−2.8	−1.7 to 1.5	3.1	−2.0	−1.9 to 1.8	3.4
65+ years old									
2080–2099	−51	−21 to 90	181	−55	−38 to 16	48	−41	−25 to 17	23
2040–2059	−43	−24 to 22	38	−48	−36 to 9.6	34	−41	−22 to 13	25
2020–2039	−35	−23 to 18	32	−41	−29 to 9.7	35	−33	−16 to 14	27

Note: Percentage change in net age-specific heat- and cold-related mortality from 2012 levels. Likely range represents 17 to 83 percent confidence band.

−38 to +16 deaths per 100,000 by late century for RCP 4.5 and −25 to +17 deaths per 100,000 for RCP 2.6.

The distribution of potential changes broadens moving forward in time for RCP 4.5 and RCP 8.5, with almost no broadening for RCP 2.6 (figure 8.2). Changes in all-age mortality rates by late century exhibit a 90 percent confidence interval spanning 16.4 deaths per 100,000 in RCP 4.5, which widens to span 35 deaths per 100,000 in RCP 8.5. Across age groups, the spread of potential outcomes is largest for ages 65+, with a 90 percent confidence interval that spans 232 deaths per 100,000 by the end of the century in RCP 8.5.

In percentage terms, the spatial distribution of projected impacts is uneven across the country (figure 8.3), with the Southeast, Southwest, southern Great Plains, mid-Atlantic, and central Midwest suffering the largest increases in mortality rates (deaths per 100,000 people) because the number of high-temperature days in these regions increases substantially, but the baseline climate in these locations is too warm for them to experience many lethally cold days initially, minimizing the gains from warming. In contrast, the Rockies, Appalachia, Northwest, northern Great Plains, northern Midwest, and Northeast (except New York City and New Jersey) experience net reductions in mortality rates for the median RCP 8.5 projection because the gains from a smaller number of cold days outweigh the losses during additional hot days. For a sense of scale, we compute changes in total mortality assuming no population growth from the 2010 U.S. Census and see how these gross national patterns translate into changes of national aggregate mortality (figure 8.3, right column). Because many of the regions that benefit from climate change have less-dense populations (e.g., northern Great Plains) than many of the regions that see increasing mortality rates (e.g., Gulf states), the net change in national outcomes is for the risk of mortality to rise on average.

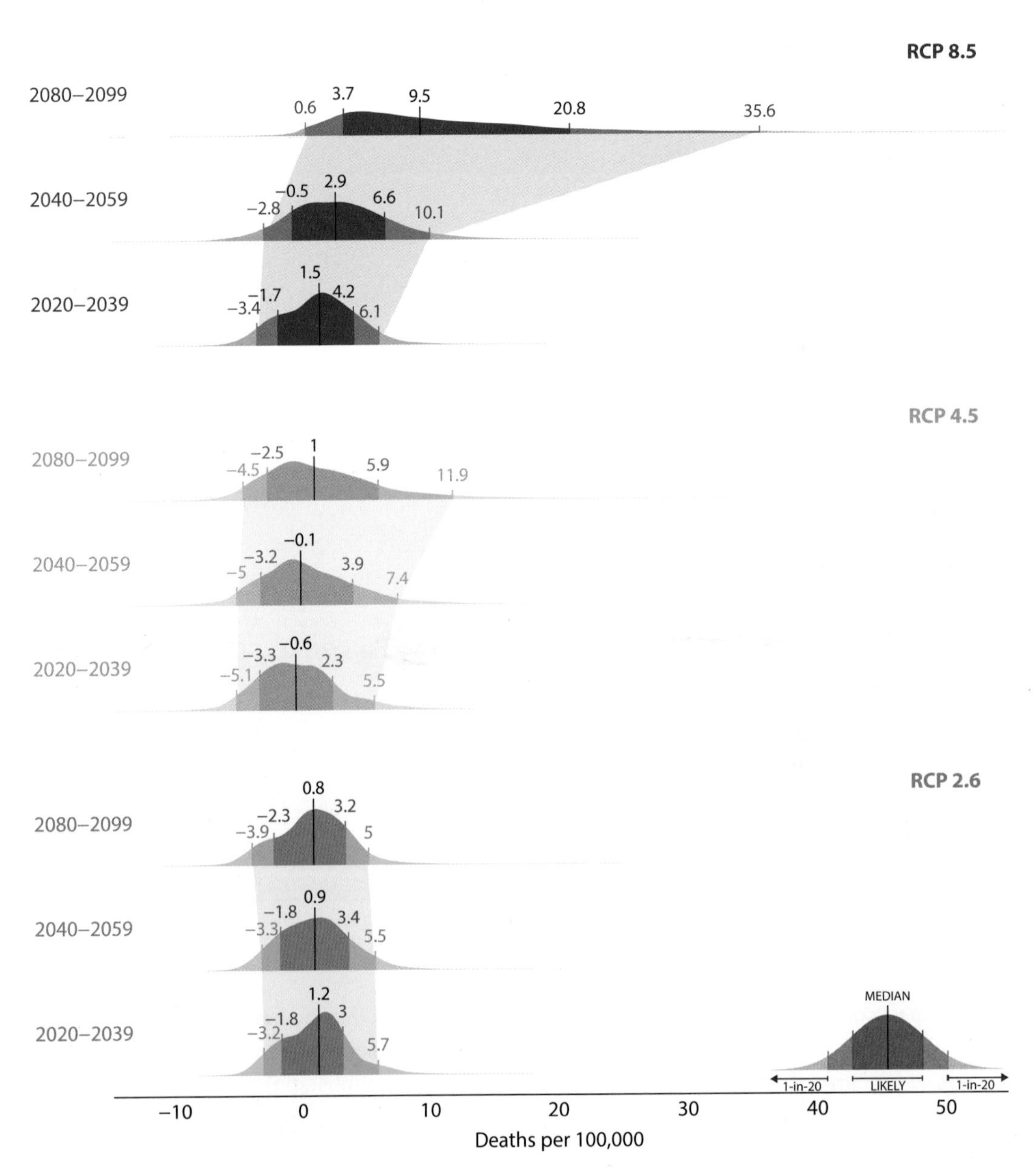

FIGURE 8.2. **Change in National Mortality Rate**

Net percentage change in all-age, heat- and cold-related mortality (deaths per 100,000)

The impacts are described in terms of average changes over 20-year intervals. These averages are useful for describing persistent economic changes in future periods, but they mask short-lived events that may only last a year or two but have substantial economic consequences. Within each 20-year window, the likelihood of extreme annual events, such as a very high mortality year, evolves with the climate. One way to describe how the likelihood of extreme events changes is to examine how frequently we expect to experience years that are as deadly as the worst year experienced during two decades of recent history, a so-called 1-in-20 year event. In figure 8.4, we plot the estimated number of years that will have temperature-related mortality larger than historically observed 1-in-20 year mortality events. For each year, we plot the expected number of these extreme years that will be experienced in the 20 years that follow; that is, we plot what the immediate future appears like to an individual in a given year.

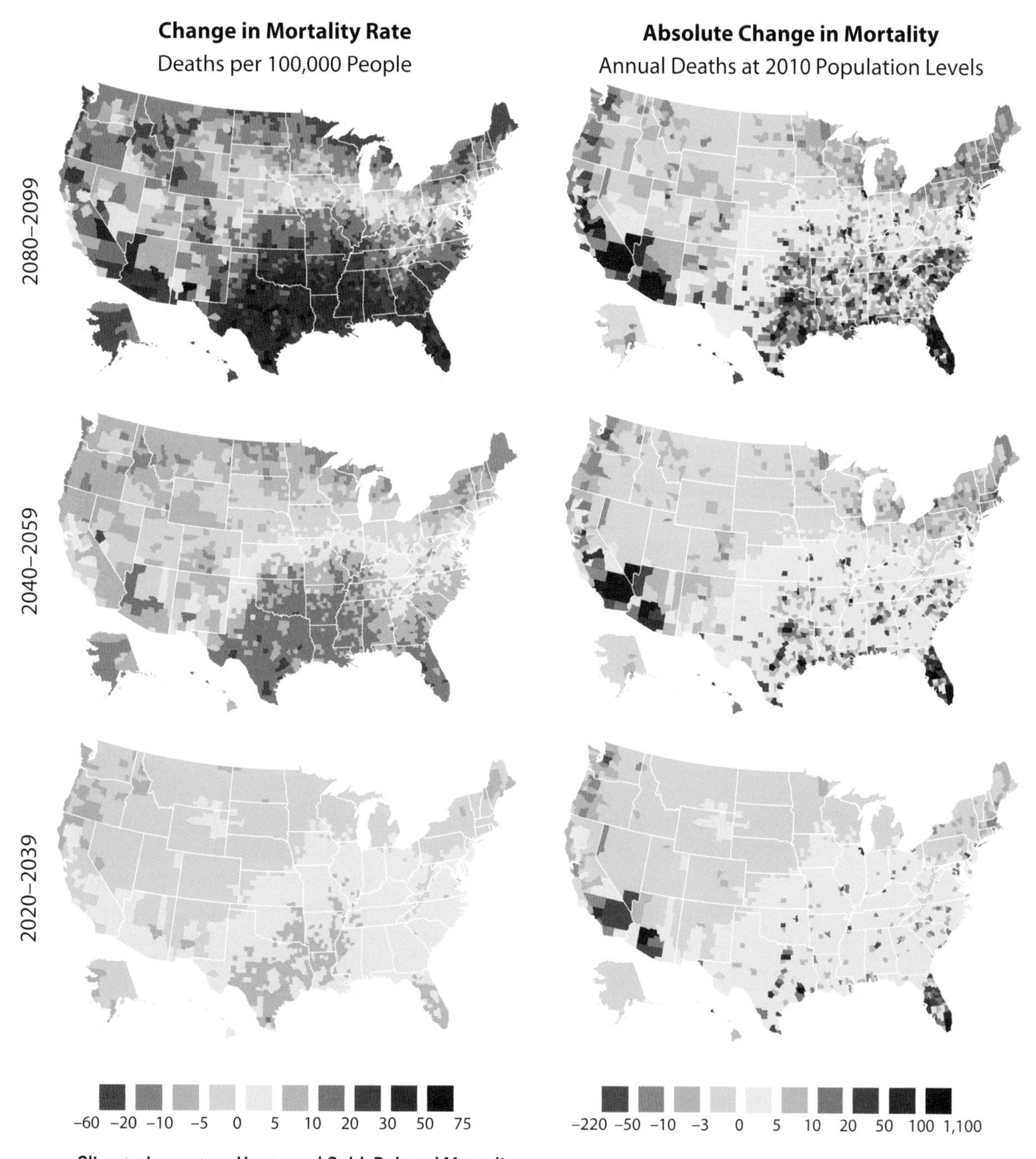

FIGURE 8.3. **Climate Impact on Heat- and Cold-Related Mortality**

RCP 8.5 median

By 2030, mortality events that used to occur only once every 20 years will be expected to occur at roughly the same rate during the following 20 years for RCP 8.5, with the frequency falling slightly for RCP 4.5 and RCP 2.6. The frequency of these extreme mortality events falls slightly in the first half of the century because reductions in cold-related mortality more than offset increases in heat-related mortality. However, after 2040 the effect of extreme heat events drives the likelihood of extreme mortality events steadily upward in RCP 8.5, causing these events to occur roughly 11 times every 20 years by the end of the century. The frequency of extreme mortality events remains

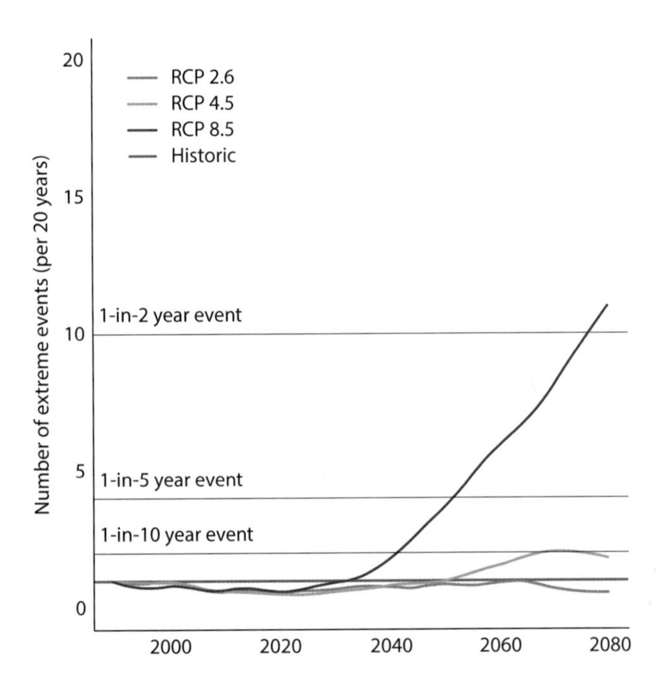

FIGURE 8.4. **Projected Change in Frequency of Temperature-Related Mortality Events Equal to or Worse than Historical 1-in-20 Year Events**

Expected frequencies across climate models and weather; frequencies are plotted in the first year of each 20-year period

slightly depressed in RCP 2.6 throughout the century and rises to twice every 20 years by 2070 in RCP 4.5.

It is important to note that these estimates assume the national distribution of individuals remains fixed relative to 2010 (U.S. Census estimates) and that other health factors that interact with heat-related mortality also remain fixed. If populations migrate for health-related reasons or invest in other protective investments, such as air-conditioning, then these projections will overstate future impacts. We explore the impact of some of these potential adaptations on our projections in part 5.

OTHER IMPACTS

Air-Quality Impacts (Ozone, Allergens, Particulate Matter)

Climate change will affect air quality through several pathways, including temperature-related effects on regional ambient concentrations of ozone, fine particles, and dust. Air pollutants such as ozone (O_3), carbon monoxide (CO), sulfur dioxide (SO_2), nitrogen oxide (NO_x), volatile organic compounds (VOCs), and particulate matter (PM) have both acute and chronic effects on human health, including respiratory irritation, chronic respiratory and heart disease, lung cancer, acute respiratory infections in children and chronic bronchitis in adults, and aggravation of preexisting heart and lung disease or asthmatic attacks (Kampa & Castanas 2008). Short- and long-term exposures have also been linked with premature mortality and reduced life expectancy.

The public-health effects of ozone pollution are significant. A large body of evidence, summarized in the EPA's most recent comprehensive assessment, provides clear evidence of the health effects of exposure to ground-level ozone (U.S. Environmental Protection Agency 2013). The strongest evidence links ozone exposure to respiratory illness, as ozone has been found to irritate lung airways, diminish lung function, and exacerbate respiratory symptoms and has been associated with increased hospital admissions and emergency room visits (U.S. Environmental Protection Agency 2013). Symptoms can arise quickly, sometimes within 1 hour of exposure, or more gradually from cumulative exposure over time, or even several days after exposure. Studies have found that nationwide ozone exposure from 2000 to 2002 resulted in not only nearly 800 deaths, but also more than 4,500 hospitalizations and emergency room visits, and 365,000 outpatient visits, which resulted in costs of more than $6.5 million (Knowlton et al. 2011). One study of ozone pollution in Los Angeles from 1993 to 2000 found that an 0.01-ppm increase in the 5-day average ozone level (the EPA National Ambient Air Quality Standard for the 8-hour daily maximum is 75 parts per billion [ppb], or equivalently 0.075 ppm) is associated with a 4.7 percent increase in hospitalizations for respiratory illness. Given ambient levels in Los Angeles during that period reached 0.05 ppm, the additional hospitalization costs from that ozone exposure amounted to $1.85 million per year (Moretti & Neidell 2009).

In some cases, ozone exposure can even lead to death. Mortality effects have been shown to be both short-term, happening within hours of exposure, and cumulative, having effects over long-term periods of exposure. Short-term exposure to an 0.01-ppm increase can lead to an increase in cardiovascular and respiratory mortality of 0.52 percent (Bell, McDermott, & Zeger 2004), and a 4 percent higher risk of death from increases in seasonal averages of 0.01 ppm (Jerrett et al. 2009). The mortality and morbidity effects are exacerbated for particularly susceptible populations, such as children, the elderly, and those with existing respiratory conditions such as asthma.

Ground-level ozone is created in reactions between NOx and VOCs in the presence of sunlight. Consequently, studies find a direct correlation between temperature and ozone formation. During summer 1990 in Atlanta, maximum ozone concentrations were found to rise with maximum daily temperatures (Bernard et al. 2001). That same summer, hospitalizations for childhood asthma and other reactive airway diseases were found to increase nearly 40 percent after high-ozone days (White et al. 1994).

With emissions at today's levels, warmer summer temperatures are expected to increase summertime ground-level ozone concentrations. One study estimated the potential future impact of climate change on ozone concentrations and public health for 50 eastern U.S. cities (Bell et al. 2007). Using global and regional climate and regional air-quality models, but keeping ozone precursor emissions flat at 1993–1997 levels, the study was able to isolate the effect of temperature on future ozone concentrations. It found that on average across the 50 cities, by midcentury the summertime daily 1-hour maximum was expected to increase by 4.8 ppb from 1990 levels, with the largest increase, at 9.6 ppb, under a business-as-usual greenhouse gas emission scenario. The average number of days per summer exceeding the national regulatory standard was found to increase by 68 percent on average, and hospitalizations for asthma increased by 2.1 percent as a result of temperature increases alone. A recent study by Pfister et al. (2014), using downscaled climate projections (using RCP 8.5) and regional chemical-transport models, found similar results for midcentury. Without emission controls on ozone precursors, they found the number of high-ozone events (with concentrations >75 ppb) show a clear increase from current levels (70 percent) over most of the contiguous United States.

Particulate matter—especially very fine particles such as those found in smoke or haze—can get deep into the lungs and cause serious health problems. While the potential impacts of temperature on particulate matter levels across the United States are not well understood (Tagaris et al. 2007; Tai, Mickley, & Jacob 2012), there is a clear relationship between particulate pollution and wildfires, which are expected to be more frequent in some areas of the United States as temperatures rise and landscape becomes drier. Smoke produced by wildfires contains particulate matter, carbon monoxide, and several ozone precursors including nitrogen oxides, non-methane organic compounds, and volatile organic compounds, which can significantly reduce air quality in local communities and areas downwind (Dennekamp & Abramson 2011). The respiratory health impacts of such exposures have been found to be significant, in line with exposures to similar levels of emissions in urban areas from vehicle tailpipes, power plants, and other industrial sources.

In many areas of the United States, air pollution from wildfires and biomass burning represents a significant share of total exposure, especially as state programs have worked to reduce emissions from vehicles and other man-made sources. In California, criteria pollutant emissions from biomass burning were found to contribute emissions equivalent to 18 and 34 percent of man-made emissions of carbon monoxide and particulate matter (PM2.5), respectively. The same study found that under a medium-high future climate-change scenario, end-of-century emissions from wildfires in California are projected to increase 19 to 101 percent above rates experienced in recent decades (Hurteau et al. 2014). They found the emissions increases to be most extreme in Northern California, an effect that was influenced very little by adjusting development patterns to control for impacts on population exposure. Emissions from wildfires in the Sierra Nevada will directly impact the San Joaquin Valley air basin, one of the most populous and fastest growing in the state, and one with a high probability of exceeding federal air-quality standards for ground-level ozone. Under future climate scenarios, degraded air quality in the basin due to wildfires is expected to affect an additional 1.5 million to 5.5 million people.

Allergies are the sixth most costly chronic disease category in the United States, and the direct medical costs of two of the main allergic diseases—asthma and hay fever—are estimated to be $12.5 billion and $6.2 billion per year, respectively (U.S. Environmental

Protection Agency 2008b). The production of plant-based allergens will also be affected by climate change. Increased pollen concentrations and longer pollen seasons, resulting from warmer temperatures and higher ambient CO_2 concentrations, which can generate allergic responses and exacerbate asthma episodes, diminish productive work and school days and incur health-care costs (Wayne et al. 2002; Ziska 2008). Studies have shown that, between 1995 and 2001, the pollen season for ragweed, a significant cause of hay fever in the United States, increased as much as 13 to 27 days in the central United States, with the largest increases observed in northern cities, including Minneapolis, Fargo, and Madison (Ziska et al. 2011).

Extreme Weather, Waterborne and Vector-Borne Diseases, and Other Impacts

A whole host of other potential risks to health may result from changing climatic conditions. Whenever there is a negative impact on the built environment from extreme events, including storms, fires, and flooding, human health is at risk and loss of life may result. Changes in the frequency of extreme precipitation events will have consequences for health hazards associated with direct damage wrought by storms and floods (including injury and mortality), as well as ensuing exposures to waterborne diseases, toxins, sewage, and contamination from mold and other respiratory irritants (Joyce et al. 2013). Floods are the second deadliest of all weather-related hazards in the United States, accounting for nearly 100 deaths each year, the highest portion of which occurs as a result of flash floods and flooding associated with tropical storms (Ashley & Ashley 2008). Persistent heavy rains and thunderstorms in summer 1993 brought flooding across much of the central United States, resulting in 48 deaths and $30 billion in damage (NOAA National Climatic Data Center 2013).

Heavy precipitation and runoff contribute to increased risk of waterborne disease from increased surface and groundwater contamination. Outbreaks of diseases caused by organisms such as *Giardia* and *Escherichia coli* and other acute gastrointestinal illnesses have been linked to heavy rainfall events, such as the one in Milwaukee, Wisconsin, in 1993, which led to 403,000 cases of intestinal illness and 54 deaths (Hoxie et al. 1997). More than half of the waterborne disease outbreaks in the United States in the last half of the twentieth century were preceded by extreme rainfall events, according to a study conducted at the Johns Hopkins Bloomberg School of Public Health (Curriero et al. 2001). In urban watersheds, more than 60 percent of the annual load of all contaminants is transported during storm events, increasing the risk to vulnerable urban populations exposed to dangerous contaminants.

Additional impacts include changes in the distribution of diseases borne by insects, changes in crop yields and quality as well as global food security, changes in the frequency and range of harmful algal blooms, and risks resulting from population displacement (due to sea-level rise and extreme weather events) (Joyce et al. 2013). Many of these impacts are difficult to study, and their causal processes and results are less easily quantified (McMichael, Woodruff, & Hales 2006).

Vulnerable Populations

It is important to take into account that climate-related risks are disproportionately higher for the most vulnerable subpopulations, including children and the elderly, low-income communities, and some people of color. Children in particular face increased impacts from heat waves (Basu & Samet 2002), air pollution, and infectious disease and impacts from extreme weather events (American Academy of Pediatrics 2007). The elderly and those with preexisting health conditions face greater risk of death from heat waves and suffer more severe consequences from air pollution and flood-related health risks (Balbus & Malina 2009). Low-income communities, already burdened by the high incidence of chronic illness, inadequate access to health services, and limited resources to adapt to or avoid extreme weather, are also disproportionately impacted by climate-related events (Balbus & Malina 2009; Reid et al. 2009).

CHAPTER 9
CRIME

CRIME is an important social force in the United States. The incidence of both violent and nonviolent crime affects individuals, households, and communities at a very personal level. The threat of crime shapes how societies organize themselves to protect their fellow citizens, their families, and their neighborhoods from the potential consequences of crime. Victims of crime know firsthand its effects on quality of life.

Crime is also an important economic force in the United States. In areas where crime is prevalent, residents notice direct effects on housing prices, education, and job availability. The opportunity costs are high, as crime removes both victims and perpetrators of crimes from the productive workforce. Crime prevention and prosecution comes at significant costs to society, as public and private expenditures are redirected from other more productive uses. In 2010, public spending on police protection, legal and judicial services, and corrections totaled more than $260 billion for all jurisdictions (Bureau of Justice Statistics 2010). Interpersonal violence each year amounts to tens of thousands of deaths across the United States, with millions more the victims of assault and rape. Property crime, including burglary and larceny, affect nearly 10 million people each year (Federal Bureau of Investigation 2012). Given the magnitude of current losses due to criminal activity, even small changes in crime rates can affect communities at very personal and economic levels and can ultimately have a substantial detrimental effect on the U.S. economy as a whole.

According to the Federal Bureau of Investigation, many factors influence crime rates including population density, age, education, family cohesiveness, divorce rates, effectiveness of law enforcement, and weather. Research efforts have long focused on understanding crime's causes and contributors in order to improve the effectiveness of crime-prevention efforts. Because the human and economic stakes are so high, every potential cause has been seriously considered. These efforts have determined that weather and climate have a consistent and significant effect on human conflict, broadly defined, including both violent and nonviolent crime. This relationship has been documented around the globe, across all types of conflict, levels of development, and all spatial scales, through all phases of human history to modern times (Hsiang, Burke, & Miguel, 2013).

Much attention has been given to climate's effect on war and civil conflict, especially in regions where the scale of conflict is large and where climate extremes are already evident. While it may be easy to imagine increased incidence of civil conflicts in hot, arid, resource-constrained countries, the empirical link to the climate applies just as readily to armed robbery in downtown Los Angeles. Findings from a growing body of rigorous quantitative research across multiple disciplines has found that weather, and in particular temperature, affects the incidence of most types of violent and nonviolent crime in American cities and rural areas alike. Of course, climate is not the primary cause of crime, but studies find clear evidence that climate variations can have substantial effects (Jacob, Lefgren, & Moretti 2007; Card & Dahl 2011; Ranson 2014).

Despite rising temperatures, the United States is in the midst of a historic decline in crime rates. Nonetheless, the impact of climate on crime in American communities is real, and crime rates could increase—or decline more slowly than they otherwise would—as temperatures rise across the United States in the coming century. With more than 1.2 million incidents of violent crime and nearly 9 million incidents of property crime in the United States last year, the potential for even a small increase relative to a world without climate change is significant enough—in both human and economic terms—to merit a serious assessment of the risk. As with all impacts in this assessment, social and economic factors may determine local or national trends in crime, but a changing climate may alter these trends substantially, imposing real costs on Americans. We assess the temperature-related impacts on both violent and property crime in the United States.

BACKGROUND

Studies across multiple disciplines—including criminology, economics, history, political science, and psychology—have found that climatic events have exerted considerable influence on crime and human conflict, even when controlling for all other possible explanations. This is true regardless of geography (whether Africa or the United States), time period (as relevant in ancient times as last year), duration of climatic events (lasting hours, days, or months), or spatial scale (global down to neighborhood or even building level) (Hsiang & Burke 2013).

The evidence is particularly strong for one climate variable: temperature. Studies from across the United States, drawn from extensive, high-quality time-series data, provide compelling evidence of the heat–crime link. Studies have repeatedly found that individuals are more likely to exhibit aggressive or violent behavior toward others if temperatures are higher (Kenrick & MacFarlane 1986; Vrij, Van der Steen, & Koppelaar 1994; Mares 2013). This has been documented for a whole range of aggressive behaviors: horn-honking by frustrated drivers, player violence during sporting events, and more serious criminal activity including domestic violence, assault, and murder (Kenrick & MacFarlane 1986; Anderson, Bushman, & Groom 1997; Cohn & Rotton 1997; Anderson et al. 2000; Rotton & Cohn 2000; Jacob, Lefgren, & Moretti 2007; Larrick et al. 2011; Mares 2013; Ranson 2014). The influence of higher temperatures on individuals has also been found to lead to increased retaliatory violence among groups. Studies have shown that police officers are more likely to use deadly force in a training simulation when confronted with threatening individuals in a hotter environment, and hot days have contributed to more rapid escalation of retaliatory violence at sporting events (Vrij, Van der Steen, & Koppelaar 1994; Larrick et al. 2011). Temperature's role is evident even when you remove the confounding effects of normal seasonal or annual fluctuations in crime rates, economic and cultural factors, enhanced crime reporting over time, and changes in law-enforcement activity (Ranson 2014).

While there is substantial evidence to support the link between warmer temperatures and the incidence of crime, studies have not been able to determine the precise physiologic mechanism(s) by which this occurs. There are several potential explanations. One suggests that individual criminal behavior is determined by rational decisions about the costs and benefits of certain actions and that weather factors into the probability of committing a crime without getting caught (Jacob, Lefgren, & Moretti 2007). Another possible explanation is based on consistent evidence that temperature affects aggression levels, affecting an individual's judgment in a way that causes loss of control and heightened propensity to commit criminal acts (Anderson, Bushman, & Groom 1997; Card & Dahl 2011). Yet another possible explanation is that the frequency of criminal acts is determined in part by opportunity; in this case, certain climate conditions allow for increased social interaction, expanding opportunities for crime to occur (Rotton & Cohn 2003). Pleasant weather, for instance, brings victims and offenders in closer proximity as people flock outdoors, resulting in increased violence, particularly robberies and assaults (Cohn & Rotton 1997). No single explanation has been able to account for all of the observed patterns, indicating it is quite likely that several of these mechanisms are at play (Hsiang & Burke 2013).

OUR APPROACH

To understand what climate change may mean for U.S. crime rates in the future, we rely on statistical studies that isolate the observed effect of temperature and rainfall on crime in the United States and apply this to projected future conditions. Because there are strong cross-county patterns in crime, as well as strong trends over time (that may differ by location) and over seasons, we rely on studies that account for these patterns when measuring the effect of climate variables on crime rates. Two published studies provide nationally representative estimates that satisfy these criteria, which we use to construct quantitative projections. Ranson (2014) examines county-level monthly crime rates during 1960–2009, and Jacob, Lefgren, and Moretti (2007) examine jurisdiction-level weekly crime rates during 1995–2001. Both studies identify the incremental influence of temperature and rainfall changes on violent crimes and property crimes using data collected by the Federal Bureau of Investigation. Ranson's analysis provides greater coverage over years and across the country (the 2010 data covers 97.4 percent of the U.S. population); however, the studies may be complementary because their sample structures and statistical approaches differ somewhat, reflecting different modeling decisions.

Figures 9.1 and 9.2 display an optimally weighted average dose-response curve for both violent crime and property crime, drawn from Ranson (2014) and Jacob, Lefgren, and Moretti (2007). Both studies largely agree that higher daily maximum temperatures strongly and linearly influence violent crime, with a somewhat weaker and probably nonlinear influence on property crime. On average, increasing a county's temperature by 10°F for a single day increases the rate of violent crime linearly by roughly 0.2 percent. Jacob and colleagues assume that all types of crime respond linearly to temperature, but when Ranson examines nonlinearity, he finds that property crime increases linearly up to 40°F to 50°F and then levels off.

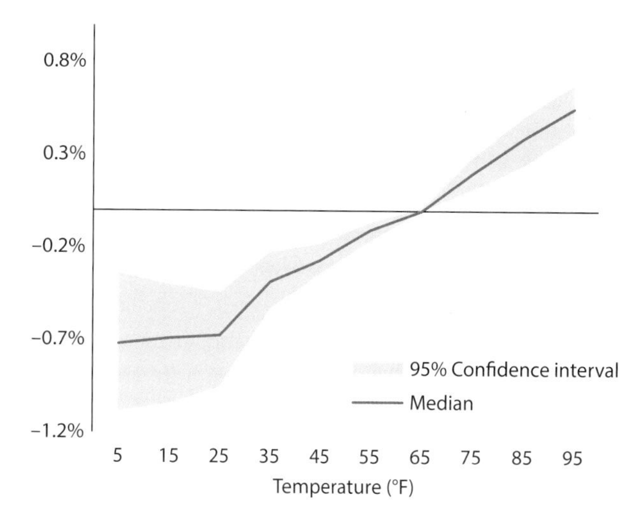

FIGURE 9.1. **Temperature and Violent Crime**

Percentage change in incidence of violent crime vs. daily maximum temperature (degrees Fahrenheit)

These findings are consistent with other work that finds aggressive behavior increases roughly linearly with temperature and that property crime is mainly constrained by opportunity (e.g., it is more difficult to steal cars and other property when it is extremely cold and snowy outdoors). Both papers find that the effects of rainfall are much smaller and less influential, with higher rainfall slightly increasing property crime and slightly decreasing violent crime.

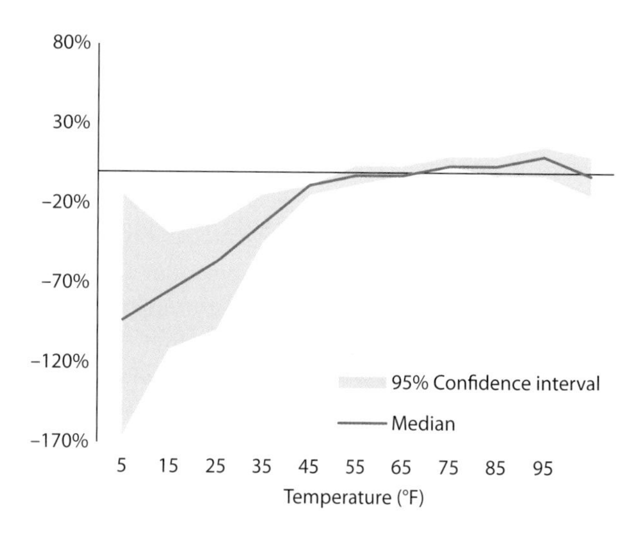

FIGURE 9.2. **Temperature and Property Crime**

Percentage change in incidence of property crime vs. daily maximum temperature (degrees Fahrenheit)

Both studies consider whether increased crime rates caused by high temperatures induce forward displacement ("harvesting") of crimes or generate new crimes that would not otherwise occur. Jacob and colleagues examine weekly crime rates and find that when a hot week triggers additional crime, roughly half of the violent crimes and a third of the property crimes would have otherwise occurred in the following 4 weeks, with no evidence of displacement beyond the fourth week. Consistent with this finding, Ranson examines monthly crime rates and finds that after an abnormally warm month, there is no evidence that crime in the following month is reduced. Thus, both studies find that temporal displacement beyond a one-month time frame is minimal, although there is evidence of displacement within that month. To account for this temporal displacement in our analysis, we only consider temperature-induced crime that would not have occurred in later periods.

Ranson examines whether there is evidence that populations adapt by examining if the sensitivity of crime to temperature declines over time or is lower in counties that are hotter on average. Ranson finds that the response of crime to temperature has remained virtually unchanged since 1960, with only suggestive evidence that the sensitivity of violent crime has fallen very slightly over the half-century. Ranson finds no evidence that hotter counties are better adapted in this respect, as the response of hotter and colder counties is indistinguishable.

To assess potential future impacts of climate change on crime, we simulate changes in violent and property crime rates under different climate scenarios relative to a future in which the climate does not drive changes in crime after 2012—although other social and economic trends are assumed to continue. Within each scenario, we account for uncertainty in climate models, weather, and statistical results, causing our projection to be a probability distribution of potential outcomes at each moment in time.

Considering the potential impact of changes in temperature and precipitation on crime, we find that crime generally increases as early as 2020–2039, and the range of likely changes is unambiguously positive by midcentury for all scenarios (figure 9.3). In RCP 8.5, we estimate violent crime is *likely* to increase 0.6 to 2.1 percent by midcentury and 1.9 to 4.5 percent by late century, with a 1-in-20 chance that late-century increases will be below 1.7 percent or exceed 5.4 percent relative to baseline crime rates. Examining property crime, we find that impacts tend to be substantially smaller in percentage terms for all cases, with late-century rates in RCP 8.5 *likely* rising 0.4 to 1.0 percent, with a 1-in-20 chance that the rise in property crime rates will be less than 0.3 percent or more than 1.1 percent. Property crime does not increase as strongly as violent crime because the impact function for property crime is nonlinear and flattens at temperatures higher than 55°F (see figure 9.2), whereas the impact function for violent crime continues to increase even at high temperatures (see figure 9.1). Thus, future warming is likely to have a smaller percentage effect on property crime because much of the warming will increase temperatures on warm or hot days that are already above 55°F, a change that does not affect property crime but does affect violent crime.

Projected changes are smaller in magnitude for RCP 4.5 and RCP 2.6, with the *likely* range of changes for violent crime spanning an increase of 0.6 to 2.5 percent for RCP 4.5 and −0.1 to 1.3 percent for RCP 2.6 by late century. The 1-in-20 outcomes span a narrower range than for RCP 8.5, with a 1-in-20 chance of crime rates rising less than 0.2 percent or more than 3.2 percent for RCP 4.5 and a decrease larger than −0.2 percent or an increase above 1.5 percent for RCP 2.6. Projections for property crime are similar in structure but are roughly one third the magnitude.

Across all RCPs, the distribution of potential changes broadens moving forward in time, and the rate of spreading increases moderately with increasing emissions for violent crime only (see figure 9.3). Changes in violent crime rates by late century exhibit a 90 percent confidence interval spanning 1.7 percentage points in RCP 2.6, which widens to span 3.0 and 3.7 percentage points in RCP 4.5 and 8.5, respectively, indicating that future uncertainty over violent crime rates increases with warming.

The spread of potential outcomes is relatively more consistent across RCP scenarios for property crime, primarily because there is no response of property crime to variation in high temperatures that generate uncertainty for violent crime projections (figure 9.4).

The "lumpiness" (or multimodality) of the distributions results from the high statistical precision for the econometric results used to generate these projections. For other impacts, statistical uncertainty causes the climate-model-specific distributions to be broader and to overlap more, making them less well defined. Thus, the lumpiness of the crime distributions indicates that

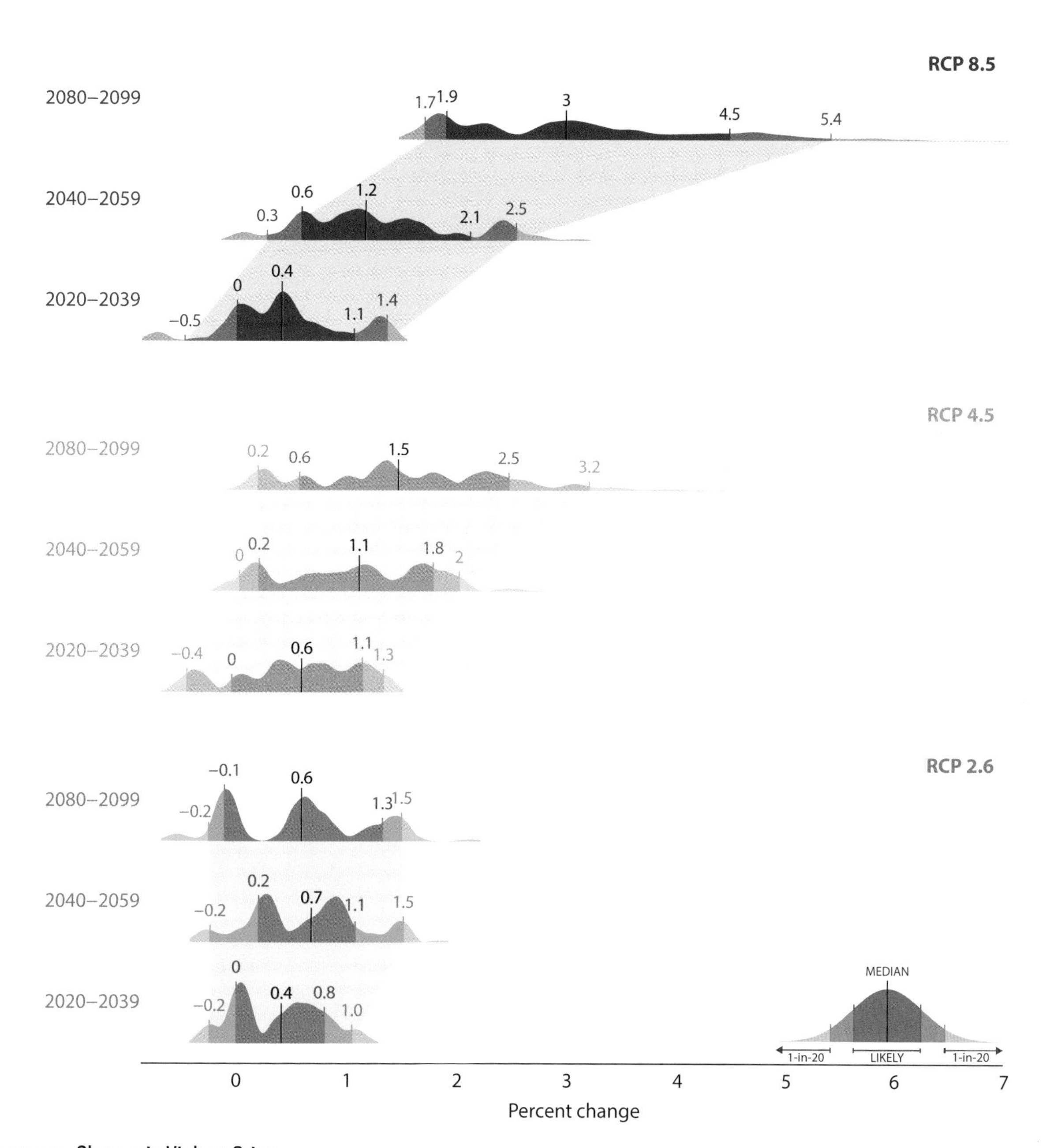

FIGURE 9.3. **Change in Violent Crime**

between-climate-model uncertainty drives the bulk of uncertainty in these specific projections.

In percentage terms, the spatial distribution of projected impacts is relatively similar across the country in the median RCP 8.5 projection, with increases in violent crime projected across all counties (figure 9.5, left column) and increases in property crime projected across almost all counties (figure 9.6, left column). There are somewhat smaller increases in property crime in the South, Southwest, and California because these locations have fewer cold days in their baseline climate, and warming during cold days drives the response of property crime. For a sense of scale, we compute changes in the total number of crimes committed. We assume the baseline distribution of crimes remains fixed relative to 2000–2005 averages, assuming no population growth, change in law enforcement, or additional social trends (figures 9.5 and 9.6, right columns). The national distribution of additional crimes is strongly influenced by the locations of urban centers (e.g., New York, Chicago, Los Angeles), although it is also

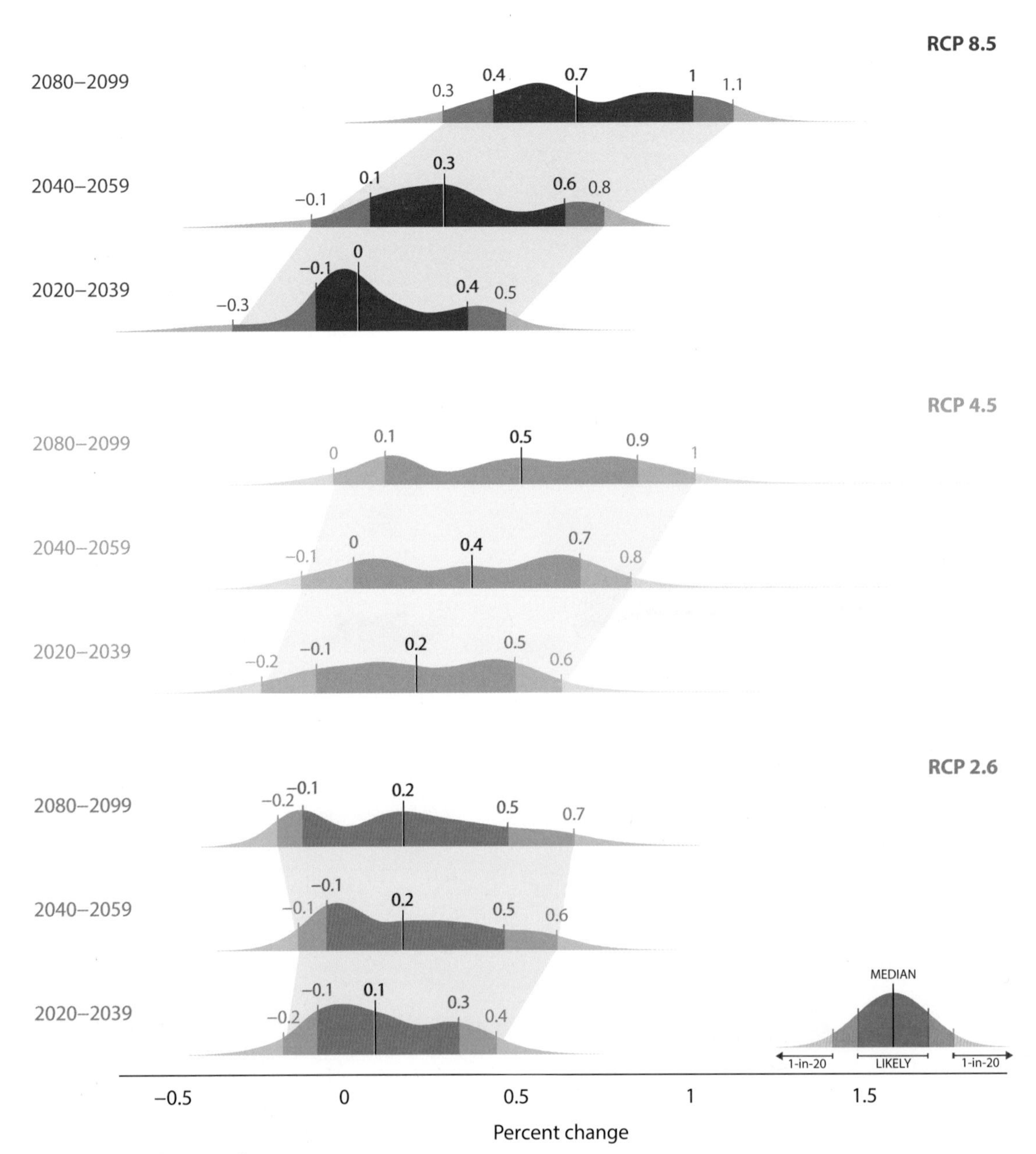

FIGURE 9.4. **Change in Property Crime**

possible to discern the larger impact in northern counties for property crime.

These impacts are described in terms of average changes over 20-year intervals. These averages are useful for describing persistent economic changes in future periods, but they mask short-lived events that may only last a year or two but have substantial consequences for well-being. Within each 20-year window, the likelihood of extreme annual events, such as a very high-crime year, evolves with the climate. One way to describe how the likelihood of extreme events changes is to examine how frequently we expect to experience years that are as damaging as the worst year over two decades of recent history, a so-called 1-in-20 year event.

In figures 9.7 and 9.8, we plot the estimated number of years that will have temperature-related crime anomalies larger than historically observed 1-in-20 year anomalies. For each year, we plot the expected number of these

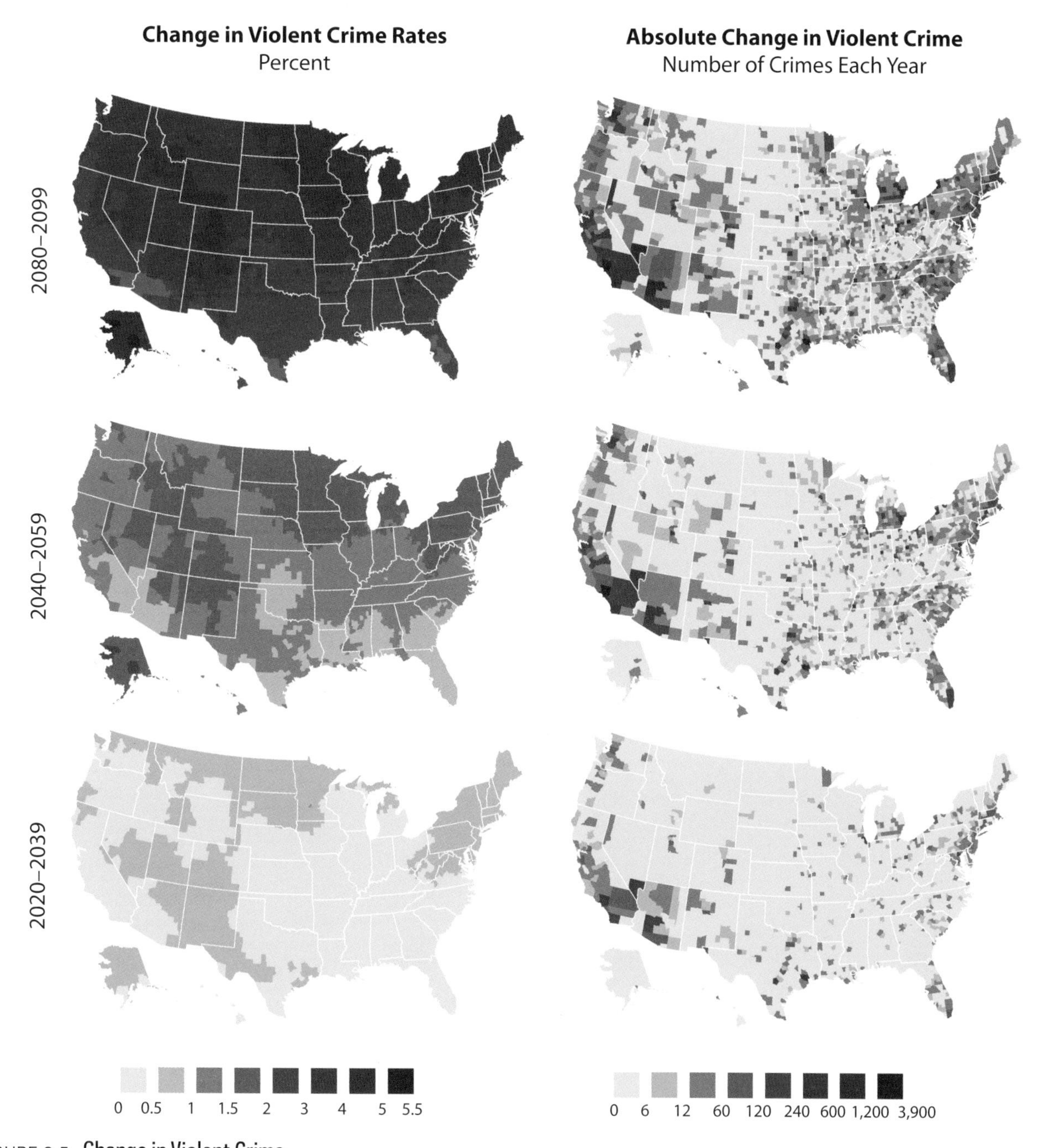

FIGURE 9.5. **Change in Violent Crime**

RCP 8.5, median

extreme years that will be experienced in the 20 years that follow; that is, we plot what the immediate future looks like to an individual in a given year. As early as 2020, climate-related violent crime anomalies that used to occur only once every 20 years will be expected to occur roughly seven times in the following 20 years for RCP 2.6 and ten times in the following 20 years for RCP 4.5 and 8.5. Extreme property crime anomalies grow in frequency more slowly and do not reach as high values as those for violent crime. By 2080, violent crime extreme events will be occurring roughly ten times every 20 years in RCP 2.6, 16 times every 20 years in RCP 4.5, and

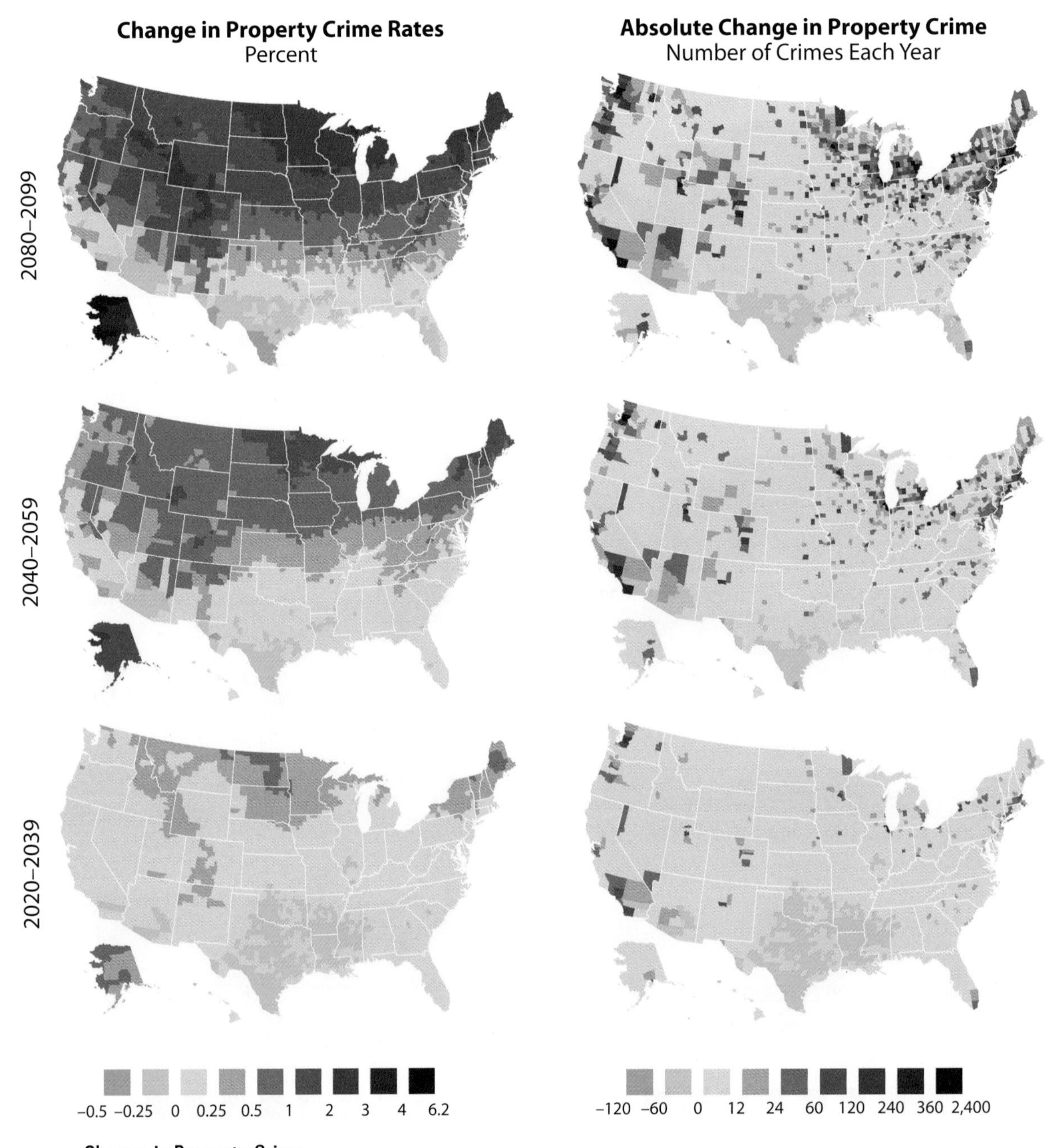

FIGURE 9.6. **Change in Property Crime**

RCP 8.5, median

every year in RCP 8.5. The frequency of extreme property crime anomalies are similar in RCP 8.5 by the end of the century but only reach ten events every 20 years in RCP 4.5 and seven events every 20 years in RCP 2.6.

It is important to note that these estimates assume the national distribution of crime remains fixed relative to the average crime rate during 2000–2005 and that geographic patterns in law enforcement remain unchanged. It is likely that if these changes in crime occur, communities will respond by expanding their law-enforcement activities. One can roughly consider how much additional resource communities would need to invest in policing activity to offset these increases by using estimates for the effectiveness of policing activity in reducing crime.

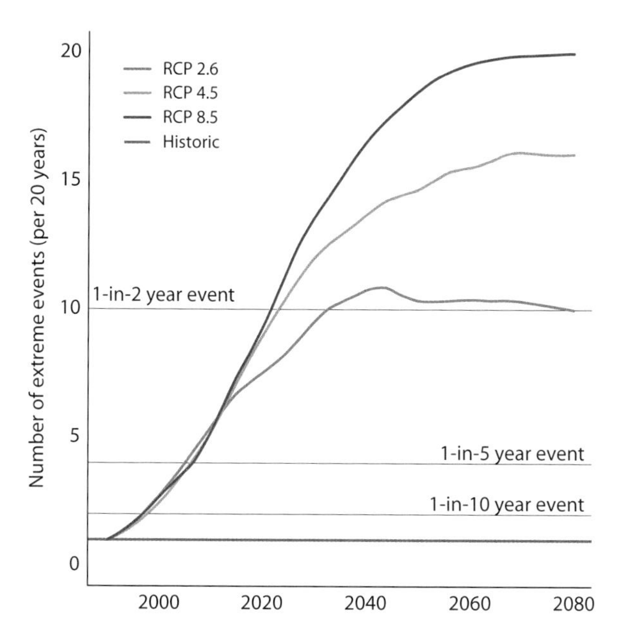

FIGURE 9.7. **Projected Change in Frequency of Temperature-Related Violent Crime Anomalies Equal to or Worse than Historical 1-in-20 Year Events**

Expected frequencies across climate models and weather; frequencies are plotted in the first year of each 20-year period

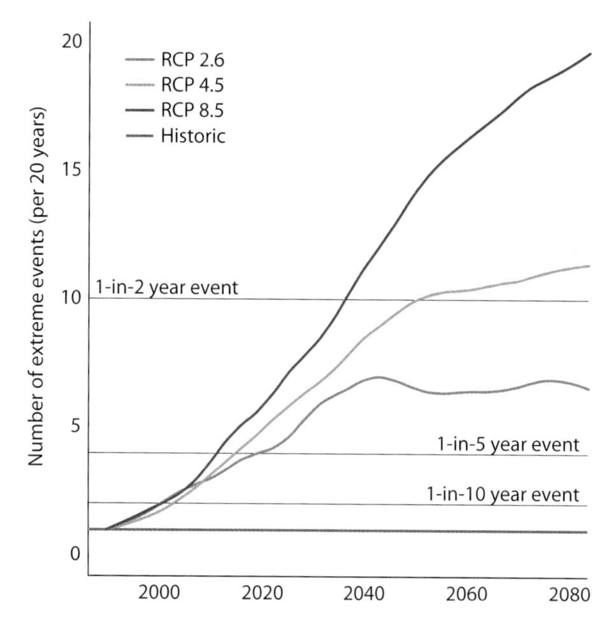

FIGURE 9.8. **Projected Change in Frequency of Temperature-Related Property Crime Anomalies Equal to or Worse than Historical 1-in-20 Year Events**

Expected frequencies across climate models and weather; frequencies are plotted in the first year of each 20-year period

Observing that each 1 percent increase in the size of the police force reduces crime by roughly 0.1 to 0.6 percent (Chalfin & McCrary 2013), our results suggest that to offset fully the *likely* range of late-century violent crime changes in RCP 8.5, police forces would have to grow by 3 to 19 percent (lower end of likely range) to 8 to 45 percent (upper end of likely range). In part 5 of this book, we explore how our projections might change if future populations continue to adapt to climate-related crime at historically observed rates.

CHAPTER 10

ENERGY

ENERGY is a key ingredient in U.S. economic growth. Ensuring a reliable supply of electricity and other sources of energy is critical to the financial security of American businesses and households and to the national security of the country as a whole. While dynamic enough to respond to the climate conditions of the past, our energy system, as currently designed, is poorly prepared for future climatic changes. Rising temperatures, increased competition for water supply, and elevated storm-surge risk will affect the cost and reliability of the U.S. energy supply. Climate change will also shape the amount and type of energy consumed. In this chapter, we quantify the demand-side impacts of the projected changes in temperature described in chapter 4 and discuss the range of supply-side risks the U.S. energy sector faces as well.

BACKGROUND

Energy demand is highly climate-sensitive in some sectors, and temperature in particular is a significant determinant of both the quantity and type of energy consumed. Demand for heating and cooling, which accounts for roughly half of residential and commercial energy use, fluctuates hourly, daily, and seasonally in response to outdoor ambient temperatures. Warmer winter temperatures as a result of climate change will reduce heating demand, particularly in northern states, which is currently met largely through the combustion of natural gas and fuel oil in boilers, furnaces, and water heaters. At the same time, hotter summer temperatures will increase demand for residential and commercial air-conditioning run on electricity. Climate-driven changes in air-conditioning can have an outsized impact on the electric-power sector, forcing utilities to build additional capacity to meet even higher peak temperatures.

OUR APPROACH

To assess the effect of the projected temperature changes discussed in chapter 4 on U.S. energy consumption, we turned first to the econometric literature. Because there are strong cross-location patterns in energy demand, as well as strong trends over time (that may differ by location) and over seasons, we focused on studies that account for these patterns when measuring the effect of climate variables on energy demand. Two studies provide estimates that satisfy these criteria, although only one is nationally representative. Deschênes and Greenstone (2011) examine state-level annual electricity demand for the country from 1968 to 2002 using data from the U.S. Energy Information Administration (EIA), and Auffhammer and Aroonruengsawat (2011) study building-level electricity consumption for each billing cycle (roughly a month) for California households served by investor-owned utilities (Pacific Gas and Electric, San Diego Gas and Electric, and Southern California Edison). Both studies identify the incremental change in electricity consumed for each additional day at a specified temperature level. Deschênes and Greenstone provide national coverage, while Auffhammer and Aroonruengsawat provide greater temporal and spatial resolution across the full range of climate zones in California; thus, the studies may be complementary.

Both studies find that electricity consumption increases during both hot days that exceed roughly 65°F and cold days that fall below roughly 50°F (figure 10.1). Incremental increases in daily temperature cause electricity consumption to rise more rapidly than incremental decreases in temperature, although both changes have substantial impacts on overall demand. Auffhammer and Aroonruengsawat further examine how the shape of this dose-response function changes with the climate zone that each household inhabits, finding that in hotter locations that are more likely to have air-conditioning widely installed, electricity demand increases more rapidly with temperature. This suggests that as populations adapt to hotter climates, they install more air-conditioning infrastructure and use air-conditioning more heavily for hot days at a fixed temperature.

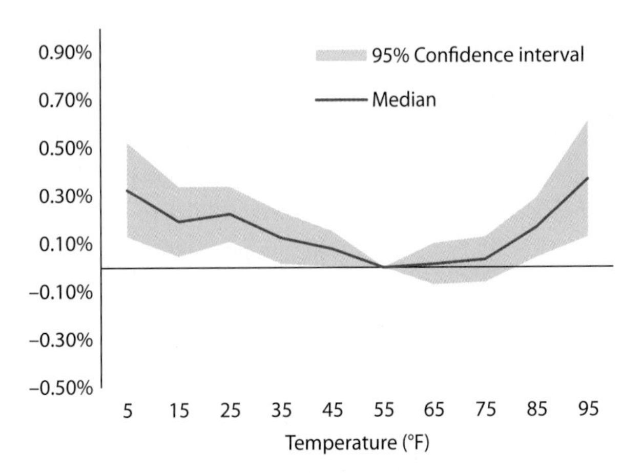

FIGURE 10.1. **Temperature and Electricity Demand**

Observed change in electricity demand (percent) vs. daily temperature (degrees Fahrenheit)

Widening the Lens

Unfortunately, the available econometric studies only capture part of the energy-demand story. While residential and commercial electricity demand rises alongside temperature, as households and businesses increase their use of air-conditioning, natural gas and oil demand in those two sectors falls. Many households and businesses use natural gas or oil-fired boilers and furnaces for heating, rather than electricity. The econometric studies mentioned earlier only cover changes in demand, not changes in price. To capture these fuel substitution and price effects, we use RHG-NEMS, a version of the EIA's National Energy Modeling System (NEMS; more information on NEMS is available at www.eia.gov/oiaf/aeo/overview) maintained by the Rhodium Group.

NEMS is the model used by the EIA to produce its *Annual Energy Outlook*, the most widely used projection of future U.S. energy supply and demand. NEMS is the most detailed publicly available model of the U.S. energy system, as it includes every power plant, coal mine, and oil and gas field in the country. Individual consumer decisions regarding how much to heat or cool one's home, which appliance to buy, and what car to drive are explicitly modeled, as are producer decisions regarding new electricity, oil, gas, and coal production. Temperature is an input into NEMS and affects heating and cooling demand in the residential and commercial sectors. The appliances and equipment used to meet this demand influences the quantity of electricity, natural gas, and oil supplied to household and business consumers.

We began by comparing the modeled impact of a given change in temperature on electricity demand in NEMS with the empirically derived dose-response function described earlier and found very similar results. We then modeled the impact of a range of regional temperature projections from chapter 4 to capture the change in total energy demand, energy prices, and delivered energy costs. NEMS only runs to 2040 but is still useful in modeling the impact of longer-term temperature changes relative to the energy system we have today. As we are measuring the impact of climate-driven changes in energy demand relative to a baseline, the baseline itself matters less. Modeling long-term temperature changes in NEMS provides a reasonable estimate of the relative change in demand, price, and costs given current economic and energy-system structures.

RESULTS

Energy Demand

Consistent with the econometric estimates, we find meaningful climate-driven increases in residential and commercial electricity demand. Under RCP 8.5, average nationwide electricity demand in the residential and commercial sectors *likely* increases by 0.7 to 2.2 percent by 2020–2039, 2.3 to 4.9 percent by 2040–2059, and 6.2 to 14 percent by 2080–2099 (figure 10.2). The largest increases occur in the Southwest, the Southeast, and southern Great Plains states (figure 10.3). Texas, Arizona, and Florida see late-century *likely* increases of 9.6 to 21 percent, 8.5 to 21 percent, and 9.6 to 22 percent, respectively. At the other end of the spectrum, most New England states and those in the Pacific Northwest see low single-digit *likely* increases, with declines possible in certain counties.

In RCP 4.5, we find a *likely* increase in average electricity demand of 0.2 to 1.9 percent by 2020–2039, 1.2 to 4.1 percent by 2040–2059, and 1.7 to 6.6 percent by 2080–2099.

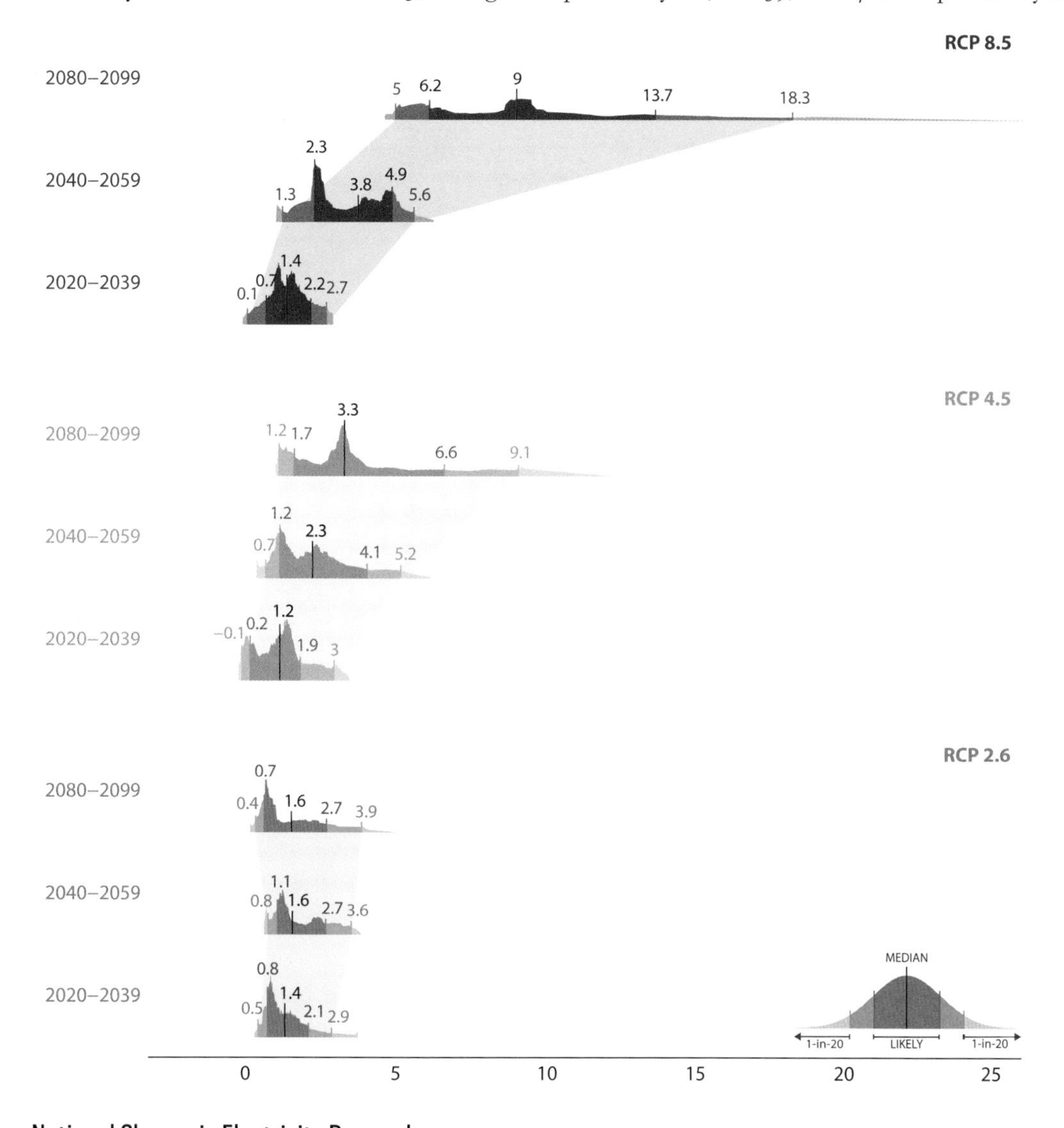

FIGURE 10.2. **National Change in Electricity Demand**

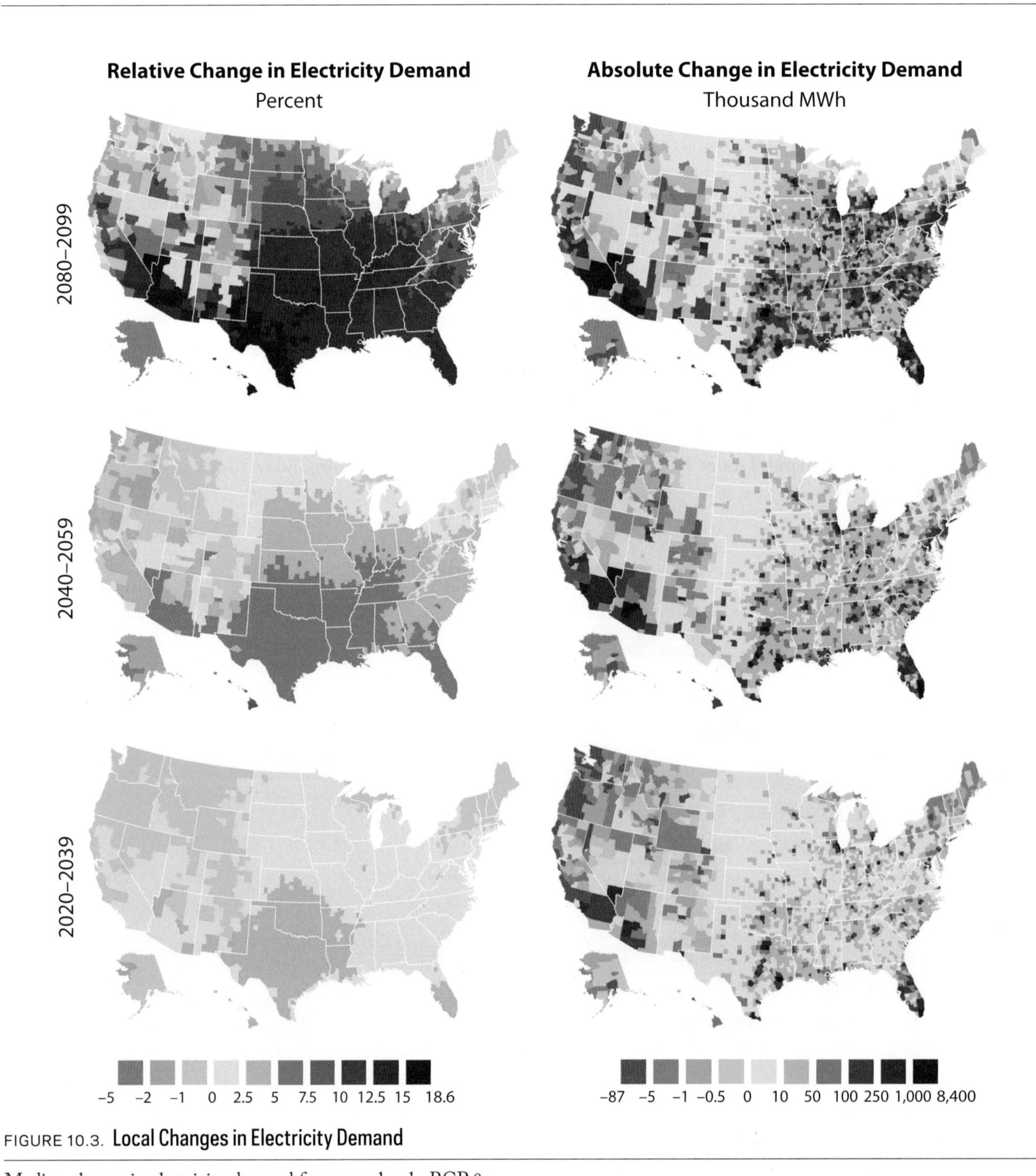

FIGURE 10.3. **Local Changes in Electricity Demand**

Median change in electricity demand from 2012 levels, RCP 8.5

In RCP 2.6, we find a *likely* increase of 0.8 to 2.1 percent by 2020–2039, 1.1 to 2.7 percent by 2040–2059, and 0.7 to 2.7 percent by 2080–2099.

Offsetting this increase in cooling-driven electricity demand, we find a significant decline in heating-driven natural gas and fuel oil demand in the residential and commercial sectors under RCP 8.5. This decline is concentrated in the Northeast, upper Midwest, northern Great Plains, and Northwest, areas with the greatest heating needs today. Total natural gas demand does not fall because demand from the power sector increases, but the net effect of changes in heating and cooling demand is a very modest change in energy consumption overall.

Energy Costs

While we find little climate-driven change in total energy demand, the switch from heating demand to cooling demand raises total energy costs. Climate-driven increases in cooling demand increase electricity consumption during the hottest times of the day and hottest periods of the year, when electricity demand is already at its peak. Higher peak demand requires the construction of additional power-generation capacity to ensure reliable electricity supply. Under RCP 8.5, we find a *likely* increase in installed power-generation capacity due to climate-driven changes in electricity demand of 8 to 95 gigawatts (GW) by 2020–2030, 73 to 212 GW by 2040–2059, and 223 to 532 GW by 2080–2099.

While most of this capacity would only operate part of the time (during peak demand periods), the capital costs as well as operating costs are passed on to electricity consumers. The resulting electricity price increases lead to a *likely* 0.1 to 2.9 percent increase in total annual residential and commercial energy costs on average by 2020–2039, 2.1 to 7.3 percent by 2040–2059, and 8 to 22 percent by 2080–2099 (figure 10.4). The greatest *likely* increases occur in the Southeast (12 to 28 percent), Great Plains (9 to 30 percent), and Southwest (11 to 25 percent) in 2080–2099. At the other end of the spectrum, the Northwest sees a *likely* change in energy expenditures of −4.5 to +3.7 percent late-century, while the Northeast sees a 4.1 to 13.6 percent *likely* increase. In RCP 4.5 and RCP 2.6, smaller increases in demand lead to less generation capacity construction and thus lower energy cost increases, though the RCP 4.5 and RCP 2.6 cost estimates do not include any increase in energy costs resulting from a change in U.S. energy supply necessary to reduce greenhouse-gas emissions consistent with either climate pathway. The cost estimates described above also exclude the supply-side climate impacts discussed later.

OTHER EFFECTS

Climate change will impact energy supply as well as energy demand. The energy supply chain is long and complex. There are a number of points in that supply chain where climate-related disruptions could interrupt delivery of electricity and heating or transport fuels.

Climate change will also have negative impacts on some sources of energy supply. For example, rising average temperatures and more frequent temperature extremes will reduce the efficiency of thermoelectric generation and transmission, while reduced sea ice in the Arctic will enable greater offshore oil and gas exploration. We describe some of the most significant supply-side climate impacts in the subsections that follow.

Thermal Generation Efficiency

Coal, natural gas, oil, nuclear, and biomass power plants all produce electricity by boiling water and using steam to spin a turbine. This steam is then recycled by cooling it back into water. Higher ambient air temperatures as a result of climate change reduce the efficiency of this process. The magnitude of the impact depends on a number of plant- and site-specific factors. For most combined-cycle plants, every 1.8°F (1°C) increase in air temperature will *likely* reduce electricity output by 0.3 to 0.5 percent (Maulbetsch & Difilippo 2006). For combined-cycle plants with dry cooling, often more sensitive to warmer ambient temperatures, the reduction can be as large as 0.7 percent (Davcock, DesJardins, & Fennel 2004). For natural gas–fired combustion turbines, which are often used for peak demand times, each 1.8°F increase in temperature will likely result in an 0.6 to 0.7 percent decline in electricity output, and for nuclear power, output losses are estimated at approximately 0.5 percent (Linnerud, Mideksa, & Eskeland 2011). Combining these reductions with the projected increase in average summer temperatures under RCP 8.5 described in chapter 4 suggests thermal efficiency declines could reduce total electricity generation by 2 to 3 percent by midcentury and 4 to 5 percent by late century, depending on the energy technology mix.

Nearly all the electric-power plants in the United States use water for cooling, and the power sector accounts for nearly half of total U.S. water withdrawals (Energy Information Administration 2011). Ambient temperatures affect surface-water temperatures. Surface-water temperatures in many U.S. rivers have risen in recent years (Kaushal et al. 2010) and are projected to continue to warm due to climate change in the decades ahead (Cloern et al. 2011; Van

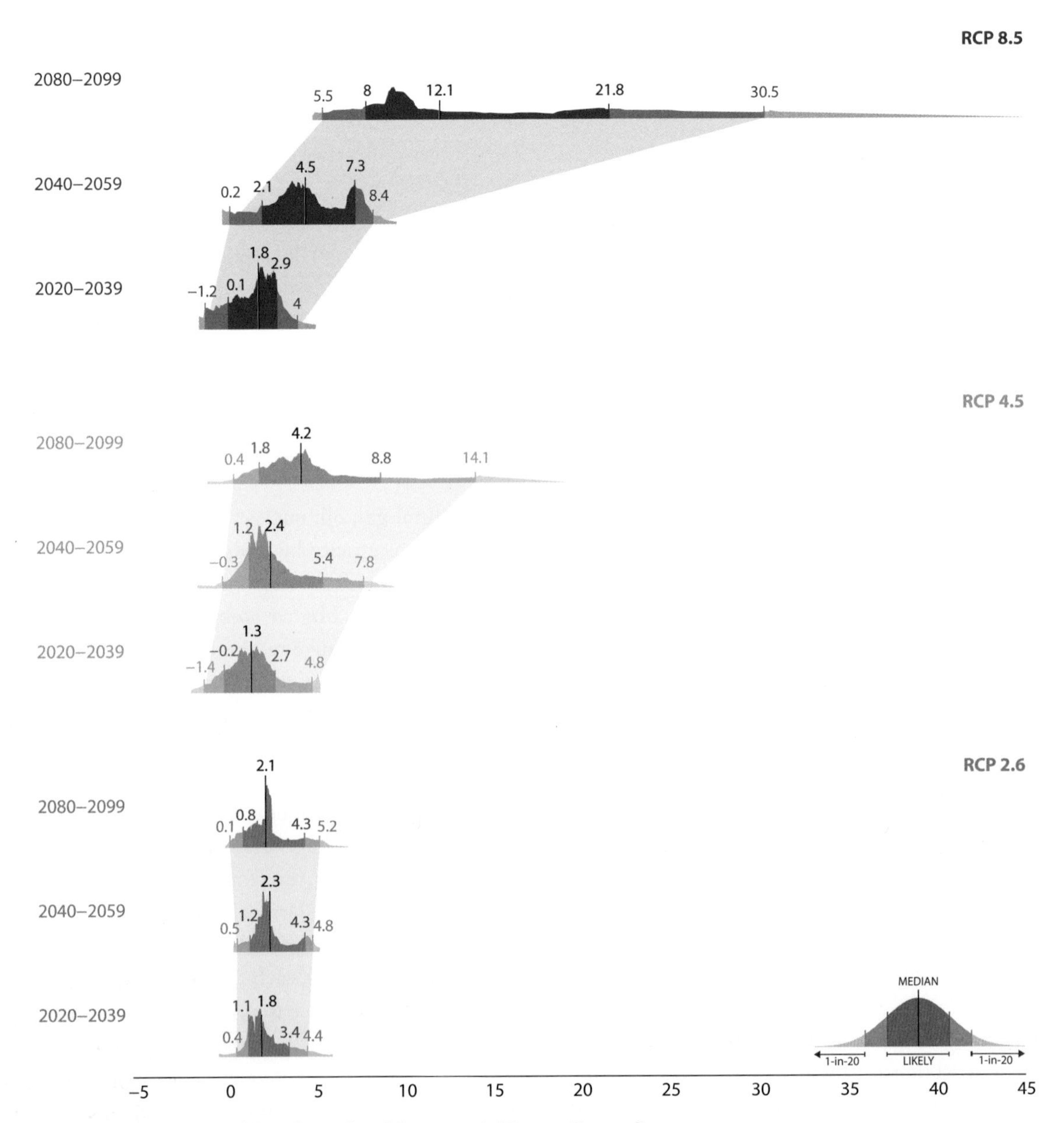

FIGURE 10.4. **Change in Annual Residential and Commercial Energy Expenditures**

Vliet et al. 2012; Georgakakos et al. 2014). Warmer water temperatures can degrade the efficiency of cooling processes and reduce electricity production as well (Van Vliet et al. 2012). In August 2012, record water temperatures in Long Island Sound shut down one reactor at Dominion Resources' Millstone Nuclear Power Station in Connecticut because the temperature of the intake cooling water exceeded the technical specifications of the reactor. While no power outages were reported, the 2-week shutdown resulted in the loss of 255,000 megawatt-hours of power, worth several million dollars (U.S. Nuclear Regulatory Commission 2012).

The majority of U.S. thermal power plants currently use once-through cooling systems, which use water from a nearby lake, river, aquifer, or ocean to cool steam and then return it to the body of water from which it was withdrawn. Because of the elevated temperatures of discharged water, thermal discharge limits have been established to protect aquatic ecosystems. Increasing water temperatures put power plants at risk of exceeding these limits, with the potential for financial penalties or forced curtailments (Skaggs et al. 2012). Indeed, large coal and nuclear plants have, in several cases in recent history, been forced to restrict operations because of higher water temperatures (Averyt et al. 2011). A recent study projected a decrease in average summer capacity of thermoelectric plants with once-through cooling of 12 to 16 percent and of those with recirculation cooling systems of 4.4 to 5.9 percent by mid-century, depending on the emissions scenario (Van Vliet et al. 2012). The study also found that the probability of extreme (greater than 90 percent) reductions in power production will on average increase by a factor of 3.

Electricity Transmission

Approximately 7 percent of generated electricity is lost during transmission and distribution (known as "line losses"), with the greatest losses occurring on hot days (Energy Information Administration 2012a). Increased average temperatures, as well as more frequent temperature extremes, will likely exacerbate these transmission and distribution losses (U.S. Global Change Research Program 2009; Sathaye et al. 2012; Wilbanks et al. 2012b). Warmer temperatures are also linked to diminished substation efficiency and life span (Sathaye et al. 2012). Current line losses are valued at nearly $26 billion (Energy Information Administration 2012b), so even small increases in loss rates can have a significant impact on electricity producers and consumers. A recent study found that a 9°F increase in average summer temperatures in the Southwest (within our projected end-of-century range under RCP 8.5) would result in a 7 to 8 percent reduction in transmission carrying capacity (Sathaye et al. 2013). Extreme heat events could result in even higher losses. Depending on the duration and intensity of the event, extreme temperatures can lead to power outages, as happened in 2006 when power transformers failed in Missouri and New York during a heat wave, causing widespread electricity supply interruptions (U.S. Global Change Research Program 2009).

Arctic Oil and Gas Production

Climate change is already shaping the energy landscape in Arctic Alaska, which has warmed faster than any other region of the United States to date, with both positive and negative impacts for the U.S. energy supply. Alaska currently accounts for more than 10 percent of U.S. crude oil production and is home to a large share of the national oil and gas resource base (U.S. Energy Information Administration 2013). Warming temperatures have already resulted in permafrost thaw, which is beginning to threaten onshore infrastructure on which oil and gas exploration and production depends. Energy pipelines built on permafrost are at increasing risk of rupture and leakage, and warmer temperatures are already resulting in shorter winter road seasons. The number of days of allowable travel on Alaskan tundra have been cut in half over the past 30 years, limiting the time during which onshore oil and gas exploration and production equipment can be used (Alaska State Legislature 2008). In a changing and unstable Arctic, the cost of maintaining existing infrastructure will likely increase, as will design and construction costs for new onshore infrastructure. Climate change is opening up new sources of oil and gas development as well. Higher temperatures are reducing sea-ice cover, which is improving access to substantial offshore oil and natural gas deposits in the Beaufort and Chukchi seas.

Water Availability

Current U.S. energy production is extremely water-intensive, and climate change will impact U.S. water supply in myriad ways (see chapter 17). Increased evaporation rates or changes in snowpack may affect the volume and timing of water available for hydropower and power-plant cooling, and changing precipitation patterns can affect bioenergy production. In regions where water is already scarce, competition for water between energy production and other uses may also increase. Regions that depend on water-intensive power generation and fuel extraction will be particularly vulnerable to changes in water availability over time.

At 40 percent of total freshwater withdrawals, thermal power generation is the largest water consumer in the United States (Kenny et al. 2009). Seasonal and chronic water scarcity has resulted in electricity supply disruptions in the past, particularly during periods of low summer flow. For example, a drought in the southeastern United States in 2007 forced nuclear and coal-fired power plants within the Tennessee Valley Authority system to shut down some reactors and reduce production at others (National Energy Technology Laboratory 2010). Similar water-driven shutdowns occurred in 2006 along the Mississippi River at the Exelon Quad Cities Illinois plant and at some plants in Minnesota. A recent assessment found that nearly 60 percent of coal-fired power plants in the United States are located in areas subject to water stress from limited supply or competing demand from other sectors (National Energy Technology Laboratory 2010).

Although annual average precipitation will likely increase across the continental United States over the next century, changes in seasonality of precipitation, timing of spring thaw, and climate-driven changes to surface runoff may affect surface and groundwater supplies in some regions. Potential future water scarcity increases the risk of electricity supply disruptions in some regions. In particular, surface water and groundwater supplies in the Southwest, Southeast, and southern Rockies are expected to be affected by runoff reductions and declines in groundwater recharge, increasing the risk of water shortages (Georgakakos et al. 2014). According to the Electric Power Research Institute, approximately one quarter of electricity generation in the United States—250 GW—is located in counties projected to be at high or moderate water supply sustainability risk in 2030 (EPRI 2011). The study found that all generation types will be affected, with 29 GW of nuclear, 77 GW of coal, and 121 GW of natural-gas generation capacity in counties with "at risk" water supplies.

Hydroelectric generation accounts for 7 percent of the total U.S. electricity supply, roughly 20 percent of electricity generation in California and the Northeast and up to 70 percent of electricity generation in the Pacific Northwest (Energy Information Administration 2013; Georgakakos et al. 2014). Projected climatic changes, including more precipitation falling as rain and less as snow, reduced snowpack, and earlier peak runoff, may decrease annual water storage and runoff. The resulting reductions in stream flow will decrease available hydropower generation capacity. The degree of impact will vary widely by region, with the western United States expected to be at greatest risk.

Water also plays a vital role in oil and gas production. Large volumes of water are used throughout the production process, including enhanced oil recovery, hydraulic fracturing, well completion, and petroleum refining. As the share of U.S. oil and gas production coming from unconventional sources (including coal-bed methane, tight gas sands, and shale oil and gas) increases, access to water will similarly increase in importance in sustaining U.S. production growth (U.S. Department of Energy 2013a). In times of water stress, oil and gas operations must compete with other water users for access, limiting availability and driving up costs. During the severe drought of July 2012, oil and natural gas producers faced higher water costs or were denied access to water for 6 weeks or more in several states including Kansas, Texas, Pennsylvania, and North Dakota (Dittrick 2012; Ellis 2012; Hargreaves 2012).

Coastal Storms and Sea-Level Rise

The sea-level rise and coastal storm dynamics discussed in chapter 4 threaten important energy assets as well as commercial and residential property. Superstorm Sandy demonstrated the extent to which coastal storms can disrupt energy supply. Storm surge and high winds downed power lines, flooded substations and underground distribution systems, and damaged or shut down ports and several power plants in the Northeast (U.S. Department of Energy 2013b). More than 8 million customers in 21 states lost power, further threatening vulnerable populations reeling from the effects of the storm (U.S. Department of Energy 2012a). Sandy also forced the closure of oil refineries, oil and gas pipelines, and oil and gas shipping terminals, impeding fuel supply in the region.

More than half of total U.S. energy production and three quarters of electricity generation take place in coastal states (U.S. Energy Information Administration 2013). The concentration of critical facilities in vulnerable coastal areas creates systemic risk not only for the region but also for the nation as a whole. The Gulf Coast is a prime example. The region is responsible for half of U.S. crude oil and natural gas production and is home to nearly half of the country's refining capacity, with nearly 4,000 active oil and gas platforms, more than 30 refineries, and 25,000 miles of pipeline (Entergy 2010; Wilbanks et al. 2012a). It is also home to the U.S. Strategic Petroleum Reserve (SPR), with approximately 700 million barrels of crude oil stored along the Gulf Coast for use in the event of an emergency (U.S. Department of Energy 2012b). With a substantial portion of U.S. energy facilities located in the

Gulf, isolated extreme weather events in the region can disrupt natural gas, oil, and electricity markets throughout the United States (Wilbanks et al. 2012a).

Outside of the Gulf Coast, other regional energy hubs are also at risk. The National Oceanic and Atmospheric Administration warns that outside of greater New Orleans, Hampton Roads near Norfolk, Virginia, is at greatest risk from sea-level rise and increased storm surge. The area is home to important regional energy facilities, including the Lamberts Point Coal Terminal, the Yorktown Refinery, and the Dominion Yorktown power plant (Wilbanks et al. 2012a). On the other side of the country, many of California's power plants are vulnerable to sea-level rise and the more extensive coastal storm flooding that results, especially in the low-lying San Francisco Bay area. An assessment performed for the California Energy Commission found that the combined threat of sea-level rise and the incidence of 100-year floods in California puts up to 25 thermoelectric power plants at risk of flooding by the end of the century, as well as scores of electricity substations and natural gas storage facilities (Sathaye et al. 2012).

Wildfires

Wildfires (see chapter 18) also pose a risk to the nation's energy infrastructure. During summer 2011, severe drought and record wildfires in Arizona and New Mexico burned more than 1 million acres and threatened two high-voltage lines transmitting electricity from Arizona to approximately 400,000 customers in New Mexico and Texas. In 2007, the California Independent System Operator declared an emergency due to wildfire damage to more than two dozen transmission lines and 35 miles of wire, with nearly 80,000 customers in San Diego losing power, some for several weeks (SDG&E 2007; Vine 2008). More frequent and severe wildfires increase the risk of physical damage to electricity transmission infrastructure and could decrease available transmission capacity.

CHAPTER 11
COASTAL COMMUNITIES

Temperate climates, attractive scenery, ease of navigation, and access to ocean food supplies have put coastlines at the forefront of human development throughout history and around the world. The United States is no exception. Today, counties touching the coast account for 39 percent of total U.S. population and 28 percent of national property by value. Coastal living carries risk, particularly on the East Coast and along the Gulf of Mexico, where hurricanes and other coastal storms inflict billions in property and infrastructure damage each year. Climate change elevates these risks. Rising sea levels will, over time, inundate low-lying property and increase the amount of flooding that occurs during coastal storms. Moreover, as discussed in chapter 4, warmer sea-surface temperatures may change the frequency and intensity of those storms.

BACKGROUND

A growing body of research assesses the potential impacts of sea-level rise (SLR) on coastal communities. Early studies focused on developing a methodology for site-specific estimates of damage from SLR that could be used as a model for nationwide assessments (Yohe 1990). Several compared the cost to coastal property of damage from mean sea-level rise with the cost of protecting that property with sea walls, structural enhancements, and other adaptive measures (Yohe et al. 1996; Yohe and Schlesinger 1998).

Subsequent work expanded to regional assessments. One of the first was conducted by the U.S. Environmental Protection Agency (Titus & Richman 2001), which identified areas vulnerable to inundation from higher sea levels along the Atlantic and Gulf coasts. A subsequent U.S. interagency assessment of the Mid-Atlantic region simulated a 1-meter SLR running from New York through Virginia and estimated the associated impacts on residential property and coastal residents (CCSP 2009). The first robust national estimate of potential inundation damage from SLR, as well as the cost of protective measures, was published in 2011 (Neumann et al. 2011) using the National Coastal Property Model (NCPM) developed by Industrial Economics, Inc., for the U.S. Environmental Protection Agency.

Permanent inundation from mean sea-level rise is only one of the risks climate change presents to coastal property and infrastructure. Higher average sea levels lead to higher storm surges and elevated flooding risks (Frumhoff et al. 2007), even if the intensity or frequency of storms remains unchanged (Frazier et al. 2010). Kemp and Horton (2013) found that while the record 13.9-foot storm tide in New York Harbor during Superstorm Sandy was primarily due to the coincidence of the strongest winds with high tide, SLR driven by historical climate change added more than 1 foot to that 13.9-foot total.

A number of recent studies have assessed coastal communities' vulnerability to future SLR-driven increases in storm surge. At a local scale, after Superstorm Sandy, the New York City Panel on Climate Change analyzed the risk to the city's property and infrastructure from future climate-driven changes in sea levels and storm activity (NPCC 2013). California conducted an assessment of the impact of SLR on the Bay Area's 100-year floodplains for coastal storms (**San Francisco Bay Conservation and Development** Commission 2011; Heberger et al. 2012), and Harrington and Walton (2008) estimated the impacts on coastal property for six coastal counties in Florida. Neumann and colleagues have incorporated projected increases in storm surge as a result of both mean SLR and potential changes in hurricane intensity and frequency into the NCPM for select cities (Neumann et al. 2014).

OUR APPROACH

Alongside the academic and policy-oriented work described earlier, private companies have developed sophisticated models to estimate potential losses from coastal storms. These models are used by the insurance industry in underwriting flood and wind insurance products, by the finance industry in pricing catastrophe bonds, and by local officials in coastal communities in preparing for and responding to hurricanes and other coastal storms. While not traditionally used in this way, they are also incredibly powerful tools for understanding how climate change will likely shape both industry and coastal community risk exposure in the years ahead.

Risk Management Solutions (RMS) is a leading provider of such tools, along with models for quantifying and managing other catastrophic risks, from earthquakes to terrorist attacks to infectious disease; RMS is a partner in this assessment. To assess the value of property at risk from future SLR, we mapped the probabilistic local SLR projections described in chapter 4 against RMS's detailed exposure data set, which covers buildings, their contents, and automobiles for all coastal counties in the United States. To analyze the impact of local SLR on storm surge and flood damage during hurricanes and nor'easters, we used RMS's North Atlantic Hurricane Model. This model combines state-of-the art wind and storm surge modeling and a stochastic event set that represents more than 100,000 years of hurricane activity and spans the range of all possible storms that could occur in the coming years (see appendix C).

The result of this analysis is the first comprehensive, nationwide assessment of the risk to coastal communities from mean SLR and SLR-driven increases in storm surge from hurricanes and nor'easters under a full range of climate futures, and at a very high level of geographic resolution. Taking this work one step further, we explore the impact of changes in hurricane frequency and intensity projected by Knutson et al. (2013) for RCP 4.5 and Emanuel (2013) for RCP 8.5 on both future storm surge and wind damage (see chapter 4). (While we capture projected change in frequency and intensity from the cyclogenesis models used by Knutson et al. and Emanuel, we do not capture projected change in landfall location. This could have a meaningful impact on the geographic distribution of hurricane-related losses and is worthy of considerable additional research.)

There is considerable uncertainty surrounding future coastal development patterns, which makes accurate cost projections challenging. Over the past few decades, population and property values in coastal counties have grown faster than the national average, putting more people and assets at risk. It is unclear the extent to which this trend will continue going forward, given constraints to further development and expansion in many coastal areas. Rather than attempt to predict how the built environment will evolve in the decades ahead, we assess the impact of future changes in sea level and storm activity relative to the American coastline as it exists today. Damage is reported in current dollars against current property prices.

RESULTS

Inundation from Mean Sea-Level Rise

While all coastal states are at risk from rising sea levels, some are much more vulnerable than others. Under RCP 8.5, for example, between 4.1 percent and 5.5 percent of total insurable residential and commercial property in the state of Louisiana will *likely* be below mean sea level (MSL) by 2050 (excluding that property already below MSL), growing to 15 to 20 percent by 2100 (figure 11.1). Florida is the second most vulnerable state in percentage terms, with 0.4 to 0.6 percent of current statewide property *likely* below MSL by 2050, growing to 1 to 5 percent by 2100. In dollar terms, between $33 billion and $45 billion worth of current Louisiana property will *likely* be below MSL by 2050, growing to $122 billion to $164 billion by 2100. The total value of current Florida property at risk is similar, with between $15 billion and $23 billion *likely* below MSL by 2050, growing to $53 billion to $208 billion by 2100 (figure 11.2).

Nationwide, we find that between $66 billion and $106 billion worth of current coastal property will *likely* be below MSL by 2050 under RCP 8.5 unless protective measures are taken (table 11.1), growing to $238 billion to $507 billion by 2100 (table 11.2). The value of current property *likely* under MSL falls to $62 billion to $85 billion by

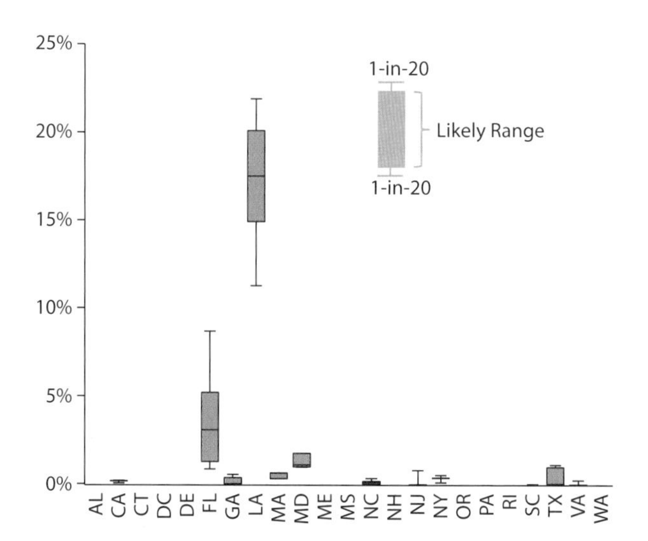

FIGURE 11.1. **Share of Current Property Below MSL in 2100 Under RCP 8.5**

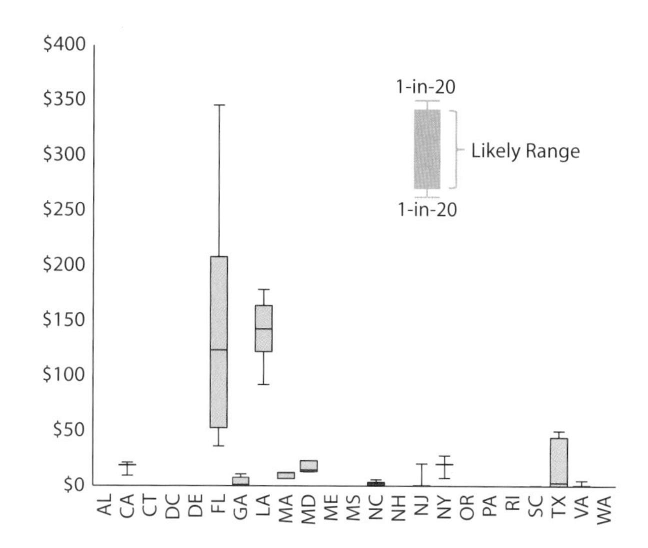

FIGURE 11.2. **Value of Current Property Below MSL by 2100**

RCP 8.5, billion 2011 U.S. dollars

2050 in both RCP 4.5 and 2.6. By 2100, nationwide property *likely* below MSL is $175 billion to $339 billion in RCP 4.5 and $150 billion to $276 billion in RCP 2.6.

Two factors explain this relatively small difference in inundation between the RCPs. First, the expanding ocean and melting ice sheets respond to both the amount of warming and the length of exposure to elevated temperatures. Temperatures begin to diverge significantly between RCPs only in the second half of the century; sea level, which integrates temperature, diverges later. Second, the largest sources of uncertainty in sea level are potential positive feedbacks in the behavior of ice sheets, particularly the West Antarctic Ice Sheet (WAIS). For example, for parts of the sea floor that are appropriately sloped, it is possible that, as a warming ocean eats away at the base of the WAIS (which unlike most of the Greenland and East Antarctic Ice Sheets largely sits below sea level), it will expose more of the ice sheet to the ocean, which will accelerate melt, exposing still more ice, and so forth. Such feedbacks are poorly understood at present; the uncertainties arising from this low level of understanding are independent of emissions and therefore cause the projected ranges of sea-level change for all the RCPs to overlap considerably.

At the tails of the SLR probability distribution, inundation damage is considerably worse than the *likely* range. For example, there is a 1-in-20 chance more than $346 billion of current Florida property (8.7 percent) could be

TABLE 11.1 Additional current property below MSL and MHHW by 2050 (in billion 2011 U.S. dollars, at current property prices)

Probability	*RCP 8.5*		*RCP 4.5*		*RCP 2.6*	
	MSL	*MHHW*	*MSL*	*MHHW*	*MSL*	*MHHW*
1-in-100 chance above	156	523	143	472	129	456
1-in-20 chance above	126	465	107	400	106	397
Likely range	66 to 106	323 to 389	62 to 85	294 to 366	62 to 85	287 to 360
1-in-20 chance below	61	256	60	240	60	226
1-in-100 chance below	52	186	51	181	50	172

MSL, mean sea level; MHHW, mean higher high water.

below MSL by the end of the century under RCP 8.5 (see figure 11.2) and a 1-in-100 chance that more than $681 billion of current Florida property (17 percent) could be lost by 2100 unless defensive measures are taken. Nationwide, there is a 1-in-20 chance that more than $701 billion of current property will be below MSL by 2100 and a 1-in-100 chance it will be more than $1.1 trillion (see table 11.2).

While roughly two thirds of all current property *likely* below MSL by 2050 is in Louisiana and Florida, which may become three quarters of all property by the end of the century, Maryland, Texas, Massachusetts, North Carolina, New York, New Jersey, and California also face meaningful inundation risk. In Maryland, for example, between $13 billion and $23 billion (0.7 and 1 percent) of current statewide property will *likely* be below MSL by 2050, with losses concentrated in Queen Anne's and Talbot counties located on the east side of the Chesapeake Bay. In Texas, up to $44 billion of current property will *likely* be below MSL by the end of the century, including important industrial and energy infrastructure.

Inundation risk from SLR extends beyond those properties underwater at average tide levels. There is currently $1.6 trillion in coastal property that is above MSL, but at or below peak high-tide levels, often referred to as mean higher high water (MHHW) levels. Most of this property is protected by shoreline defense built up over the course of decades or even centuries. As MSLs rise, the high-tide mark will rise as well, putting additional property in the line of fire. Without defensive investments (see chapter 22 for a discussion), these properties risk significant damage.

Figures 11.3 and 11.4 show the share and value of additional current property below MHHW by 2100 due to

TABLE 11.2 Additional current property below MSL and MHHW by 2100 (in billion 2011 U.S. dollars, at current property prices)

Probability	*RCP 8.5*		*RCP 4.5*		*RCP 2.6*	
	MSL	*MHHW*	*MSL*	*MHHW*	*MSL*	*MHHW*
1-in-100 chance above	1,114	1,636	719	1,433	613	1,332
1-in-20 chance above	701	1,432	495	1,135	431	990
Likely range	238 to 507	724 to 1,144	175 to 339	759 to 926	150 to 276	430 to 830
1-in-20 chance below	166	509	134	400	116	362
1-in-100 chance below	131	383	105	313	102	246

MSL, mean sea level; MHHW, mean higher high water.

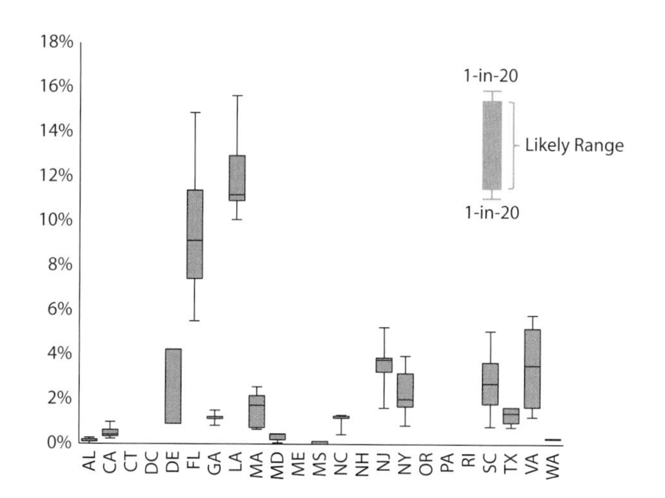

FIGURE 11.3. **Share of Current Property Below MHHW Caused by SLR by 2100 Under RCP 8.5**

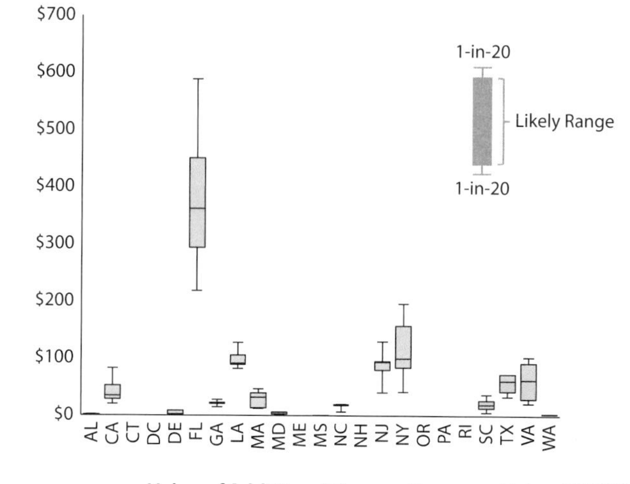

FIGURE 11.4. **Value of Additional Current Property Below MHHW Caused by SLR by 2100**

RCP 8.5, billion 2011 U.S. dollars

mean SLR. The value is two to four times larger than that for property below MSL, depending on time frame and SLR scenario (see tables 11.1 and 11.2).

To illustrate the risks presented by local SLR to coastal communities, we map inundation levels in 2100 at MSLs in the median, 1-in-100, and 1-in-200 projections for RCP 8.5 for Miami, Norfolk, Houston, and Wilmington, North Carolina, in figures 11.5 through 11.8. The inundation threat to Miami is particularly grave at a citywide level, but will also challenge the viability of several neighborhoods in

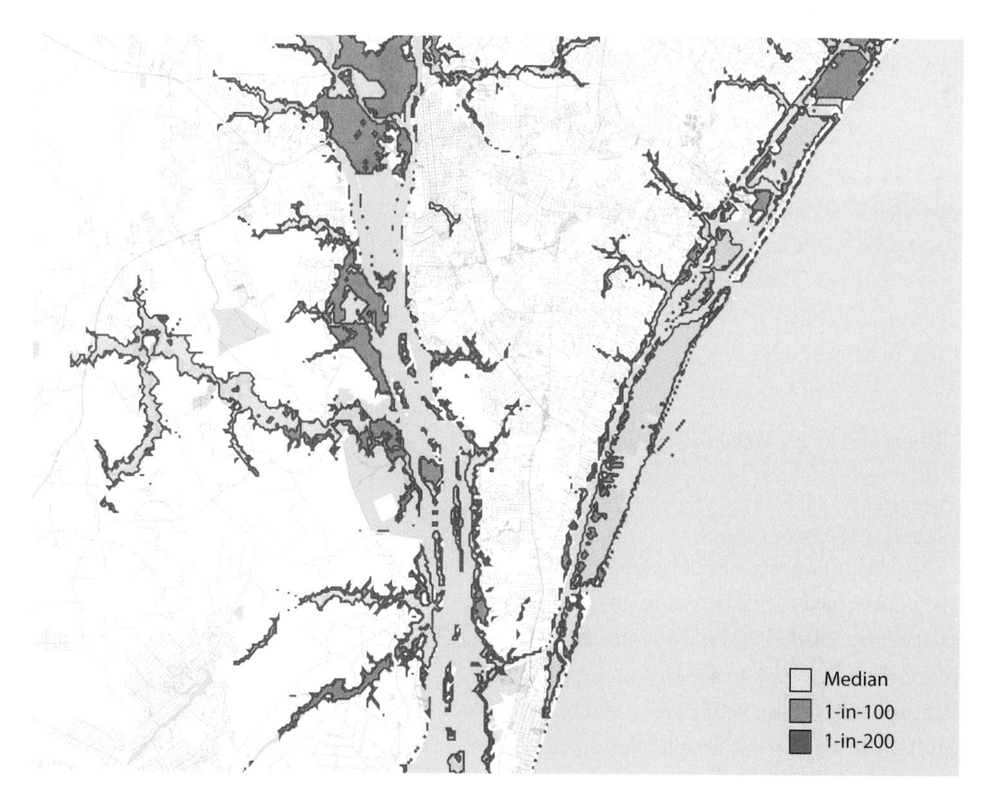

FIGURE 11.5. **Wilmington MSL Projections in 2100 Under RCP 8.5**

FIGURE 11.6. **Miami MSL Projections in 2100 Under RCP 8.5**

New York City, Wilmington, and elsewhere. In Houston, while the center of the city is reasonably safe, critical energy infrastructure is at risk. In Norfolk, major naval installations are threatened by SLR. This choice of examples is illustrative only. Many other cities in the country face significant SLR risk. These maps also do not show property at or below MHHW but above MSLs.

Storm Surge

As mentioned earlier, higher sea levels also mean greater flooding during hurricanes and other coastal storms. These storms currently result in roughly $27 billion in average annual commercial and residential property damage and business interruption costs along the East Coast and Gulf

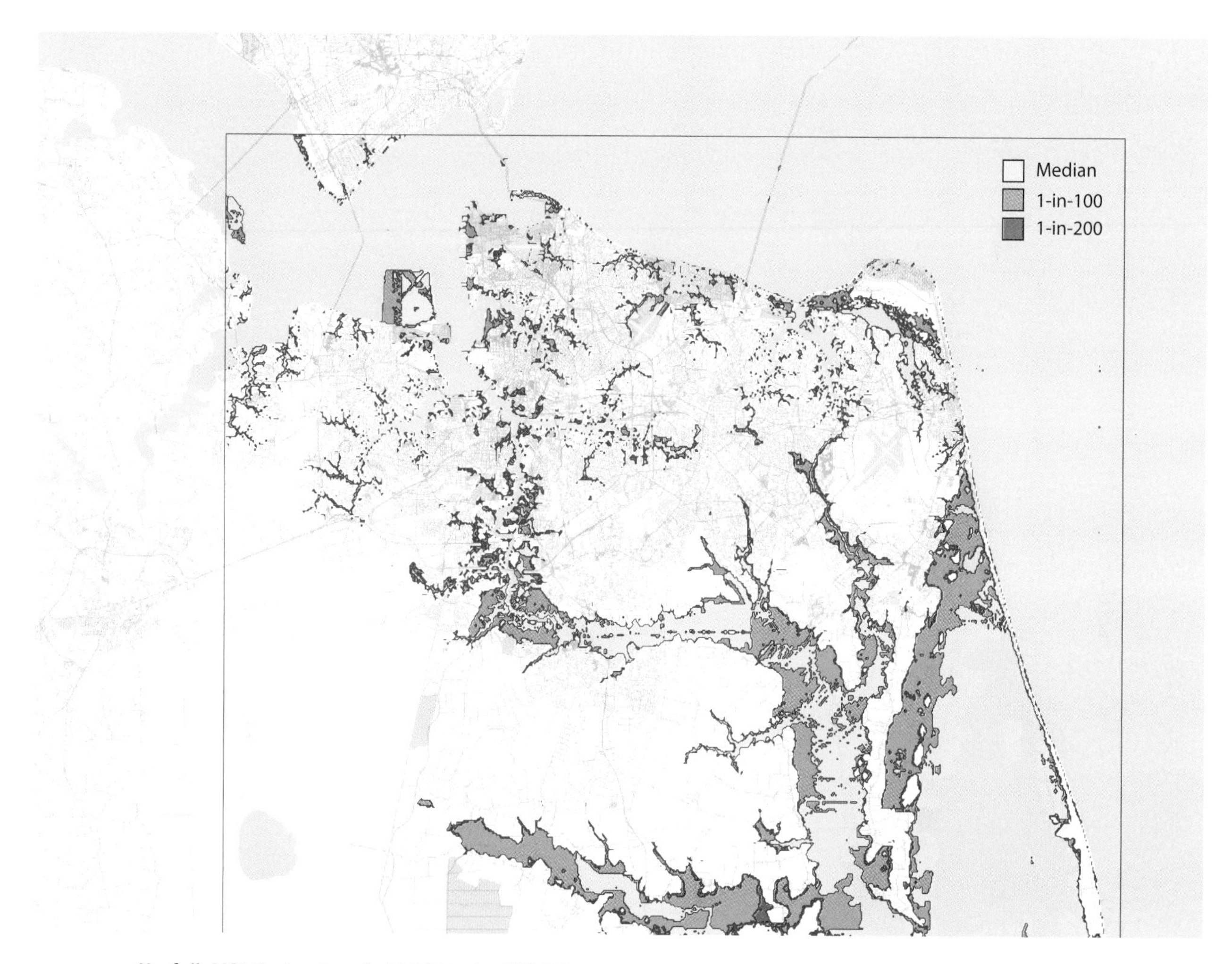

FIGURE 11.7. **Norfolk MSL Projections in 2100 Under RCP 8.5**

of Mexico, with roughly half of that occurring in Florida. The impact of climate change on flooding during coastal storms is larger and more immediate than the impact of gradual SLR-driven inundation discussed earlier. Assuming current hurricane activity continues, SLR under RCP 8.5 will *likely* increase average annual losses by $2 billion to $3.5 billion per year as early as 2030, a 7 to 13 percent increase over current levels. This increase in storm damage, like the storms themselves, will not be evenly spread across time. These numbers reflect the expected average annual loss of all storms across different scenarios for SLR.

As with inundation from SLR, this climate-driven increase in expected storm damage hits some states harder than others (figures 11.9 through 11.14). The largest relative *likely* increases occur in Delaware (16 to 39 percent by 2030), New Jersey (14 to 36 percent), New York (11 to 27 percent), and Virginia (13 to 28 percent). In absolute terms, Florida faces a far greater increase in expected storm damage due to higher sea levels than any other state. By 2030, average annual losses *likely* grow by $738 million to $1.3 billion.

By 2050, average annual losses from hurricanes and nor'easters will *likely* grow to $5.8 billion to $13 billion nationwide under RCP 8.5, a 21 to 48 percent increase from current levels, due just to mean SLR. Average annual losses in New Jersey will *likely* increase by between 64 and 174 percent, by 53 to 155 percent in Delaware, and by 45 to 110 percent in Virginia. In absolute terms, Florida will *likely* see an additional $1.9 billion to $4 billion a year in storm damage by 2050 unless protective measures are taken, while New York will *likely* see an additional $658 million to $3 billion in coastal storm-related costs each year.

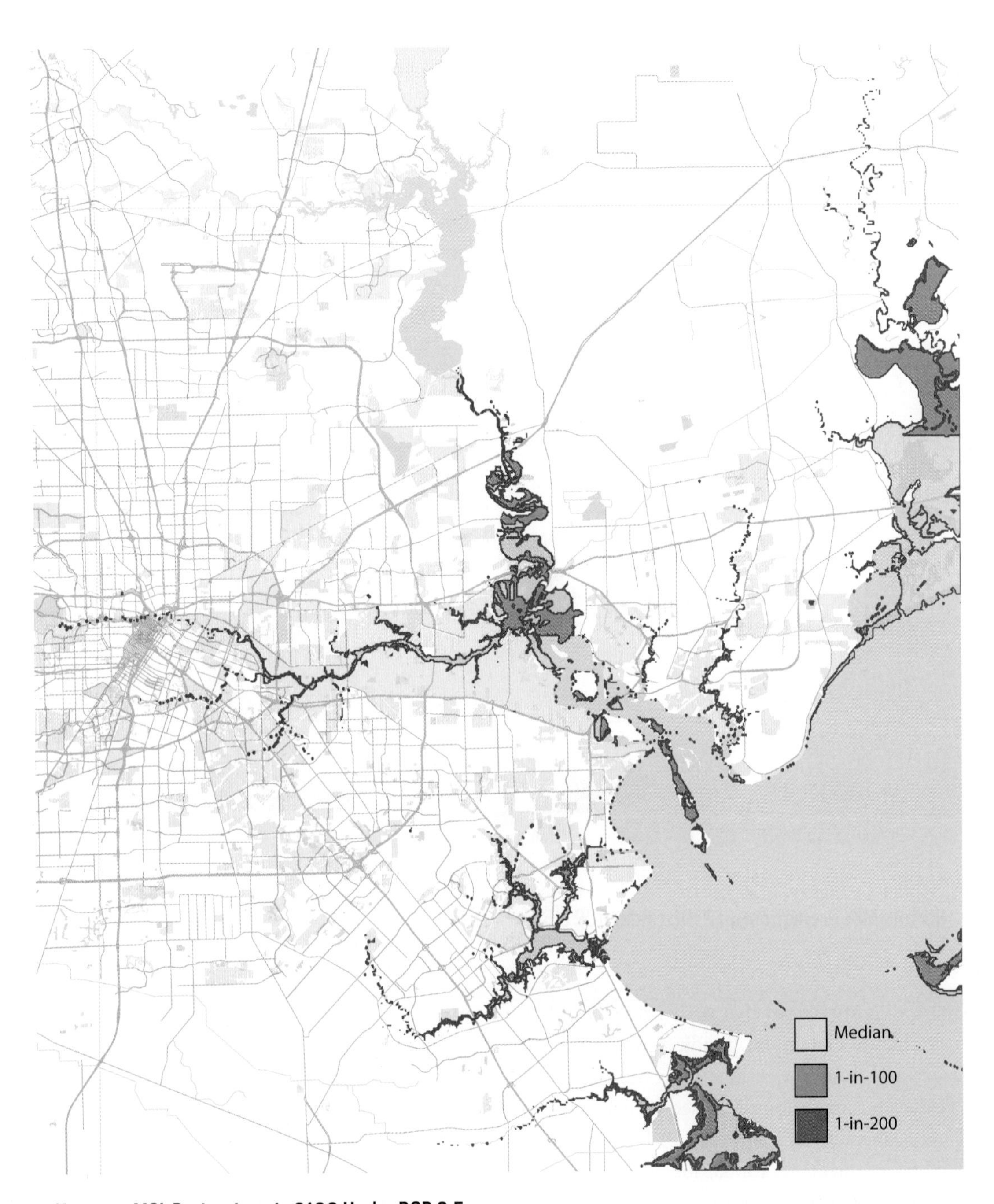

FIGURE 11.8. **Houston MSL Projections in 2100 Under RCP 8.5**

By 2100, SLR-driven increases in average annual hurricane and nor'easter damage will *likely* grow by $19 billion to $33 billion under RCP 8.5, a 71 to 122 percent increase from current levels. There is a 1-in-20 chance that damage could grow by more than $42 billion by 2100 and a 1-in-100 chance that damage could grow by more than $50 billion. Conversely, there is a 1-in-20 chance that average annual losses will only grow by $15 billion or less and a 1-in-100 chance of a less than $8.6 billion increase. Florida will *likely* see a $7 billion to $14 billion, or 60 to 104 percent, increase above current levels. New York will *likely* see a $2.6 billion to $5.2 billion increase, or 159 to 313 percent, and New Jersey will *likely* see a $1.4 billion to $3.7 billion increase, or 208 to 414 percent.

Averaged over the two-decade intervals used for other impact categories, the *likely* SLR-driven increase in average annual coastal storm damage is $2 billion to $3.6 billion on average by 2020–2039, $5.7 billion to $12 billion

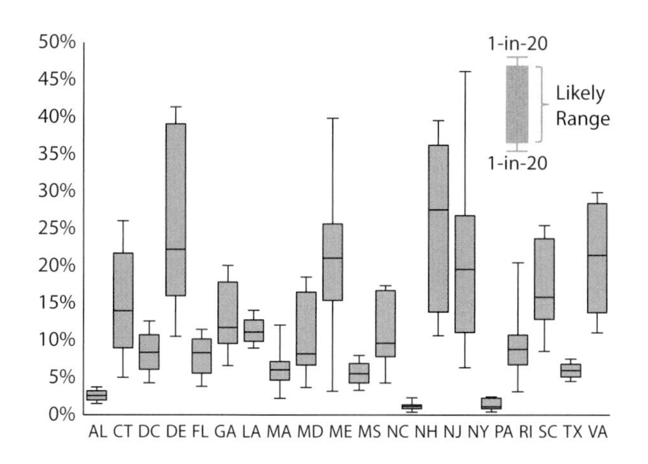

FIGURE 11.9. **Relative Increase in Average Annual Coastal Storm Damage Caused by Higher Sea Levels in 2030**

Percent change from 2010 expected average annual losses

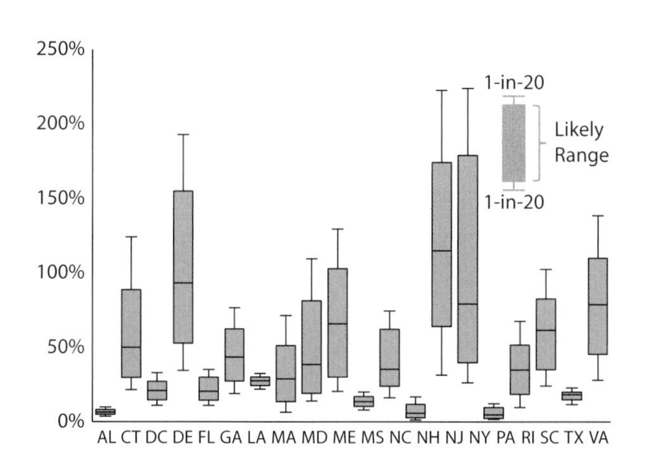

FIGURE 11.10. **Relative Increase in Average Annual Coastal Storm Damage Caused by Higher Sea Levels in 2050**

Percent change from 2010 expected average annual losses

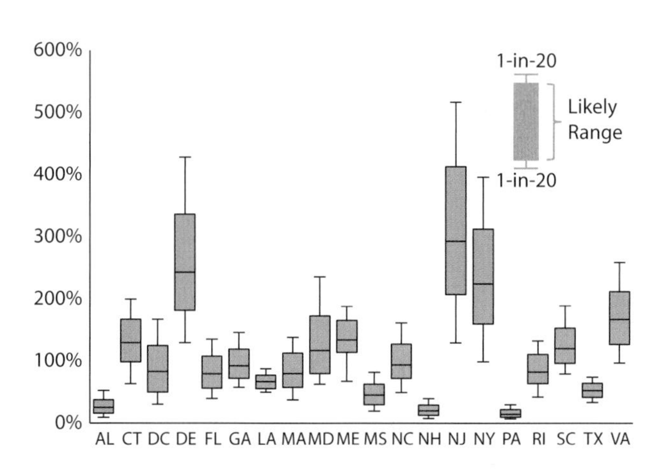

FIGURE 11.11. **Relative Increase in Average Annual Coastal Storm Damage Caused by Higher Sea Levels in 2100**

Percent change from 2010 expected average annual losses

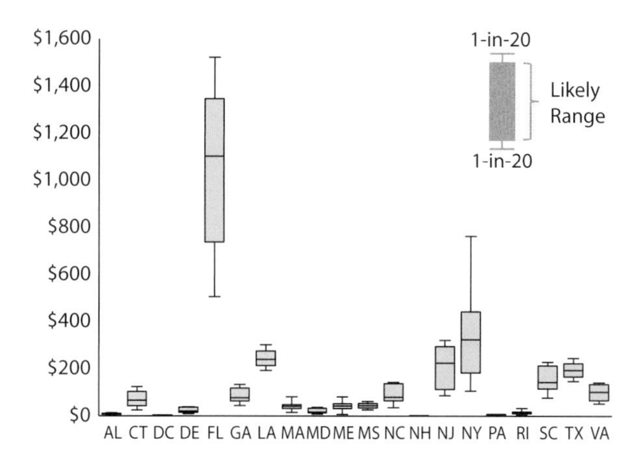

FIGURE 11.12. **Absolute Increase in Average Annual Coastal Storm Damage Caused by Higher Sea Levels in 2030**

Million 2011 U.S. dollars relative to 2010 expected average annual losses

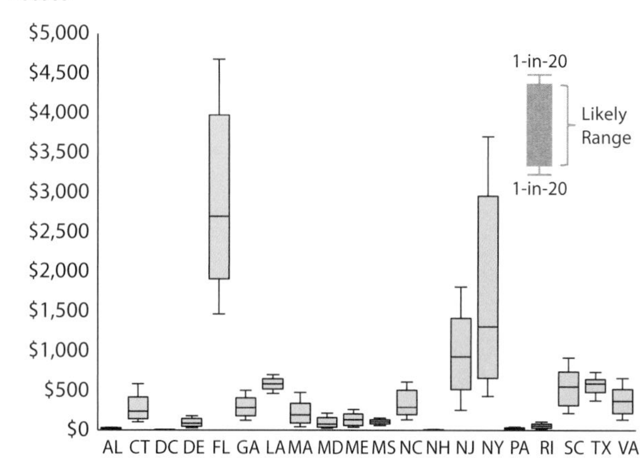

FIGURE 11.13. **Absolute Increase in Average Annual Coastal Storm Damage Caused by Higher Sea Levels in 2050**

Million 2011 U.S. dollars relative to 2010 expected average annual losses

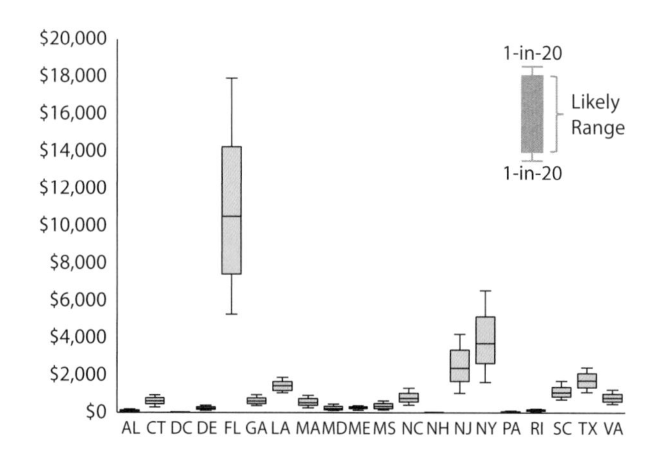

FIGURE 11.14. **Absolute Increase in Average Annual Coastal Storm Damage Caused by Higher Sea Levels in 2100**

Million 2011 U.S. dollars relative to 2010 expected average annual losses

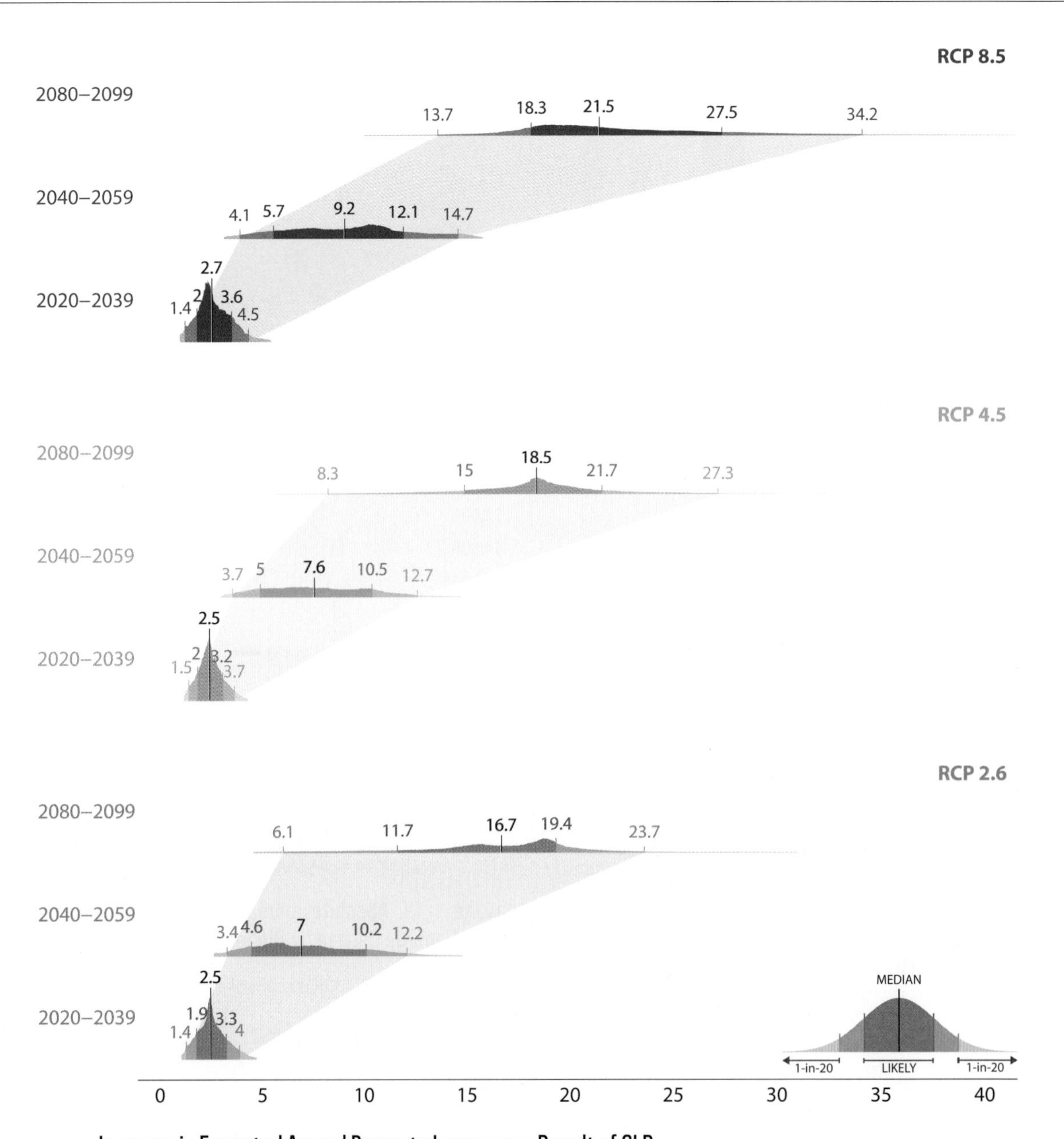

FIGURE 11.15. **Increase in Expected Annual Property Losses as a Result of SLR**

Assuming no change in hurricane activity (billion 2011 U.S. dollars)

on average by 2040–2059, and $18 billion to $28 billion on average by 2080–2099 (figure 11.15).

The relatively small difference in SLR in this century between RCPs translates into a relatively small difference in SLR-driven surge damage between RCPs as well. In RCP 4.5, the *likely* increase in average annual coastal storm damage due to mean SLR is between $5 billion and $11 billion by 2040–2059 and between $15 billion and $22 billion by 2080–2099. In RCP 2.6, the *likely* range falls to $4.6 billion to $10 billion on average by 2040–2059 and $12 billion to $19 billion on average by 2080–2099.

Another way to think about the risk to coastal property from SLR-driven increases in storm surge is to map the change in extent of flooding during a 1-in-100 year flood; or, put another way, areas with a 1 percent chance of being flooded in any given year. Buildings within the 100-year floodplain are generally required to purchase flood insurance by the federal government. Projected

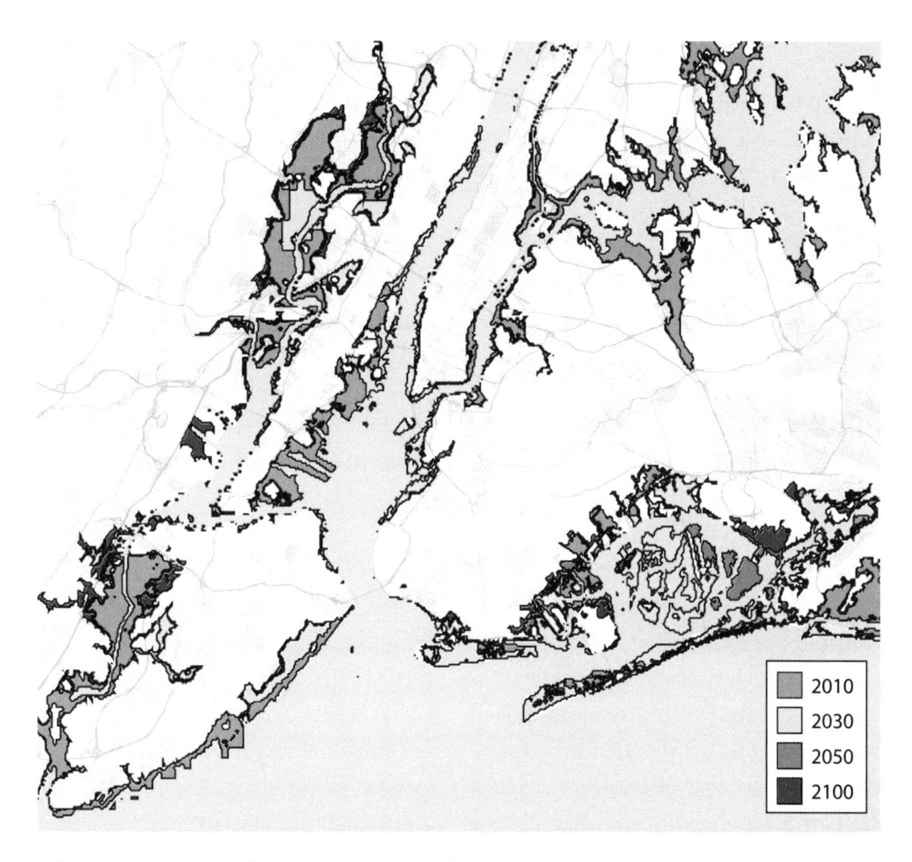

FIGURE 11.16. **The New York City 100-year Floodplain Under Median RCP 8.5 SLR**

Assumes historical hurricane activity

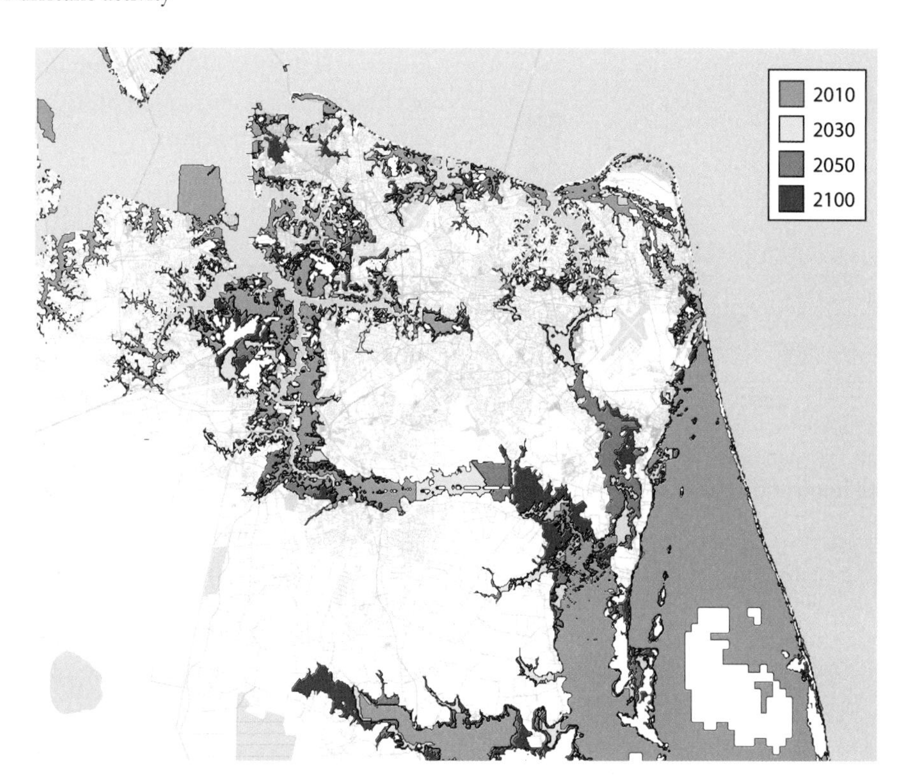

FIGURE 11.17. **The Norfolk 100-year Floodplain Under median RCP 8.5 SLR**

Assumes historical hurricane activity

local SLR will materially change the 1-in-100 year floodplain in many communities, and as soon as the next 10 to 20 years. Figures 11.16 and 11.17 show the change in the 100-year floodplains of New York City and Norfolk, respectively, as a result of projected SLR in our median RCP 8.5 scenario.

CHANGES IN HURRICANE FREQUENCY AND INTENSITY

There is considerable uncertainty about how climate change will influence the frequency and intensity of hurricanes going forward, but the impact of potential hurricane activity change is significant. For example, using ensemble projections from Emanuel (2013) for changes in hurricane frequency and intensity under RCP 8.5, average annual damage from East Coast and Gulf of Mexico hurricanes and nor'easters will *likely* grow by \$3.0 billion to \$7.3 billion by 2030, an 11 to 22 percent increase from current levels (figure 11.18). By 2050, the combined impact of higher sea levels and modeled changes in hurricane activity *likely* raise annual losses by \$11 billion to \$23 billion, roughly twice as large of an increase as from changes in local sea levels alone. By the end of the century, the combined *likely* impact of SLR and modeled changes in hurricane activity raise average annual losses by \$62 billion to \$91 billion, three times as much as higher sea levels alone.

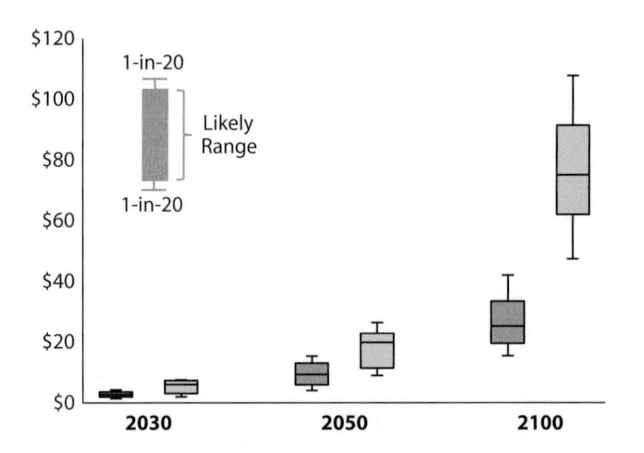

FIGURE 11.18. **Increase in Average Annual Losses with Historical and Projected Hurricane Activity**

Billion 2011 U.S. dollars: RCP 8.5 ensemble tropical cyclone activity projections from Emanuel (2013). Blue is historical, green is projected.

Under RCP 4.5, using changes in hurricane activity projected by Knutson et al. (2013), the increase in average annual commercial and residential property damage as a result of climate change is *likely* \$2.7 billion to \$7.0 billion by 2030, \$11 billion to \$22 billion by 2050, and \$56 billion to \$80 billion by 2100. Averaged over the two-decade intervals used for other impact categories, the increases are \$3.6 billion to \$5.7 billion by 2020–2039, \$11 billion to \$22 billion by midcentury, and \$47 billion to \$65 billion by late century (figure 11.19). The increase in damage resulting from either Emanuel's or Knutson and colleagues' projections for future changes in hurricane activity are due to both greater storm surge (even without climate-driven SLR) and greater wind damage.

While examining different RCPs, both Emanuel and Knutson and colleagues find significant changes in hurricane activity as a result of warmer sea-surface temperatures. Should this finding turn out to be correct, changes in storm activity could be a more important determinant of climate-driven changes in hurricane damage than SLR alone in the years ahead.

KEEPING OUT THE SEA

There are a number of steps individual building owners, community organizations, and policy makers at the local, state, and national levels can take to guard against some of these coastal effects. These include strengthening buildings, constructing sea walls, and nourishing beaches. In part 5 of this book, we analyze the extent to which these adaptive measures can reduce the risk that coastal communities face.

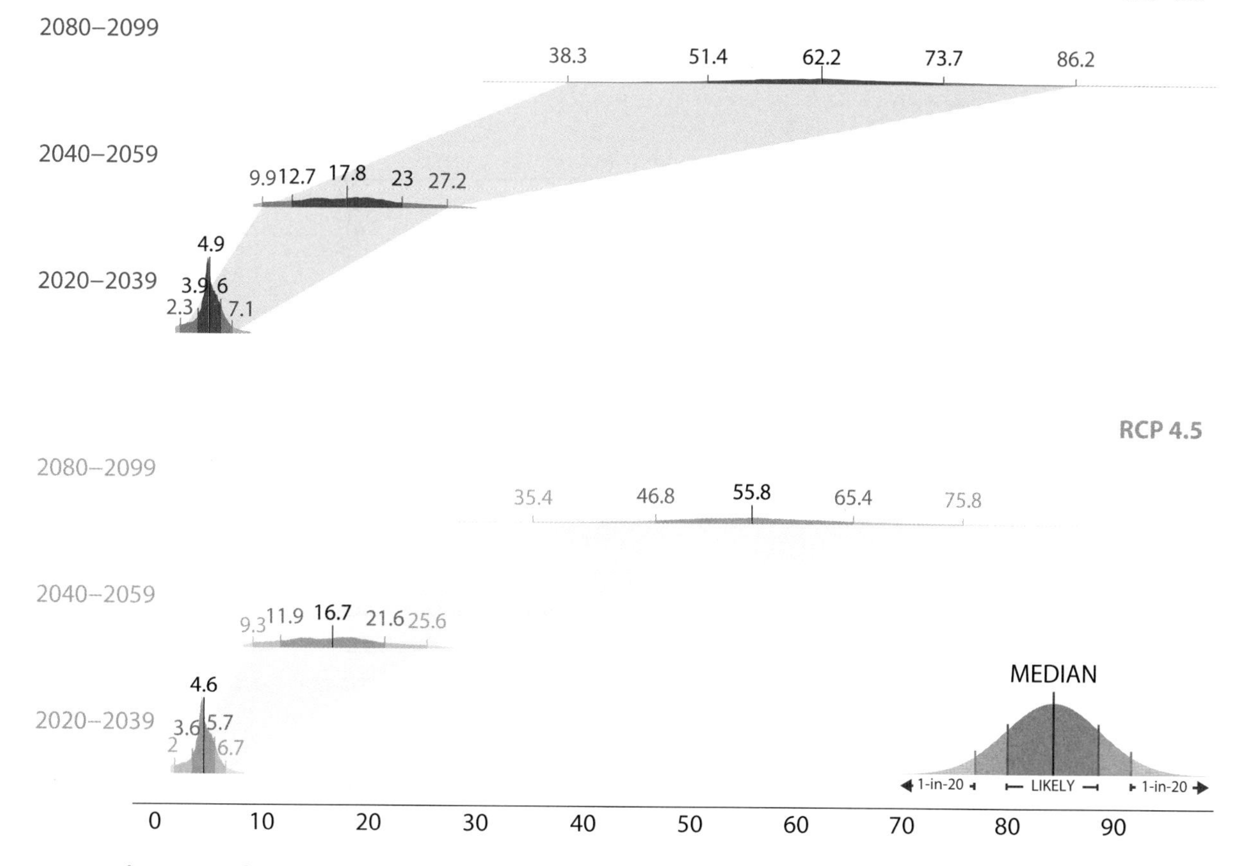

FIGURE 11.19. **Increase in Average Annual Losses with Projected Hurricane Activity**

Billion 2011 U.S. dollars.

Source: RCP 8.5 tropical cyclone activity projections from Emanuel (2013); RCP 4.5 projections from Knutson et al. (2013)

PART 3

PRICING CLIMATE RISK

OPENING COMMENTARY

GEOFFREY HEAL

COLUMBIA BUSINESS SCHOOL

THE set of chapters in part 3 attempts to put the effects of climate change in the United States into an economic context by placing dollar values on them. This certainly makes sense, as dollars are what we generally use for comparing and adding up very different types of goods and services. At the same time, we should be aware that there may be dimensions of climate effects that are not well-reflected in dollar values, no matter how ingeniously these are computed. Consider deaths from heatstroke: government agencies commonly value these at a few million dollars, a number known technically as the "value of a statistical life," but this clearly does not reflect the real impact of the loss of a family member. Who would sacrifice their children for a contribution to their retirement fund? The same issue arises with an issue the report does not cover—extinction of species. How do we set a cost to the loss of a species? What is the cost of the loss of a life-form that has been a part of our world since way before we became human? We have to recognize that there are possible consequences of climate change that are simply not adequately captured by the utilitarian calculus involved in reduction to dollar values. (The same can surely be said of the costs of crime: There is an unpleasantness associated with living in a crime-ridden area that goes beyond the expected loss from crime.) And we need to be aware that these are often the really important and irreversible consequences—as the examples of death and extinction indicate. So, any dollar estimate of the costs of climate change is necessarily a lower bound on the true costs.

But, given that we are in the game of evaluating climate change economically, this report does a good job. This is a game we need to play: Although we are frequently told that the costs of preventing climate change—the costs of moving away from fossil fuels—are prohibitive and will destroy our prosperity, a trove of recent work suggests just the opposite. A complete response has two elements: to demonstrate that in fact the costs of moving away from fossil fuels are low, and that the costs of business as usual, of allowing the climate to change, are very great and dominate the costs of moving to clean energy. Thus we do need a dollar figure for the climate change effects that can be valued in money terms, even though we must realize that there is a lot that can never be captured in the dollar figures.

There are a couple of points that stand out in part 3 and bear repetition. One is the massive geographic variation in climate effects: Some states are barely affected, and others are affected quite significantly. The northwestern states escape largely unscathed from most aspects of climate change, whereas the Southeast and the Great Plains suffer some serious losses. The Northeast is harder hit than the Northwest but is also one of the regions least affected. There is some irony here: The regions least affected are those that are most aware of and concerned about climate change, whereas those most affected are blasé to skeptical about the issue. Florida, with a climate-skeptic governor, comes out badly from all analyses. This geographic variability is reflected at the global level: Some countries will be little affected, some may even gain, and others will be destroyed. In general, hotter countries will lose a lot and cooler countries will lose a little from an altered climate. As rich countries tend to be cool and poor ones hot, this will worsen the world's inequality (Park & Heal 2014). The report finds a similar phenomenon domestically: Climate change is regressive in its domestic effects.

Another of the analysis's observations that bears repetition is that the time horizon matters: There are big differences between the effects 25, 50, 100, and 200 years out. In the next decade or so, we may encounter no dramatic consequences of climate change, but this will certainly not imply that there will be none: No dramatic changes for decades is quite consistent with very harmful changes toward the end of the century and beyond. It is worth emphasizing that climate change will not stop in 100 or 200 years just because our simulations stop there: It will progress and probably get progressively faster as various positive feedbacks come into operation. Bad outcomes late in the century will mean disastrous outcomes next century. My great-grandchildren will still be around then, so this is not an irrelevant consideration.

Now I want to switch to some more focused and technical comments on the analysis of part 3. For many reasons—a lot of them mentioned in the report—the report underestimates the economic costs of climate change. There are effects about which we do not know enough for quantifying and modeling, even though we have a strong intuition that they will be important. But even effects that are included are likely to be underestimated. As an example, take the case of food production. There is little question that agriculture in the United States—and in other countries, too—will be negatively affected by the changes that are likely in our climate. How do we value this drop in agricultural output? The answer that the report gives, an obvious one, is to take the drop in output and multiply it by the price of output. This gives us the dollar value of the impact. But this misses an important point, which is that consumers may well value food at more than they pay for it. This raises the issue of what economists call consumer surplus: I may be willing to pay more for something than the price that it sells for. That is particularly true of food: We are not going to stop eating just because the supply of food goes down and its price up. So, we should really value a drop in agricultural output not by market price but by what consumers would be willing to pay for food, which could be very much more than the market price. The same is true for property lost to sea-level rise or storms: we by default value that at its market price, whereas it may be worth far more to its owners.

The analysis studies four other categories of economic losses in addition to losses of agricultural output and of property from storm or sea damage: these are changes in labor productivity, in energy use, in heat-related mortality, and in crime rates. I have already noted that there are aspects of death and of crime that are clearly not reflected in conventional assessments of their costs, so that the report is underestimating the costs of these. Higher temperatures and humidity levels reduce labor productivity, and the report does a good job of trying to measure this, but there is another aspect of heat and humidity that goes with reduced productivity but is additional, which is the amenity value of climate. People don't like heat and humidity, which is one reason why they are less productive under these circumstances. The disutility from heat and humidity has an economic cost, possibly a large one, but this is not reflected in the report's calculations (although there are other studies that try to measure this amenity value). The report indicates how significant this effect could be when it discusses in chapter 4 the growth of wet-bulb temperatures in excess of 80°F—temperature and humidity combinations that pose a serious threat to human health.

The analyis is open about the issues that it omits because there is no adequate research on which to base conclusions, but again it should be emphasized that some of these are potentially highly influential. One example is water shortages (such as those now occurring in the West), which can disrupt both industry and agriculture. Although the report does factor in possible changes in rainfall when considering the effects of climate change on agriculture, more extreme rainfall changes could make current predictions seem wildly optimistic. Another omitted issue is the death of forests, something we are already seeing in parts of the Rockies, where bark beetle populations are multiplying in response to higher winter temperatures. This is tied to the rapid growth of wildfires in the West, caused by a combination of heat, dryness, and forest dieback resulting from the beetle infestations. Storm incidence is another aspect of climate change that could have very direct effects on the lives of many Americans and is not fully captured by the report's calculations: while coastal floods and hurricanes are considered, inland floods, windstorms, and snowstorms are not. As we have seen in the past few years, all of these can cause massive harm. So far, they have eluded the skills of modelers.

An innovative aspect of the analysis is its attempt to calculate a "risk premium" and a "fairness" or "equity" premium, the former related to the increase in risk facing Americans and the latter an attempt to value the change in income distribution wrought by an altered climate. I haven't seen a calculation of an equity premium before, and the report convincingly argues that it is large, possibly doubling the damage from climate change. Although I think the equity premium calculations are good and innovative, I think the risk calculations leave something out. Climate change doesn't just introduce risk; it introduces uncertainty. Here I'm using a distinction due to Frank Knight. Risk involves random outcomes whose probabilities are known, as in throwing fair dice: uncertainty, random outcomes with unknown probabilities. The latter is the case with climate change: We really don't know the probabilities that the increase in temperature in, say, the Southeast in 2075 will be 4 or 5 or 6 degrees Celsius. We have some ideas about the relative likelihoods of different outcomes but not a well-defined probability density function of the type beloved of teachers of statistics 101. Knight's term "uncertainty" has now been replaced by "ambiguity," a term introduced by Daniel Ellsberg in 1961 (Ellsberg 1961). We are generally averse not only to risk but also to ambiguity, and even more averse to the latter than to the former. Future work might investigate how accounting for ambiguity alters these results.

Another complex issue that the analysis raises is whether Gross Domestic Product (GDP) or economic output is the right baseline from which to measure deviations due to climate change. The report expresses most impacts as a percent change from an output measure. In fact, many researchers now argue that output-based measures of economic activity are not good indicators of welfare or of sustainability and that we should be relating the consequences of climate change to a different baseline. This is too complex an issue to explore here but is still worth mentioning. It is not clear how allowing for this would affect the report's calculations.

A strategic issue in any study of the economics of climate change is the choice of a discount rate, the rate at which future costs and benefits are discounted (devalued) relative to the present because of their futurity. The choice of a discount rate is highly controversial, with heated arguments about whether the choice should be zero or something closer to commercial rates of return. The report avoids discounting climate damage explicitly, but in the instance of discounting worker earnings it uses a 3 percent rate. I've referred so far to "the" discount rate, but in fact there are two concepts here and it is important to distinguish between them. One is the pure rate of time preference (PRTP), the rate at which we discount the future just because it is the future. It is generally agreed that the choice of a PRTP is a value judgment, an expression of one's personal ethical perspective. The second is the consumption discount rate (CDR), the rate at which the valuation of an increment of consumption falls over time. This can differ from the PRTP because for example future people may be richer (or poorer) and so less (or more) deserving of extra income—again a value judgment. It is this CDR that is relevant for cost-benefit studies of the type that the report is conducting, and the CDR is given by a formula called the Ramsey rule, after the famous Cambridge economist, mathematician, and philosopher Frank Ramsey:

$$\text{CDR} = \text{PRTP} + \text{Elasticity} \times \text{Income growth}$$

Here, income growth is the rate of growth of income, and elasticity is the elasticity of the marginal utility of income, which is the formal name for the coefficient of inequality aversion that the report uses in calculating the cost of inequality associated with climate impacts. (If the future is uncertain—which is universally the case—then there should be additional terms in the expression for CDR reflecting uncertainty and risk aversion.) The PRTP should probably be a small number, less than 1 percent. The report considers a range of values for inequality aversion up to 8: It is not clear what the future growth rate of the U.S. economy will be, and indeed more relevant here is the future growth rate of the income of the average American, which has been zero to negative in the past few decades. Taking these factors into account, it is possible that 3 percent is high for the CDR, and it might have been better for the report authors to try several different discount rates and report results for all of them when considering worker earnings. Also, as the growth rate might vary over time, the CDR might also vary (see Arrow et al. 2014).

A final comment is that the analysis considers the impacts of climate change *in* the United States: This is not the same as the impacts of climate change *on* the United States, as the United States can and indeed will be affected by changes in the climate elsewhere. World food markets are interconnected, so a drop in food production resulting from climate change in India or China will affect world market prices and so the prices faced by U.S. buyers. And the submergence of small island states in the Pacific or the Caribbean will inevitably generate fluxes of refugees, and it is likely that the United States will have to accept some responsibility for them.

The bottom line here is that this is a great analysis. It pulls together the emerging literature on climate impacts better than anything else I know of, at least for the United States. It would be wonderful to have similar studies for other regions, particularly some of those that are less affluent and less able to cope with climate disruptions. I think the report underestimates the impact of climate alterations on the United States, but I also think this is almost inevitable for any serious scholarly study, as there is so much about climate impacts that we don't yet understand. And in an interconnected world, studying climate change in one country can only give us a partial picture of the forces that our climate experiments are letting loose on the planet.

CHAPTER 12

FROM IMPACTS TO ECONOMICS

WHAT are the economic consequences of the climate impacts described in the preceding chapters? Rising sea levels, increased flooding, and more frequent and intense coastal storms damage capital that must be rebuilt. Changing yields affect the financial health of both agricultural producers and farming communities. Climate-driven changes in mortality rates shape overall labor supply, and temperature influences the productivity of that labor. Higher energy prices reduce real household income and raise business costs. Changes in crime rates affect property values and public expenditures on police and other security services. The costs of climate change will not be evenly spread throughout the country. The nature and magnitude of the economic risks Americans face depends very much on who they are and where they live.

Economists began studying the impact of climate change on modern economies in the early 1990s, starting with the pioneering work of Yale professor William Nordhaus (1991), Peterson Institute for International Economics fellow William Cline (1992), and London School of Economics professor Samuel Fankhauser (1993). Research focused on combining climate and economic models to enable an integrated assessment of the relationships between (a) economic activity and greenhouse-gas emissions, (b) greenhouse-gas emissions and global temperature increases, and (c) global temperature increases and economic activity. The first of these "benefit-cost integrated assessment models" (IAMs) were developed by Nordhaus (1994), Cambridge professor Chris Hope (1993), and University of Sussex professor Richard Tol (1995). These three models continue to be among the most often used, although a few others have joined their ranks (Revesz et al. 2014).

IAMs are primarily used to conduct cost-benefit analysis of emission-reduction strategies at the global level (Mastrandrea 2010). The cost of climate change is quantified through one or more climate "damage functions," which provide monetary estimates of climate impacts associated with different increases in global average temperatures, often expressed as a percentage loss of Gross Domestic Product (GDP). This is mapped against an "abatement cost function" that provides a monetary estimate of the cost of reducing greenhouse-gas emissions, also generally expressed as a percentage loss of GDP, to estimate the economically optimal level of emissions reduction.

Because their scope of coverage is so broad, IAMs necessarily rely on simplified representations of individual components of both climate and economic systems (Kopp & Mignone 2012). For example, most models explore changes in global mean temperatures and sometimes sea level, but not how these changes shape temperature, precipitation, and sea level at a local scale. Economic costs are generally assessed and presented as global aggregates or aggregates for a small number of multinational regions, with no subnational geographic detail and often no sectoral detail. Most IAMs assume the economy naturally adapts to climate change to the extent it makes economic sense to do so. Those climate costs that can't be addressed through adaptation are weighed against the cost of reducing emissions based on a single representative decision-maker's attitude toward risk and level of concern regarding future economic liabilities.

These features make IAMs less useful for the type of risk assessment we seek to provide with this report than the national or global benefit-cost analysis for which they have traditionally been used. American businesses and households experience climate change in the form of shifts in local daily temperature and precipitation patterns, rather than global annual averages. Global or nationwide economic cost estimates are useful in international or national policy-making but do little to inform local risk-management decisions. Indeed, local decision-makers, whether state infrastructure planners, property developers, agricultural producers, or individual households, need more tailored information in order to make the adaptation investments the IAMs assume will occur. These individuals and institutions differ both in risk tolerance and in planning and investment time horizons, making economy-wide risk aversion and time-preference assumptions irrelevant to their respective decision-making processes.

Over the past decade, researchers have begun taking a more granular approach to assessing the economic cost of climate change, including in the United States. For example, in 2004, economist Dale Jorgenson used a computable general equilibrium (CGE) model of the U.S. economy and the best climate projections and impact estimates available at the time to assess national costs at a sectoral as well as economy-wide level (Jorgenson et al. 2004). In 2009, David Abler, Karen Fisher-Vanden, and colleagues conducted a similar assessment for Pennsylvania, providing a greater level of geographic resolution (Abler et al. 2009). A 2010 report by Sandia National Laboratory analyzed potential economic impacts in the United States across a wider range of climate scenarios than past assessments (Backus et al. 2010).

A category of IAMs distinct from the benefit-cost IAMs discussed earlier, known as process-based IAMs, contain detailed representations of the energy and agriculture sectors. These process-based IAMs have traditionally been used for analyses of the cost-effectiveness of climate-change mitigation strategies rather than for assessments of the risks of climate change. Recent work, however, has begun to incorporate feedbacks from climate change onto these models' representation of the national and global economy, laying the groundwork for assessments of the economic costs of climate change (Calvin et al. 2013; Reilly et al. 2013).

Building on this work, in part 3 we quantify the economic consequences of the climate impacts described in part 2 by sector, state, and region and across a full range of potential climate futures. We assess the impact on those sectors, states, and regions directly affected (chapter 13), how those impacts ripple throughout the region and the country, and how climate impacts in a given year affect the rate of economic growth in subsequent years (chapter 14). Finally, we explore how the differences in time preference, risk appetite, and concern about inequality shape the national significance of these economic impacts (chapter 15).

CHAPTER 13

DIRECT COSTS AND BENEFITS

All assessments of the economic consequences of developments that may occur decades in the future must grapple with uncertainty about how the economy will evolve going forward absent those developments. This assessment of the economic risk of climate change is no different. Researchers are left with two choices: (a) assess the consequences of future developments relative to current economic structure and population distribution or (b) try to predict how the structure of the economy will change and assess the consequences of future developments relative to a hypothetical economic future. To quantify the direct costs and benefits of the climate-driven changes in agricultural production, labor productivity, mortality, crime, energy demand, and coastal storm damage described in previous chapters, we take approach (a). As stated from the outset of this report, our goal is to provide decision makers with a better understanding of the risks they face, and their views on the likely evolution of particular sectors or the U.S. economy as a whole in the years ahead may well differ from ours—but we can all agree on the structure and size of the current economy.

Of course, some of the following costs may be ameliorated by populations adapting to changes in the climate (an issue we discuss and explore quantitatively in chapter 22), and one notable form of adaptation will be the out-migration of individuals away from increasingly adverse climates, a response that directly contradicts the earlier assumption of a fixed population distribution. If this out-migration occurs, it will slightly dampen the effects of climate change and reduce the magnitude of costs (and benefits) estimated later, but it is unlikely that it will cause our direct estimates to be off by a large amount. The size of the difference between our estimates (when populations are fixed) and actual costs (when some populations migrate) is equal to the product of two relatively small numbers: the fraction of the population that migrates and the fraction of the sector that is lost to climate. Because the product of two small numbers is a very small number, the distortion in our estimates introduced by assuming no migration is small (Hsiang 2011). For example, if 1 percent of a population migrates out of a region because of climate change and the mortality rate in that region rose by 9.5 deaths per 100,000 because of climate change (our median estimate for the nation), the mortality that is averted by this migration would be roughly 0.095 deaths per 100,000 in the original population (1 percent times

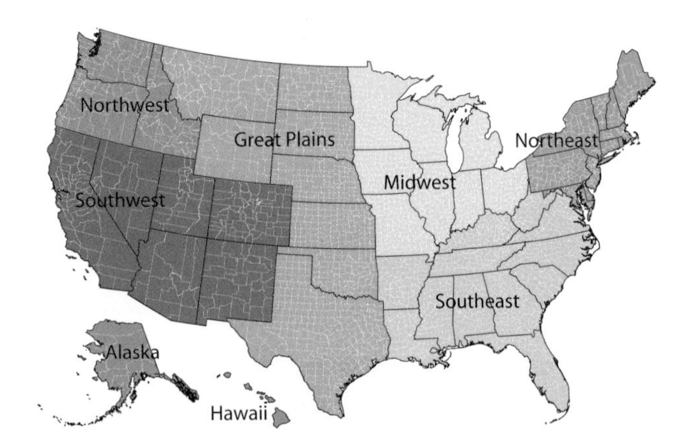

FIGURE 13.1. **U.S. Regional Definitions**

Regions as defined for the U.S. National Climate Assessment

9.5 deaths per 100,000). This means our baseline mortality estimate of 9.5 deaths per 100,000 is 0.095 deaths per 100,000 too large and should be corrected to 9.4 deaths per 100,000, a very minor adjustment given the scale of other uncertainties in our analysis.

All results presented in this chapter are for RCP 8.5. In part 5 of this book, we assess the extent to which global emission reductions consistent with an RCP 4.5 or RCP 2.6 scenario, as well as investments in adaptation, reduce these costs (and benefits). Direct costs are reported at the county level (either for illustrative counties or binned by decile) and at the state level on an annual basis, grouped by National Climate Assessment region (figure 13.1). (We have less confidence in many impact estimates for Hawaii and Alaska than in those for the lower 48 states and exclude them from our analysis in this chapter. That does not, however, mean they do not face economic risks from climate change.)

AGRICULTURE

We quantify the direct costs and benefits of climate-driven changes in commodity crop yields described in chapter 6 using the National Income Product Accounts of the Bureau of Economic Analysis and more detailed input-output tables from the Minnesota IMPLAN group (see appendix D). We assume any future change in commodity crop yields (maize, wheat, soy, and cotton) translates into a change in overall agricultural output proportional to each product's current share of total agricultural output (using current crop prices) for the county, state, or region in question. We count the change in agricultural output as the direct cost or benefit in that geographic area. At the national level, assuming current farming practices continue, the *likely* change in yields under RCP 8.5 range from an average annual direct cost of –$8.5 billion (i.e., an $8.5 billion benefit) to +$9.2 billion by 2020–2039, –$8.2 billion to +$19 billion by 2040–2059, and –$12 billion to +$53 billion by 2080–2099.

These are relatively modest effects in the context of today's $17 trillion U.S. economy against which they are measured—as expected, because these four crops account for less than one third of U.S. agricultural output by value, and for just 0.2 percent of total national economic output. As described in chapter 6, climate change will likely result in an increase in yields in some parts of the country and a decrease in others, the combination of which results in relatively modest net changes at the national level. The local economic significance of this regional heterogeneity in agriculture impacts is exacerbated by the wide variation in agriculture's importance in different state economies (figure 13.2). The four commodity crops included in our analysis accounted for 2.6, 2.2, 2, and 1.2 percent of total economic output in Nebraska, South Dakota, Iowa, and North Dakota, respectively, in 2011. These states, and key counties within them, face economically significant changes in agricultural output (both positive and negative) over the course of the century.

Likely nationwide, agricultural impacts per capita range from a $27 benefit to a $29 cost on average by 2020–2039 and from a $26 benefit to a $61 cost by 2040–2059. Figures 13.3 and 13.4 compare these national estimates to the *likely* range for individual counties, ranked by median projected per capita cost and binned by decile. Within each decile bin, critical values for the probability distribution of impacts (e.g., medians) are averaged across counties. The most vulnerable 10 percent of counties fare considerably worse than the national average, with *likely per capita* costs of –$722 to +$1,793 (median of +$793) by 2020–2039 and –$244 to +$3,382 (median of +$1,632) by 2040–2059. The most positively affected counties see *likely* per capita costs of –$638 to +$164 (median of –$261) by 2020–2039 and –$1,102 to +$8 (median of –$574) by 2040–2059.

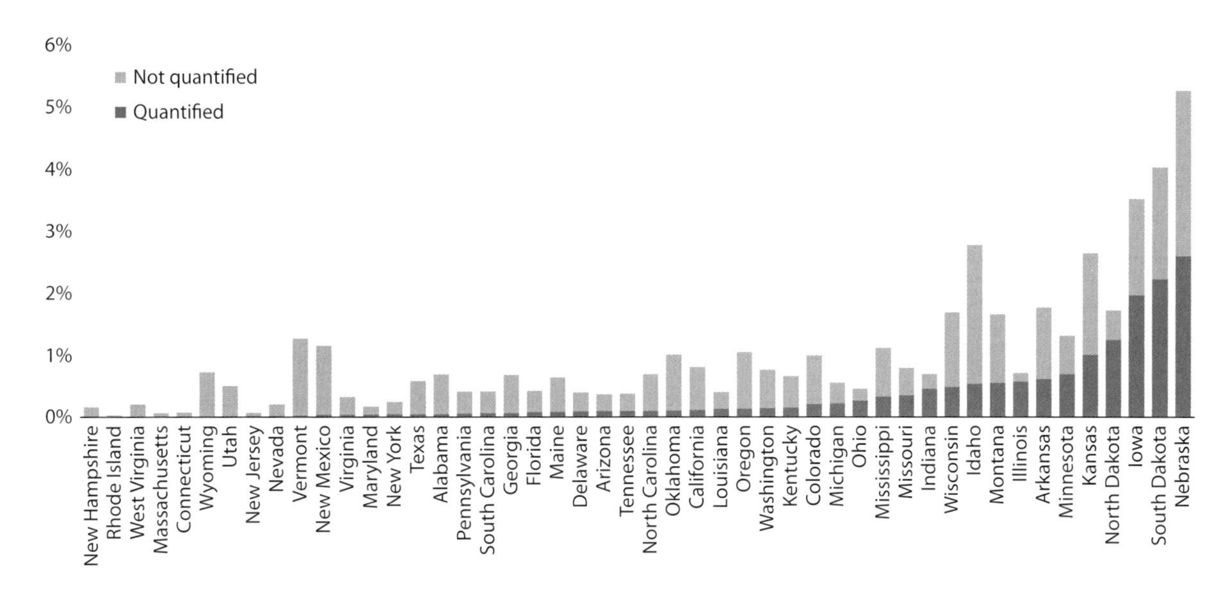

FIGURE 13.2. **Agricultural Production as a Share of State Output**

2011, broken down by crops quantified in this report and those excluded

By late century, the impacts described in chapter 6 translate into *likely* per capita costs of −$37 to +$169 (median of +$68) nationwide, with regional disparities getting even larger (figure 13.5). Iowa and Nebraska see the largest *likely* costs on a per capita basis at +$275 to +3,882 (median of +$1,996) and +$550 to +$3,416, respectively. North Dakota sees the largest *likely* net benefit per capita, with Montana, Oklahoma, and the Pacific Northwest seeing more modest *likely* gains.

LABOR

We assess the direct costs and benefits of the climate-driven changes in labor productivity described in chapter 7 by mapping projected percentage changes both in high-risk and low-risk sectors against the value added by those sectors in the 2011 IMPLAN input-output tables by geographic area. We assume that a 1 percent change in the

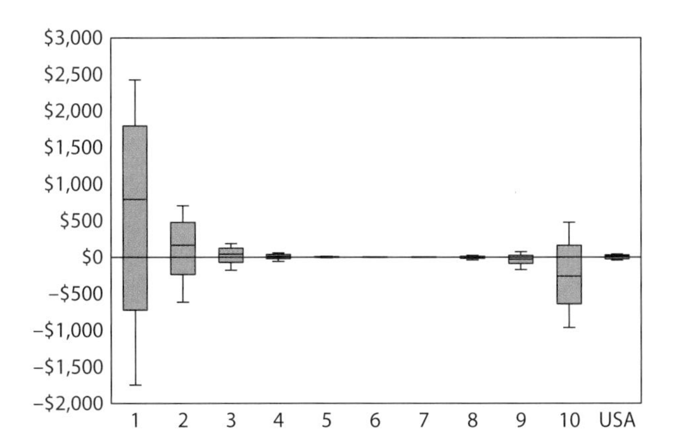

FIGURE 13.3. **County-Level per Capita Direct Costs from Changes in Agricultural Yield by Decile, 2020–2039**

RCP 8.5; 2011 U.S. dollars per capita, assumes current economic structure, crop mix, and agricultural prices; negative values indicate net benefits

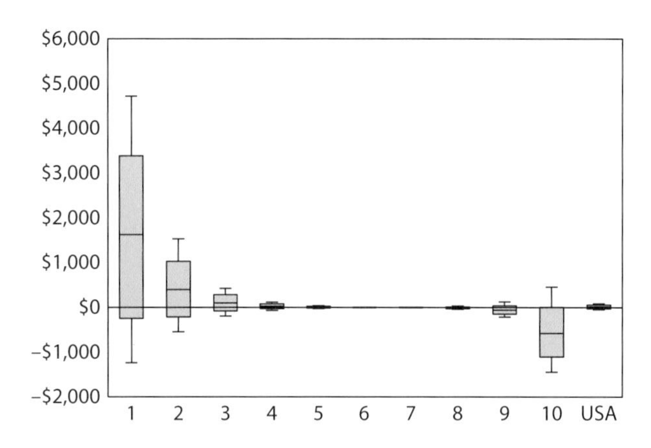

FIGURE 13.4. **County-Level per Capita Direct Costs from Changes in Agriculture Yield by Decile, 2040–2059**

RCP 8.5; 2011 U.S. dollars per capita, assumes current economic structure, crop mix, and agricultural prices; negative values indicate net benefits

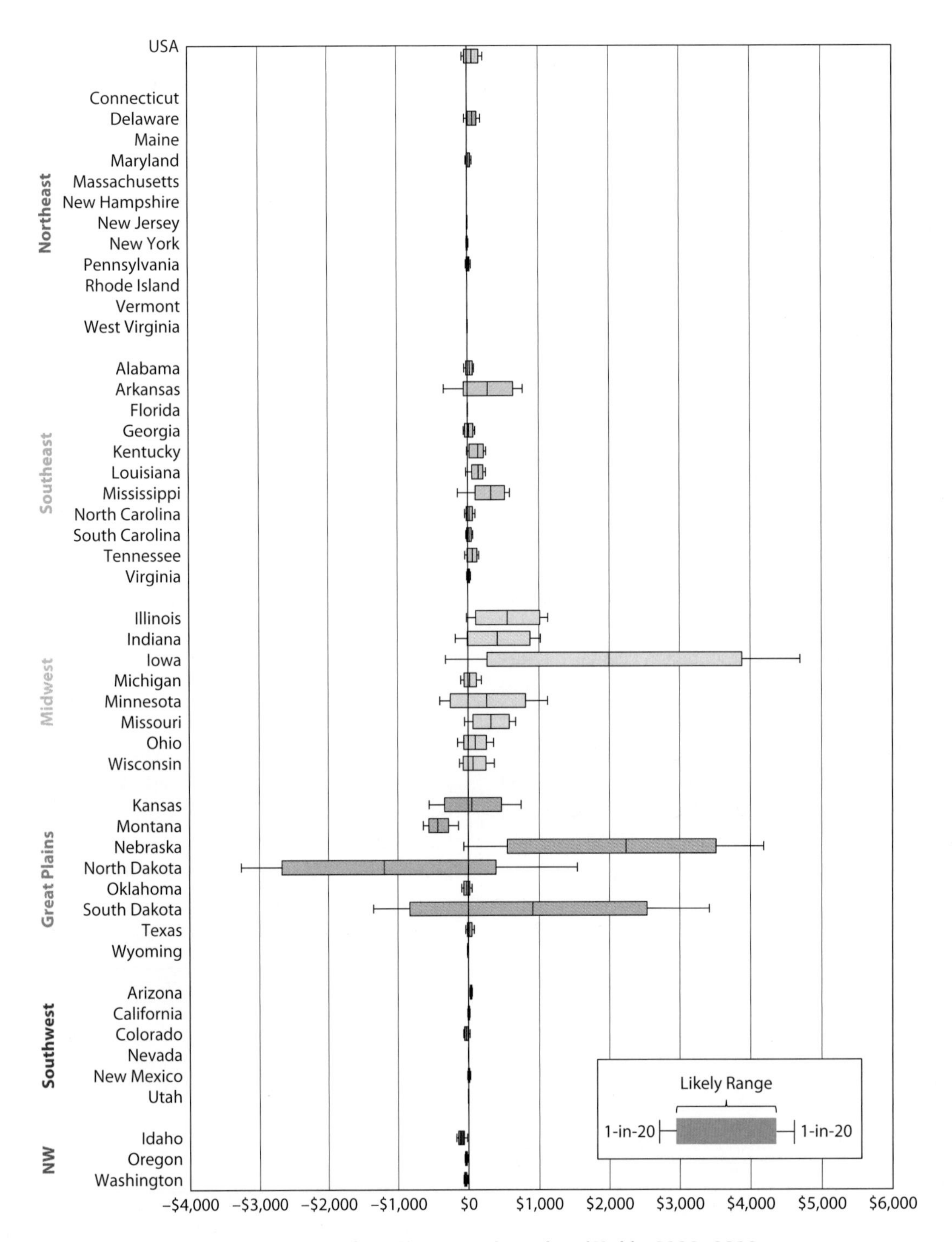

FIGURE 13.5. **State-Level per Capita Direct Costs from Changes in Agricultural Yields, 2080–2099**

RCP 8.5; 2011 U.S. dollars, assumes current economic structure, crop mix, and agricultural prices; negative values indicate direct benefits

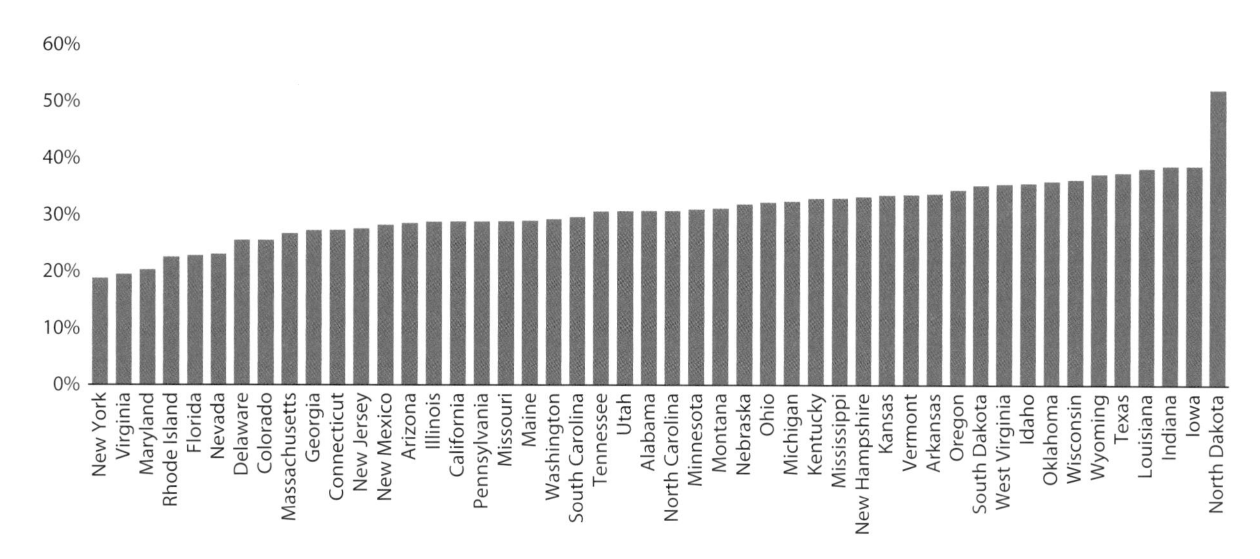

FIGURE 13.6. **Share of State Employment in High-Risk Sectors, 2011**

high-risk sector labor productivity results in a 1 percent change in the high-risk sector value-added and calculate the direct cost of that 1 percent of change in value-added at 2011 prices. At the national level, assuming that the current sectoral mix in the economy remains constant, *likely* average annual direct labor productivity costs (high-risk and low-risk combined) under RCP 8.5 are +$0.1 billion to +$22 billion by 2020–2039, +$10 billion to +$52 billion by 2040–2059, and +$42 billion to +$150 billion by 2080–2099—considerably larger than the nationwide agricultural impacts.

The regional variation of the impact of changes in temperature on labor productivity in high-risk sectors is not as geographically varied as the impact on agricultural productivity, nor is the economic importance of those high-risk sectors. That said, there is still a meaningful amount of variation in the high-risk sectors' share of total state employment, ranging from 53 percent in North Dakota to 19 percent in New York in 2011 (figure 13.6). Combined with modest variation in the climate-driven rate of high-risk labor productivity decline, this variation produces meaningful differences across counties and states. By 2020–2039, climate-driven changes in labor productivity will *likely* cost between +$0.3 and +$69 per capita on average nationwide (figure 13.7). For the most vulnerable counties, binned by decile, the *likely* range is −$19 to +$270. By 2040–2059, *likely* national average per capita costs grow to +$36 to +$171, with the most vulnerable 10 percent of counties *likely* seeing +$94 to +$473 in per capita costs (figure 13.8).

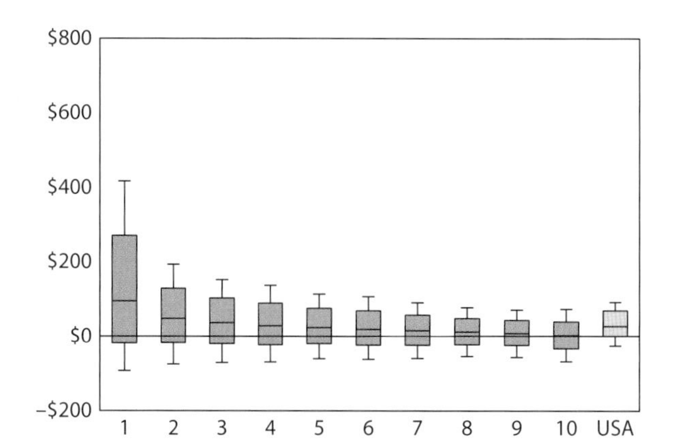

FIGURE 13.7. **County-Level per Capita Direct Costs from Changes in Labor Productivity by Decile, 2020–2039**

RCP 8.5; 2011 U.S. dollars; negative values indicate net benefits

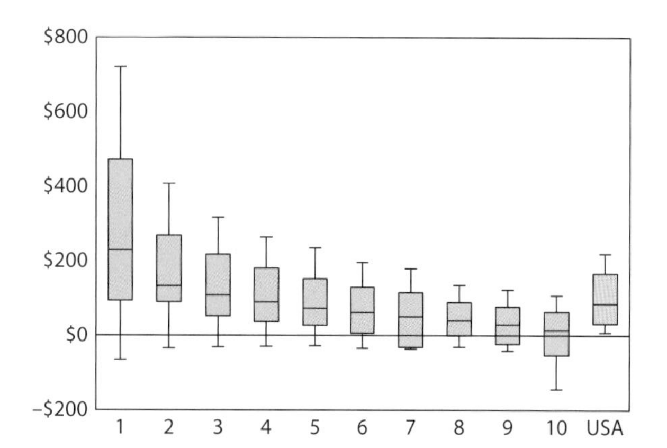

FIGURE 13.8. **County-Level per Capita Direct Costs from Changes in Labor Productivity, 2040–2059**

RCP 8.5; 2011 U.S. dollars; negative values indicate net benefits

By late century, national average per capita costs from climate-driven labor productivity declines grow to +$133 to +$479. The Northeast and the Northwest see smaller *likely* costs than the national average, while most states in the Southeast, Great Plains, Southwest, and Midwest see larger *likely* costs. At a state level, the largest direct per capita cost of climate-driven changes in labor productivity are in Texas (+$276 to +$1,040), Louisiana (+$262 to +$962), and North Dakota (+$233 to +$957) and are due in large part to the relatively large share of the workforce in high-risk sectors in these states and, in the case of Texas and Louisiana, larger percent reductions in productivity (figure 13.9).

The climate-driven changes in temperature-related mortality discussed in chapter 8 directly impact economic activity by changing available labor supply. We measure these "market" costs and benefits by calculating the change in state-level labor supply resulting from climate-driven changes in temperature-related mortality. We assess mortality by age cohort, as discussed in chapter 8, using IMPLAN socioeconomic data and labor-force participation rate estimates by age cohort from the Bureau of Labor Statistics (available online at www.bls.gov/emp/ep_table_303.htm). As with labor productivity, we assume a 1 percent change in labor supply translates into a 1 percent change in value-added.

Unlike climate-driven changes in labor productivity, however, the direct cost or benefit of mortality changes compounds over time. Workers who die in one year do not return to the labor force in the next year, during which time additional workers may die from climate-driven temperature increases. This leaves two options for assessing market costs. The first is to assess the lost or gained lifetime labor productivity of a projected climate-driven change in mortality in a given year and discount (using a 3 percent discount rate) those future losses to the year in which the death occurred. The second is to use a population model to assess changes in the composition of the workforce over time as climate-driven mortality impacts evolve. In this chapter, we use the former technique (described in the appendix D). We use the second technique as part of our exploration of how macroeconomic effects might shape direct climate costs and benefits in the next chapter.

At a national level, the present value of lost lifetime labor supply from *likely* annual climate-driven mortality under RCP 8.5 is +$3.4 billion to +$14 billion on average by 2040–2059, but with a net benefit likely in a significant number of states in the Northeast, Upper Midwest, Upper Great Plains, and Northwest due to fewer cold-related deaths. By late century, many of these states still see *likely* benefits, but not enough to offset a sharp increase in heat-related mortality across the Southeast and in many Midwest, Great Plains, and Southwest states (figure 13.10). Nationwide *likely* mortality costs rise to +$13 billion to +$41 billion, or +$42 to +$130 per person. Louisiana, Texas, Oklahoma, and Florida see the highest *likely* mortality increases, more than twice the national average, while New England, Oregon, and Washington see the largest *likely* mortality declines.

Changes in lifetime labor income is, of course, a somewhat narrow measure of the value of a human life. In analyzing the benefits and costs of policies with an impact on mortality rates, governments often look to a population's "willingness to pay" for small reductions in their risk of dying (Viscusi & Aldy 2003). This is often referred to as the "value of statistical life" (VSL). In the United States, the Environmental Protection Agency (EPA) uses a central estimate of $7.9 million per person (in 2011 dollars) regardless of age, income, or other population characteristics (EPA 2010). Aldy and Viscusi (2007) have found important differences in willingness to pay by age cohort, and as temperature-related mortality affects older Americans more than the population on average, the EPA VSL could be an overestimate of the willingness to pay to avoid these impacts. As it is the standard used by the U.S. government, we include it here as an upper-bound estimate.

Using the EPA central VSL estimate, we find *likely* average annual nationwide mortality costs under RCP 8.5 of –$12 billion to +$161 billion (median estimate of +$69 billion) by 2040–2059. Late century, this grows to +$90 billion to +$506 billion, more than twice as high as the market costs of climate-driven mortality. These values translate into +$287 to +$1,617 on a nationwide per capita basis (figure 13.11). As with the market costs of the mortality impacts described earlier, there is considerable variation among states. Florida sees the highest *likely* costs, at +$2,163 to +$5,979 per capita, while Maine sees the lowest, at –$2,080 to –$1,015.

CRIME

We assess the direct costs of the climate-driven increase in violent and property crime described in chapter 9 using average cost estimates for specific types of crimes, such as homicide or larceny (Heaton 2010). The costs of specific

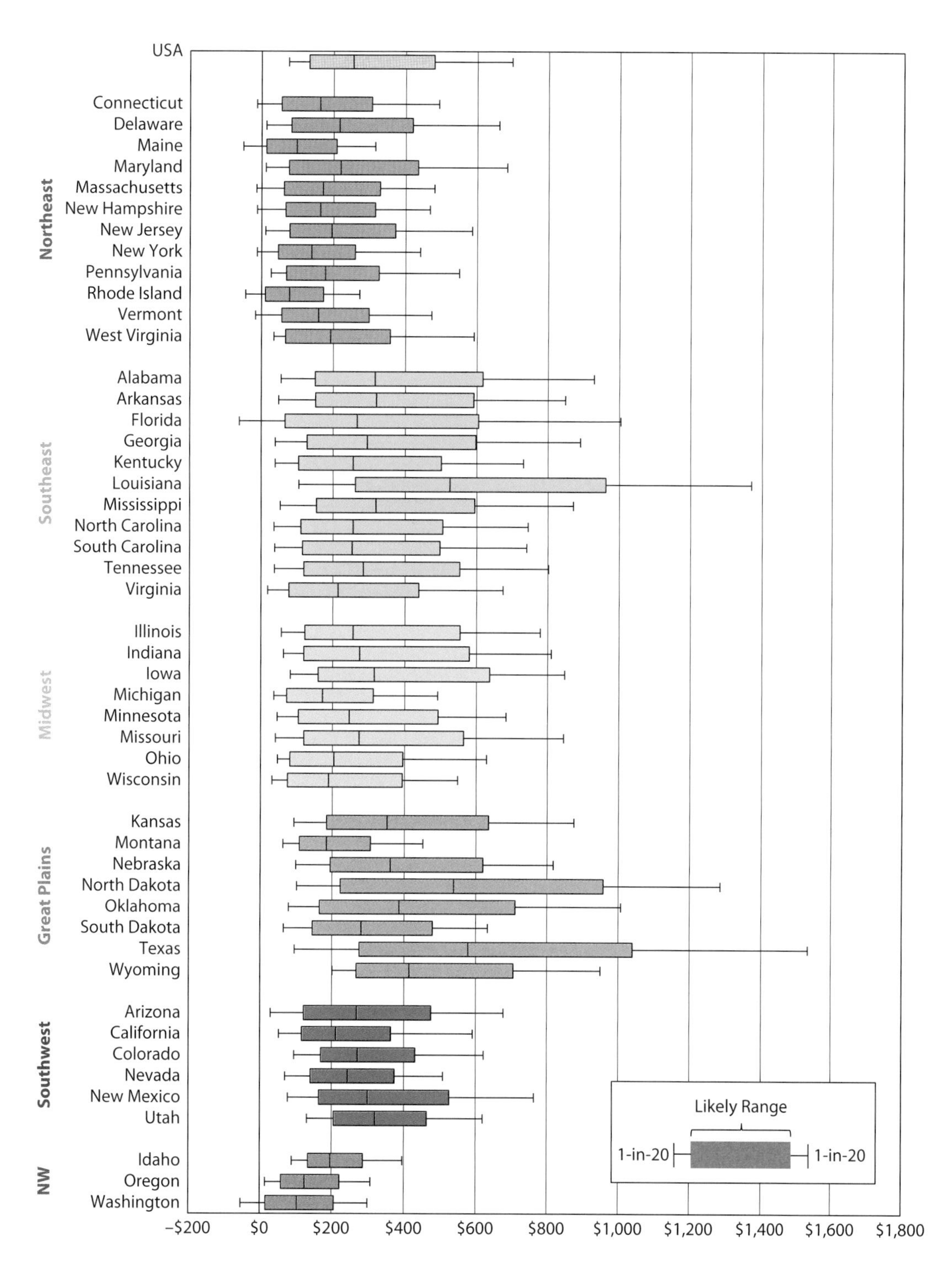

FIGURE 13.9. **State-Level per Capita Direct Costs from Changes in Labor Productivity, 2080–2099**

RCP 8.5; 2011 U.S. dollars; negative values indicate net benefits

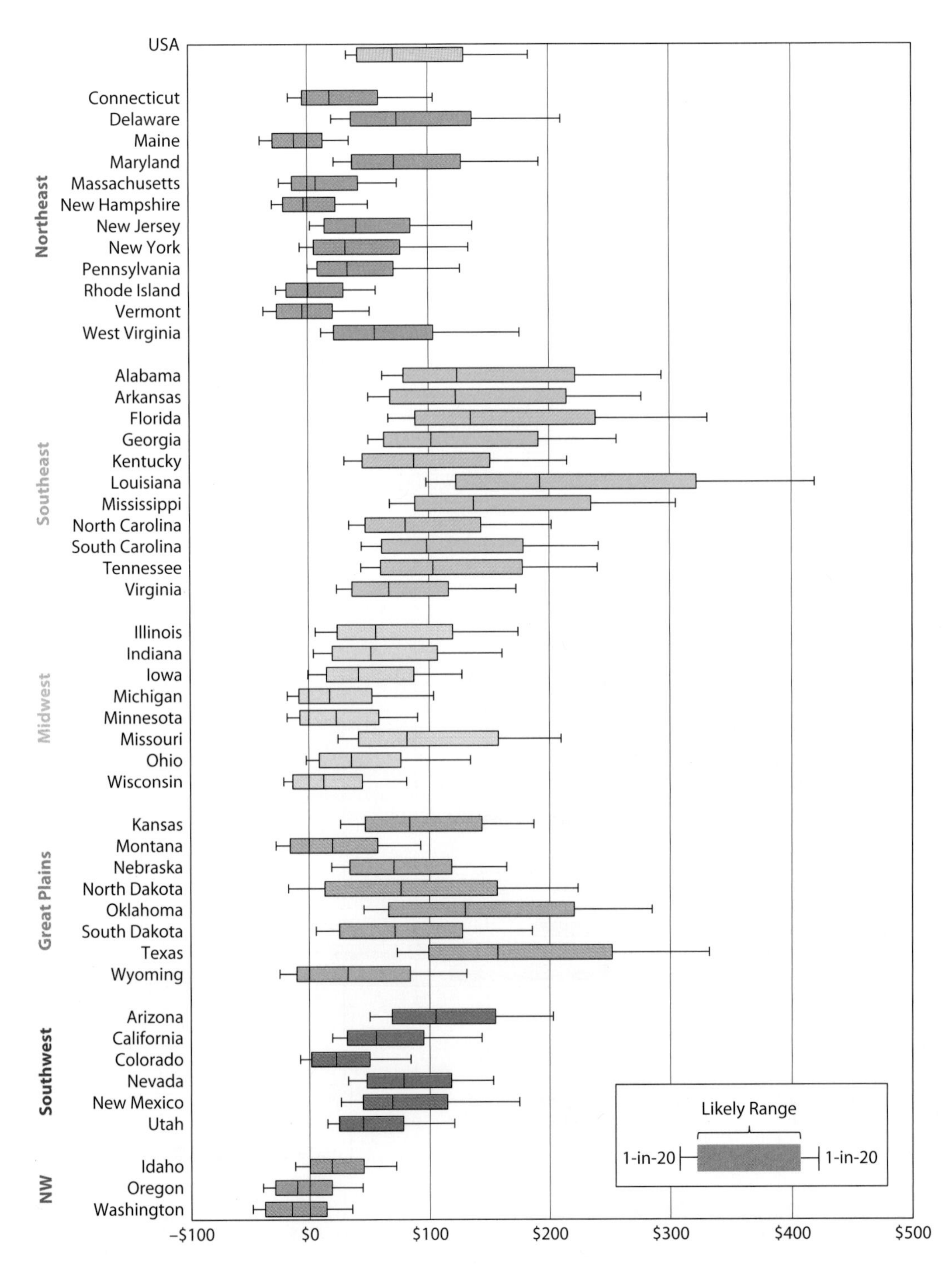

FIGURE 13.10. **State-level per Capita Direct Costs from Changes in Mortality, 2080–2099**

Using market estimates in RCP 8.5; 2011 U.S. dollars; negative values indicate direct benefits

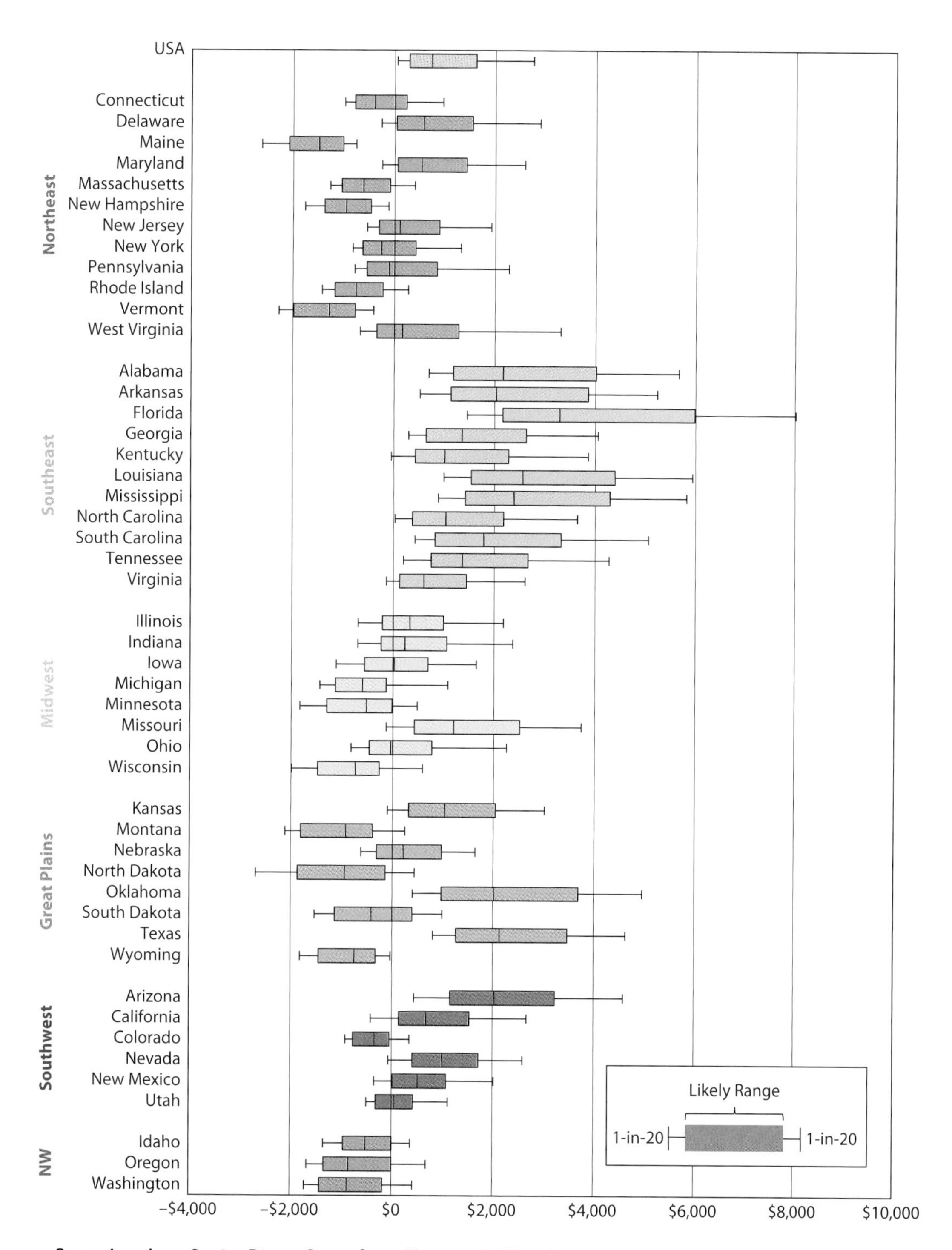

FIGURE 13.11. **State-Level per Capita Direct Costs from Changes in Mortality, 2080–2099**

Using a VSL of $7.9 million in RCP 8.5; 2011 U.S. dollars; negative values indicate direct benefits

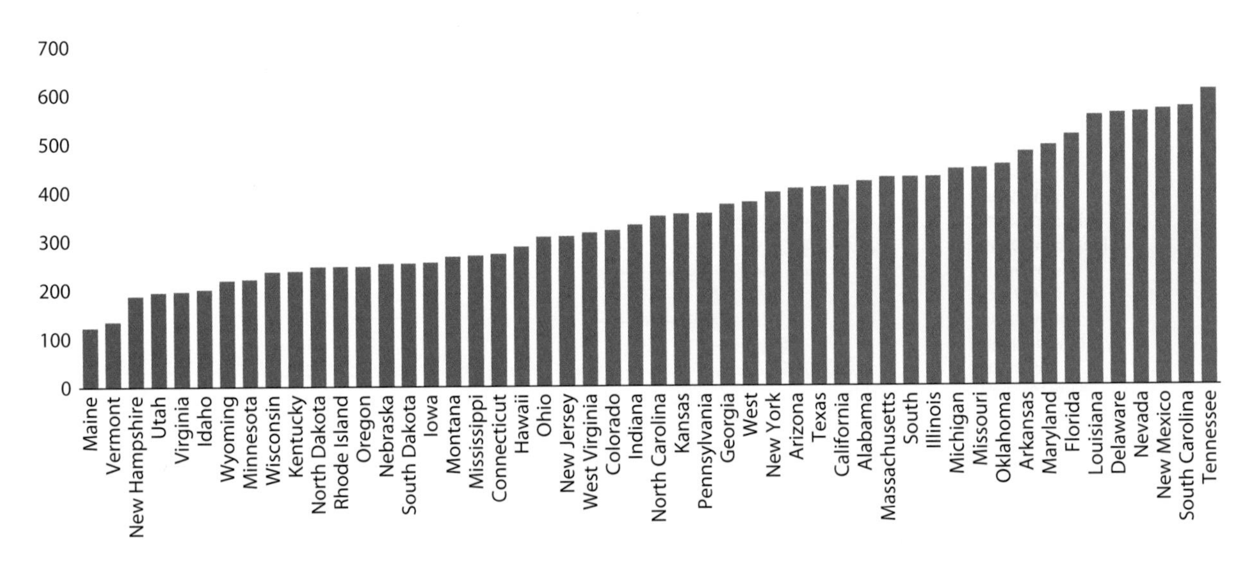

FIGURE 13.12. **Violent Crime Rates by State**

Crimes per 100,000 people, 2011, FBI Uniform Crime Reporting

crimes are estimated by combining accounting-based estimates, which attempt to enumerate costs incurred by victims (such as doctors' bills or lost assets), and contingent valuations of specific crimes, which try to elicit individuals' willingness to pay to avoid specific crimes using surveys. We assume that, in the future, the relative frequency of specific violent crimes and specific property crimes remains fixed within each state, but that the overall rate of these two classes of crimes evolves with the climate. At the national level, the *likely* change in direct property and violent annual crime costs under RCP 8.5 is $0 to $2.9 billion on average by 2020–2039, $1.5 billion to $5.7 billion by 2040–2059, and $5.0 billion to $12 billion by 2080–2099, making crime the least economically significant impact quantified in this report at a national level.

There is meaningful regional variation in climate-driven crime costs due to both differences in local climate projections and underlying crime rates (figures 13.12 and 13.13).

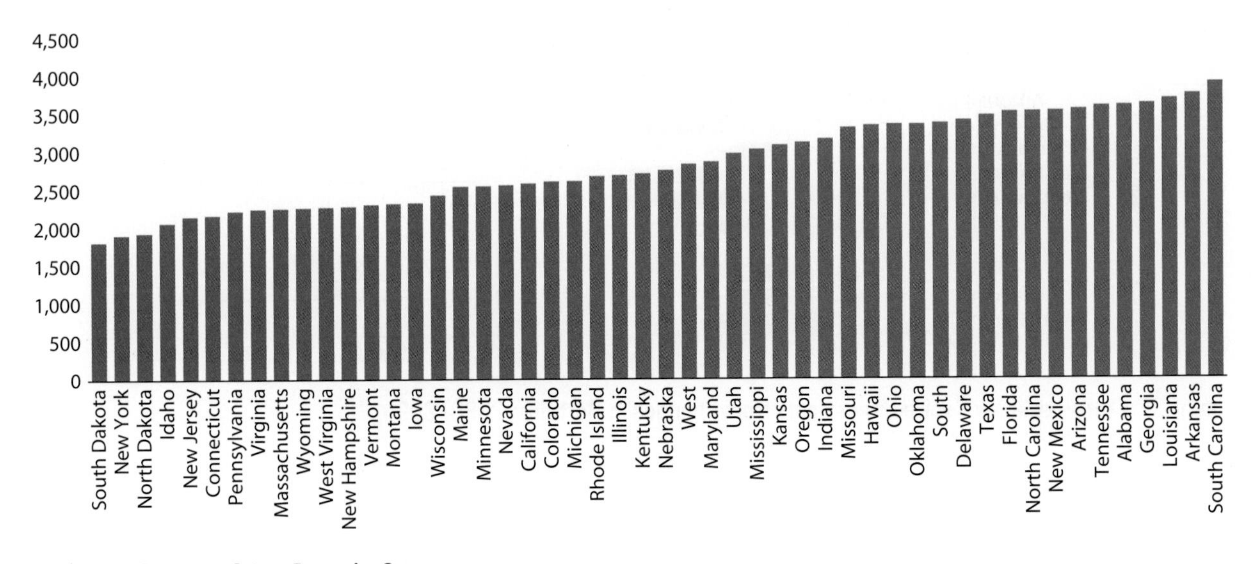

FIGURE 13.13. **Property Crime Rates by State**

Crimes per 100,000 people, 2011, FBI Uniform Crime Reporting

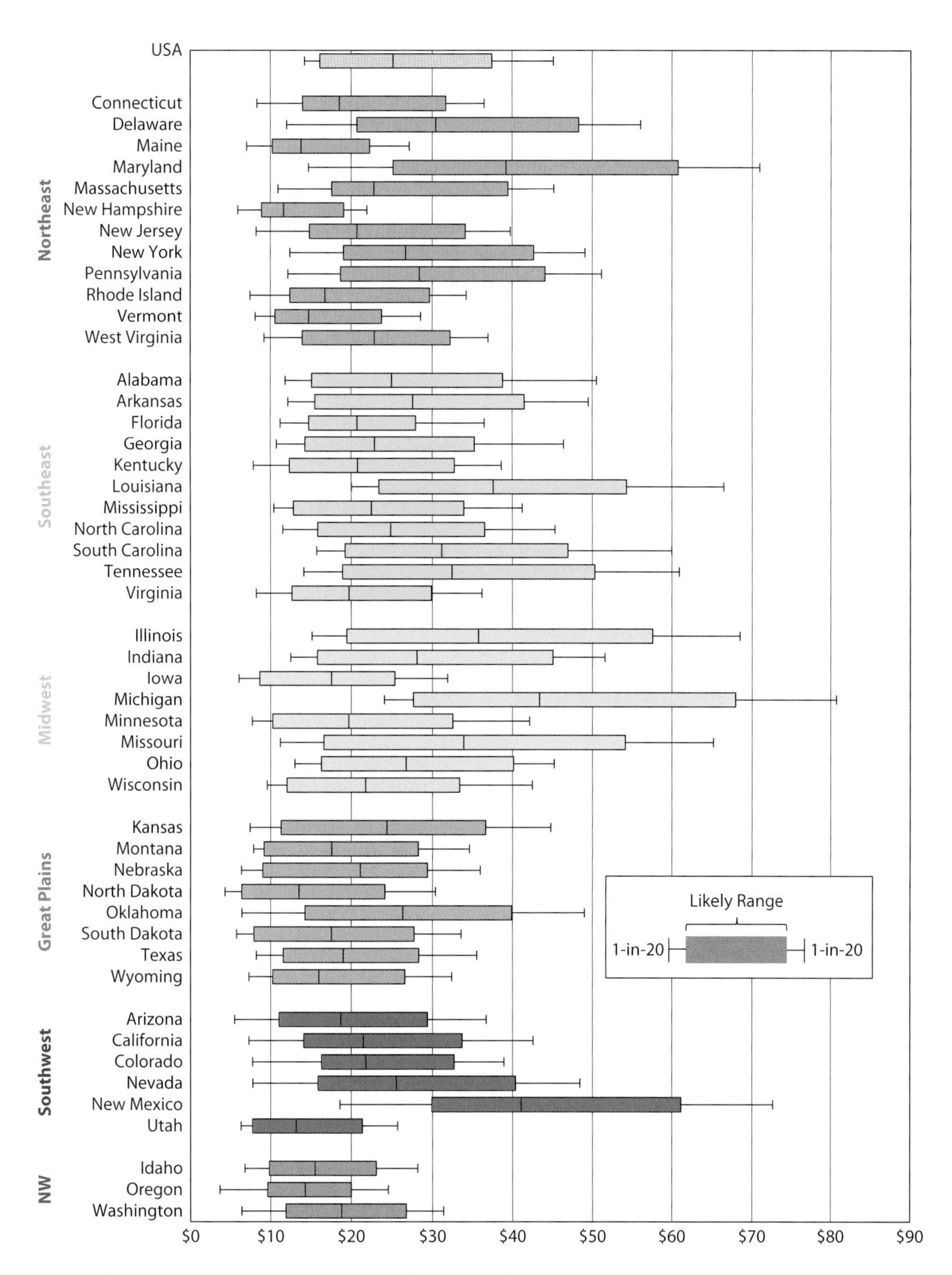

FIGURE 13.14. **State-level per Capita Direct Costs from Changes in Crime Rates, 2080–2099**

RCP 8.5; 2011 U.S. dollars

Likely national direct crime costs average $16 to $37 on a per capita basis in 2080–2099. Michigan, New Mexico, Maryland, Louisiana, and Illinois see the largest increases, though still relatively modest on a per capita basis. The per capita increases in crime costs are lowest in Utah, New England, and the Pacific Northwest (figure 13.14).

ENERGY

We assess the direct costs and benefits of climate-driven changes in energy demand using the estimates of percentage change in energy expenditures outlined in chapter 10 relative to current energy expenditure levels. At the national level, future changes in temperature mapped against today's U.S. energy market *likely* increase average annual energy expenditures under RCP 8.5 by $0.5 billion to $11 billion on average by 2020–2039, $8.3 billion to $29 billion by 2040–2059, and $32 billion to $87 billion by 2080–2099. Local changes in energy expenditures vary based both on local climate projections and local energy market conditions. Nationwide *likely* average annual energy expenditure increases by 2020–2039 are $1.5 to $37 on a per capita basis (figure 13.15), growing to $27 to $94 on average by 2040–2059 (figure 13.16). For the most vulnerable 10 percent of counties, however, the *likely* average per capita increase is $4 to $119 in 2020–2039 and $78 and $229 in 2040–2059. At the other end of the spectrum, 10 percent of counties see a *likely* combined decrease in energy expenditures of $3 to $60 per capita by 2020–2039 and $15 to $78 by 2040–2059. By late century, annual per capita energy expenditures *likely* increase by $102 to $279 (figure 13.17). The Northeast and Northwest see a much more modest increase (with some states seeing decreases in the median projection), as declines in heating costs offset much (and some places all) of the increase in cooling costs. Expenditures rise most in the Southeast and the more southern states in the Great Plains and Southwest regions, where temperatures reach their highest levels and states currently have little heating demand to lose. In Arizona and Florida, for example, per capita energy expenditures rise by more than twice the national average.

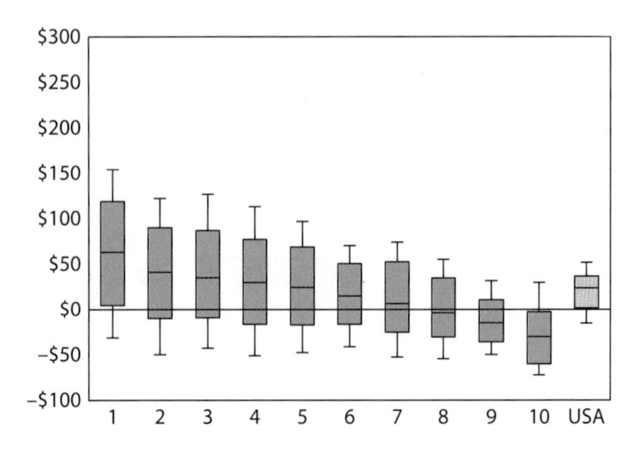

FIGURE 13.15. **County-Level per Capita Direct Costs from Changes in Energy Expenditures, 2020–2039**

RCP 8.5; 2011 U.S. dollars; negative values indicate net benefits

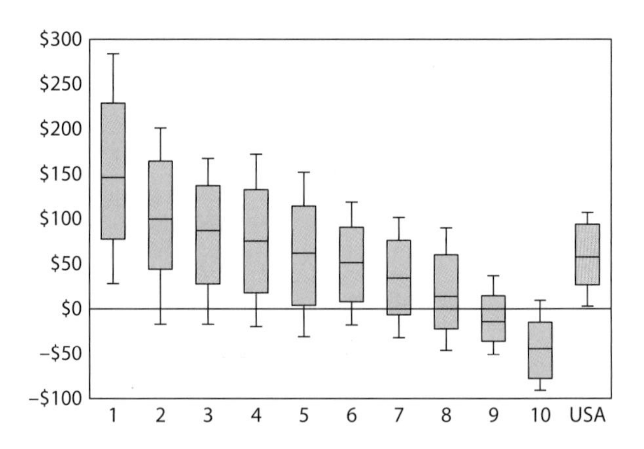

FIGURE 13.16. **County-Level per Capita Direct Costs from Changes in Energy Expenditures, 2040–2059**

RCP 8.5; 2011 U.S. dollars; negative values indicate net benefits

COASTAL COMMUNITIES

We assess the direct cost of climate-driven changes in coastal storms using the average annual loss estimates described in chapter 11. At the national level, assuming coastal property exposure remains unchanged, projected sea-level rise (SLR) increases average annual losses from hurricanes and other coastal storms by $2 billion to $3.7 billion by 2020–2039 under RCP 8.5, $6 billion to $12 billion by 2040–2059, and $18 billion to $27 billion by 2080–2099.

As discussed in chapter 11, the impact of these SLR-driven changes in storm damage is not evenly spread. Potential costs vary, not only between coastal and noncoastal states

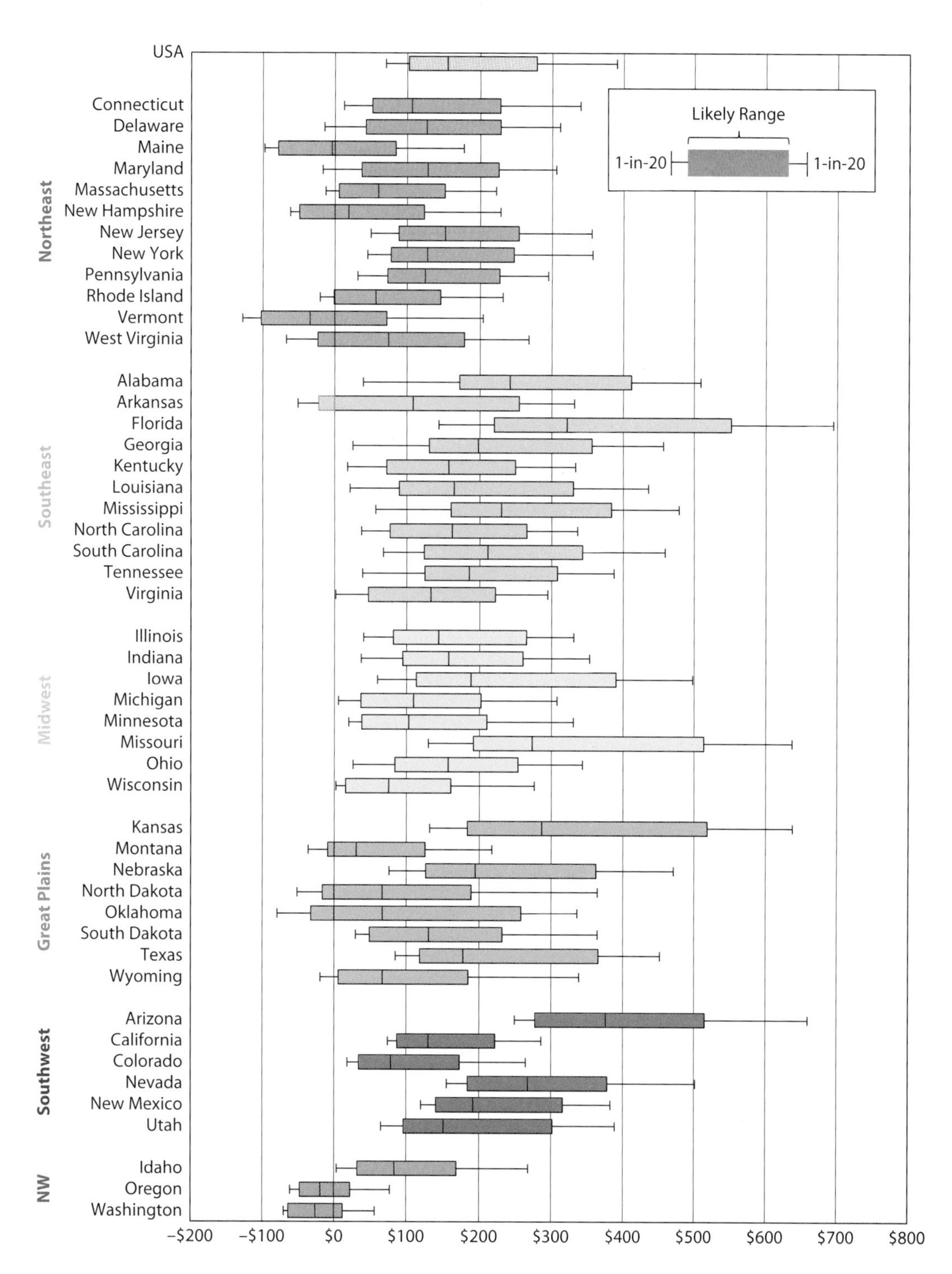

FIGURE 13.17. **State-Level per Capita Direct Costs from Changes in Energy Expenditures, 2080–2099**

RCP 8.5; 2011 U.S. dollars; negative values indicate net benefits

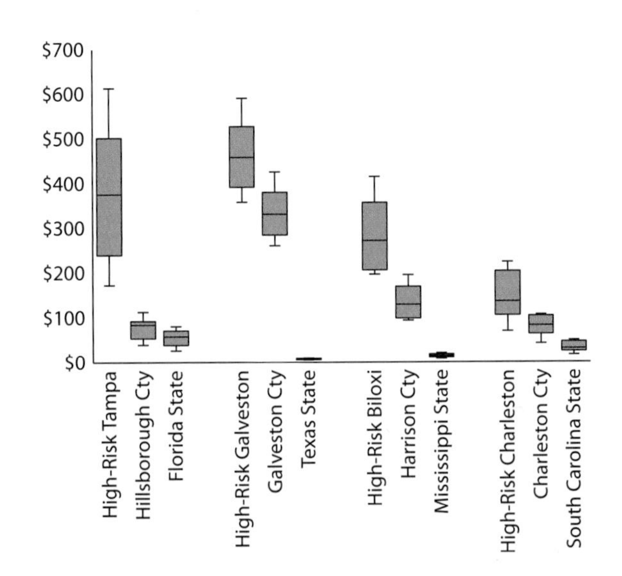

FIGURE 13.18. **Per Capita Increase in Average Annual Coastal Storm Damage due to SLR, 2030**

RCP 8.5, 2011 U.S. dollars

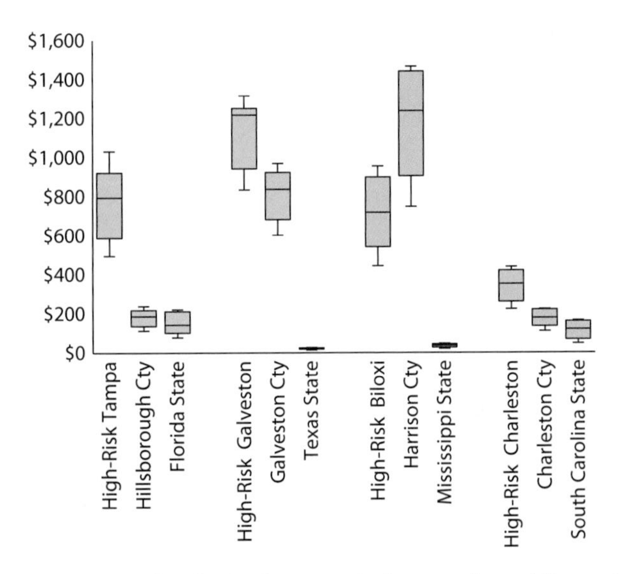

FIGURE 13.19. **Per Capita Increase in Average Annual Coastal Storm damage due to SLR, 2050**

RCP 8.5, 2011 U.S. dollars

but also within coastal states. The direct risk of SLR-driven changes in storm damage is concentrated in particularly vulnerable coastal communities over the next few decades, broadening to entire coastal states by midcentury.

Take, for example, a single-family wood home on the coast in Tampa, Florida. In the RMS exposure data set, such a home has an insurable value of $222,000, with an average annual hurricane loss of $5,005. By 2030, higher sea levels in Tampa will *likely* raise this structure's annual average loss by $627 to $1,310, growing to increases of $1,534 to $2,404 by 2050. Based on the average number of people per household in Tampa, that translates into a per-resident *likely* increase in average annual losses of $240 to $502 by 2030 and $588 to $921 by 2050. In figures 13.18 and 13.19, we compare that to the increase in per capita average annual loss in Hillsborough County (in which Tampa resides) as a whole and in the entire state of Florida. In Hillsborough County and Florida, per capita average annual losses *likely* grow by $54 to $93 and $39 to $72, respectively, by 2030 and $137 to $218 and $101 to $211, respectively, by 2050. We provide similar comparisons for Galveston, Texas, Biloxi, Mississippi, and Charleston, South Carolina.

In Galveston, a typical single-family coastal home worth $191,000 faces $4,752 in average annual hurricane losses today that *likely* grow by $1,035 to $1,392 by 2030 and $2,488 to $3,303 by 2050. In Biloxi, a typical single-family coastal home worth $194,000 faces $10,800 in average annual hurricane losses today that *likely* grow by $527 to $915 by 2030 and $1,384 to $2,299 by 2050. In Charleston, a typical single-family coastal home worth $180,221 faces $2,329 in average annual hurricane losses today that *likely* grow by $254 to $492 by 2030 and $629 to $1,016 by 2050. In figures 13.18 and 13.19, these increases are translated into per capita terms using the average number of people per household in those cities and compared to the *likely* per capita increase in losses for the counties and states that house those cities as a whole. (Note that for all these estimates, we assume home values remain unchanged. Over time, home values will appreciate, increasing damage in dollar terms. Incomes will also rise, however, so the relative impact of the damage on household budgets will be less than the absolute damage. As with all costs discussed in this chapter, we compare future climate impacts to current incomes, asset prices, and economic output.)

Of course, much of the ultimate cost for these concentrated damages will not be borne by the affected households themselves, but will be spread more broadly through private insurance to other policyholders and through state-backed and federal government–backed insurance and disaster relief to other taxpayers.

While specific communities within coastal states continue to bear considerably more risk from SLR-driven increases in storm damage than each state as a whole, over time statewide costs in certain parts of the country begin to mount, particularly when combined with *likely* inundation damage from mean SLR and rising mean higher high water (MHHW) levels (see chapter 11). By late in the century (2080–2099 average), the *likely* SLR-driven increase in average annual inundation (at the MHHW level) and storm loss for the country as a whole translates into $85 to $138 per American based on current population and property prices. For Florida, however, the increase is more than five times that amount, at $520 to $931 per person (figure 13.20). The *likely* per capita increase in damage in New Jersey is roughly three times the national average at $294 to $437 per year. For the average New Yorker, the direct annual costs of an SLR-driven increase in inundation and storm damage late in the century is $185 to $332 per capita.

In exploring the potential impact of modeled changes in hurricane activity, we only looked at changes in frequency and intensity of different storms, not changes in the geographic distribution of where hurricanes make landfall. Such changes may well occur but were beyond our ability to analyze for this report. If storm geography remains the same but storm intensity and frequency change as modeled by Emanuel (2013) for RCP 8.5, the *likely* increase in national average annual storm losses grows to $4 billion to $6 billion on average by 2020–2039, $13 billion to $23 billion by 2040–2059, and $51 billion to $74 billion by 2080–2099. Adding average annual inundation from MHHW rise, the *likely* increase in average annual losses by 2080–2099 grows to $59 billion to $89 billion, or $193 to $287 on a per capita basis nationwide. The *likely* increase in per capita losses in Florida grows to $1,530 to $2,280. In New Jersey and New York, the *likely* per capita increase in losses grows to $423 to $679 and $280 to $504, respectively (figure 13.21).

SUMMING UP

This assessment of direct costs and benefits helps illuminate which of the climate impacts quantified in this report matter most from an economic standpoint and for which parts of the country. At a national level, mortality (using the VSL), labor productivity, coastal damage, and energy demand are the most significant, both at midcentury (figure 13.22) and late century (figure 13.23). Notably, commodity agricultural impacts are relatively small at a national level, though they have received the most research attention in the academic community. Climate-driven changes in labor productivity, which have received comparatively scant research focus, likely pose a much more substantial nationwide economic risk.

At a state level, the relative importance of each impact varies greatly (figure 13.24). Coastal damage ranks relatively high in the Northeast and Southeast, while agricultural impacts play a considerably larger role in the Midwest and Great Plains than for the country as a whole. Labor productivity costs are meaningfully sized and relatively evenly spread, while mortality costs, which are large, are positive for some states and negative for others.

The combined *likely* direct cost of these six impacts in the late century, using changes in labor income estimates for mortality and historical hurricane activity, is 0.7 to 2.4 percent of GDP, with a 1-in-20 chance of costs less than 0.4 percent or more than 3.4 percent. The Southeast sees *likely* combined direct costs considerably higher than for the county as a whole. The *likely* range for Florida is 2.3 to 6.0 percent of economic output, and for Mississippi it is 1.8 to 5.7 percent (figure 13.25). A number of Midwest and Great Plains states also see *likely* combined impacts considerably higher than the national average, while most Northeast and Northwest states *likely* see relatively little direct costs.

Switching from changes in labor income to the VSL as the measure of mortality costs significantly increases the total. At a national level, the combined *likely* direct cost of our six quantified impacts is 1.2 to 5.4 percent of economic output late in the century (figure 13.26). In Florida and Mississippi, *likely* direct costs rise to 7.6 to 20.6 percent and 5.7 to 17.8 percent. The difference between impacts in the South and in the Northwest and Northeast grows using this measure, with net benefits in the median projections for eight Northwest and Northeast states.

At the upper bound of our estimates, using projected hurricane activity and the VSL for mortality costs, late-century *likely* combined direct costs at a national level are 1.4 to 5.7 percent of economic output (figure 13.27). For Florida, the most at-risk state, combined *likely* direct costs rise to 10.1 to 24 percent of state economic output, while Vermont sees combined *likely* direct benefits of 0.8 to 4.5 percent of total output.

Finally, note that because we are taking an enumerative approach and many known impacts are not quantified, the numbers presented in this chapter should not be viewed as a comprehensive portrait of all economic costs and benefits (Pindyck 2013; Stern 2013). Many of these unquantified impacts are discussed in part 4 of this book, and some of the methods for evaluating the costs of "deep" structural uncertainties in climate impacts (Weitzman 2009; Heal & Millner 2014) are discussed in chapter 15.

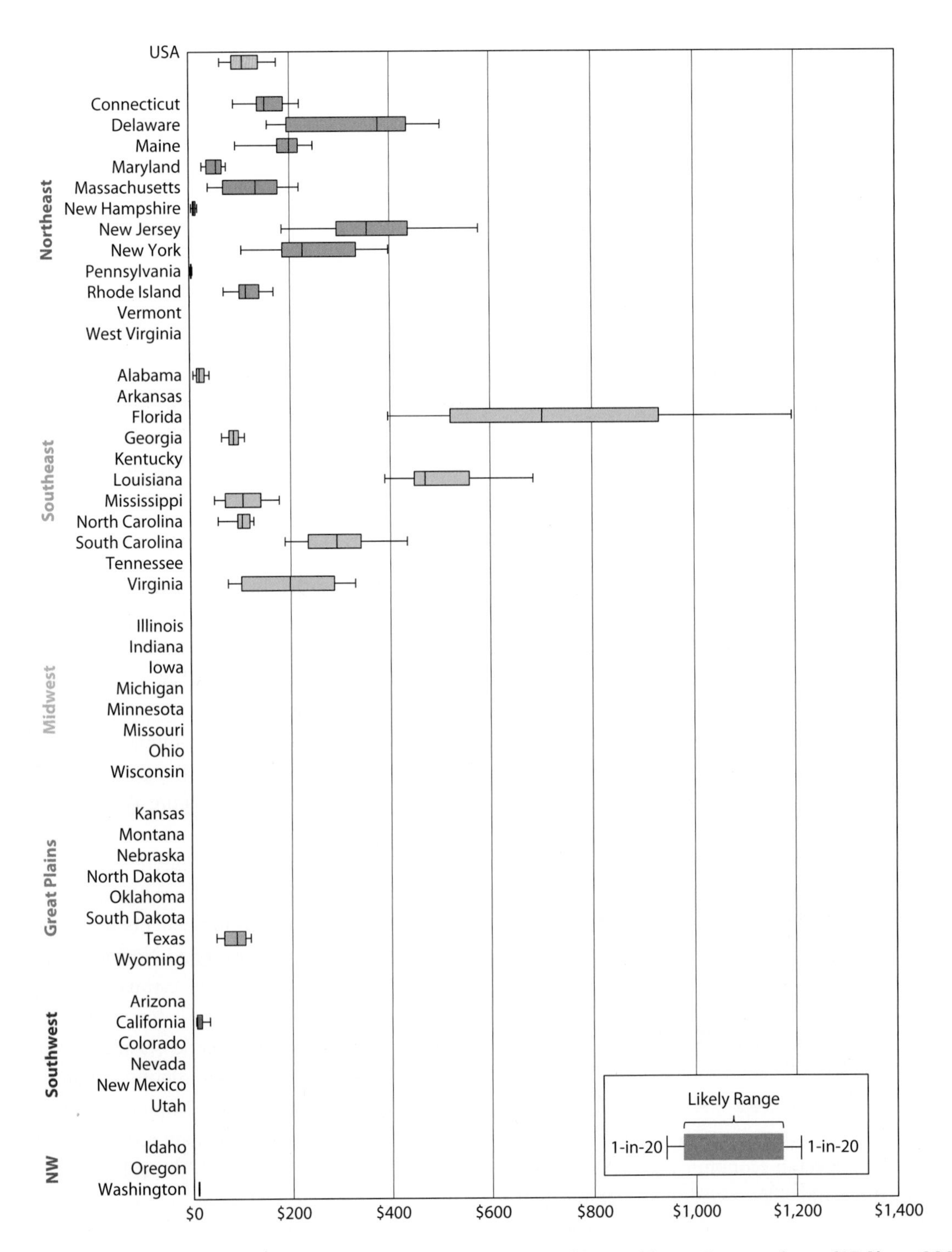

FIGURE 13.20. **Per Capita Inundation Damage and Increase in Average Annual Coastal Storm Damage due to SLR Alone, 2080–2099**

RCP 8.5; 2011 U.S. dollars

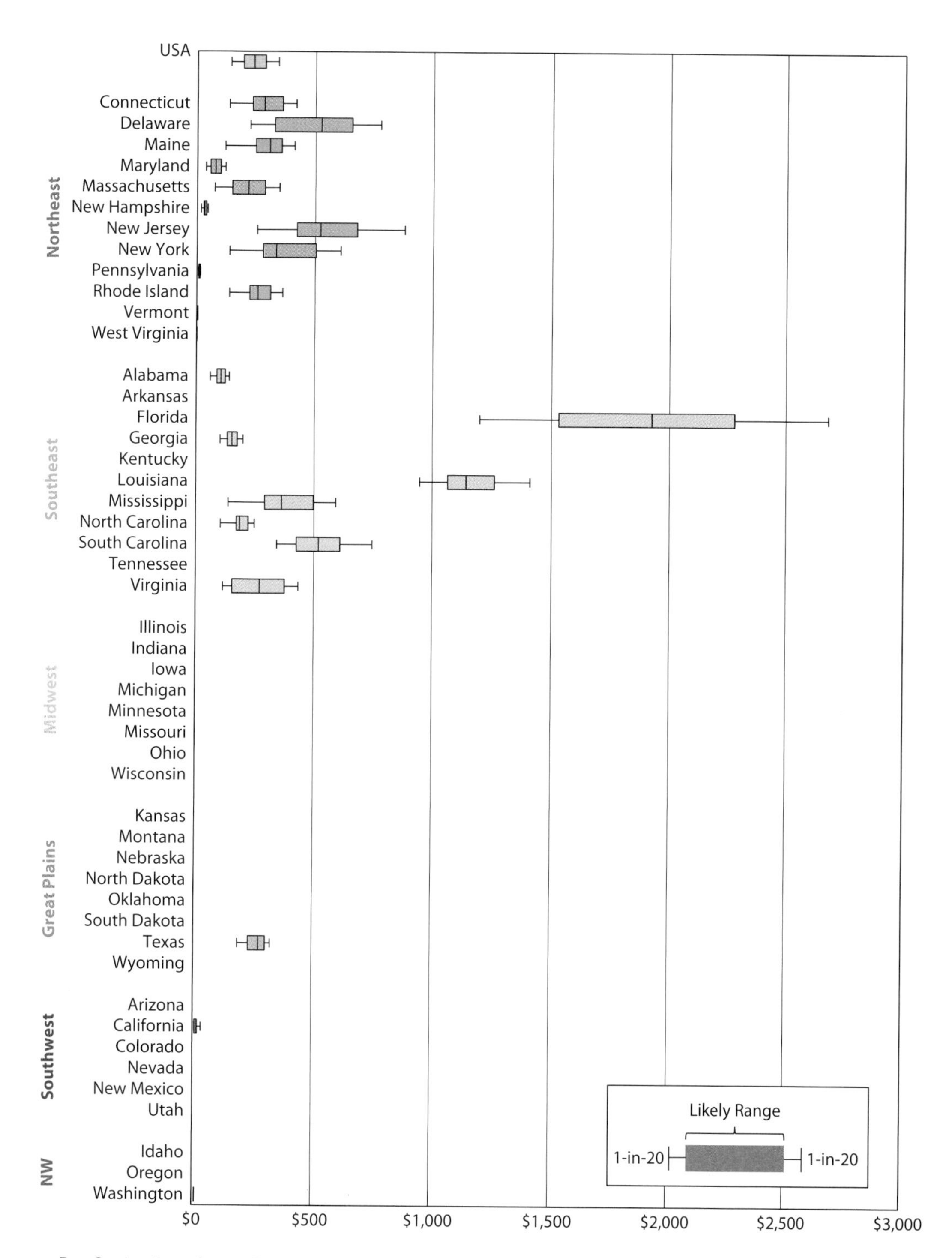

FIGURE 13.21. **Per Capita Inundation Damage and Increase in Average Annual Coastal Storm Damage due to SLR and Potential Hurricane Activity Changes, 2080–2099**

RCP 8.5; 2011 U.S. dollars; hurricane activity projections from Emanuel (2013)

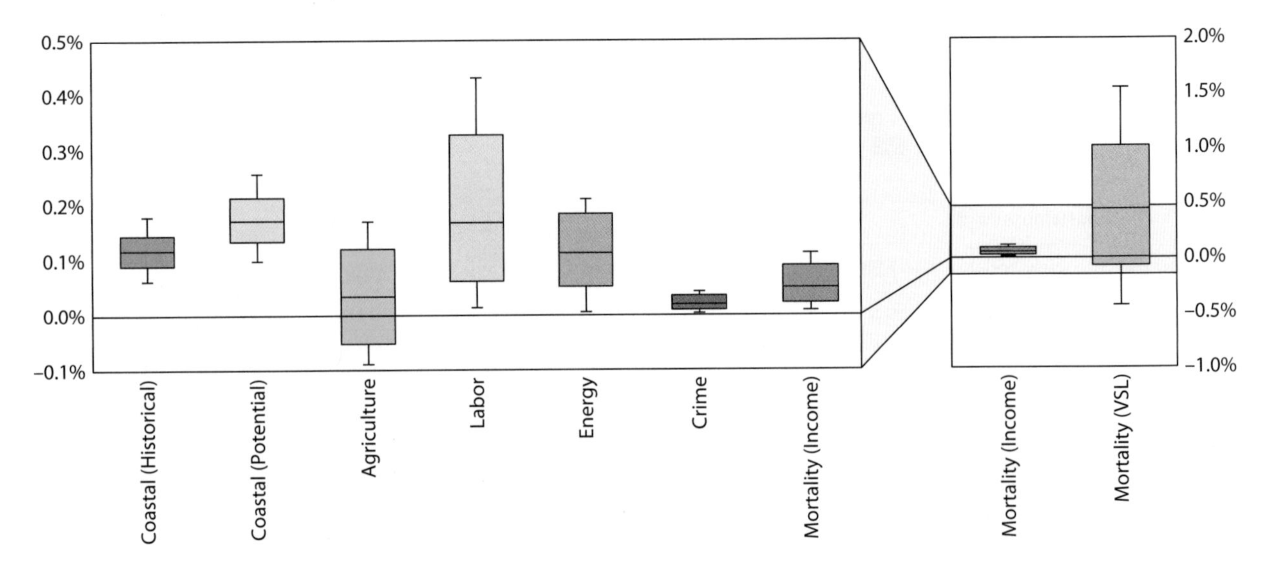

FIGURE 13.22. **Direct Costs and Benefits as a Share of GDP, 2040–2059**

RCP 8.5

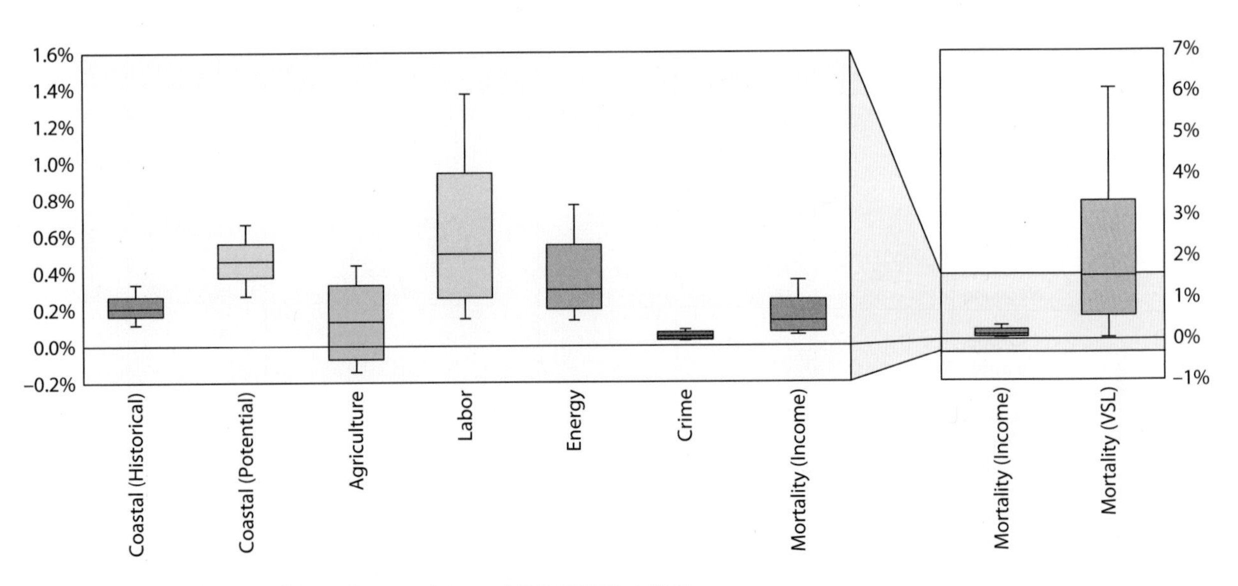

FIGURE 13.23. **Direct Costs and Benefits as a Share of GDP, 2080–2099**

RCP 8.5

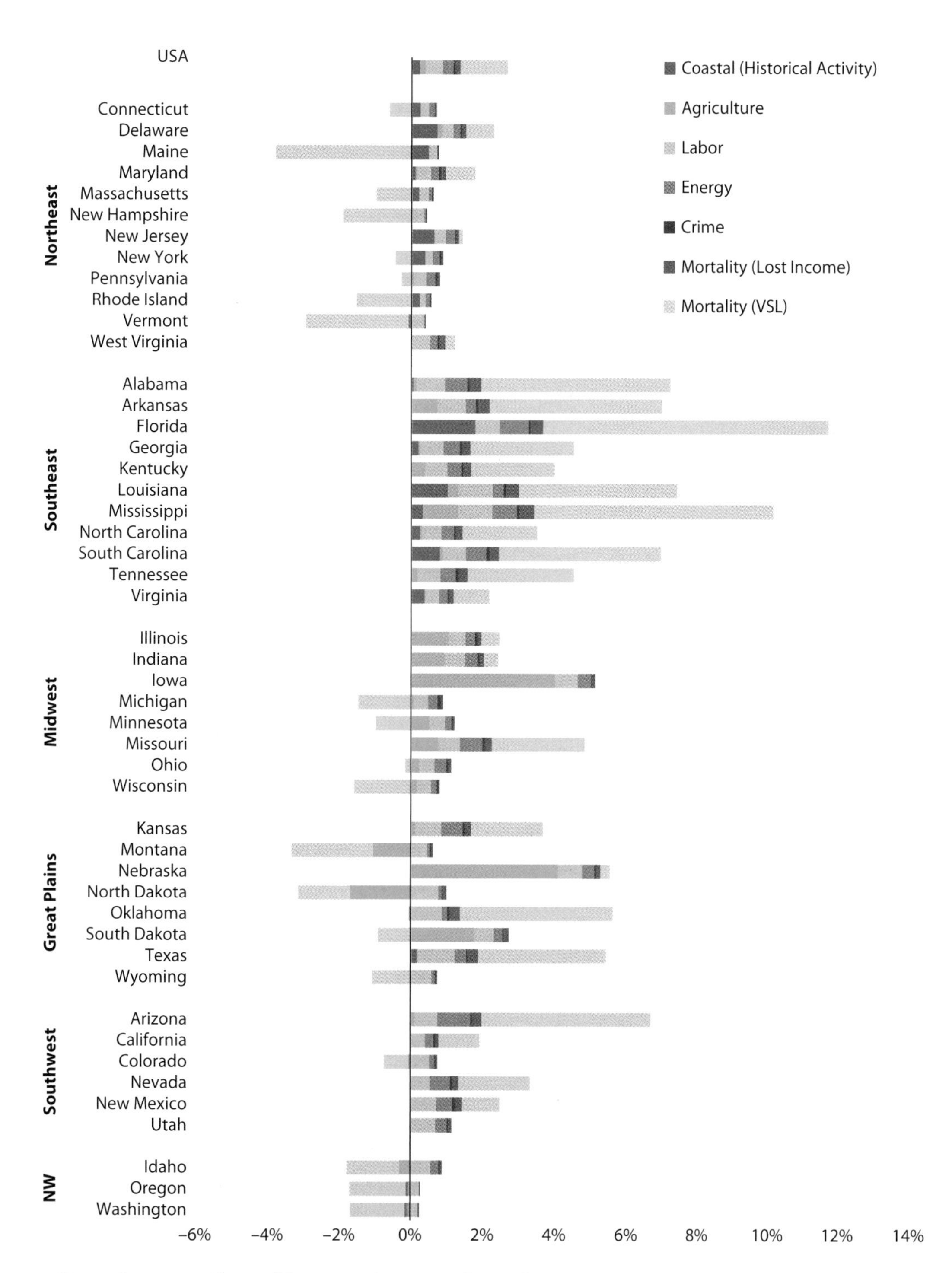

FIGURE 13.24. **Direct Costs as a Share of Economic Output at the Median, 2080–2099**

RCP 8.5

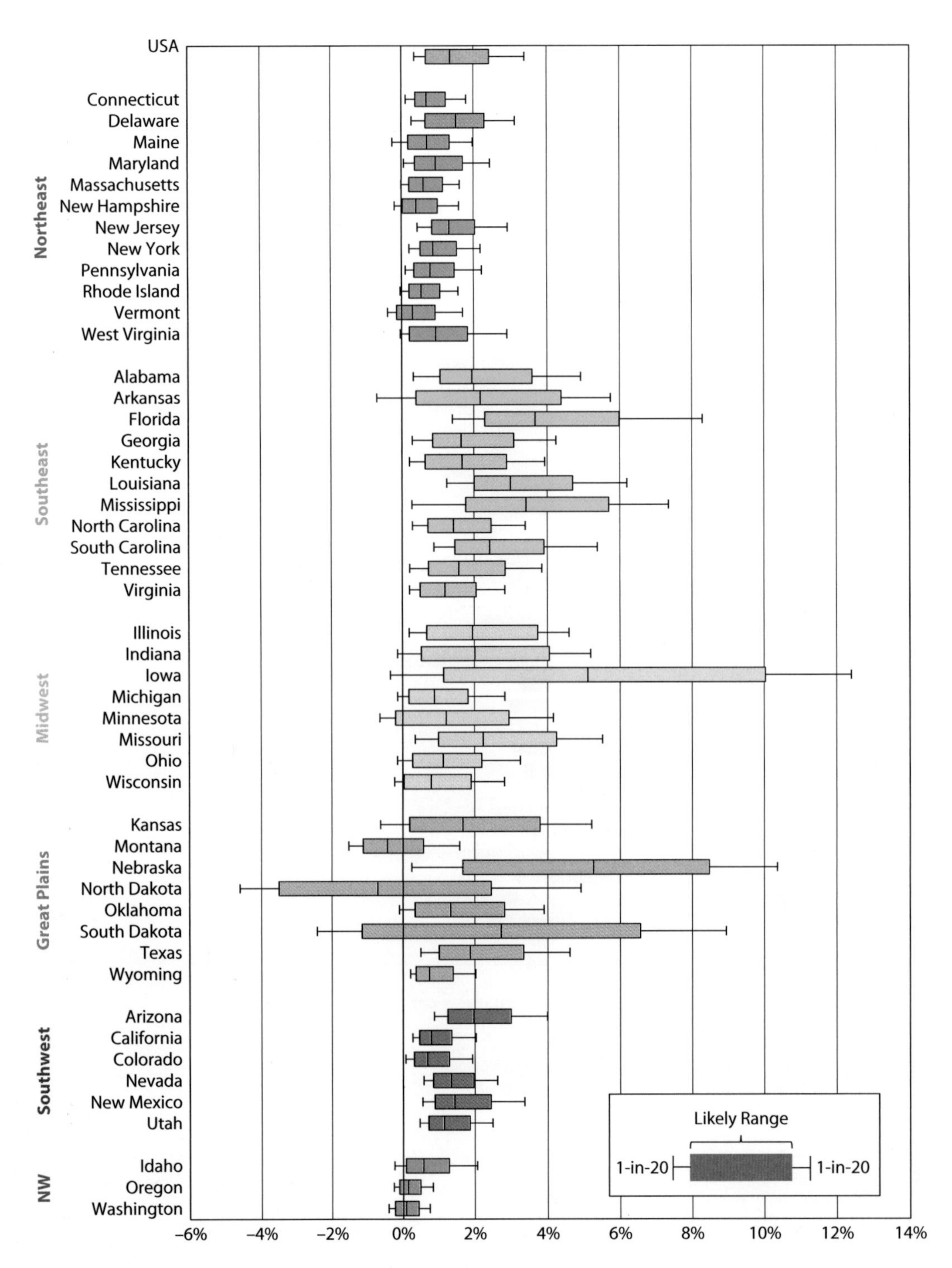

FIGURE 13.25. **Six Quantified Impacts with Historical Hurricane Activity and Market Mortality Cost, 2080–2099**

RCP 8.5; percent of economic output; negative values indicate net benefits

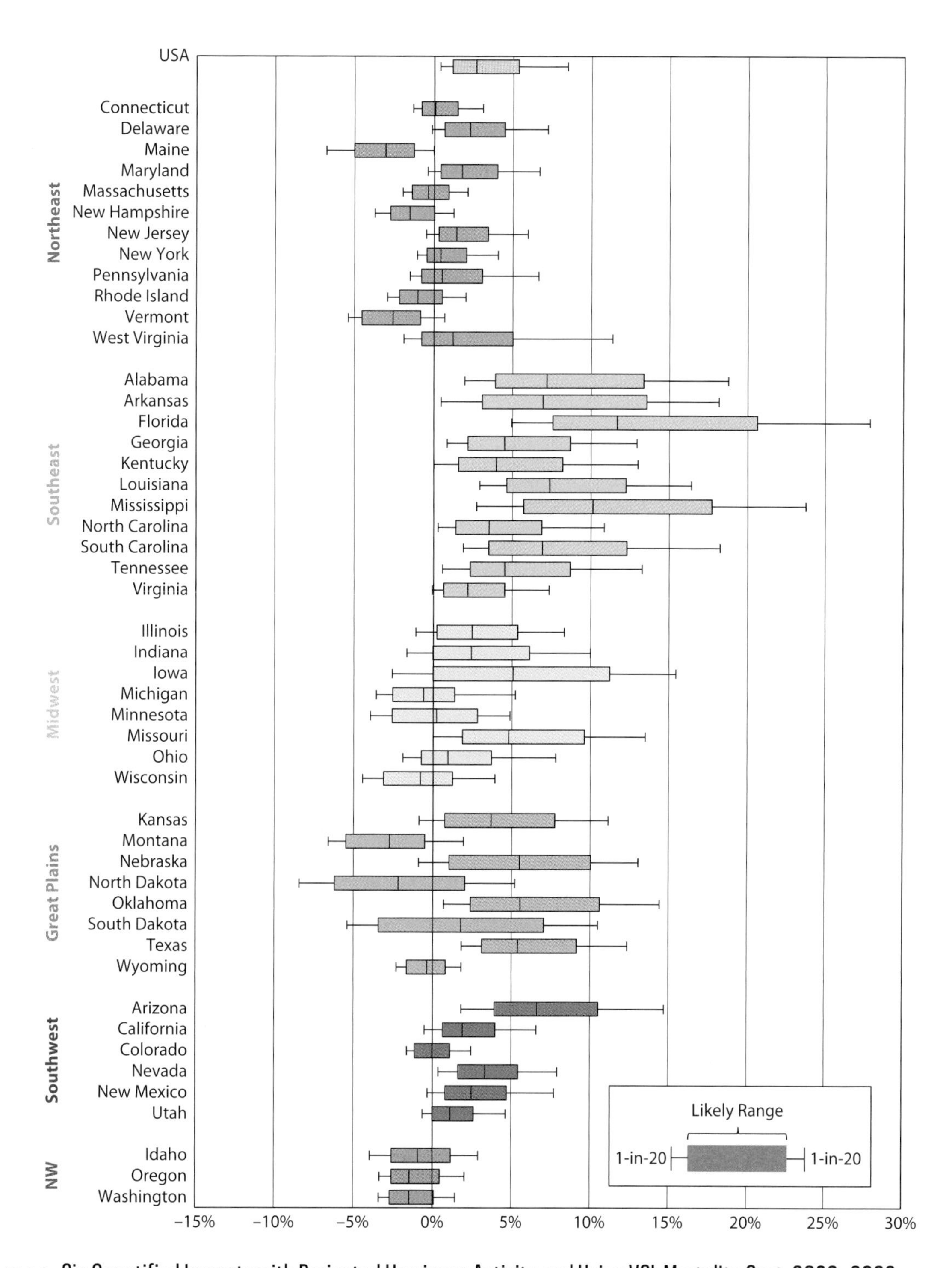

FIGURE 13.26. **Six Quantified Impacts with Projected Hurricane Activity and Using VSL Mortality Cost, 2080–2099**

RCP 8.5; percent of economic output; negative values indicate net benefits

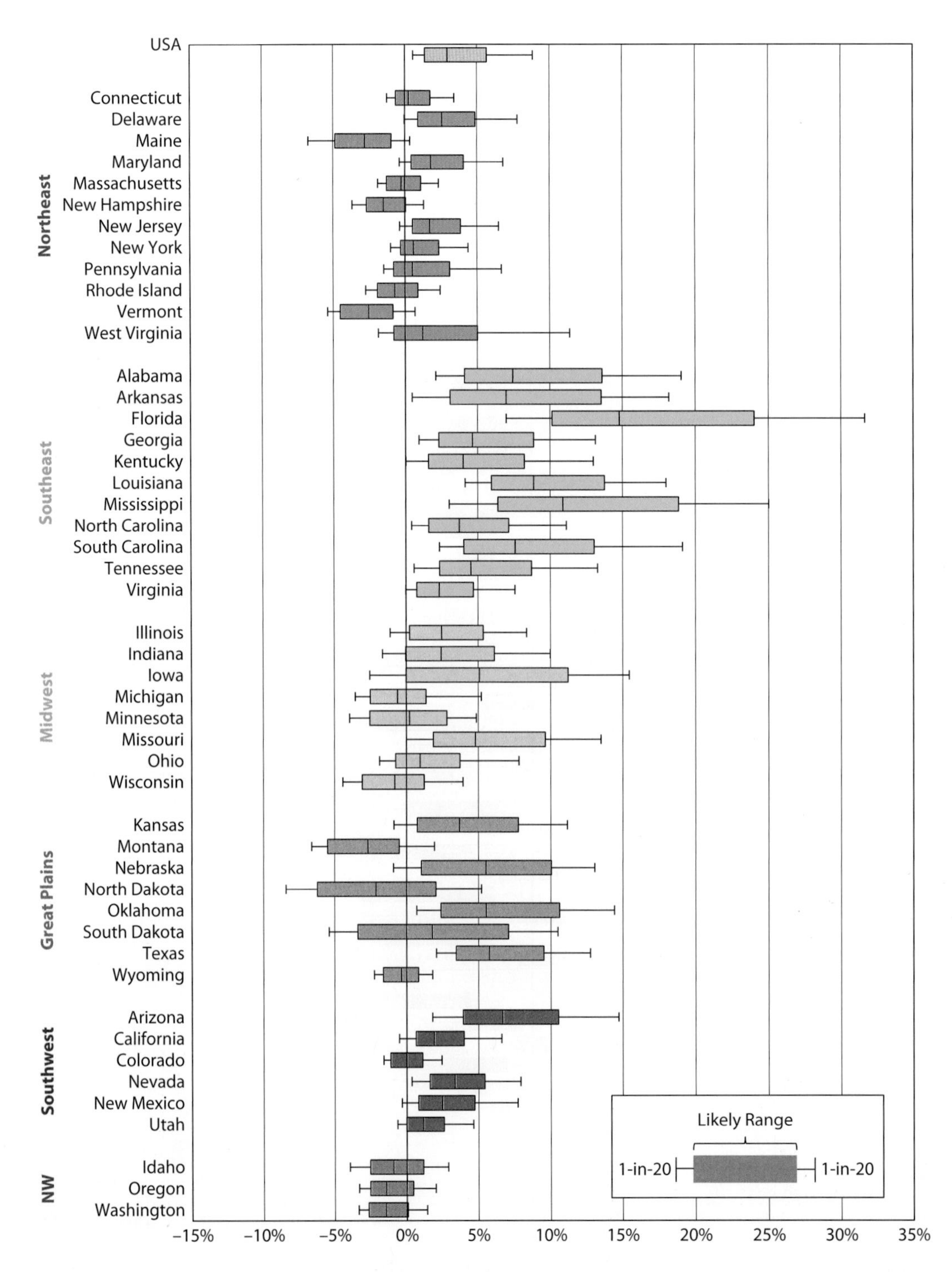

FIGURE 13.27. **Six Quantified Impacts with Projected Hurricane Activity and Using VSL Mortality Costs, 2080–2099**

RCP 8.5; percent of economic output; negative values indicate net benefits

CHAPTER 14

MACROECONOMIC EFFECTS

IN the preceding chapter, we assessed the direct costs and benefits of the climate impacts covered in this report and found significant variation by state. U.S. states are all part of the same national economy, however, and direct impacts in one sector or region have implications for other sectors and regions as well. For example, a decrease in agricultural output in Iowa affects food prices nationwide (and globally). Damage to coastal property raises borrowing costs in noncoastal regions because of the national (and global) nature of capital markets. Higher energy costs flow through the economy and can increase the price of a wide range of goods, and changes in labor productivity alter what people do for a living and where they work. This chapter examines the extent to which some of these macroeconomic effects shape the overall magnitude and the regional variation in the direct costs and benefits of the climate impacts quantified in this report.

METHODOLOGY

To illustrate some of these concepts, we use RHG-MUSE, a dynamic recursive computable general equilibrium (CGE) model of the U.S. economy developed and maintained by the Rhodium Group and integrated into the SEAGLAS platform. RHG-MUSE represents sectoral and regional relationships as they exist in the economy today, based on a framework developed by Rausch and Rutherford (2008) and similar to CGE models used in other climate-change assessments, such as those of Jorgenson et al. (2004) and Abler et al. (2009). RHG-MUSE is solved annually from 2012 to 2100 and simulates the growth of the U.S. economy through changes in labor, capital, and productivity. The model is calibrated using state-level social accounting matrices (SAMs) from the Minnesota IMPLAN

Group (more information on IMPLAN is available at www.implan.com). For computational simplicity, we aggregated the 440 sectors in the IMPLAN SAMs to create nine sectors tailored to the impacts covered in this report. A full description of RHG-MUSE is available in appendix D.

Our hesitation in predicting how the U.S. economy will evolve between now and 2100, expressed in the preceding chapter, still holds. Yet as some of the macroeconomic consequences of the direct climate impacts unfold over time, projections are necessary. We calibrate RHG-MUSE to maintain the country's current economic structure and demographic profile in the baseline scenario throughout the modeling time frame. Both population and economic output grow, but the sectoral and geographic shares of both employment and output remain roughly the same. This allows for a relatively apples-to-apples comparison with the direct economic impacts described in chapter 13 and maximizes consistency with our empirically based impact estimates.

We represent the climate impacts covered in this report in RHG-MUSE as follows (with a detailed discussion available in appendix D).

Agriculture

We represent climate-driven changes in agricultural productivity as a percent change in total output productivity affecting the baseline productivity in that year. This means that for a given level of capital, labor, and intermediate goods use, a state's agricultural output will be equal to the baseline output given the same level of inputs multiplied by the productivity change. The model propagates this change across the economy through price and quantity effects.

Labor Productivity

We represent climate-driven changes in labor productivity through the productivity of labor inputs to high-risk and low-risk sectors by state. New production activities are able to respond to this change and substitute labor for capital, or vice versa, but extant production suffers a proportional reduction in output and a loss of productive capital.

Health

Unlike in the preceding chapter, we represent climate-driven changes in mortality by tracking their impact (by age cohort) on the size and composition of the U.S. population using a population model incorporated into RHG-MUSE and reducing available labor supply accordingly.

Crime

We exclude crime from the CGE model because of the mixture of market and nonmarket factors in our direct economic impact estimates.

Energy

We increase residential and commercial energy costs by state based on the energy expenditure estimates laid out in chapter 10.

Coastal Communities

We represent climate-driven inundation, flood, and wind damage to coastal property as a direct reduction of capital stock. This is implemented before each run of the model's static core, such that some fraction of the depreciated capital stock will be unavailable for earnings and use in the coming period. Because rates of return must equalize across states and sectors in RHG-MUSE, new capital formed by savings/investment replaces lost capital stock until rates are equal. We also capture business interruption by reducing industrial productivity in affected states consistent with RMS business-interruption estimates for that particular sea-level rise (SLR) scenario.

RESULTS

The macroeconomic dynamics in RHG-MUSE alter the direct costs and benefits of the climate-driven changes in agricultural production, labor productivity, mortality, energy costs, and coastal property in several ways.

The ability of firms to substitute factors of production in response to changes in prices, capital stock, or labor supply reduce direct costs. For example, the impact of climate-driven reductions in labor productivity on economic output is decreased over time through increased application of capital. Likewise, SLR-driven damage to coastal capital stock is offset in part through greater application of labor. When changes in prices as a result of climate impacts reduce demand for goods from a sector, labor and capital are freed up for other sectors, offsetting the direct costs.

There are other macroeconomic effects in the model that amplify direct costs. Most important in our analysis is the impact of damaged coastal capital stock in a given year on economic growth in subsequent years. The need to rebuild damaged coastal property redirects investment that would have otherwise occurred elsewhere in the economy, reducing economic output in the process. We find that over the course of decades, the cumulative impact on growth of single-year coastal capital stock damage is several times larger than the direct cost to the coastal property receiving the damage.

On net, the economy-wide cost estimates from RHG-MUSE are slightly higher than the direct costs presented in chapter 13. Under RCP 8.5, the likely late-century combined direct cost for climate-driven changes in coastal damage (assuming historical activity rates), labor productivity, energy demand, mortality (using market estimates), and agricultural production are 0.7 to 2.4 percent of GDP nationwide (figure 14.1). The *likely* range from RHG-MUSE is 1.0 to 3.0 percent. We may be underestimating costs in RHG-MUSE because of the different treatment of mortality. In calculating direct climate-driven mortality costs and benefits in a given period, we estimate the net present value (at a 3 percent discount rate) of lifetime earnings lost by deaths in that period. In RHG-MUSE, the late-century mortality costs are the cumulative impact of all climate-driven mortality occurring up until that point. Thus, the late-century estimates of direct mortality costs include lifetime earnings lost after 2100, while the RHG-MUSE estimates do not. Especially as net national climate-driven mortality increases sharply in the second half of the century, this difference in approach leads to higher mortality-related estimates in the direct cost approach.

For some impacts, the macroeconomic effects captured in RHG-MUSE reduce regional variation in costs, due primarily to free movement of capital and

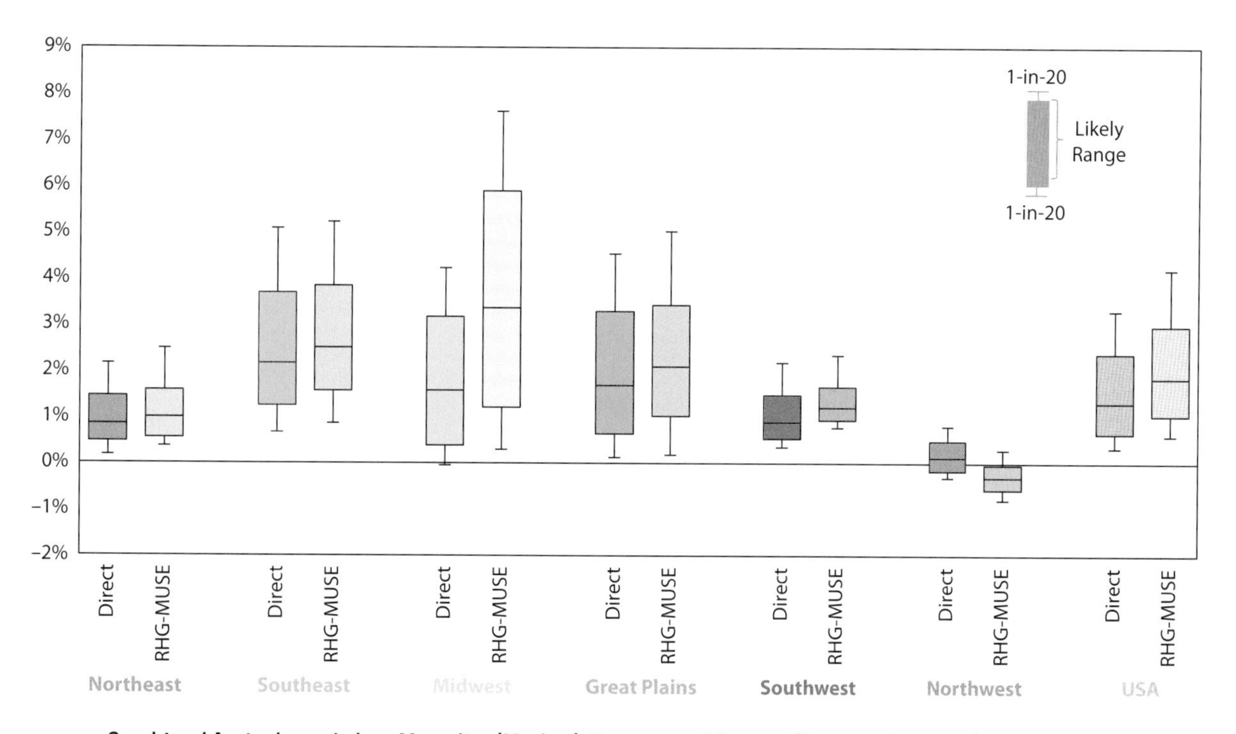

FIGURE 14.1. **Combined Agriculture, Labor, Mortality (Market), Energy, and Coastal (Historical Activity) Impact**

Calculated as direct costs and modeled using RHG-MUSE under RCP 8.5 in 2080–2099; percent of output; negative values indicate net benefits

goods across state borders in the model (agriculture is a notable exception, as discussed later). For example, the increase in investment demand in coastal states as a result of SLR-driven damage draws investment away from other states equally. This is likely an optimistic assumption; the price of investment goods like cement and steel will probably rise more in storm-damaged areas than the national average, even if borrowing costs rise equally nationwide.

At the same time, however, we assume that labor is fixed by state, which prevents additional regional smoothing from occurring. As a result, the relative return on labor rises in more heavily affected states, which could attract labor from other parts of the country. In contrast, if the climatic changes causing that damage also reduce relative livability, the state-level reduction in labor supply could be even greater than the direct labor productivity and mortality costs suggest.

The Southeast and Great Plains fare better once the macroeconomic effects captured in RHG-MUSE are included, while the Midwest does considerably worse. This is due to the indirect economic impacts in the region of a climate-driven decline in commodity agricultural production. Notably, the Pacific Northwest sees even greater net benefits in RHG-MUSE, despite being part of the same national economy, due largely to the long-run growth benefits of increased labor supply from a decline in cold-related deaths.

This macroeconomic modeling exercise should serve primarily as a conceptual exercise. Predicting how markets will respond to climate impacts over the course of eight and one-half decades is extremely challenging. In addition, national macroeconomic conditions are heavily shaped by events around the world, and other countries will also be affected by climate change. International climate impacts in turn comprise a macroeconomic impact to the U.S. economy—an effect not assessed in this report. Nonetheless, this illustration shows that the macroeconomic dynamics we do capture modestly decrease the combined national cost of the five modeled impacts and that, while they reduce regional inequality somewhat, significant differences remain.

There are a number of research groups actively working to build more sophisticated economic models that can capture a broader range of national and international dynamics, such as the Project on Integrated Assessment Model Development, Diagnostics and Inter-Model Comparisons (PIAMDDI). In addition to continuing to improve the SEAGLAS platform in the months and years ahead, we have designed the analysis to be modular and open source in the hopes that other researchers can integrate those components they find useful into their own work and build upon the findings of this prospectus.

CHAPTER 15

VALUING RISK AND INEQUALITY OF DAMAGES

Two central contributions of this report are to characterize the uncertainty associated with the economic impacts of climate change and to estimate the extent to which these impacts will be borne unequally among Americans. In both cases, we note that average impacts gloss over an important aspect of the story: If the climate changes, there is a sizable chance that different types of impacts will be larger or smaller than the central estimate, and in many cases specific regions of the country experience impacts that differ substantially from the national average. While the primary purpose of this report is to provide empirically based, spatially explicit information about the risks businesses, investors, and households in different parts of the United States face from climate change, these insights are also important in how we price climate risk at a national level.

Both risk and inequality can increase the perceived costs of climate change above the expected cost of climate change; that is, the average impact we expect to see across possible futures and across regions of the United States. Uncertain outcomes and unequal impacts increase our perception (or valuation) of these costs because as individuals and as a society, we generally dislike uncertainty in our futures (e.g., individuals buy home insurance in part because the risk that a catastrophe could bankrupt a family is worrisome) and we dislike strong social inequalities (e.g., individuals donate money to charity in part to alleviate the hardships of poorer individuals). *How much* we dislike uncertainty and inequality affects how much these factors should influence our decision-making process, and they inform us of how much we should focus on future uncertainty or inequality in climate-change impacts relative to the average impact of climate change. In economics, the extent to which we are concerned about risk and inequality can be described by two factors:

Risk aversion: How averse are we to the uncertain possibility of bad future outcomes?

Inequality aversion: How much do we dislike having some individuals suffer greater losses than others, especially if proportionately greater losses fall on poorer individuals?

Both of these types of aversion reflect our underlying preferences and can thus be measured empirically, although it is possible that a decision maker may be more risk averse or inequality averse than one would estimate by observing individuals in a population. This might be true, for example, if increasing inequality has indirect effects on the economy or a population's social well-being that are not understood or internalized by individuals within a population; it might also be true because the preferences of individuals acting collectively through democratic processes may differ from those of individuals acting individually in a market. It is worth noting that in many assessments of climate-change impacts, risk aversion and inequality aversion are assumed to be the same, although recent work suggests that the two need not, and very likely should not, be treated that way (Fehr & Schmidt 1999; Engelmann & Strobel 2004; Bellemare, Kröger, & Soest 2008; Crost & Traeger 2014).

Here we use our new results describing the probability distribution of impact across states within the United States to illustrate through example how decision makers could adjust their valuations of the damages from climate change to account for aversion to risk and inequality (see appendix E for mathematical details). In both cases, we summarize the additional costs imposed by risk and inequality as a *premium*, which is the additional cost that we would be willing to bear to avoid the inherent risk and additional inequality imposed by climate-change impacts. We assume the well-being of all Americans should be treated equally and consider how the value of mortality (using the VSL) and direct agricultural losses could be adjusted to account for the structure of their risk and their unequal impact, in large part because these two example sectors have nonlinear response functions that generate some of the largest variations in damage.

RISK AVERSION

Accounting for risk aversion stems from the observation that individuals and society at large dislike uncertainty in future costs. For example, suppose Anna has a salary of $40,000 this year. Further suppose Anna knows that if she stays at her current job, there is a 95 percent chance that she will get a 10 percent raise (a gain of $4,000) and a 1 percent chance that she will be fired (a loss of $40,000). The expected value of staying at her current job is therefore $41,800 (the sum of 95 percent times $44,000 and 5 percent times $0). Further suppose she has the opportunity to switch to a new job that also pays $40,000 but guarantees her employment next year (with no raise). If she is risk-neutral, then her current job is worth $1,800 more to her than the alternative; if she is risk-averse, she might nonetheless opt for the more certain alternative because she wished to avoid the possibility of being fired.

Following conventional practice, we measure risk aversion with a coefficient of relative risk aversion (RRA). An RRA of zero reflects risk neutrality; higher values reflect higher levels of risk aversion. Studies of the relative rates of return of safe investments (such as U.S. treasury bonds) and risky investments (such as stocks) suggest that the RRA reflected in U.S. financial investments is between 2.5 and 6, although it could be as low as 1 or as high as 12 (Ding et al. 2012). Another study of investments, aimed at separating risk aversion from preferences between current and future consumption, suggests an RRA of 9.5 (Vissing-Jørgensen and Attanasio 2003). Experimental results from a survey of individuals in the United States, the United Kingdom, Canada, Australia, and Mexico similarly suggest that the central third of individuals surveyed have values between 3 and 5, although one third of individuals surveyed have values less than 3 (half of these individuals are between 1.5 and 2.0), and one fifth of individuals surveyed have values greater than 7.5 (Atkinson et al. 2009).

We can use the RRA to turn the projected probability distributions of losses in each state in each period into *certainty-equivalent* losses per capita in that period; in other words, we find the losses that an individual would bear with certainty that have the same welfare impact as the distribution of losses characterized in this report (see appendix E). The *risk premium* is the difference between the certainty-equivalent loss and the expected actual loss: It is the hypothetical quantity one would be willing to pay just to avoid the uncertainty in climate effects (Kousky, Kopp, & Cooke 2011).

Because the production of commodity crops (maize, wheat, soy, and cotton) represents a small fraction (about 0.8 percent) of total economic output, it can have only a small effect on total income: In the 1-in-20 worst case for RCP 8.5 in 2080–2099 in the hardest-hit state, Iowa, agricultural losses constitute a 9 percent loss of overall output. The risk premium is therefore small relatively, ranging up to 7 percent of the value of the lost output for a high RRA of 10 (see the first row of table 15.1, where inequality aversion [IA] = 0). By contrast, because the mortality effects can be quite large—the 1-in-20 chance loss for RCP 8.5 in 2080–2099 is equivalent to 20 percent of output in Florida if measured using VSL—the risk premium can be significant. Even in the absence of

TABLE 15.1 Combined inequality-risk premiums for agricultural impacts, 2080–2099

		RRA					
		← *Low Risk Aversion*			*High Risk Aversion* →		
		0	2	4	6	8	10
IA: High Inequality Tolerance →	0	0%	1%	3%	4%	6%	7%
	2	7%	8%	10%	11%	13%	14%
	4	9%	11%	12%	13%	15%	16%
IA: ← Low Inequality Tolerance	6	10%	11%	13%	14%	15%	16%
	8	13%	14%	15%	16%	17%	18%
	10	20%	21%	22%	23%	24%	25%

Note: RCP 8.5; premium as a percentage of expected losses for maize, wheat, cotton, and soy output

inequality aversion, strong risk aversion can add as much as 16 percent to the value of the mortality losses (see the first row of table 15.2, where IA = 0).

INEQUALITY AVERSION

Accounting for inequality aversion is important because most individuals dislike the notion that some individuals bear far more of a group's cost than other members of a group. For example, in team efforts, most individuals usually find themselves unhappy if some members of the team do not "pull their weight," thereby forcing others to do additional work to make up for this shortfall. In a more extreme example, if a foreign country were to invade a single U.S. state, Americans throughout the rest of the country would not simply stand by and let the population of that one state fend for itself; rather, the whole country would come to the aid of the invaded state. There are many similar cases, such as natural disasters, where the nation spends both effort and money to protect and support small groups of individuals because we do not believe those individuals should be left to suffer tremendous costs alone—instead, the country expends additional resources to share these burdens.

TABLE 15.2 Combined inequality-risk premiums for mortality impacts, 2080–2099

		RRA					
		← *Low Risk Aversion*			*High Risk Aversion* →		
		0	*2*	*4*	*6*	*8*	*10*
IA: High Inequality Tolerance →	0	0%	3%	6%	9%	13%	16%
	2	33%	37%	41%	46%	51%	56%
	4	72%	78%	84%	90%	97%	104%
IA: ← Low Inequality Tolerance	6	114%	121%	129%	138%	147%	157%
	8	155%	164%	173%	184%	195%	207%
	10	190%	200%	211%	223%	236%	250%

Note: RCP 8.5; premium as a percentage of expected losses, applying value of a statistical life

The degree of inequality aversion can be measured with a coefficient of inequality aversion (IA), analogous to the RRA. An IA of zero reflects inequality neutrality, implying there is no cost to inequality, while higher values reflect increasing levels of inequality aversion. Experimental results suggest that different individuals have a very broad range of IA values, with the central third of individuals having values between 2.0 and 7.5, one quarter having values between 0.5 and 1.5, and nearly a third having values greater than 7.5 (Atkinson et al. 2009).

In any given period, we can use the IA to turn the projected distribution of losses into an equivalent national, *inequality-neutral* loss (Gollier 2013). The inequality-neutral loss is a hypothetical economic loss that has the same welfare impact as the actual loss but, as a fraction of income, is shared evenly by all Americans. In other words, we find the level of loss that, if equalized across states, would yield the same welfare as the unequal cross-state distribution of output per capita (see appendix E). The *inequality premium* is the difference between the inequality-neutral loss and the expected actual loss; it is the hypothetical quantity one would be willing to pay just to avoid the additional inequality imposed by climate change impacts. Inequality-neutral losses will be larger if individuals who are initially poorer are harmed relatively more by climate change; they may be smaller if individuals who are initially richer are harmed relatively more.

Unlike the risk premium, it is possible for the inequality premium to be negative if climate change reduces initial wealth disparities, which can happen if climate change imposes sufficiently larger damage on wealthy populations than on poorer populations. In this case, the unequal distribution of climate impacts would lower the perceived cost of those impacts relative to their expected value. Thus, it is not obvious *ex ante* that accounting for inequality aversion will necessarily increase the perceived cost of climate change.

We note that we do not resolve differences in damage across counties within a state—accounting for such differences would likely increase the inequality premium—although cross-state impacts tend to be more unequal than cross-county impacts within each state. We also do not account for the distributional effects of climate change within a state by income or demographic group, which are likely more important than differences across counties.

For both agriculture and mortality, the inequality premium can be significant. Strong inequality aversion alone can increase the value of agriculture losses by up to 20 percent (see the first column of table 15.1, where RRA = 0), although the macroeconomic effects described in the preceding chapter dampen the inequality of direct agricultural impacts to some extent.

Strong inequality aversion can more than double the value of mortality losses, adding a 190 percent premium for an IA of 10 (see the first column of table 15.1, where RRA = 0). The large magnitude of the inequality premium for mortality arises because the mortality increase is highest in some of the nation's poorest states and least in some of the richest. Among the poorest ten states, the per capita median mortality increase is 28 deaths per 100,000 people under RCP 8.5 in 2080–2099 (with additional deaths among the poorest states exceeding 30 per 100,000 in Florida and Mississippi); among the ten richest states, the average is a decrease in deaths of 3 per 100,000 people (with the reduction in deaths among the richest states exceeding 10 per 100,000 in Alaska, North Dakota, and Washington).

PUTTING IT TOGETHER

Thus far in this chapter, we separately analyzed the risk and inequality premiums for two types of impacts. However, we can combine both risk aversion and inequality aversion to compute an *inequality-neutral, certainty-equivalent damage* (see appendix E). This value is the hypothetical cost that, if shared equally among all individuals with certainty, would have the same welfare impact as the actual unequal distribution of state-specific risks. The combined *inequality-risk premium* is the difference between this hypothetical cost and expected damage; it is the cost of having unequal economic risks imposed by climate change. Calculating this premium helps us conceptualize how inequality in expected losses, inequality in the uncertainty of losses, and inequality in baseline income together increase the perceived value of climate-change damage.

For both agricultural losses and mortality, combining risk aversion with inequality aversion yields a higher inequality-risk premium. The magnitude of the increment from the combination partially reflects the magnitude of the individual effects. For agriculture, the increased premium from layering risk aversion (which is small in this impact category) on top of inequality aversion is small;

for strong risk and inequality aversion, it amounts to a 25 percent increase in the value of losses compared to 20 percent from strong inequality and no risk aversion. For mortality, in contrast, strong risk and inequality aversion combined can add a premium of 250 percent compared to 190 percent for inequality aversion in the absence of risk aversion. If we focus on values most frequently observed in experiments (RRA and IA of roughly 4 [Atkinson et al. 2009]), which for RRA also coincides with the central Ding et al. (2012) estimate based on comparison of the prices of U.S. stocks and bonds, then the inequality-risk premium on agricultural loss and mortality are 13 and 90 percent of the expected loss. If instead we use an RRA of 10, close to the Vissing-Jørgensen & Attanasio (2003) estimate of 9.5, the mortality premium rises to 104 percent of the expected loss.

Overall, for the example impact categories we have assessed here, the inequality premium is substantially larger than the risk premium. While on its face this finding may be surprising—risk aversion is, after all, a key motivator for many policies and measures to manage climate-change risk—it is not when considered in the broader context of this analysis.

DECISION MAKING UNDER UNCERTAINTY

While our estimated probability distributions for the impacts we quantify represent a rigorous effort to assess probabilities in a framework that is both internally consistent and consistent with the best available science, they do not represent the only valid estimates. Among other factors, alternative climate-model-downscaling techniques, alternative probability weightings of climate models, alternative priors for the impact functions, and alternative assumptions about the changing structure of the economy would all change the estimates. There is no single correct approach.

Under such conditions of "deep" uncertainty, economists and decision scientists have developed a range of alternatives to the classical von Neumann-Morgenstern expected utility paradigm of estimating a single probability distribution that is interpreted as "true" and using this distribution to weight possible outcomes (as we did earlier when we estimated our risk premiums). One finding from this work is that many decision-makers are *ambiguity averse*: they view the non-uniqueness of the probability distribution as imposing a cost premium on top of the risk premium (Heal & Millner 2014). Another finding is that if catastrophic outcomes are possible and decision makers cannot rule them out, the remaining "fat tail" of the probability distribution of potential losses imposes an exceptionally large risk premium (Weitzman 2009).

Alternative approaches to the expected utility paradigm include a "maxmin" approach (i.e., choosing a course of action that among all possible worst-case outcomes is the least bad, an approach closely related to the "precautionary principle") and an "α-maxmin" approach (i.e., making a decision based on a weighted mixture of the most likely outcome and the worst possible outcome). One might also choose to minimize regret—to find a strategy that across all possible futures minimizes the difference between the realized outcome and the best one could have done in the absence of uncertainty. These three approaches could all be applied using the impacts we estimate in this report to characterize the worst possible outcomes and the most likely outcomes. There are also additional alternative approaches that address the cost of ambiguity by estimating multiple probability distributions, each of which is assigned a probability of being correct (Kunreuther et al. 2012; Heal & Millner 2014).

Finally, we note that in this analysis, we have quantified only a subset of the potential costs of climate change, many of which cannot yet be rigorously assessed in an economic framework. Without the inclusion of the missing effects described in part 4, any evaluation of the worst possible outcomes would be incomplete. The true worst-case future is characterized not just by lost labor productivity and widespread heat-related deaths but also by (among other changes) international conflict and ecological collapse (Stern 2013). Part 4 surveys the gaps in our coverage.

PART 4
UNQUANTIFIED IMPACTS

OPENING COMMENTARY

NICHOLAS STERN AND BOB WARD

THIS book provides an important and careful contribution to the understanding of the economic risks posed by climate change to some key sectors of the United States. But, as discussed in part 4, it does not cover all the risks and so necessarily underestimates the potential consequences. A number of those omitted risks are not easily modelled or measured, but they are real and potentially of great importance.

The analysis quantitatively estimates the effect of climate change in the United States on some aspects of agriculture, labor, health, crime, energy, and coastal communities. Separate chapters in part 4 are dedicated to qualitative assessments of how climate change affects water, forests, and tourism, all of which could have huge repercussions on lives and livelihoods. Water, for instance, includes both river and surface flooding and drought. Some of the difficulties involved in assessing these impacts is that they will vary from region to region, future projections are uncertain, and they depend on other factors influenced by human action, such as water demand and urban run-off. But the current evidence is that the risks are very large, particularly in the rather likely case in which flood and drought events become more frequent and intense.

For instance, extended drought promotes desertification, causing a potentially irreversible change in the environment with drastic implications for where and how people live. Similarly, floodplains of rivers can grow to cover much wider areas, rendering some locations essentially uninhabitable without the development of strong and extensive defenses, and making agricultural land unproductive.

The damage to forests caused by unmanaged climate change can also cause extensive economic damage, not just for local communities, but also by fundamentally altering the local environment. Forested areas, for instance, can be better at soaking up rainwater and fixing the soil than other areas. But the die-back of forests due to climate change in the United States can also have global economic consequences, because forests are an important absorber of carbon dioxide emissions, and so die-back accelerates the atmospheric accumulation of carbon dioxide from human emissions. Essentially, the die-back of forests due to climate change is a form of positive feedback in the climate system. In the case of tropical and boreal forests, die-back could constitute a "tipping point" in the global climate system, as chapter 3 acknowledges.

Many potential impacts of future climate change are difficult to assess and quantify, not least because of the deep uncertainty associated with unmanaged global warming of 2°C (3.6°F) or more above pre-industrial, or mid-nineteenth-century, levels. As described in chapter 2, global temperatures that are 3°C (5.4°F) higher than pre-industrial levels have not occurred on Earth for millions of years, well outside the evolutionary experience of modern *Homo sapiens* (no more than a quarter of a million years). On current emissions trajectories for the world, it is likely that a rise of 3°C (5.4°F) would be exceeded within 100 years.

While the report has attempted to quantify the direct impacts of sea-level rise on coastal communities, its focus on projections up to the end of the century (with projections up to 2200 appearing only in appendix A) misses out the potential long-term continuation over many centuries that will fundamentally alter the coastline of the United States.

There is no reason to suppose that the biggest impacts of sea-level rise will occur by 2100; warming and expansion of the oceans is an extended and inexorable process. Indeed, the report's projections indicate a likely global mean sea-level rise of 2 to 3 feet (60 to 100 cm) by 2100, increasing to 4 to 9 feet (130 to 280 cm) by 2200. But, if a threshold is passed beyond which the melting of the major land-based polar ice sheets becomes unstoppable and irreversible, increases in sea level, with potential sliding and melting of land ice, could accelerate. The projections in this report indicate a 1-in-200 chance of more than 5.8 feet (1.8 m) of global mean sea-level rise by 2100 and 21 feet (6.3 m) by 2200.

The focus on 2100 for sea-level rise reflects the current lack of comfort of climate scientists and economists in determining the long-term risks of climate change in the face of uncertainty about both future atmospheric concentrations of greenhouse gases and the sensitivity of the climate. But its use excludes very serious risks of consequences that are both irreversible and unstoppable.

Climate scientists also acknowledge that their models do not capture all of the feedbacks that can accelerate climate change and make its management extremely difficult. Chapter 3 notes the potential for such "tipping points" in the climate system. One very important example is the possible release of large volumes of methane, a powerful greenhouse gas, due to the thawing of permafrost layers in polar regions. These effects are omitted, despite being regarded as large by climate scientists, because they cannot be made precise or formal enough to be included in the models, or because their probability is considered to be below some lower threshold. While they might be excluded from this report and other models for understandable reasons, the consequences are potentially immense, in terms of their impact, and it would be most unwise to exclude such risks from the argument and from assessment of policy.

But, as Stern (2013) points out, the global integrated assessment models used by some economists to turn impacts (such as those in this report) into losses add further gross underassessment of the risks because many of them contain key and unjustified assumptions which build in the expectation that unmanaged climate change will have only modest consequences.

First, the global models assume that underlying economic growth is driven by factors that will not be affected very much by climate change. This has the effect of making the assumption that these factors work to increase global wealth and incomes powerfully over a century, no matter what the level of climate change or its impact. The labor productivity and coastal capital stock damages described in the report are examples of impacts that could decrease the overall rate of economic growth.

Second, economic losses in the global models are represented by functions that restrict damage in a given period to current output only and further relate only to temperature in that period. They disregard the fact that capital, labor, and land during the period could be damaged both by previous impacts and by current impacts. Hence, climate change can damage stocks, not just output flows, and cast the impacts long into the future. Notably, this report is the first analysis of this type to account for stocks of damage from hurricanes and sea-level rise, but there are other potential impacts with similar dynamics that it omits.

Third, the damage functions used by the global models are quantitatively weak, and so cannot incorporate very large, and potentially catastrophic, impacts. They are so weak as to be thoroughly implausible: for example many of them embody damages of just a few percent of GDP for warming of 5–6°C (9.0–10.8°F) to reach temperatures that have not existed on Earth for tens of millions of years, and which would likely transform the relationship between human beings and the planet, with many densely populated parts of the world likely to become uninhabitable through effects such as the extreme heat and humidity discussed in chapter 4. It is extremely difficult to extrapolate accurately from empirical measurements of the impacts experienced so far after less than 1°C (1.8°F) of warming to those created by a rise in temperature of 5°C (9.0°F) or more.

This combination of factors leads to the unwarranted assumption in the global models that populations in the future will be much wealthier than current generations, regardless of the amount of climate change that takes place. This report attempts to avoid these shortcomings, but—as it acknowledges—it does not provide a comprehensive analysis of the economic impacts of climate change. Simply summing up the six impacts it quantifies would yield estimates of total risk that are likely to be very conservative.

Dietz and Stern (2014) explored how a conventional global economic model that addresses just some (but far from all) of the biases and shortcomings, and takes into account a wider range of risks, can imply that unmanaged climate change would lead to a strong decline in living standards by the end of the century.

A further limitation of the global models is that they measure economic effects only in terms of aggregate consumption or gross domestic product (in other words have just one consumption or income dimension), and hence exclude the consequences of, for instance, injury or loss of life, except in terms of the supply of labor (although this report separately quantifies loss of life). The global economic models also exclude explicit analysis of environmental goods and services (as the report notes), as well as other non-market impacts.

In addition, the analysis excludes related non-climate impacts of the burning of fossil fuels, such as local air pollution. As the New Climate Economy Report (Global Commission 2014) points out (based on recent analyses by the World Health Organisation, the Organisation for Economic Cooperation and Development and the International Monetary Fund), poor air quality due to the burning of coal and diesel has severe consequences for human health in cities, with particles measuring 2.5 microns or less causing losses equivalent, for instance, to about 4 percent of gross domestic product in the United States, 6 percent in Germany, and 11 percent in China.

One of the most important omissions from the analysis is the potential for climate change, through shifts in the pattern of extreme events and other effects, such as desertification, to cause migration within the United States, across its borders, and around the world. Unmanaged climate change could lead to the movement of hundreds of millions of people from affected areas. We know from history that such migrations would result in likely extended conflict and war, with hugely negative consequences for prosperity and well-being.

Again, the analysis acknowledges this potential impact, but the single paragraph in chapter 20 devoted to it hardly reflects the seriousness of the risks involved. Conflict across the world caused by the impacts of climate change, disrupting the lives and livelihoods of hundreds of millions of people, is likely to have devastating consequences for every country and for world security. The U.S. Department of Defense and the U.S. military have already begun to highlight the way in which climate change can be a risk multiplier in politically sensitive regions in the immediate future, and are also emphasizing the security risks of climate change in the medium and longer term. And given the long lags in climate processes, one could not negotiate a "peace treaty," with rapid effect, with the climatic causes of such conflict and insecurity. There is mounting evidence about the potential for extreme events, such as droughts, both to be affected by climate change and to act as major "stressors" in human societies—as demonstrated in recent work by one of the report's lead authors (Hsiang, Burke & Miguel 2013) but left out of the quantitative analysis. And it should be noted that we are witnessing such major effects with global average temperatures currently only about 0.9°C (1.5°F) higher than they were in the late nineteenth century (the useful benchmark).

This report is a careful, very timely, and fundamentally important contribution. It provides an extremely important summary of some of the economic risks of climate change in the United States. But decision makers should realize that the potential consequences of an unchecked rise in greenhouse-gas concentrations are likely to be substantially more devastating that the report describes.

Its legacy should be continued research to assess and, where possible, quantify those impacts that cannot now be avoided, and which must be adapted to, as well as to gain a better grasp of just how huge the risks of unmanaged climate change are likely to be, even for a superpower.

CHAPTER 16

WHAT WE MISS

OUR analysis is broad, but it is far from complete. As discussed in chapter 1, we have only been able to quantify a subset of the economic risks of climate change in the United States. Figure 16.1 highlights some of the gaps. These gaps can be subdivided into several categories: incomplete coverage within included market impacts, omitted categories of market impacts, interactions between impacts, omitted nonmarket impacts, effects on international trade and security, out-of-sample extrapolation, and potential structural changes. Many of these limitations parallel those of benefit-cost "integrated assessment models" that lack the empirical calibration and spatial detail of our analysis; some of the literature discussing the limitations of the damage estimates of these models applies here as well (Yohe & Tirpak 2008; Warren 2011; Howard 2014).

MARKET IMPACTS

Incomplete Coverage Within Included Impacts

In the seven impact categories we have examined, we have focused on a subset of effects most amenable to quantitative analysis. These limitations of scope are described in the sectoral chapters (chapters 6 through 11); we summarize them here.

In the agricultural sector, we have assessed the impacts of temperature and precipitation changes on the largest commodity crops but not on fruits, vegetables, or nuts. These so-called specialty crops dominate the agricultural sectors of some states, such as California. We also do not include the effects on livestock, which, like humans, will suffer from humidity as well as heat. Nor do we include

Science

 Temperature: averages and extremes

 Precipitation: averages and extremes

 Local sea-level rise

 Humidity: wet-bulb temperature

 Strong positive carbon-cycle feedbacks

 Ice-sheet collapse

 Ocean temperature and acidification

 Ecosystem collapse

 Unknown unknowns

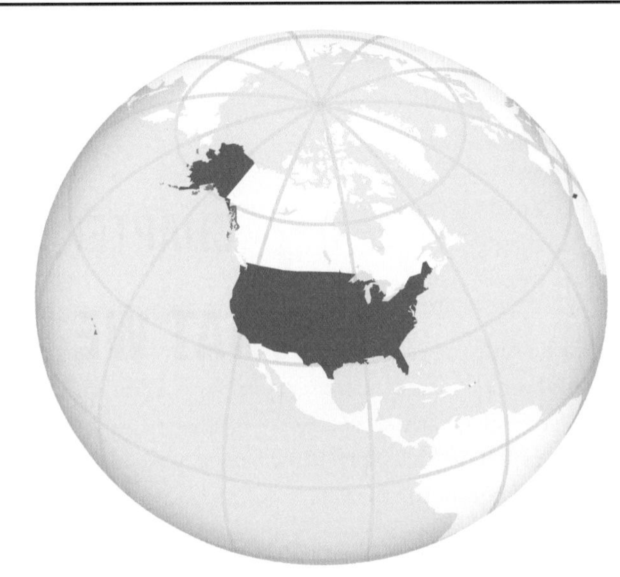

Methodology

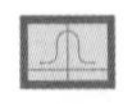 Full probability distribution, tail risks

 Market impact

 Quality of life

 Biodiversity, ecosystem loss

 Ecosystem services

 International trade

Other Impacts

 Water supply and demand

 National security

 International civil conflict

 Aid and disaster relief

 Tourism, outdoor recreation

 Fisheries

 Forests

 Wildfire

Impacts

Coastal Damages

 Inundation from sea-level rise

 Hurricanes and nor'easters

Changes in hurricane activity

 Transportation

 Infrastructure

Health

 Heat/cold-related mortality

 Respiratory effects

 Extreme weather

 Vector and water-borne disease

Agriculture

 Grains, soy, cotton yields

 Other crops: fruit, vegetables, nuts

 Livestock

Energy

Energy demand

Energy supply

Labor Productivity

Hours worked

Labor quality, health impacts

Crime

 Property crime

 Violent crime

Included

Limited

Excluded

FIGURE 16.1. **Scope of Assessment**

the effects of potentially expanded weed, pest, and disease ranges.

While we consider the effects of temperature on the number of hours worked, we do not assess the effects on the intensity of labor during working hours. Nor do we include the effects on labor productivity of the nonlethal health impacts of climate change, whether mediated by respiratory illness, vector-borne disease, or the consequences of extreme weather events.

For health impacts, we include heat- and cold-related deaths. We do not include the respiratory effects of temperature-aggravated air pollution, the health impacts of disease spread by extreme weather, the effect of temperature or weather disasters on birth weight, or the expansion in the range of vector-borne diseases such as Lyme disease. We include humidity-related heat stress only to the extent it is indirectly captured in the empirically calibrated temperature impacts; the effects of increasingly frequent, extremely dangerous category III and record-breaking, extraordinarily dangerous category IV Humid Heat Stress Index days are not included.

In the energy sector, we include changes in energy demand but not supply-side effects such as reductions in the efficiency of thermoelectric generation or electricity transmission. For coastal impacts, we include damage to capital and the cost of business interruption, but we do not include the network effects caused by damage to critical infrastructure.

Omitted Categories of Impacts

Other types of market impacts we miss entirely in our quantitative analysis. Changes in the availability of water will affect the agricultural sector and electricity generation. Like coastal storms, inland flooding driven by intense precipitation events destroys capital and interrupts businesses. Forests, which both serve as an essential resource for the forestry industry and provide less-easily monetized ecosystem services, are threatened by changes in temperature and precipitation, more frequent wildfires, and expansion of pest and disease ranges. As climate changes, the desirability of different areas as tourist destinations will change. We qualitatively address water, forest, and tourism impacts in chapters 17 through 20.

Interactions Between Impacts

Although we estimate the direct effects in each impact category independently, there are important linkages between them that extend beyond the market interactions captured by the computable general equilibrium model. For example, energy supply and agriculture compete for limited water resources. Similarly, estimates of heat-associated mortality and labor productivity reductions include implicit assumptions about the use of air-conditioning (and therefore energy) to offset some of the heat and humidity.

NONMARKET IMPACTS

Many of the most important risks associated with climate change fall outside the market economy. In this report, we have quantified mortality caused by heat and cold: while mortality affects the labor supply and therefore the market, it also directly affects human well-being. Omitted health impacts, discussed earlier, do as well.

Humans depend upon the planet's ecosystems in myriad ways, most not valued by the market. To name just a few: ecosystems absorb CO_2 from the atmosphere, recycle nutrients, pollinate plants, serve as storm barriers, and prevent soil erosion (Millennium Ecosystem Assessment 2005). Though placing a dollar value on these services is extremely challenging and highly sensitive to assumptions, the annual value of global ecosystem services has been estimated at twice global GDP (Costanza et al. 2014).

Climate change threatens to disrupt ecosystems both on land and in the ocean, which also face serious threats from land-use change, nutrient pollution, and overexploitation. The oceans also face another CO_2-related threat, that of ocean acidification, which makes it more difficult for calcifying organisms—ranging from corals to shellfish—to produce their skeletons. Ecosystem disruption related to climate change has occurred many times in Earth's past (Barnosky et al. 2012) and may represent one of the most serious climate-change risks. Given the complexity of the problem, however, efforts to understand the economic consequences of future ecosystem changes are still at an early stage.

More generally, throughout this analysis, we measure impacts in terms of their effect on GDP. But GDP is a metric of economic output; it is not a measure of human welfare. Agricultural production constitutes only about 6 percent of world GDP, but the effect on human welfare of an agricultural collapse would be much larger. For many, communities and ecosystems have a value that extends beyond their productive capacity. For a parent, the welfare impact of losing a child to heat-related mortality is much greater than the net present value of that child's expected future earnings. The risks posed by climate change should be viewed in this broader context; in some cases, this may lead to policies or investments that would not be merited based on a monetary benefit-cost analysis alone.

INTERNATIONAL TRADE AND SECURITY

Our analysis focuses on the effects of climate change in the United States, but the United States is not a world unto itself. Its fate is bound economically and politically to that of the nearly 7 billion people outside its borders. For globally traded goods, such as crops, trade effects may dominate domestic changes. If the agricultural sectors of other countries are more severely affected than our own, demand for U.S. crops may rise even at elevated prices. Similarly, if the labor productivity impacts in other countries are more severe than in the United States, the United States may gain a competitive advantage even if the world economy as a whole suffers. Quantifying these linkages would require the extension of our analysis to a full model of the global economy.

Climate change could also prove to be an important factor affecting global security, which is qualitatively discussed in chapter 20. Extreme weather events and longer-term climate shifts can promote migration both within and between countries. There is significant support in the academic literature for a relationship between climate and civil conflict. The 2014 Quadrennial Defense Review concluded that climate change may increase the "frequency, scale, and complexity" of future missions, while also posing a threat to defense installations (U.S. Department of Defense 2014).

OUT-OF-SAMPLE EXTRAPOLATION

The impact sectors we consider all are calibrated either directly (as in the case of the five sectors with empirical models) or indirectly (as in the case of the process-model-based analyses) against historical behavior. While history provides us the only data set against which to test and calibrate our projections, climate change will increase the frequency of record-breaking weather that falls outside past experience. Because the structure of empirically derived dose-response functions beyond the limits of historical experience is unknowable, there is no foolproof technique for estimating how impacts will look in these out-of-sample cases. Thus, for simplicity and clarity, throughout this report we have used the conservative assumption that record-breaking temperatures will have impacts similar to the estimated impact of the hottest days on record. We examine the importance of this assumption in a sensitivity test (see appendix B) in which we instead linearly extrapolate all dose-response functions beyond the hottest conditions observed historically. We find that, in general, this adjustment to our modeling approach has only minor impact, primarily because most days over the next century will be hotter than historical averages but will not exceed the temperature of historical national maxima and thus are well-described by the in-sample structure of our dose-response functions.

More generally, the likelihood that the climate will produce unexpected surprises will increase as temperatures rise outside the realm of past human experience. The appearance in the eastern half of the countries of summer days so hot and humid that short periods of moderate, shaded outdoor activity can induce heat stroke in healthy individuals is an example of a known phenomenon outside the realm of past experience (see chapter 3). Some of the tipping points discussed in chapter 3 represent known unknowns, and in a complex system like Earth, there almost certainly will be "unknown unknowns" that are entirely beyond our current knowledge.

STRUCTURAL CHANGES

In our analysis, we assume that the structure of the U.S. economy remains as it is today—an assumption almost guaranteed to be wrong. GDP will grow in different regions at different rates due to a combination of factors ranging from demography, to policy, to climate. By the end of the century, some of the dominant industries may be ones that—like the IT and biotech industries today—were unknown eight decades previously.

As we discuss in chapter 22, social and technological innovations may reduce some of the damaging effects of climate change. The efficiency of air-conditioning may increase significantly faster than demand for cooler air. Genetically engineered crops, different planting schedules, and more efficient irrigation may offset effects on the agricultural sector. Defensive structures, relocation away from threatened coastlines, and structures designed for periodic flooding may all reduce the impacts of coastal storms and sea-level rise. Extreme heat and humidity may not be a problem if the people of 2100 spend their entire lives in climatically controlled domed cities like those envisioned in the science-fiction novels of the 1950s.

The statistician George Box famously observed that "all models are wrong, but some are useful." Our analysis provides a projection of what today's economy would look like in the face of twenty-first-century climate change, not a prediction of what the economy of 2100 will look like. The structural changes that can reduce climate risk are more likely if policy makers, business leaders, and citizens are equipped with knowledge about the risks posed by climate change. We have tried to craft our analysis to address this need.

CHAPTER 17

WATER

WATER is a fundamental resource for our society, our economy, and the health of our communities and ecosystems. It is critical not only for our own consumption but also for food production, electricity generation, and many industrial activities. Although we have dealt with them throughout history, droughts and floods continue to pose significant risks to the U.S. economy, our health, and our way of life.

Climate change affects water resources through multiple pathways, changing risks from water scarcity and overabundance, affecting water quality, and shifting patterns of water availability within and among regions and communities. Climate change can affect water supply by altering precipitation, surface runoff, and stream-flow patterns, as well as by increasing evaporation from lakes, reservoirs, soils, and plants. It can affect water demand directly by increasing irrigation and landscape watering needs and indirectly through increased energy use for air-conditioning and thus water use for cooling of thermoelectric power plants. Shifting precipitation patterns and heavier storms can intensify droughts and floods. Rising water temperatures and saltwater infiltration of near-shore groundwater reservoirs can affect water quality.

In concert with demographic, land-use, and other socioeconomic changes, climate change poses novel challenges for water planning and management. Existing water infrastructure and legal frameworks, created assuming an unchanging climate, may not be adequate to address these challenges. Managing water risk in a changing climate requires reevaluating strategies to meet our diverse water needs and protect natural ecosystems, with this informed by projections of both supply and demand.

WATER DEMAND

Water is a fundamental input to nearly every sector of the U.S. economy, creating competing demands across a wide range of users. Linkages with agriculture and energy production dominate water use in the United States. In 2005, freshwater withdrawals from surface water and groundwater sources totaled nearly 350 million gallons per day (Kenny et al. 2009). Freshwater withdrawals for thermoelectric power generation (41 percent) and irrigation (37 percent) are largest, followed by municipal and residential uses (14 percent), industry and mining (5.5 percent), and livestock and aquaculture (3 percent). Most water for thermoelectric power generation is returned to its original source after use (at a higher temperature), while most water for irrigation is consumed during use. Western states account for most irrigation withdrawals, while eastern states account for most thermoelectric generation withdrawals. Water withdrawal estimates do not include water for in-stream uses such as hydropower production, a significant source of electricity generation in the Northwest, California, New England, and Alaska (Energy Information Administration 2013; Georgakakos et al. 2014). Minimum in-stream flow requirements have also been established in many places to protect freshwater ecosystems.

From 1960 to 1980, water withdrawals rose dramatically (Kenny et al. 2009), but they have since been relatively stable due to increases in the efficiency of irrigation and thermoelectric cooling processes and declines in industrial water withdrawals across the United States and in irrigated acreage in western states. These have offset increases in municipal and residential use and in livestock and aquaculture (Kenny et al. 2009; Foti, Ramirez, & Brown 2012).

Projections of future water demand are dependent on assumptions about future population growth, socioeconomic development, and technological change, as well as the effects of climate change on water use. One research effort, assuming a continuation of the historical trends described earlier and the A1B socioeconomic scenario (in which U.S. population growth declines slowly, with total population nearing 500 million by 2100), found that demand for withdrawals in the absence of climate change would increase only 3 percent from 2005 levels by 2060 and 13 percent by 2090 (Foti, Ramirez, & Brown 2012; Brown, Foti, & Ramirez 2013). With climate change, however, the same effort projects that demand for withdrawals will increase substantially, mainly due to increased irrigation and landscape watering needs and to a lesser extent to increased water use for electricity production to meet assumed growth in air-conditioning (Foti, Ramirez, & Brown 2012; Brown, Foti, & Ramirez 2013). Under the A1B emissions scenario (intermediate between RCP 8.5 and RCP 6.0), demand is projected to increase 12 to 41 percent by 2060 and 35 to 52 percent by 2090, depending on the climate model used, with greater increases projected in the western states. One uncertainty is the effects on crop water use of climate change–induced changes in growing season and increased CO_2 in the atmosphere, which may compensate to some extent for the effects of higher temperatures reflected in the irrigation demand increases presented here (Elliott et al. 2013; Prudhomme et al. 2013; Wada et al. 2013; Brewer et al. 2014).

WATER-SUPPLY AND WATER-MANAGEMENT CHALLENGES

Balancing climate-driven changes in supply and demand across water uses poses challenges for water management. Managers are increasingly recognizing the need for adaptive responses. Analyses of potential challenges have been undertaken at the national scale, for specific river basins, for specific municipalities, and for specific water uses. All regions of the United States face water-management risks, and the Southeast and Southwest including California are seen as most likely to experience water shortages (Barnett & Pierce 2009; Rajagopalan et al. 2009; Foti, Ramirez, & Brown 2012; Roy et al. 2012; Georgakakos et al. 2014; Romero-Lankao et al. 2014).

Changes in Precipitation, Runoff, and Stream Flow

The primary source of freshwater is precipitation—falling rain and snow—either through runoff when soils are saturated into rivers, lakes, and other surface-water bodies or through recharge of groundwater. Annual average precipitation has increased over the past century in much of the continental United States, with notable increases in parts of the Northeast, Midwest, and Great Plains over recent decades (Walsh et al. 2014). Precipitation is very likely to increase in the Northeast and likely to increase in the Midwest, with increases particularly in the winter and spring. Springtime precipitation decreases are likely in the Southwest, and summer precipitation decreases are likely in the Great Plains and Northwest.

Surface runoff has increased in the Northeast and the Mississippi basin and decreased in the Northwest during the past half-century, and a decreasing trend is emerging for the Colorado River basin (Luce & Holden 2009; McCabe & Wolock 2011; U.S. Department of the Interior, Bureau of Reclamation 2012; Georgakakos et al. 2014). In the future with continued high emissions of greenhouse gases, surface-water and groundwater supplies in parts of the Southwest, Southeast, and southern Rockies are expected to be affected by runoff reductions and declines in groundwater recharge, increasing the risk of water shortages (Seager et al. 2013; Georgakakos et al. 2014). Annual runoff is projected to decrease in some river basins in the Southwest, including the Colorado and Rio Grande, with mean and median runoff reductions of ~10 percent projected for California, Nevada, Texas, and the headwaters of the Colorado River over the next few decades, with greater model agreement for the Colorado headwaters and Texas than for California and Nevada (Cayan et al. 2010; Seager et al. 2013; Georgakakos et al. 2014). Annual runoff is also projected to decrease in the Southeast, driven by temperature-induced reductions in soil moisture (Zhang & Georgakakos 2012; Brewer et al. 2014). Annual runoff is projected to increase in the second half of the century (with little change through the middle of the century) in river basins in the Northwest and north-central United States such as the Columbia and Missouri (Brewer et al. 2014; Georgakakos et al. 2014).

The U.S. Bureau of Reclamation has conducted a series of western river-basin-level assessments of climate risk, which provide local illustrations of the broader trends described earlier. For example, the Colorado River supplies drinking water for almost 40 million people across seven western states, water for irrigation of 5.5 million acres producing 15 percent of U.S. crops and 13 percent of U.S. livestock, and water for hydropower facilities totaling 4,200 MW of electric-generating capacity (U.S. Department of the Interior, Bureau of Reclamation 2012). Over the past century, there have been multiple years when water use was greater than supply, with resulting shortages in the upper parts of the basin (that rely more on annual stream flow) rather than in water storage in the river system. Basin-level projections of water supply and demand by the Bureau of Reclamation indicate that decreasing annual flows and decreased snowpack result in decreased spring/summer runoff in the upper basin. At the same time, demand is projected to increase because of development and climate factors. Comparing median projections of water supply with median projections of water demand yields a 3.2 million acre-foot imbalance in the Colorado River basin by 2060 (U.S. Department of the Interior, Bureau of Reclamation 2012; note that the confidence bands on these estimates are wide and year-to-year demand and supply are variable). This imbalance represents about 20 percent of total average annual Colorado River consumptive use over the past 10 years, or roughly equivalent to estimated national water demand for municipal and industrial uses in 2015, which is projected to grow over time.

Timing of Water Availability

Changes in the timing of runoff can also challenge water-management efforts. They can increase the need for storage infrastructure and reevaluation of complex allocation and operation frameworks in response to shifts in seasonal stream flow. They can increase the risk of water shortages by creating or widening mismatches between the annual cycles of supply and demand, for example, where peak stream flow is shifting earlier in the spring and demand is generally highest later in the summer (Georgakakos et al. 2014). Increases in the amount of precipitation falling as rain rather than snow, decreases in the amount of water stored in spring snowpack, and earlier melting has changed stream-flow patterns in many rivers in the western United States (Mote 2006; Pierce et al. 2008; Fritze, Stewart, & Pebesma 2011; Hoerling et al. 2012a).

Due in part to continued decreases in spring snowpack and earlier snowmelt (Diffenbaugh, Scherer, & Ashfaq 2012; Georgakakos et al. 2014), increases in winter runoff are projected in north-central U.S. basins, as well as in basins along the West Coast such as the Sacramento and San Joaquin. Decreases in winter runoff are projected in Texas (Seager et al. 2013). Significant decreases in spring runoff are also projected for basins in the Southwest, including in California and the Southern Rockies (Seager et al. 2013; Brewer et al. 2014; Georgakakos et al. 2014). Declining snowpack can also affect groundwater in many mountainous areas of the United States where snowmelt is an important component of recharge (Earman et al. 2006; Earman & Dettinger 2011; Georgakakos et al. 2014).

SECTORAL IMPACTS

Agriculture

Year-to-year variation in water availability is a key determinant of agricultural production. The changes in precipitation due to climate change and the amount and timing of water supply described earlier will affect management of water for agriculture and competition with other water uses during times of scarcity. For example, agricultural water use is often curtailed to lessen shortages for municipal/household, commercial, and industrial users, as was the case in Texas in 2012 and 2013, when the Lower Colorado River Authority cut most Coastal Bend rice farmers' water to limit water restrictions in Austin (Phillips, Rodrigue, & Yücel 2013). Reduced water availability for agriculture may also lead to contraction in irrigated acreage in some areas, particularly in the western United States (Elliott et al. 2013). As discussed earlier, warmer temperatures will increase the water needs of crops and the demand for irrigation, although higher atmospheric concentrations of CO_2 can also increase water-use efficiency of some crops (Elliott et al. 2013; Prudhomme et al. 2013; Wada et al. 2013; Brewer et al. 2014; Hatfield et al. 2014).

Climate-change influences on patterns of climate extremes will also have direct effects on agricultural production. In 2011, flooding of the Mississippi River caused an estimated $1.3 billion in agricultural damage in Arkansas and Mississippi (NOAA 2013b). Droughts, often in concert with heat waves, cause significant agricultural damage. In the past 3 years, drought has had major regional effects in the United States. Drought and heat-wave conditions in 2011 in the southern Great Plains and Southwest caused $12 billion in damage across sectors (NOAA 2011; NOAA 2013a). The 2012 drought was the most extensive since the 1930s, with moderate to extreme drought conditions affecting more than half the country from the summer through the end of the year and causing $30 billion in damage across sectors (NOAA 2012; NOAA 2013a). The 2012 drought hit U.S. corn and soybean production particularly hard, with low rainfall and high temperatures during the growing season substantially reducing crop yields (NOAA 2012). Drought conditions receded in many Midwest and Great Plains states in 2013 but remained or expanded in western states (NOAA 2013a). California, for example, experienced the driest calendar year in the 119 years of recorded observations (NOAA 2013b). (See chapter 2 for a discussion of drought projections. For further discussion of climate-related changes in precipitation and water availability and impacts to agriculture, see chapter 5.)

Impacts on Electricity Generation

Cooling for thermoelectric power plants is dependent on water availability and water temperature, with power generation particularly at risk during periods of low summer flow and high water temperatures. For example, a 2007 drought in the southeastern United States forced nuclear and coal-fired power plants within the Tennessee Valley Authority system to reduce production, pushing up electricity prices (Kimmell & Veil 2009; National Energy Technology Laboratory 2010).

Water temperature has increased in some U.S. rivers (Kaushal et al. 2010), and temperatures are projected to continue to warm because of climate change (Cloern et al. 2011; Van Vliet et al. 2011; Georgakakos et al. 2014). Higher water temperatures are expected to degrade the efficiency of cooling processes and electric generation and inhibit release of heated water from once-through cooling systems because of regulations protecting ecosystems (Van Vliet et al. 2012; Georgakakos et al. 2014). A recent study projected a decrease in average summer capacity of 12 to 16 percent for once-through cooling systems and 4.4 to 5.9 percent for recirculation cooling systems by the 2040s, dependent on the emissions scenario (Van Vliet et al. 2012).

Hydropower supplies 20 percent of electricity generation in California, Alaska, and the Northeast and up to 70 percent of electricity generation in the Northwest (Energy Information Administration 2013; Georgakakos et al. 2014). Runoff projections suggest hydropower production may decrease in the southern United States (particularly the Southwest) and increase in the Northeast and Midwest (Georgakakos et al. 2014). But seasonal changes in the timing and amount of stream flow are also projected to affect the operation of hydroelectric plants, with actual future production dependent on the capacity of facilities, competition with other water uses, and basin-level changes in runoff amount and timing. For example, a study of hydropower production in the Pacific Northwest projected increases in winter of approximately 5 percent, decreases in summer of 12 to 15 percent, and overall annual reductions of 2 to 3 percent by the 2040s, with larger decreases in summer production of 17 to 21 percent by the 2080s (Hamlet et al. 2010).

These supply-side energy impacts will impose costs on energy consumers above and beyond the climate-driven changes in energy demand discussed in chapter 8.

Water Quality

Changes in air and water temperature, precipitation intensity, drought, and stream flow due to climate change directly affect water quality, as do changes in land use and other aspects of use and management of land and water resources. Worsening water quality can affect ecosystems and downstream water users, and several studies project decreasing quality in the future due to the combined effects of climate change and development (Tu 2009; Praskievicz & Chang 2011; Wilson & Weng 2011; Romero-Lankao et al. 2014). Increases in precipitation intensity along with changes in wildfire activity due to climate change can also affect sediment, nutrient, and contaminant loads and water quality, with negative effects for downstream water use (Emelko et al. 2010; Osterkamp & Hupp 2010; Georgakakos et al. 2014). Increasing air and water temperature is increasing thermal stratification in lakes and reservoirs, which can release nutrients and pollutants from bottom sediments (Sahoo & Schladow 2008, 2011; Schneider & Hook 2010; Georgakakos et al. 2014; Romero-Lankao et al. 2014).

FLOODING

Flooding causes fatalities and significant damage to property and agriculture, with average annual damage between 1981 and 2010 estimated at $7.8 billion (in 2011 dollars) (NOAA 2013b). Floods in 2011, including in the Northeast and along major river basins in Mississippi, Missouri, and Ohio, caused 108 fatalities and $8.4 billion in damage (NOAA National Climatic Data Center 2013). Flash floods, urban flooding, and coastal flooding are all strongly tied to heavy precipitation events, while river floods are also dependent on basin topography and existing levels of soil moisture. All floods are also affected by human land-use and land-management decisions.

In most of the United States, heavy precipitation events have become more frequent and intense over the past several decades, with the amount of precipitation during such events increasing in all regions of the continental United States except the Southwest and Northwest (Walsh et al. 2014). These trends have not yet been linked to changes in flood frequency, but heavy precipitation increases are projected to continue across the United States (Kunkel et al. 2013b; Wehner 2012; Wuebbles et al. 2013; Walsh et al. 2014).

Flood frequency and severity may increase in the Midwest and Northeast based on climate and hydrological projections (Georgakakos et al. 2014). Future flood risks across the United States are difficult to estimate, given their dependence on land-use trends such as urbanization, but such trends including development in coastal areas and floodplains can exacerbate the effects of increased flooding (Hejazi & Markus 2009; Doocy et al. 2013; Georgakakos et al. 2014; Romero-Lankao et al. 2014). For example, one study estimated 30 to 40 percent increases by midcentury in flood discharge associated with a 100-year flood in the West and Northeast, with 50 to 60 percent increases by the end of the century in areas of the Northeast, in the Pacific Northwest, and in other urbanized areas of the West, due to the combined effects of climate change, population growth, and land-use change (Kollat et al. 2012). Coastal flooding is described in chapter 4.

ECONOMIC DAMAGE AND ADAPTATION COSTS

A small number of studies have estimated water-related economic damage and adaptation costs associated with climate change. A study examining national economic damage from changes in water supply and demand for agricultural, public and domestic, and commercial and industrial use, as well as for hydropower and in-stream flow requirements, estimated total damage from climate change by 2100 (in 2007 dollars) of $4.2 billion per year under a business-as-usual scenario that falls between RCP 8.5 and 6.0 and $3.6 billion per year under a policy scenario similar to RCP 4.5 (Henderson et al. 2013). Damage estimates for 2025 in this study are $734 million and $690 million, respectively, and are largest in the West and Southeast. Such cost estimates are highly dependent on assumptions about future runoff and evaporation, the categories of water use included in the analysis, and the handling of reallocation of water among uses (e.g., shifting from agricultural use to other uses during times of scarcity). The cost estimates above, for example, do not include livestock, mining, and cooling for thermoelectric power generation, which may exclude some of the damage from climate change. These estimates also do not include damage due to flooding and changes in water quality, which were large in earlier studies (Frederick & Schwarz 1999; Hurd et al. 1999; Henderson et al. 2013). Water transfers may themselves involve substantial transaction costs as well as follow-on social and economic impacts (Hurd & Coonrod 2012). The analysis also does not include adaptation, which could reduce some damage at an associated cost, nor the potential reductions in agricultural water use associated with carbon fertilization that could reduce agricultural damage.

The estimated investment needs for water infrastructure over the next few decades without considering climate change are quite large. The EPA estimates that U.S. water infrastructure faces 20-year capital investment needs without climate change of $335 billion for drinking-water systems and $298 billion for wastewater and stormwater treatment and collection (U.S. Environmental Protection Agency 2008a, 2009). But changing infrastructure needs imply additional costs and make it important to consider climate change in such decisions in order to spend money wisely. A preliminary analysis of the costs of adaptation by the National Association of Clean Water Agencies (NACWA) and the Association of Municipal Water Agencies (AMWA) estimated total adaptation costs for infrastructure and operations and maintenance through 2050 of $325 billion to $692 billion for drinking-water systems and $123 billion to $252 billion for wastewater systems, with the largest costs in the Southwest followed by the Southeast and ranges dependent on assumptions about temperature and runoff changes and stringency of inland and coastal flood-protection measures (NACWA & AMWA 2009).

CHAPTER 18

FORESTS

FROM the boreal forests of Alaska, to the California redwoods, to northeastern deciduous trees and southeastern pines, forests span a third of total U.S. land area (about 750 million acres) and provide important natural and economic benefits. In economic terms, they provide valuable commodities like timber and bioenergy, recreational opportunities, and employment for local communities. The U.S. forest-products industry produces $200 billion in sales per year and employs about 1 million workers, generating an additional $54 billion each year in payroll (USDA 2013b). Although less easily quantified, forests also provide important ecosystem services including wildlife habitat, clean drinking water, flood control, and carbon storage, as well as other social, cultural, and aesthetic benefits (Joyce et al. 2013; Scholes et al. 2014).

U.S. FOREST HEALTH IS HIGHLY CLIMATE DEPENDENT

The health of U.S. forests and forest-related industries is directly influenced by the climate. Gradual changes in temperature and precipitation patterns, as well as extreme weather events like drought, affect forest growth, species distribution, and overall condition. Climate factors also affect the incidence and extent of damage from forest disturbances such as wildfire, pests, and disease (Anderegg, Kane, & Anderegg 2013).

A 2012 USDA assessment determined that climate change has already significantly affected the nation's forests through a host of mechanisms (Vose, Peterson, & Patel-Weynand 2012). For example, earlier snowpack melt in spring paired with warmer summer temperatures and extended drought in some regions has led to tree die-off, and more frequent and intense forest fires have caused extensive damage in increasingly dry areas. Milder winter temperatures have contributed to the arrival of bark beetles and other pest outbreaks at higher elevations. Changes in the distribution of tree and plant species, as well as the timing of their natural cycles, have been attributed to rising temperatures, with many plant, insect, and animal species shifting northward over the past century (Joyce et al. 2013). In some areas where tree growth has been limited by cold temperatures and short growing seasons, the warming climate has resulted in acceleration of forest growth (under 1 percent per decade) (Boisvenue & Running 2006).

Climate is just one among many factors that influence the health of U.S. forests. Some of the most significant changes in U.S. forests over the past few decades are a result of land-use changes such as increased urbanization and conversion for agriculture, harvest of forest products and bioenergy development, fire suppression and prevention programs, and air and water contamination. The interaction between changes in these factors and changes in climate make isolation of their individual effects difficult, especially in cases where climate and socioeconomic drivers are related (e.g., as domestic and global demand for forest products, bioenergy, and agriculture drive land-use change and climate change simultaneously).

CLIMATE-DRIVEN DISTURBANCE PUTS U.S. FORESTS AT RISK

The U.S. National Climate Assessment, based on observed changes over the past 30 years, found with high confidence that future climate change will further shift forest-disturbance patterns (Joyce et al. 2013). The type and magnitude of such disturbances will differ regionally and will likely be more variable going forward, posing significant challenges for state and local resource managers. By the end of the century, nearly half of the western U.S. landscape will experience climate profiles never before seen by forest species currently inhabiting that region, making it difficult to predict how ecosystems will respond (Rehfeldt et al. 2006). Changes in temperature and precipitation patterns are expected to trigger dangerous disturbances, potentially doubling the area burned by midcentury and increasing by an even greater amount the proportion of western forests affected by insect infestations (Vose, Peterson, & Patel-Weynand 2012). Increased drought and warmer temperatures are expected to exacerbate these stresses, leading to higher tree mortality, slow regeneration in some species, and altered species composition.

Climate change is expected to affect U.S. forest health in other ways, including shifting habitat and species composition, changing invasive plant species distribution and success rates, and altering the hydrological cycles that affect local and regional water quality. These impacts, while important to overall forest health and potentially significant when taken as a whole, are complex and subsequently quite difficult to characterize across American forests.

In the following sections, we go into more depth on the likely influence of a changing climate on the frequency and intensity of forest disturbances from wildfires, drought, and pest and pathogen infestation.

WILDFIRES

Impacts from U.S. Wildfires Are Large and Growing

In 2013, more than 47,500 wildfires burned more than 4.3 million acres, with the highest incidence in California, Nevada, New Mexico, Oregon, Idaho, Colorado, and Arkansas according to the National Interagency Fire Center. On June 30, 2013, nineteen firefighters were killed while working to contain the Yarnell Hill Fire in Arizona, the third highest firefighter death toll attributed to wildfires in U.S. history. On August 17, 2013, the third largest fire in California's history was sparked, eventually burning more than 250,000 acres near Yosemite Park (CAL FIRE 2013).

Fire is a leading source of forest disturbances in the United States (Flannigan, Stocks, & Wotton 2000). Since the mid-1980s, large wildfire activity in North America has been marked by increased frequency and duration and longer wildfire seasons. The annual area burned by large forest wildfires (greater than 400 hectares) between 1987 and 2003 was more than six times as large as the area burned between 1970 and 1986 (Westerling et al. 2006).

Fire plays an important role in ensuring forest equilibrium, but wildfires can also have significant economic, social, and environmental costs. The U.S. Forest Service and the U.S. Department of the Interior spend an average of $3.5 billion a year to fight fires, three times what they spent annually in the 1990s. State governments spend another $2 billion annually on wildfire protection (Bracmort 2013). Lloyds of London estimates that direct losses from catastrophic wildfires in the United States totaled $28.5 billion between 1980 and 2011 (Lloyds 2013). Nearly half of that cost came in just the past decade. An average of 47 percent of average losses over the past three decades were insured. In 2012, catastrophic fires caused $595 million of insured losses across the United States.

A full accounting of the immediate and long-term costs of wildfires should also take into account a range of impacts on ecosystems, infrastructure, businesses, and individuals that are not easily quantified. These include impacts to human safety and health, loss of human life, effects on regional economies from the loss of livelihood and property, and the expense of settlement evacuations.

Fire Activity in the United States Is Strongly Influenced by Climatic Factors

Fires require biomass to burn; dry, hot, or windy atmospheric conditions conducive to combustion; and ignitions. Climate can affect all three of these factors in complex ways and over multiple timescales (Moritz et al. 2012). Climate—including temperature, precipitation, wind, and atmospheric moisture—is a critical determinant of fire activity. Climate controls the frequency of weather conditions that promote fire, whereas the amount and arrangement of fuels influences fire intensity and spread. Climate influences fuels on longer timescales by shaping species composition and productivity (Marlon et al. 2012), and large-scale climatic patterns are important drivers of forest productivity and susceptibility to disturbance (Vose, Peterson, & Patel-Weynand 2012).

Despite marked impacts by human activities, climate conditions were the primary factor in twentieth-century wildfire activity in the American West. Between 1977 and 2003, temperature and precipitation provided the dominant controls on wildfire (Littell et al. 2009). Historical fire records going back as far as 500 CE show that biomass burning in the western United States generally increased when temperatures and drought area increased and decreased when temperatures and drought declined (Marlon et al. 2012). The greatest increases in fire activity have occurred in mid-elevation, Northern Rockies forests, where land-use histories have relatively little effect on fire risks and are strongly associated with increased spring and summer temperatures and an earlier spring snowmelt (Westerling et al. 2006).

Wildfire Impacts Are Expected to Increase

Future trends of fire severity and intensity are difficult to determine because of the complex and nonlinear interactions between weather, vegetation, and people (Flannigan et al. 2009). Uncertainty in the link between climate and forest fire increases as climate conditions move outside historical ranges. Without historical analogs, and considering the highly nonlinear climate-fire relationship, it is difficult to predict how potential climate futures—and the forest fuel conditions governed by these climate drivers—will affect fire intensity and activity (Westerling et al. 2011).

Despite these limitations, the U.S. National Climate Assessment determined that there is very high confidence that, under projected climate changes, there is "high risk that western forests in the United States will be impacted increasingly by large and intense fires that occur more frequently" (Joyce et al. 2013). Several studies have found that fire activity will increase substantially with warming temperatures, in combination with an increase in the frequency and severity of drought, pests, and pathogens across much of the western United States (Westerling & Swetnam 2003; Keane et al. 2008; Littell et al. 2009; Williams et al. 2010). Eastern forests are less likely to experience near-term increases in wildfire as warmer temperatures are less likely to coincide with seasonal dry periods or more protracted drought.

Climate variables—primarily temperature and precipitation—can affect fire impacts through changes to fire area, fire severity, and length of fire seasons. Fire seasons are lengthening for temperate and boreal regions, and this trend is expected to continue in a warmer world (Flannigan et al. 2009). National Park Service data indicate that fire ignitions are now occurring both earlier and later in the season, and the average duration of wildfires has increased from less than 10 days to more than a month (Frost 2009).

Impacts on fire activity are different for each region, due in large part to regional variations in hydrology. Conditions are expected to become more humid and rainy in the East and drier in the West, resulting in fire activity declining in the eastern United States, while rising considerably in the western United States (Pechony & Shindell 2010). Western U.S. forests are particularly vulnerable and will be increasingly affected by large and intense fires that occur more frequently (Joyce et al. 2013). Climate change is expected to increase wildfire risk during the summer and fall on the southeast Pacific Coast, Northern Plains, and the Rocky Mountains (Liu, Stanturf, & Goodrick 2010). Fire area in the West is projected to increase significantly in most ecological zones, with estimated future increases in annual area burned ranging from less than 100 percent to greater than 600 percent, depending on the region, time frame, methods, and future emissions and climatic scenario (Littell et al. 2009).

One study found that a temperature rise of 3.2°F by midcentury would produce a 54 percent increase in annual area burned in the western United States relative to the present day (Spracklen et al. 2009). As burn area is ecosystem dependent, the study found that forests of the Pacific Northwest and Rocky Mountains will likely experience the greatest increases (78 and 175 percent, respectively). A study looking at fire risk in California and Nevada predicts a 10 to 35 percent increase by midcentury, depending on the greenhouse-gas emissions scenario and global climate model used (Westerling & Bryant 2008). More dramatic increases in temperature (such as those expected under RCP 8.5) when accompanied by drought are likely to produce a response in fire regimes that are beyond those observed during the past 3,000 years (Marlon et al. 2012).

DECLINING FOREST HEALTH AND TREE DIE-OFF

In recent decades, warming temperatures, intense droughts, and insect outbreaks have contributed to decreasing tree growth and increasing mortality in many forest types throughout the United States, affecting 20 million hectares and many tree species since 1997 from Alaska to the Mexico border (Bentz et al. 2010). Average annual mortality rates have increased by a factor of 2 to 3, from less than 0.5 percent of trees per year in the 1960s to 1.0 to 1.5 percent today (van Mantgem et al. 2009).

It is difficult to isolate individual causes of tree death, as factors such as drought, higher temperatures, and pests and pathogens are often interrelated (Allen et al. 2010; Joyce et al. 2013). However, according to the 2014 National Climate Assessment, rates of tree mortality have increased with higher temperatures in the western United States (van Mantgem et al. 2009; Williams et al. 2010; Joyce et al. 2013). This effect is less direct in eastern forests, where forest composition and structure appear to drive impacts in recent decades. As the National Climate Assessment notes, although the extent to which recent forest disturbances can be directly attributed to climate change is uncertain, recent research provides a clear indication that climatic variables will affect ecosystems and alter the risks U.S. forests face today.

Tree mortality and forest die-off triggered by dry and hot conditions have been documented in most bioregions of the United States over the past two decades, with increases in wildfires and bark beetle outbreaks in the 2000s likely related to extreme drought and high temperatures in many western regions (Williams et al. 2010). Coniferous tree species have seen widespread and historically unprecedented die-off in recent years, mainly as a result of drought and pests such as bark beetles (Adams et al. 2009). Forests within the southwestern United States have been particularly sensitive to drought and warmth; from 1984 to 2008, as much as 18 percent of southwestern forest area experienced mortality due to bark beetles or drought (Joyce et al. 2013). In Alaska, more than 1 million hectares of several spruce species experienced die-off. Such die-off events can create significant additional risk to surrounding forests and local communities, as tree death and the accompanying increase of dead wood will influence the fire risk of forests (Anderegg, Kane, & Anderegg 2013).

Changes in temperature, precipitation, pest and pathogen dynamics, and more extreme climate events such as drought are expected to lead to increased instances of widespread forest die-off in the future (Anderegg, Kane, & Anderegg 2013). Western forests have experienced the greatest impacts, even more severe than recent estimates, and with projected increases in temperature and aridity out to 2100, substantial reduction in tree growth and increased mortality is expected, in particular in the Southwest (Dale et al. 2001; Allen et al. 2010; Scholes et al. 2014). As temperatures increase to levels projected for midcentury and beyond, eastern forests may be at risk of die-off or decline similar to recent die-offs experienced in the western United States.

Climate influences the survival and spread of insects and pathogens directly, as well as the vulnerability of forest ecosystem infestation. Epidemics by forest insects and pathogens affect more area and result in greater economic costs than other forest disturbances in the United States (Dale et al. 2001). Native and non-native insect pest species and pathogens can greatly alter forest habitat and modify ecological processes, often leading to extensive ecological and economic damage (Dukes et al. 2009). In the United States, insects and pathogenic disturbances

Regional Wildfire Impacts

Yellowstone

Large fires have increased in the northern Rockies in recent decades in association with warmer temperatures, earlier snowmelt, and a longer fire season (Westerling et al. 2011). Although human activity—through fire suppression, forest thinning, and fuel treatment—plays a role, climatic variables were found to be of primary importance in most forests, especially at higher elevations where human activity is less prevalent. Recent studies indicate that the greater Yellowstone ecosystem, a large conifer forest ecosystem characterized by infrequent, high-severity fire, is approaching a temperature and moisture-level tipping point that could be exceeded by the mid-twenty-first century. Westerling et al. estimate that with climate-related increases in fire occurrence, area burned, and reduced fire rotation (down to 30 years from the historical 100 to 300 years), there is a real likelihood of Yellowstone's forests being converted to nonforest vegetation during the mid-twenty-first century (Westerling et al. 2011).

California

Wildfire in California comes at a very high price. Seven of the ten costliest U.S. wildfires in history before 2011 occurred in California. Wildfire risks and their associated costs pose significant challenges to state and local governments, with state fire-suppression costs of more than $1 billion each year. The risk to private property has also increased over recent years as development along the wildland-urban interface has increased, with now more than 5 million homes in more than 1,200 communities at risk. The largest changes in property damage occurred in areas close to major metropolitan areas in coastal southern California, the Bay Area, and in the Sierra foothills northeast of Sacramento. In 2003, more than 4,200 homes were destroyed by wildland fires in southern California, resulting in more than $2 billion in damage (Radeloff et al. 2005).

affect more than 20 million hectares on average each year, with an annual cost of $2 billion (Dale et al. 2001). Shifts in climate are expected to lead to changes in forest infestation, including shifts of insect and pathogen distributions into higher latitudes and elevations and increased rates of development and number of generations per year (Waring et al. 2009; Bentz et al. 2010). The National Insect and Disease Risk Map (NIDRM), prepared by the U.S. Forest Service to provide a nationwide strategic appraisal and spatial mapping of the risk of tree mortality due to insects and diseases from both endemic and non-endemic forest pests, estimates that by 2027, 81 million acres (more than 10 percent of total U.S. forest land) will be in a hazardous condition for insects and diseases. This assessment does not take into account the potential impacts from climate change but concludes that climate change will significantly increase the number of acres at risk, including elevated risk from already highly destructive pests (Krist et al. 2014).

CHAPTER 19

TOURISM

For many travel destinations across the United States, climate is the main attraction. Drawn to the nation's coasts by sunshine, sand, and sea and to mountain ranges by snow and lush forests, tourists, it seems, are the quintessential fair-weather friends. The modern tourism industry in the United States has been built to satisfy the highly climate-sensitive desires of the millions of American and foreign visitors. The climate itself, and the amenities it provides—snow in the mountains, abundant water and fish stocks in rivers and lakes, and healthy coral and marine ecosystems—is a natural resource upon which the tourism industry depends. Mountain resorts in the Rockies, for example, depend on regular and abundant snow to support more than 20 million visitors each winter. In Hawaii, hotels, restaurants, and tour operators rely on the state's year-round sunshine and sandy beaches to draw tourists from all over the world, accounting for a full fifth of the state's economic output.

Climate change will likely significantly reshape the tourism industry nationwide. Tourist "demand" will be influenced over time as tourists take into account changing conditions when weighing destination options. Climate change will also affect tourism "supply" as some destinations experience loss of or greater instability in the climate resources on which they depend. For example, sea-level rise and increased storm surge may damage beach resorts, low-elevation mountain resorts may have trouble maintaining adequate snow, and water scarcity may limit the season for whitewater rafting in areas facing drought. The risk of these potential impacts creates significant implications for local businesses and communities that depend on tourism and the climate-sensitive resources that attract visitors. Some regions will also gain, as climate change makes certain parts of the country more attractive tourist destinations.

CLIMATE IS ALREADY A MAJOR FACTOR INFLUENCING U.S. TOURISM SUPPLY AND DEMAND

Climate conditions affect the supply of tourism opportunities in several ways. First, climate determines the length and quality of the tourist season. In many areas of the United States, warming temperatures have shifted the onset of spring and summer conditions to earlier in the year. Tourist activity has been shifted as well, with peak park attendance in areas of increased average temperatures coming 4 days earlier in the year, on average (Buckley & Foushee 2012). Historical examples of year-to-year variability have shown that warmer, longer summers can provide a significant boost to tourist activity in national parks in northern latitudes. Warmer average temperatures can also mean abbreviated winter seasons, reducing opportunities for winter-sport activities. High-altitude locations (including the Colorado Rockies), often thought to be more protected from these effects, have experienced substantial shifts in the timing of snowmelt and snowmelt runoff (Clow 2010). Winter tourism has experienced noticeable changes in snow season length and quality, as the western United States and parts of the northern Great Plains, Midwest, and Northeast see earlier spring melting (Mote 2006; Pierce et al. 2008; Fritze, Stewart, & Pebesma 2011; Hoerling et al. 2012b).

Tourist destinations can also experience direct impacts from climate-related events that affect their ability to attract visitors. Extreme wet or dry years, for example, can make specific destinations unsuitable for the outdoor activities upon which they depend. Wildfires can block access to outdoor recreation areas, and coastal storms and flooding can drive away beachgoers. In the spring and summer of 2002, for example, severe drought in Colorado created dangerous wildfire conditions that kept summer visitors at bay, with a 30 percent reduction in reservations at state-park campgrounds (Butler 2002). On more rare occasions, storms and other extreme events can wipe out an entire tourist season or even multiple seasons, depriving communities of tourist-related income on top of direct weather-related damage. Louisiana experienced a 24 percent drop in visitor spending from 2004 to 2006 after Hurricane Katrina, and the number of visitors to New Orleans did not return to pre-Katrina levels until 2013 (University of New Orleans & LSU 2009).

Finally, climate is a significant factor in determining the operating expenses of many tourist destinations, including heating and cooling, snow-making, irrigation and water supply, and insurance costs. The tourist industry has long been exposed to seasonal and interannual climate variability and to date has developed tools to help manage the challenges such uncertainty creates for business planning and operation.

Climate also plays a role in tourism demand across the United States. Unlike tourism supply, which is somewhat fixed as destinations are unable to pick up and move to more suitable climes in times of climate variability or extreme events, tourists are by nature fair-weather and flexible in their choice of destination. Studies have shown that economic development is the principal determinant of the *level* of tourist demand: more disposable income means greater travel (Bigano, Hamilton, & Tol 2005). However, once people decide to travel, climate is a significant influence on *where* tourists choose to spend their vacations (Scott et al. 2008). Studies of tourist destination preferences have identified a universal preference for moderate temperatures (with an optimal temperature of about 70°F) and have found that American tourists in particular display a strong preference for specific precipitation levels (Lise & Tol 2002). As a result, seasonal travel patterns shift toward warmer temperatures and sunny skies in temperate regions of the United States. Perceptions of environmental quality—sufficient stream flow and fish stocks, for example, or thriving coral and beach ecosystems—are also important determinants of tourist demand.

CLIMATE CHANGE IMPACTS ON U.S. TOURISM

The sensitivity of U.S. tourist demand to climate and the natural resources it affects means that climate change will expose the industry to a wide range of potential risks. Businesses that have been built to take advantage of local climates will need to adapt to these changes over time. Such changes include rising sea levels and increased storm surge from hurricanes and other coastal storms that may damage tourist infrastructure, disrupt travel in coastal communities, and put beaches and other environmental attractions at risk. Changing hydrological patterns will affect river flows and lake levels that draw tourists for water-sport activities and affect water availability and competition among water users. Activities that require large volumes of water to sustain, such as golf (a single golf course requires the same amount of water as a city of 12,000 people), waterparks, and pools, will be most affected by changes in availability and price (Scott et al. 2008).

Future changes in temperature will have a wide array of impacts. Warmer average temperatures nationwide mean that "ideal" tourist temperatures will shift northward and to higher elevations, with potential impacts on tourist destination preference. Along with warmer temperatures comes growth in insect populations and the associated vector-borne diseases they carry, which may also affect the quality of tourist activity in some areas. More severe droughts and wildfires may limit or curtail tourist activities in affected areas. Changes in the migration patterns of fish and animals will affect fishing and hunting, and warmer ocean temperatures and ocean acidification will affect coral reefs in popular diving destinations.

The dynamic nature of tourism demand and the wide array of tourist destination types and locations across the United States make it difficult to predict how the sector as a whole will be affected by a changing climate. The very strong substitution effect on tourist demand makes it difficult to assess the impact of climate change on overall tourism levels in the United States (Lise & Tol 2002).

Several studies that consider the isolated impact of temperature rise on tourism found that the U.S. tourism industry as a whole may actually benefit as northern temperate regions become more attractive destinations for travelers globally (Bigano et al. 2006, Scott et al. 2006, Ehmer & Heymann 2008). Tourists, finding traditional southern destinations increasingly hot, are expected to go north following more ideal recreational temperatures. One study determined that with an increase of 1.8°F of global mean temperatures by midcentury, the United States will see a modest decline in foreign travelers (as they stay home or select other destinations), but more Americans (by a factor of 3) are expected to choose to stay in the United States as a result of milder weather (Berrittella et al. 2006). With domestic tourism making up the vast majority of tourist activity in the United States, contributing nearly 90 percent of total travel and tourism sales in 2012, the net economic impact of warmer temperatures was found to be positive. In general, global tourism demand models find that countries with larger shares of domestic tourism are less affected by climate change, a finding that holds for climates that are currently cool but which may warm over time, like the northern latitudes of the United States (Berrittella et al. 2006). It is important to note that existing studies of global tourism impacts have only explored changes in temperature, omitting potential sea-level rise (SLR), changing precipitation patterns, or ecosystem impacts.

Nationwide assessments also obscure important consequences for specific communities. Climate-change impacts will be experienced differently from region to region and may even vary among communities within the same state, depending not only on local climate but also shifting tourism industry dynamics. An analysis of the attractiveness of 143 North American cities, based on seven climate variables associated with tourist demand, found that by the late twenty-first century, the number of U.S. cities with "excellent" or "ideal" climate ratings in the winter months is likely to increase (Scott, Mcboyle, & Schwartzentruber 2004). In contrast, Mexico's ratings decline as temperatures rise even further, suggesting that more winter Sun-seekers will opt instead to go north, bringing additional revenue to U.S. states. However, as temperatures rise, southern U.S. states may also see losses of these sunbird tourists to northern states, altering the competitive dynamics within the U.S. market.

Although tourists are flexible enough to respond to climate change, the same cannot be said of local providers of tourist services and local economies dependent on tourism revenues. Areas where tourism constitutes the major livelihood of local communities and where such tourism is strongly climate-dependent will be the most affected. Changes in the length and quality of the tourism season will also have considerable implications for the long-term profitability and competitive relationships between destinations. In general, greater variability in climate creates uncertainty for how tourism demand will respond, making it more difficult for the tourism industry to plan and maintain business from year to year.

Winter Sports

Climate has long ruled the fortunes of winter destinations dependent on snow for skiing and other winter sports. Across the United States, winter temperatures have warmed 0.16°F per decade on average since 1895, and this rate more than tripled to 0.55°F per decade since 1970 (Burakowski & Magnusson 2012). The unpredictability of winter seasons, as warmer temperatures bring increased variability in snow quantity, quality, and season length, has made it increasingly difficult for winter destinations dependent on steady revenue from snow-seeking tourists. The unique vulnerability of the winter tourism industry to climate makes it an important area for studying the near- and long-term impacts of a changing climate on winter tourism in the United States.

The businesses and communities that depend on winter sports as a significant source of annual revenues (more than $53 billion in annual spending on gear and

Tourism Facts

In 2012, tourist-related output generated $1.46 trillion and 8 million jobs (U.S. Bureau of Economic Statistics 2013). The Outdoor Industry Association estimates recreational activities including hiking, camping, and fishing contribute nearly $650 billion in spending and $80 billion in tax revenue to federal, state, and local governments and support more than 6 million jobs. National parks see more than 280 million visitors, generating $12 billion in visitor spending and supporting nearly 250,000 jobs (Outdoor Industry Foundation 2012).

trip-related sales) are paying close attention to current and future climate trends (Outdoor Industry Foundation 2012). The expectation that climate change will bring even warmer winters, increased rainfall and reduced snowfall, and shorter snow seasons has raised concern that the U.S. winter-sports industry could face significant losses. The picture is more complicated, however, as experience to date shows that winter tourists and ski operators have proved able to adapt, to some extent, to these changing conditions, making up for lost snow through artificial production. Tourists have adjusted as well, varying the timing and frequency of winter travel. The key question over time will be whether and how tourists and the winter-sports industry react to future climate changes and at what cost.

Looking back at past impacts associated with warmer temperatures can provide some insight into how this single variable may affect future winter-sports seasons. The U.S. 2011–2012 winter season was the fourth warmest winter on record since 1896, with the third smallest winter snow cover footprint in the 46-year satellite record. An assessment for the National Ski Areas Association (NSAA) found that winter-sports visits were down nearly 16 percent, despite significant efforts by ski resorts to supplement the lack of snow with snowmaking (RRC Associates 2013). Snowpack was particularly limited across areas in the West, where parts of California, Nevada, and Arizona had snowpack less than half of average, translating into a 25 percent drop in visitors. This can have real implications for states that rely on winter tourism and for local resorts and communities that experience declining revenues. One study analyzing the winter snowfall data across the United States from 1999 to 2010 found that lower-snowfall winters were associated with fewer skier visits in nearly all states, with a total revenue difference in low-snow years of more than $1 billion (Burakowski & Magnusson 2012).

The ski industry has come to rely heavily on snowmaking in order to reduce vulnerability to variability in snow levels and season length and maintain business from year to year. Ski areas have invested millions of dollars in snowmaking capabilities, and by 2001 all ski areas in the Northeast, Southeast, and Midwest had snowmaking systems covering 62 to 98 percent of skiable terrain (Scott et al. 2006). Across the rest of the United States, by 2012 nearly 90 percent of ski resorts report snowmaking was used to supplement natural snow cover. The ability to adapt to variability in snowfall is limited, however, as it requires energy to run equipment, significant volumes of water, and below-freezing temperatures. Adaptation in the form of snowmaking, therefore, comes at a high cost, often the biggest expense for ski resorts and at times as much as half of total expenses (Burakowski & Magnusson 2012). In drought-stressed regions, water scarcity may be a limiting factor. Making an acre-foot of snow requires more than 160,000 gallons of water; a typical ski run (200 feet wide by 1,500 feet long) would take nearly 7 acre-feet of water (or approximately 1 million gallons) to make 1 foot of snow (Ratnik Industries 2010).

Recreational Fishing

A recent study by Lane et al. (2014) assessed the potential climate-change impacts on recreational freshwater fishing across the coterminous United States. They found that higher air temperatures, and to a lesser extent changes in stream flow, will alter fish habitat, resulting in a decline in more desirable recreational fish species (i.e., cold-water species like trout) and a shift toward less desirable warm-water fisheries. Under their "business as usual" scenario (coinciding with a radiative forcing of 10 W/m^2 by 2100), warmer temperatures are expected to result in more than a 60 percent loss in current cold-water fishery habitat, which will virtually disappear in Appalachia, while habitat in substantial portions of Texas, Oklahoma, Kansas, Arizona, and Florida will shift from warm-water fisheries to species of even lower recreational priority. The analysis suggests that such shifts could result in national-scale economic losses associated with the decreased value of recreational fisheries.

Impacts on Coastal Tourism

Coastal areas, and the tourist destinations they support, are some of the most vulnerable to climate change. Many of the beaches, wetlands, estuaries, coral reefs, and kelp forests that attract visitors from across the United States and internationally are managed by the U.S. National Park Service, with more than 5,100 miles of coast and 3 million acres of coastal lands under the service's management. These parks attract more than 75 million visitors every year and generate more than $2.5 billion in economic benefits to local communities. Rising sea levels are expected to change shorelines and park boundaries, resulting in a net loss where parks cannot migrate inland. Everglades National Park, which brings in more than 1 million visitors each year, is uniquely vulnerable as even slight increases in sea level are expected to lead to disproportionate increases in inundation periods for broad areas in the park and have already influenced both surface and subsurface saltwater intrusion (Stabenau et al. 2011). Because of a lack of suitable habitat, species are prevented from migrating upland, resulting in significant changes to the composition of wetland and other forest communities and the species they support.

The potential impacts of climate change on coastal tourism activity across the United States will be highly location specific, but localized studies provide examples of the type of impacts communities may face. SLR alone will likely change the coastal tourist dynamic as beach destinations become altered. One result of SLR is coastal erosion, which decreases beach width over time without intervention, and in some instances eventually eliminates a beach altogether. One study estimates the impacts of SLR-induced reductions in beach width on beach recreation demand in several southern beach communities in North Carolina (Street et al. 2007). Using estimates of likely SLR in 2030 and 2080 based on local conditions at beach sites in southern North Carolina, the study found that the lost recreational value to beachgoers is $93 million in 2030 and $223 million in 2080, a reduction of 16 and 34 percent of recreational value, respectively. Although some of the affected beachgoers, finding their preferred beaches diminished or gone entirely, will simply choose beach sites further afield (which are also likely to be impacted), tourist-dependent businesses in the area will be affected.

Climate change will also accelerate coral bleaching and disease caused by increased sea-surface temperatures in the Caribbean, which has already led to the loss of more than 50 percent of reef-building corals in the Virgin Islands park units since 2005 (Hoegh-Guldberg 1999; Buddemeier, Kleypas, & Aronson 2004). A recent study by Lane et al. (2014) found that, even under low-emission scenarios, it is likely unavoidable that South Florida and Puerto Rico will experience multiple bleaching and mortality events by 2020. The same study found, however, that low-emission scenarios (associated with a radiative forcing of 3.7 W/m^2 by 2100) may reduce the potential midcentury impact on Hawaii's coral reefs, where sea-surface temperatures are cooler and coral cover is greater and more robust. Low-emission scenarios only delay the extensive bleaching of Hawaii's corals, however, and Hawaii is still expected to see substantial reductions in coral cover by late century. The discounted loss of recreational benefits of a "business-as-usual" climate scenario (associated with radiative forcing of 10 W/m^2 by 2100) compared to the low-emission scenario is estimated at $17.4 billion (with a confidence interval of approximately $9 billion to $26 billion) (Lane et al. 2014).

CHAPTER 20

NATIONAL SECURITY

People are saying they want to be perfectly convinced about climate science projections. . . . But speaking as a soldier, we never have 100 percent certainty. If you wait until you have 100 percent certainty, something bad is going to happen on the battlefield.

—Gen. Gordon R. Sullivan (Catarious et al. 2007)

THE U.S. national security establishment is accustomed to making decisions in the face of uncertainty. In an unpredictable world, assessing potential global risks is essential to making our homeland more secure. Climate change is such a risk. The global conditions associated with climate change, including potential changes in tropical cyclone activity, additional drought and flooding, and rising sea levels, present serious risk factors that could trigger mass migration, elevate border tensions, increase demands for rescue and evacuation efforts, and heighten conflicts over essential resources, including food and water (Catarious et. al. 2007). In recent years, the U.S. military has come to recognize climate change as a direct threat to national security and has developed a risk-based approach to prepare for and manage the potential impacts both at home and abroad.

In 2006, a panel of 11 retired three-star and four-star admirals and generals formed a military advisory board to assess the impact of global climate change on U.S. national security. They concluded that "climate change can act as a threat multiplier for instability in some of the most volatile regions of the world, and it presents significant national security challenges for the United States" (Catarious et. al. 2007). The following year, in response to calls from Congress and shifting national strategic priorities, the United States Intelligence Community produced the *National Intelligence Assessment on the National Security Implications of Global Climate Change to 2030*, which highlighted "wide-ranging implications for U.S. national security interests" (Fingar 2008).

In recent years, the U.S. military and security establishment has moved beyond exploration and begun integrating climate-change risk assessment and management into normal national security planning. In its 2010 Quadrennial Defense Review, the U.S. Department of Defense (DOD) called for a strategic approach to climate to manage the effects on its operating environment, missions, and facilities and regularly to evaluate risks as new science becomes available (U.S. Department of Defense 2010). This was the first time the Pentagon addressed climate in a comprehensive planning document. Not long after, individual branches of the military began to develop their own assessments of likely impacts and plans for dealing with near- and long-term threats from climate change (U.S. Navy 2010; U.S. National Research Council Committee on National Security Implications of Climate Change for

Naval Forces 2011). The most recent Quadrennial Defense Review reinforced the need to incorporate climate risks into planning, stating that "the impacts of climate change may increase the frequency, scale, and complexity of future missions, including defense support to civil authorities, while at the same time undermining the capacity of our domestic installations to support training activities" (U.S. Department of Defense 2014).

These assessments group climate-related risks to U.S. national security into two categories. The first covers direct, physical impacts to the homeland, including threats to U.S. military installations from flooding and storms, threats to nuclear power plants or oil refineries, and the risk that critical U.S. defense forces may be diverted from core national security objectives to aid in the management of domestic extreme weather events, such as Hurricane Katrina. The second, and much larger, category covers the indirect risk that climate change will exacerbate existing conflicts abroad and heighten humanitarian and political crises in vulnerable states and populations (U.S. Department of Defense Science Board 2011). In this chapter we provide a general overview of these indirect international impacts, followed by a more in-depth discussion of the direct impacts within the United States.

INDIRECT INTERNATIONAL IMPACTS

Assessments by the Pentagon and the United States Intelligence Community have concluded that climate change could have significant geopolitical impacts around the world, contributing to environmental degradation and food and water scarcity, exacerbating poverty, increasing the spread of disease, and spurring or exacerbating mass migration. This is likely to lead to increased demand for defense support to civil authorities for humanitarian assistance or disaster response.

The U.S. military often has a unique ability to respond to large-scale extreme weather or natural disasters. In the wake of Typhoon Haiyan, which struck the Philippines on November 8, 2013, the United States not only provided $37 million in humanitarian aid but also deployed more than 14,000 U.S. military personnel to help stabilize the area and provide relief. The U.S. military gets a request for humanitarian assistance and disaster response about once every 2 weeks (Gensler 2014). As climate change heightens and exacerbates humanitarian emergencies, the demand for U.S. assistance will further strain U.S. military capacity, limiting readiness for homeland defense or combat operations that may arise (Fingar 2008).

While resource scarcity and natural disasters associated with climate change are significant threats in and of themselves, an associated risk is their potential to weaken already fragile governments, providing opportunity for increased instability and conflict, creating an additional burden on the U.S. military to respond to prevent further destabilization. The National Intelligence Assessment for 2030 concluded that climate change alone is unlikely to trigger state failure in that time frame, but the exacerbation of existing problems could be enough to endanger domestic stability in some states, giving rise to threats of regional conflict or creating openings for criminal activity or terrorism (U.S. Department of Defense Science Board 2011). For example, a dysfunctional government response to water stress—of a sort expected to become more common under climate change—is generally agreed to be one of the contributing factors to the current humanitarian disaster in Syria (de Châtel 2014).

Recent work has applied the same econometric techniques used elsewhere in this report to measure quantitatively the dose-response function linking climatic events to various forms of modern intrastate social conflict, ranging from ethnic riots (Bohlken & Sergenti 2010), land invasions (Hidalgo et al. 2010), local political violence (O'Loughlin et al. 2012), leadership changes (Burke 2012), and coups (Kim 2014) to full-scale civil conflict (Hsiang, Meng, & Cane 2011) and civil war (Burke et al. 2009). Overall, the body of econometric analysis provides consistent and strong evidence that elevated temperatures tend to increase the risk of intergroup conflict in a location by roughly 13 percent for each standard deviation of warming, with somewhat weaker evidence that rainfall extremes affect conflict in a quantitatively similar way (Hsiang, Burke, & Miguel 2013). For perspective on the size of these effects, historically observed oscillations in the global climate have been implicated in contributing to 21 percent of civil conflicts since 1950 (Hsiang, Meng, & Cane 2011). In contrast to earlier theories that populations fought over increasingly scarce resources (i.e., "water wars"), this new body of evidence suggests that more complex dynamics are responsible for these social conflicts—the leading theory argues that climatic changes cause economic conditions and local labor markets to deteriorate, reducing the opportunity cost of engaging in violence and extractive activities (Miguel, Satyanath, & Sergenti 2004; Chassang 2009; Hidalgo et al. 2010; Dal Bó & Dal Bó 2011).

In contrast to this recent progress on intrastate social conflict, there is no general empirical evidence as to whether modern interstate conflict may be affected by climate, but this may be due to an absence of studies on this topic (Hsiang & Burke 2013).

Climate will also have an impact on strategic resources, including fuels, minerals, and food supplies, as well as the security of international transport routes essential to ensuring open access. One important example is the rapid evolution of the Arctic as accelerating sea-ice melt opens the region to changing transport routes, competing territorial and resource claims, and potential conflicts. As the United States is one of five nations bordering the Arctic (with more than 1,000 miles of Arctic coastline) and has a seat on the Arctic Council, this example is of particular importance to the nation. With no overarching political or legal structures to oversee the orderly development of the region or to mediate political disagreements over Arctic resources or sea-lanes, the potential risk of conflict in the region is meaningful.

By the end of summer 2012, the area covered by sea ice shrunk to about 400,000 square miles smaller than it was the previous summer, leaving the Arctic ice cap less than half the size it had been 30 summers previously (National Snow and Ice Data Center 2014). Warming temperatures have resulted in a rapidly evolving Arctic landscape, exposing sea routes that did not previously exist, thus opening access to transport and resource extraction. Further expected warming could open up shipping shortcuts on the Northern Sea Route (over Eurasia) and the Northwest Passage (over North America), cutting existing oceanic transit times by days. Both American and other vessels (including other navies or smugglers) would have greater access, making the overall effects on national security hard to predict. It is likely that these Arctic routes would also allow commercial and military vessels to avoid sailing through politically unstable Middle Eastern waters and other pirate-infested waters, thus mitigating other threats (Borgerson 2008).

The U.S. military has long had a presence in the Arctic, but the greater access afforded by melting sea ice will require a shift in the nature of its role and the resources required to sustain it, such as increased capacity for search-and-rescue and border patrolling, alterations of naval vessels, and increased monitoring. Responding to these challenges may require investments in ice-capable technologies and military training; greater resources for management of maritime traffic, search-and-rescue, and accident clean-up capacities; and building an ice-capable commercial, scientific, and naval fleet, an investment some have suggested is approximately $11 billion for icebreakers alone (Ebinger & Zambetakis 2009).

DIRECT IMPACTS TO THE U.S. HOMELAND

Military Installations

The most direct national security threat is the potential impact of extreme weather and sea-level rise on domestic military installations and the physical infrastructure that supplies them and on our international installations of strategic importance.

The U.S. military manages property in all 50 states, seven U.S. territories, and 40 foreign countries, comprising almost 300,000 individual buildings around the globe worth roughly $590 billion (U.S. Department of Defense 2012). About 10 percent of DOD coastal installations and facilities are located at or near sea level and are vulnerable to flooding and inundation (Strategic Environmental Research and Development Program 2013). The National Intelligence Council estimated that 30 U.S. military installations were already facing elevated risk from rising sea levels in 2008, jeopardizing military readiness, which hinges on continued access to land, air, and sea for training and transport (Fingar 2008). Because of a combination of natural and human-caused factors, Norfolk, Virginia, home to the world's largest naval station, has experienced one of the fastest rates of sea-level rise in the United States.

Several recent disasters have highlighted the vulnerability of military installations to hurricane-related flooding and wind damage, as well as the increase in sea-level averages. In 1992, Hurricane Andrew damaged Homestead Air Force Base in Florida to the point that it never reopened, while Hurricane Ivan knocked out Naval Air Station Pensacola for a year in 2004, and Hurricane Katrina destroyed 95 percent of Keesler Air Force Base in Mississippi (Foley & Holland 2012). As demonstrated in all three cases, military bases in the United States are important drivers of local and regional economies, and when they are destroyed by natural disasters, there is considerable collateral economic damage. As discussed in chapter 2, scientists have a high degree of confidence that global sea levels will continue to rise as a result of current greenhouse-gas emissions trends and that higher sea levels alone increase damage from hurricanes and other coastal storms. If climate change increases the frequency and severity of the most intense Atlantic basin hurricanes, as many cyclogenesis models predict, the risks are even higher.

The Pacific Coast is not invulnerable: An uncharacteristic tropical storm ripped through Fort Irwin, California, in 2013 bringing monsoon rains, wind, and hail, leaving homes and facilities flooded. Wildfire also poses a risk in the western United States. In 2013, a 2,500-acre wildfire forced evacuations at Marine Corps Base Camp Pendleton in San Diego County.

In recognition of the growing risk posed by these and other climate-related disruptions linked to climate change, the 2010 Quadrennial Defense Review called for a climate-impact assessment at all of DOD's permanent installations (U.S. Department of Defense 2010). Several studies have been completed or are currently under way by the individual military service branches, and DOD's Strategic Environmental Research and Development Program (SERDP) launched a comprehensive research project to examine climate-change effects on coastal installations (Strategic Environmental Research and Development Program 2013). SERDP is using case studies, such as the Norfolk Naval Station, to quantify the potential effects of near-term sea-level rise and storm activity on coastal infrastructure and Pacific islands and atoll systems that are home to critical U.S. military installations.

Critical Infrastructure

National security extends beyond protecting the homeland from outside threats. The U.S. Department of Homeland Security has affirmed that protecting and ensuring the resilience of critical domestic infrastructure is essential to the nation's security and that natural hazards can disrupt the functioning of government and business and result in human casualties, property destruction, and broader economic effects (U.S. Department of Homeland Security 2009).

The daily functioning of most critical infrastructure systems is sensitive to changes in precipitation, temperature, wind, and, for coastal cities, rising sea levels (Love, Soares, & Püempel 2010). Extreme weather events can destroy or temporarily debilitate critical physical infrastructure upon which the country depends, making us less secure as a nation. Electricity transmission and distribution systems face elevated risks of physical damage from storms and wildfires, and fuel transport by rail and barge is susceptible to increased interruption and delay, disrupting U.S. economic activity and affecting millions of Americans who depend upon electricity and fuels to power and heat their homes (U.S. Department of Energy 2013b). Roads, airports, and bridges are at risk from coastal storms and flooding, potentially incapacitating entire regions. Damage to coastal homes could leave thousands homeless or without access to basic amenities, threatening public health and safety. Impairment of critical infrastructure also hampers disaster-response efforts where communications and transportation systems are needed for preparedness, evacuation, or provision of food or to provide water and other emergency services to affected populations.

A significant portion of U.S. transportation infrastructure is in poor shape or deteriorating. The U.S. Department of Transportation estimates that approximately 11 percent of all U.S. bridges are structurally deficient, 20 percent of airport runways are in fair or poor condition, and more than half of all locks are more than 50 years old (U.S. Department of Energy 2013b). More than $3.6 trillion is needed to bring infrastructure in the United States up to "good condition" (American Society of Civil Engineers 2013). Absent additional investment in climate resilience, the changes in temperature, precipitation, sea levels, and storm patterns described in chapter 2 will exacerbate these infrastructure vulnerabilities. Climate-related infrastructure risks are concentrated in large cities and low-lying coastal areas. The Gulf Coast, home to critical energy and transport infrastructure, is particularly vulnerable, with 27 percent of major roads, 9 percent of rail lines, and 72 percent of ports at or below 4 feet in elevation (see chapter 4).

Climate-driven changes in extreme events can also compromise U.S. military capacity by diverting resources from core national security objectives. The military mounted a massive response to Hurricane Katrina, for example, with more than 70,000 military personnel involved at the peak (U.S. Government Accountability Office 2006).

Migration

As Hurricane Katrina demonstrated, extreme weather events can trigger large, unplanned population movements. Although displacement is often only temporary and may not result in long-term migration, even short-term movements can exacerbate a range of problems, including increased stress on resources and ecosystems and strain on governance and security systems (McLeman & Hunter 2010). In the long term, regions across the United States may be destabilized by rising sea levels, increased storm surge, and rising temperatures, among other effects, increasing migration flows from affected areas (Fingar 2008; Feng, Krueger, & Oppenheimer 2010). Rising sea levels and extreme weather events in coastal zones could contribute to humanitarian disasters and potential refugee flows (Youngblut 2009).

In the long term, even gradual changes in climate may induce migration from the most affected areas. Evidence has shown that weather extremes (i.e., extreme temperatures, extreme precipitation, and storm frequencies) have a negative influence on where Americans choose to live, and climate-driven changes may influence regional migration within the United States (Fan, Klaiber, & Fisher-Vanden 2012). In coastal areas, sea-level rise and coastal inundation will lead to gradual land loss, and episodic flooding and permafrost melt will contribute to increasingly marginalized land, all of which may contribute to migration and/or require resettlement (Adger et al. 2014). Some migration flows are sensitive to changes in resource availability. For example, rising temperatures and changing precipitation patterns will lead to relative changes in agricultural production, possibly spurring rural to urban migration or migration across borders to seek more favorable conditions (Feng, Oppenheimer, & Schlenker 2012). Climate change is projected to increase the rate of immigration to the United States from Mexico as the country's already marginal water situation deteriorates and as a result of climate-driven changes in agricultural yields (Catarious et al. 2007; Feng, Krueger, & Oppenheimer 2010).

PART 5

INSIGHTS FOR CLIMATE-RISK MANAGEMENT

OPENING COMMENTARY

KAREN FISHER-VANDEN

PENNSYLVANIA STATE UNIVERSITY

PART 5, "Insights for Climate-Risk Management," provides an important extension to the sectoral-impact results presented in part 2 by exploring the implications for mitigation policy and potential adaptation responses. Similar to part 2, chapter 21 ("Mitigation") provides estimates of climate change–induced impacts on labor productivity, agricultural yields, mortality, crime rates, energy expenditures, and coastal damage under three alternative climate futures: a no-policy case (RCP 8.5), a modest climate policy case (RCP 4.5), and an aggressive climate policy case (RCP 2.6). The main message from this analysis is that climate policies will appreciably reduce the mean and variance of climate change–induced impacts in each of these sectors, with certain regions seeing greater reductions in impacts than others. Chapter 22 ("Adaptation") contains new analysis where, through thought experiments, the authors attempt to estimate how adaptation responses will mitigate sectoral impacts (agricultural yields, labor productivity, mortality rates, crime rates, energy expenditures, and coastal damage) from climate change. In each case, the authors find measurable reductions in losses as a result of adaptation responses.

The results of part 5 reflect cutting-edge science and are an important contribution to understanding the extent of climate change–induced sectoral impacts in the United States and how adaptation responses could alleviate these impacts. The authors strategically chose to focus on providing results for which there was ample empirical evidence, while being forthright about the gaps in their analysis. The authors emphasize that their analysis is not a cost-benefit analysis—their estimates only capture the benefits of policies as represented by the alternative RCP 4.5 and RCP 2.6 scenarios. Thus, to evaluate policies to achieve these two stabilization scenarios, the monetized values of these impacts would need to be compared to the cost of the policies, as is done in the integrated assessment model studies discussed in chapter 12.

A shortcoming of this analysis—and other assessments such as those of the IPCC and the National Climate Assessment—is that impacts, adaptation, and mitigation are considered separately, without regard to their interaction. Why is this important? For one, climate impacts could alter the feasibility of emissions-mitigation options. For instance, water required for thermal cooling in the case of nuclear power and stream flow required for hydroelectric power could face severe shortages as a result of climate change. Both are important carbon-free sources of electric power. Also, climate change could negatively affect biofuel crop productivities, another important low-carbon source of energy. Adaptation responses can also affect mitigation options by changing baseline emission levels. This may require greater reductions in greenhouse-gas (GHG) emissions to achieve atmospheric stabilization. For instance, although warming is expected to reduce energy demand for heating in certain regions of the United States, higher energy demand due to greater extensive and intensive air-conditioning use throughout the United States is expected to swamp these reductions in heating demand. This higher energy demand may make stabilizing atmospheric concentrations more costly to achieve.

Most of the studies that attempt to account for climate feedbacks and adaptation responses involve the use of integrated assessment models (IAMs)—computational models that formally couple the economic system with the natural system. A few key points emerge from reviewing the set of recently published integrated assessment modeling studies (see Fisher-Vanden et al. [2013] for a review of these IAM studies). First, none of these studies account for the results of climate impacts and adaptation responses on the set of viable mitigation strategies to reach stabilization targets. Therefore, current model results are likely to underestimate the cost of reaching stabilization goals. Second, no existing study examines how investments in adaptation measures could crowd out investments in mitigation options to reach stabilization targets, which will also lead to an understatement of the cost of meeting stabilization targets. Although a few studies account for investment crowding out, fewer studies account for changes in baseline emissions as a result of adaptation responses. The report also falls short in these areas.

Figure 1 offers a framework for thinking about the interlinkages of climate impacts, mitigation, and adaptation responses in integrated assessment. Emissions-mitigation strategies are targeted at reducing emissions generated by human activities (A). Sequestration strategies, another form of mitigation, target the reduction in GHG concentrations as a result of emissions from human activities (B). Geoengineering strategies attempt to decouple GHG concentrations from climate variables such as temperature (C) and climate variables from physical impacts such as drought and hurricanes (D). These physical impacts result in changes to sectoral productivities (E) and ultimately economic losses (F), both of which are targets of adaptation strategies. The authors of this report attempt to capture a number of the interlinkages depicted in figure 1. In the case of RCP 8.5, the report attempts to capture the full causal chain (A through F in figure 1), while the report only addresses a portion of the chain (A through D) for RCP 2.6 and RCP 4.5.

It is useful to distinguish between three types of adaptation strategies. Type I captures general equilibrium responses to price changes that, through substitution, can mitigate economic losses from productivity shocks. Type II includes adaptive and coping expenditures that mitigate impacts further down the chain, similar to type I, by reducing economic losses due to productivity shocks (e.g., insurance). Lastly, type III includes protective and defensive adaptation responses that reduce shocks to productivity due to physical impacts (e.g., seawalls, protective barriers).

As represented by the blue dashed lines in figure 1, these strategies and responses compete for investment and R&D resources, leading to potential trade-offs as discussed further later. Also, as captured by the red dashed lines, climate-change feedbacks will affect the set of available mitigation and adaptation options and thus optimal decision-making in a cost-benefit framework.

Few studies, including this book, distinguish explicitly between the three types of adaptation depicted in figure 1. Although many IAMs are capable of capturing adaptation responses of the type I (general equilibrium response) variety, only a few capture adaptation responses of the type II or type III varieties. A hybrid modeling approach has been used by some to capture other types of adaptation. In this report, the MUSE results capture some of the type I adaptation strategies, while the thought experiments touch on type II. Similar to other studies, this study does not adequately address these important type II and type III adaptation strategies.

There are a number of reasons for the lack of models with explicit representation of adaptation responses. First, adaptation responses are inherently regional and sectoral, and many models do not have the regional and sectoral detail to capture the variation in climate impacts and responses. Second, proactive adaptation decisions are inherently intertemporal, which explains why a number of models that include adaptation also include intertemporal decision making. Lastly, there has been a desperate lack

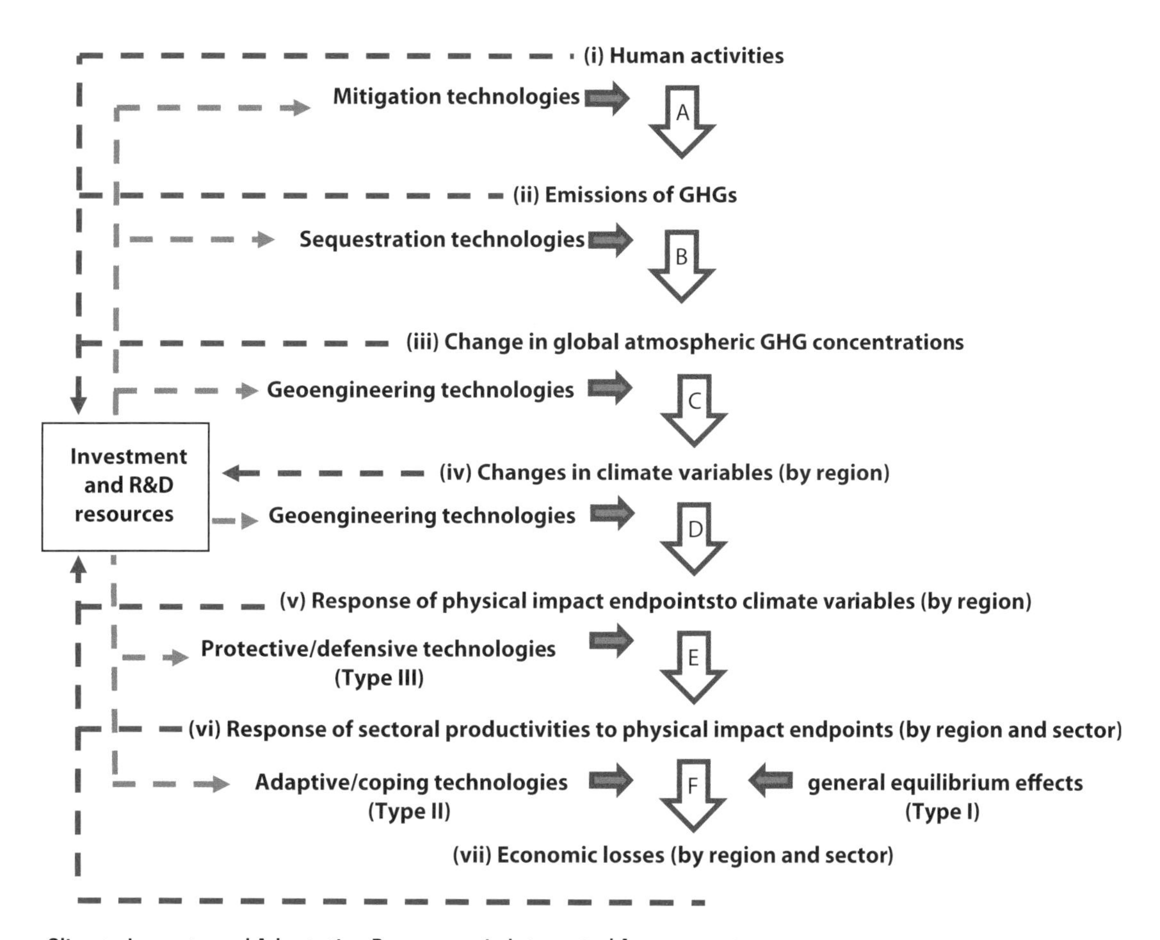

FIGURE 1. **Climate Impacts and Adaptation Responses in Integrated Assessment**

Source: Fisher-Vanden et al. (2013)

Note: GHG, greenhouse gases; R&D, research and development.

of data and empirical evidence on impacts and adaptation necessary for model calibration. Although there has been an uptick in the number of empirical studies on impacts and adaptation recently, these previous studies were not done with the intent of being incorporated into IAMs. As a result, they lack the regional and sectoral coverage to be useful for model calibration and typically collapse the (E) and (F) linkages in figure 1, instead regressing temperature on productivity or economic losses. This disconnect between empirical work and models has necessitated heroic efforts on the part of the modeler to bring empirical knowledge on impacts and adaptation responses into IAMs used for policy analysis. This report, however, attempts to address this research gap by creating IAM-compatible empirical estimates that could be incorporated into IAMs. (Planned future work by the report's research team includes the incorporation of these empirical results into an IAM.)

In sum, there are a number of important research areas to improve our estimates of the economic costs of climate change and options to mitigate. On the empirical side, the analysis has provided a valuable set of results on the direct physical sectoral impacts of climate change under different climate-mitigation scenarios that could be incorporated into an IAM to capture the interactions between mitigation options, climate feedbacks, and adaptation responses. Thus, this report provides an important first step to address the shortage of empirical work on the economic costs of climate impacts and adaptation responses that is desperately needed, along with the translational work that integrates this empirical work into IAMs for policy analysis.

CHAPTER 21

MITIGATION

RISK is the probability of an event occurring multiplied by the impact of that event should it occur. Climate risk can be managed by reducing the probability of costly climate futures by lowering global greenhouse-gas (GHG) emissions and by minimizing the impact of those futures through defensive investments and behavioral adaptation. Like the climate risk itself, the right combination of these two will depend on who you are, where you live, and the time period of concern. It will also depend on the relative cost of each option, which we do not quantify in this report. The fact that our assessment covers a range of global emissions pathways, however, allows us to assess in this chapter the extent to which global efforts to reduce GHG emissions (referred to as "mitigation") can reduce the risks we describe. In the next chapter, we discuss some of the available strategies for adapting to those changes in the climate not avoided through mitigation.

BACKGROUND

As discussed in chapter 3, the scientific community has developed a set of four harmonized Representative Concentration Pathways (RCPs) spanning the plausible range of future atmospheric GHG concentrations (figure 21.1; see also figure 3.1). RCP 8.5 represents a continuation of recent global emissions growth rates, with atmospheric concentrations of CO_2 reaching 940 ppm by 2100. RCP 2.6 reflects a future only achievable by aggressively reducing global emissions (even achieving net negative emissions by this century's end) through a rapid transition to low-carbon energy sources. Two intermediate pathways (RCP 6.0 and RCP 4.5) are consistent with a slowdown in global economic growth and/or a shift away from fossil fuels and other sources of GHG emissions more gradual than in RCP 2.6 (Riahi 2013).

Under RCP 2.6, global GHG emissions peak around 2020, while under RCP 4.5 and RCP 6.0, they peak around 2040 and 2080, respectively. Under all pathways except RCP 8.5, projected emissions for 2020 are below those that actually occurred in 2012 (Le Quéré et al. 2014). This overshoot implies that future emissions reductions need to be faster than those projected in RCPs 2.6, 4.5, and 6.0 to achieve comparable levels of cumulative emissions and therefore comparable climate outcomes.

Moving from RCP 8.5 to RCP 2.6 (as well as to RCP 4.5 or RCP 6.0) will come at a cost. While we do not quantify

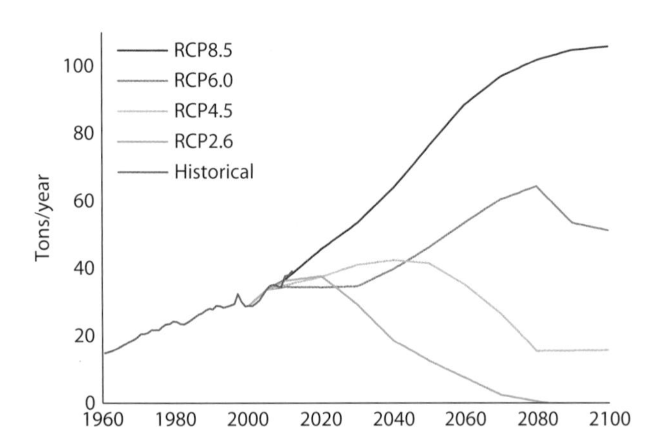

FIGURE 21.1. **Global Net Human-Caused CO_2 Emissions in the Representative Concentration Pathways**

Billion metric tons of CO2 per year

Source: Historical: LeQuere et al. (2014); RCPs: Meinshausen et al. (2011)

these costs in this assessment, there is extensive literature on this topic (Weyant & Kriegler 2014), including a recent summary from Working Group III of the IPCC (Clarke et al. 2014). Moving from RCP 8.5 to RCP 2.6 will also require coordinated global action, and we do not evaluate the prospects of such coordinated action occurring. What our analysis provides, however, is a better understanding of the potential for such action to mitigate the risks to the United States of continuing on the current global emissions pathway, by region of the country and sector of the economy, as well as its limitations.

STEERING THE SHIP

The differences in GHG emissions between RCPs emerge almost immediately. By 2030, global CO2 emissions are 25 percent below actual 2012 emissions in RCP 2.6, 5 percent higher than 2012 emissions in RCP 4.5, and 37 percent higher than 2012 emissions in RCP 8.5. Because of quirks in the way the RCPs are calculated, emissions in RCP 6.0 are below those in RCP 4.5 until the 2040s.

Inertia in the climate system, however, means that this broad range in emissions does not translate immediately into significant differences in temperature. The global mean temperature increase between 1981–2010 and 2020–2039 is *likely* 0.9°F to 1.6°F in RCP 2.6, 1.0°F to 1.6°F in RCP 4.5, and 1.1°F to 1.8°F in RCP 8.5. Because of natural variability and the greater uncertainty in how global changes translate into regional changes, the *likely* temperature projections for the contiguous United States overlap to an even greater extent: 1.2°F to 2.8°F in RCP 2.6 and 1.5°F to 3.2°F in RCP 8.5. The 1-in-20 chance projection for the three RCPs is identical: 3.6°F.

By midcentury, the *likely* global mean temperature changes for RCP 2.6 and 8.5 (1.1°F to 2.2°F in RCP 2.6 and 2.2°F to 3.7°F in RCP 8.5) no longer significantly overlap, although the changes in contiguous U.S. temperature (1.9°F to 3.5°F and 2.6°F to 5.8°F) continue to do so. Only in the second half of the century do temperature differences between the RCPs fully emerge (with a *likely* contiguous U.S. temperature increase of 1.0°F to 2.6°F in RCP 2.6 and 4.7°F to 8.8°F in RCP 8.5 by late century).

When the effects of climate inertia, physical projection uncertainty, and natural variability are combined with the statistical uncertainty in impact projections, the economic benefits of mitigation do not start to be felt until midcentury and are most obvious in the second half of the century. Figure 21.2 illustrates the time evolution of one impact, change in high-risk labor productivity, over the course of the century under RCPs 2.6, 4.5, and 8.5. Over the next 2 decades, the projected labor productivity decline is essentially independent of RCP. By midcentury, the median projection for RCP 2.6 still lies within the *likely* range for RCP 8.5, but differences between RCPs start to be clear in the tails: the 1-in-20 worst-case projection for RCP 2.6 is comparable to the median projection for RCP 8.5. By late century, the differences are large: the

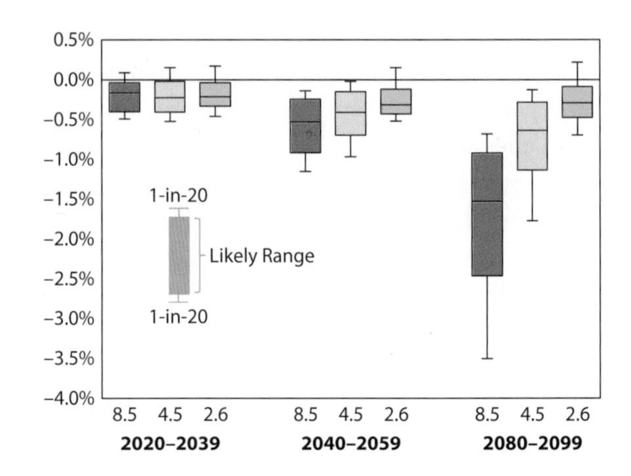

FIGURE 21.2. **Change in High-Risk Labor Productivity**

By time period and RCP

1-in-20 worst-case projection for RCP 2.6 is only slightly below the median projection for RCP 4.5, and the 1-in-20 worst-case projection for RCP 4.5 is only slightly below the median projection for RCP 8.5.

As we highlight in the remainder of this chapter, mitigation today is a crucial tool for managing some types of climate risk in the second half of this century. For the next three decades, however, the climate outcomes are largely already baked into the system. Accordingly, adaptation, as described in the next chapter, is critical for managing climate risk in the near term.

AGRICULTURE

For most of the country, agriculture on average benefits from mitigation, despite the positive effects of CO_2 fertilization. In addition to shifting the average, a major benefit of mitigation for agriculture is to truncate the large tail risk of extremely bad outcomes in major agricultural regions under RCP 8.5. By the end of the century at the national level, extreme events in annual yield losses that were historically 1-in-20 year events become 1-in-2 year events under RCP 8.5, but they are restricted to be only 1-in-5 year events under RCP 2.6 (figure 6.5). Taking the Midwest, the agricultural heartland of the country, as an example, the *likely* range of losses in RCP 8.5 extends from −8.6 percent (a small gain) down to 61 percent, whereas losses under RCP 2.6 can be constrained, spanning the much narrower *likely* range of −4.0 to 14 percent. Similar benefits of mitigation accrue for the Northeast, Southeast, and to a lesser extent the Great Plains. However, mitigation reduces potential agricultural benefits for the Southwest: the *likely* gain shifts from -5.3 percent (a small loss) to +17 percent in RCP 8.5 to -4.3 to +3.9 percent in RCP 2.6. A more exaggerated reduction in potential benefits is clear for the Northwest, although there is very limited production in that region (figure 21.3).

LABOR

The impact of climate change on labor productivity is more evenly spread geographically than for agriculture, as are ubiquitous benefits of mitigation. Midcentury *likely* declines in high-risk labor productivity nationwide are 0.2 to 0.9 percent in RCP 8.5, 0.1 to 0.7 percent

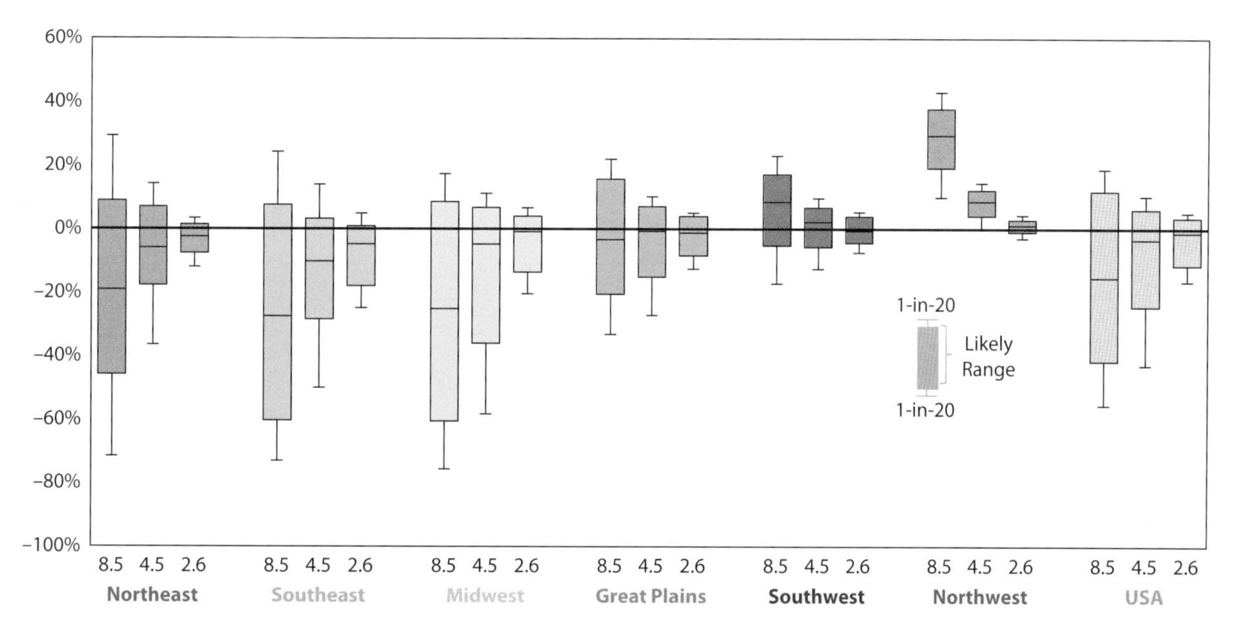

FIGURE 21.3. **Change in Maize, Soy, Wheat, and Cotton Yield, 2080–2099**

By NCA region and RCP

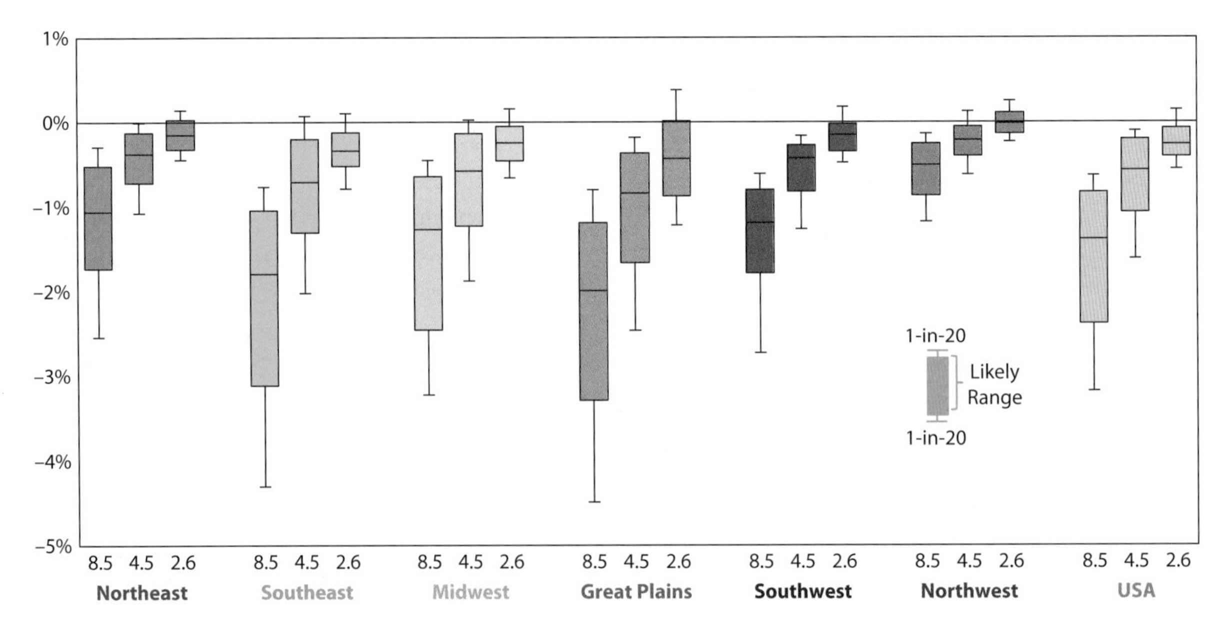

FIGURE 21.4. **Change in Labor Productivity, 2080–2099**

By NCA region and RCP

in RCP 4.5, and 0.1 to 0.4 percent in RCP 2.6. Late century, *likely* declines fall from 0.8 to 2.4 percent in RCP 8.5, to 0.2 to 1.1 percent in RCP 4.5 and 0.1 to 0.4 percent in RCP 2.6 (figure 21.4). Projected labor productivity declines in late century exhibit a long tail, especially in RCP 8.5, with a 1-in-20 chance of declines below 3.2 percent in RCP 8.5 and 1.6 percent in RCP 4.5. All regions see less labor productivity decline in RCP 4.5 and RCP 2.6, though these changes are more economically important for states like Texas, North Dakota, and Louisiana, where high-risk sectors account for a greater share of state employment.

HEALTH

As discussed in chapter 13, the effect of climate change on temperature-related mortality is one of the most economically significant impacts we quantify, as well as one of the most geographically varied. The nonlinear relationship between temperature and mortality has a strong influence over mortality's response to mitigation. Because the heat-related deaths associated with small amounts of warming roughly offset the same number of cold-related deaths, midcentury and late-century outcomes tend to look very similar in RCP 4.5 and RCP 2.6. However, late-century mortality rises rapidly in RCP 8.5 as average temperatures rise and the higher frequency of extremely hot days causes many more heat-related deaths than the number of cold-related deaths that are avoided. This pattern in mortality creates a strong incentive to avoid RCP 8.5 through mitigation, but it suggests little gain in aiming for RCP 2.6 relative to RCP 4.5.

In RCP 8.5, the *likely* increase in temperature-related mortality is 3.7 to 20.8 deaths per 100,000 people on average between 2080 and 2099. In RCP 4.5 the *likely* range falls to −2.5 to +5.9, and in RCP 2.6 it falls to −2.3 to 3.2 (figure 21.5). Projected mortality increases in late century exhibit a long tail, especially in RCP 8.5, with a 1-in-20 chance of increases greater than 36 under RCP 8.5, 12 under RCP 4.5, and 5 under RCP 2.6.

The differences between RCPs is even more significant for certain regions than for the country as a whole. The Southeast, southern Great Plains states, and parts of the Southwest will *likely* see steep declines in temperature-related mortality in RCP 4.5 or RCP 2.6 compared to RCP 8.5, while the Northwest will *likely* see an increase. The mortality benefits of mitigation in the Northeast and Midwest are more mixed.

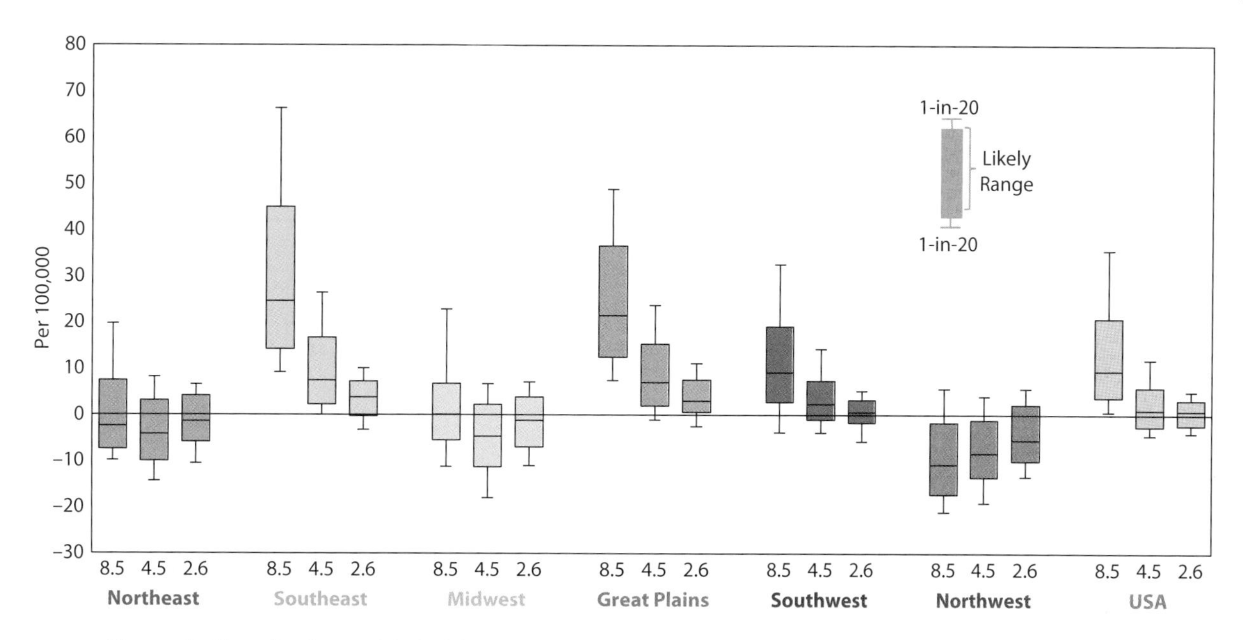

FIGURE 21.5. **Change in Mortality Rates, 2080–2099**

Deaths per 100,000, by NCA region and RCP

CRIME

The overall impact of climate change on crime rates is unambiguous but modest, particularly when compared to other factors. Climate-driven increases in crime rates are lower under RCP 4.5 and RCP 2.6 than under RCP 8.5, but not significantly so until late century. Between 2080 and 2099, the *likely* increase in the violent crime rate is 1.9 to 4.5 percent under RCP 8.5, 0.6 to 2.5 percent under RCP 4.5, and −0.1 to 1.3 percent under RCP 2.6 (figure 21.6). As with labor productivity, the reduction in climate-driven crime rate increase is comparable across regions, though the economic benefit is concentrated in states with higher baseline crime rates.

ENERGY

The impact of mitigation on climate-driven increases in nationwide energy expenditures is ambiguous until mid-century, at which point projected increases are roughly half as high under RCP 2.6 as RCP 8.5, with little difference between RCP 2.6 and RCP 4.5. By the end of the century, nationwide cost increases are considerably lower in RCP 2.6 than RCP 4.5, with *likely* ranges of 0.8 to 4.3 percent and 1.8 to 8.8 percent, respectively, which in turn are considerably lower than for RCP 8.5, with a *likely* range of 8.0 to 22 percent (figure 21.7). The largest declines in energy expenditures between RCP 8.5, RCP 4.5, and RCP 2.6 are in the Southeast, Great Plains, and Southwest—the regions that see the largest increases under RCP 8.5. Energy expenditures decline much more modestly in the Northeast and remain relatively unchanged in the Northwest.

COASTAL

Because sea-level rise (SLR) responds more slowly than temperature to changes in emissions, the impact of climate change on coastal communities is, in this century, the least responsive to changes in global emissions of the impacts we assessed. In 2050, additional current coastal property value *likely* below mean higher high water (MHHW) levels due to SLR is $323 billion to $389 billion in RCP 8.5. In RCP 2.6, this only falls to $287 billion to $360 billion. The *likely* SLR-driven increase in average annual hurricane damage in 2050 is $5.8 billion to $13 billion

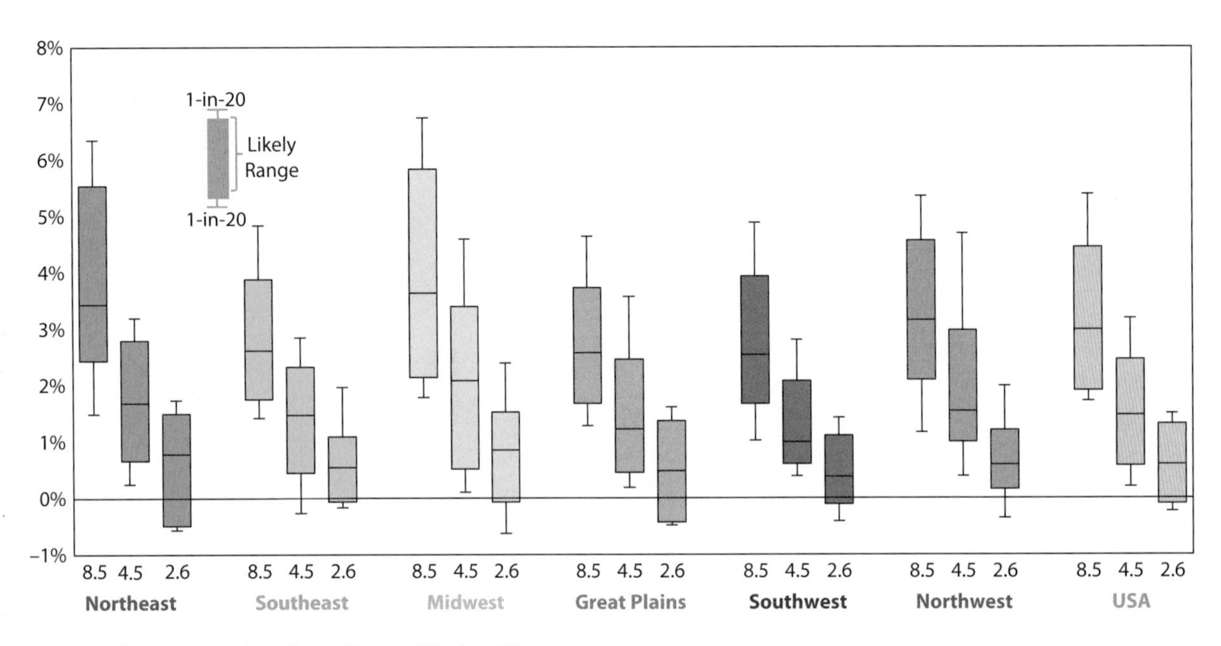

FIGURE 21.6. **Change in Violent Crime Rates, 2080–2099**

By NCA region and RCP

in RCP 8.5, $5.0 billion to $11 billion in RCP 4.5, and $4.6 billion to $10 billion in RCP 2.6.

By the end of the century, at which point the median projected increase in global sea level differs between RCP 2.6 and RCP 8.5 by about 1 foot (from a total of about 2.6 feet in RCP 8.5), there is a slightly greater difference between RCPs. *Likely* average annual inundation from SLR and SLR-driven increases in average annual hurricane damage combined are $26 billion to $43 billion between 2080 and 2099 in RCP 8.5. In

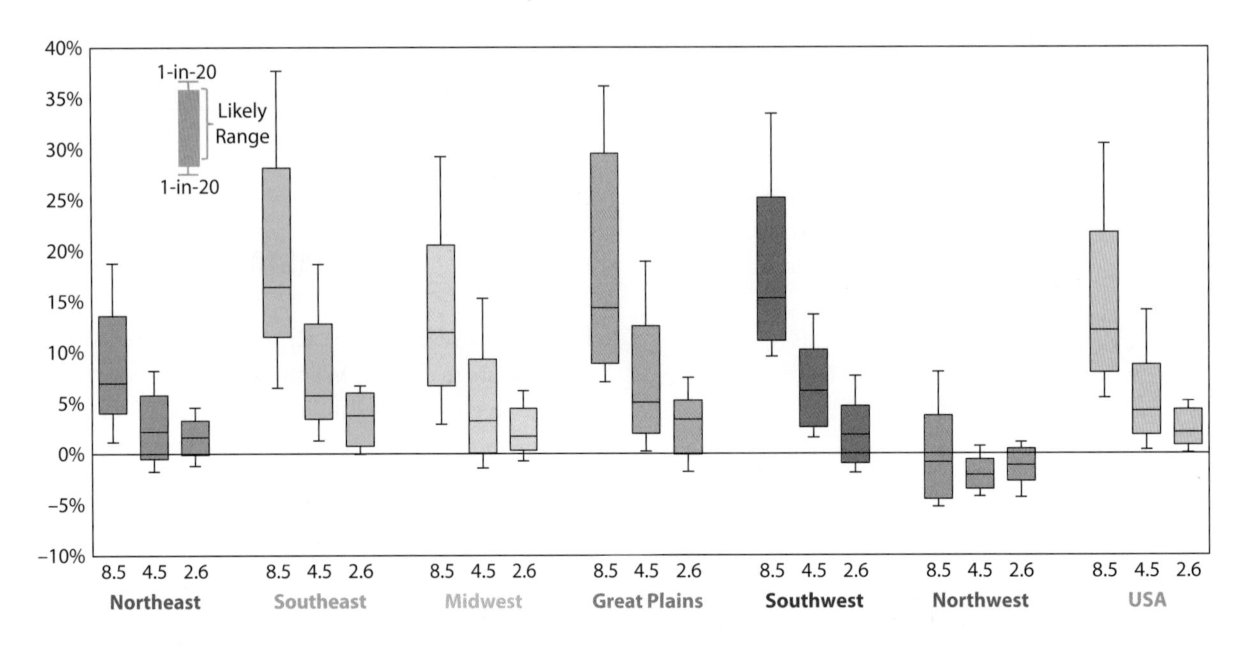

FIGURE 21.7. **Change in Energy Expenditures, 2080–2099**

By NCA region and RCP

RCP 4.5 the damage is reduced to $20 billion to $33 billion and in RCP 2.6 to $15 billion to $29 billion. This benefit is concentrated in the Northeast and Southeast, where most of the coastal inundation and hurricane risk exists (figure 21.8).

OTHER RISKS

Not all the risks that mitigation can help manage or avoid are quantified in our economic analysis. As climate conditions pass further outside the realm the planet has experienced for the past several million years, the odds of passing tipping points like those discussed in chapter 3 or of triggering unexpected planetary behaviors increases. Under RCP 8.5, the magnitude of the *likely* global warming by the first half of the next century (about 9°F to 18°F since pre-Industrial items by around 2150) will be unprecedented in the past 56 million years (see discussion of the Paleocene-Eocene Thermal Maximum in chapter 5). Under RCP 4.5, the *likely* global temperatures at the end of the century (about 2.2°F to 5.5°F higher than 1981–2010) will be comparable to those the planet last experienced about 3 million years ago (Hill et al. 2014). Under RCP 2.6, the *likely* increase in global mean temperature is limited to about 0.9°F to 2.6°F, maintaining temperatures close to a range last experienced about 125,000 years ago (Turney & Jones 2010).

As described in chapter 4, by late century under RCP 8.5, about a third of the American population (assuming the geographic distribution of population remains unchanged) is expected to experience days so hot and humid that less than an hour of moderate, shaded activity outside can trigger heat stroke (category IV on the ACP Humid Heat Stroke Index) at least once a year on average. Under RCP 4.5, only one-eighth of the population is expected to experience such a day at least once a decade on average; under RCP 2.6, the risk of such conditions is negligible for almost all Americans.

Given these and other risks not included in our economic analysis, as discussed in part 4, the results in this chapter should be viewed as highlighting the capacity of mitigation to reduce the specific set of climate risks that we have evaluated—and thus a near-certain underestimate of the overall benefit of mitigation for managing climate risk, especially in the second half of the century and beyond.

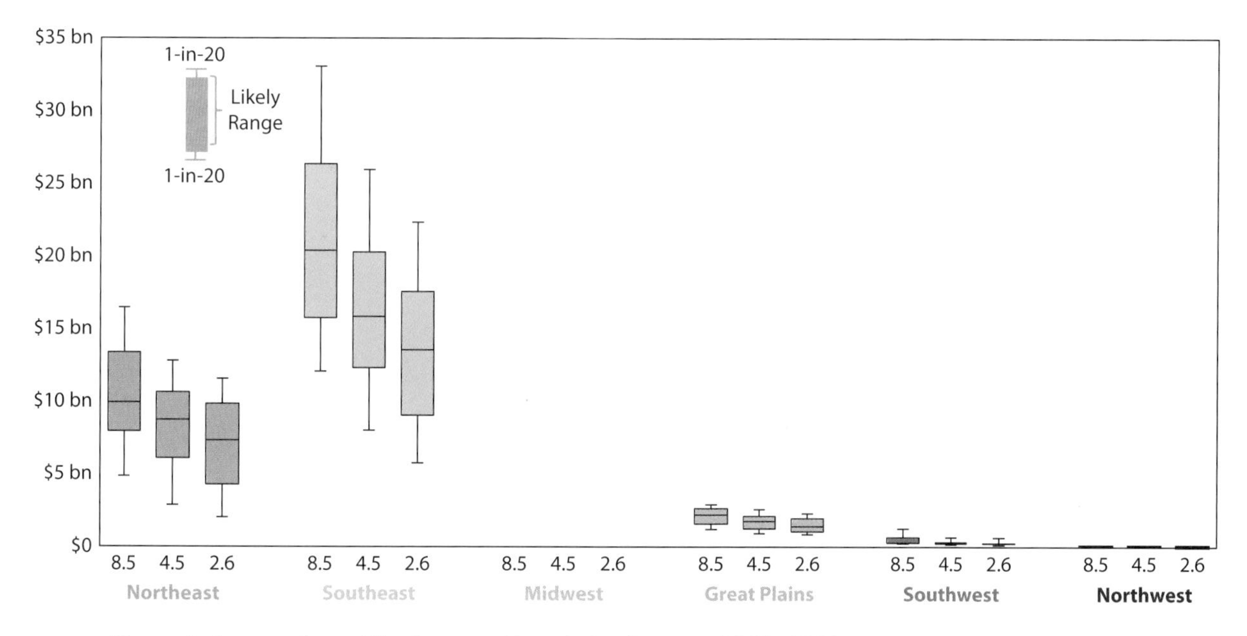

FIGURE 21.8. **Change in Average Annual Hurricane and Inundation Damage, 2080–2099**

Billion 2011 U.S. dollars

CHAPTER 22

ADAPTATION

WHILE global emission reductions can mitigate much of the climate risk Americans face, as shown in the preceding chapter, there are some climatic changes that are already "baked in" as a result of past greenhouse-gas emissions and will occur regardless of how emission levels change. In addition, many decision makers, from individual businesses and homeowners to local governments, have limited ability to affect global emissions directly and need to prepare for a range of plausible climate futures. Armed with the kind of forward-looking information provided in this report, decision makers can, however, reduce their risk exposure through adaptation.

This report provides new information on the relative economic impact of climate change on different sectors, allowing decision makers to consider where they might focus adaptation efforts. In this chapter, we consider how future populations might adapt, we demonstrate how the modeling approaches used in this report could be used to understand better the potential gains from adaptation, and we highlight weaknesses in our current understanding of the costs and benefits of adaptation.

BACKGROUND

When the climate changes and imposes economic costs on populations, these populations will respond to these changes in an effort to cope. Americans will adapt to climate change, changing how they live and how they do business in ways that are better suited for their altered environment. In general, populations adapt to climatic changes in two different ways: they change their behavior or they make "defensive investments" in new capital that mitigates the effect of climate. Behavioral changes may involve small or large changes in the actions people undertake, whether they change the time of day that they exercise, plant crops earlier in the season, or move to a different city. Defensive investments are capital investments that individuals or firms make to minimize the effects of climate that they would not have undertaken in a less adverse climate, such as the purchase of air conditioners, the building of seawalls, or the installation of irrigation infrastructure. Both behavioral adaptations and defensive investments are visible in the modern economy. For example, residential air-conditioning penetration is nearly complete in the South, where summers are already uncomfortably warm and humid, and irrigation is extensive in the West, where climates are already arid. Our ability to observe these adaptations demonstrates both their technical and economic feasibility, a notion that encourages us to believe that these adaptations will play an important role in the future of the American economy.

Importantly, however, all of these adaptations have some economic costs and may not be suitable for all future contexts. At present, we do not have a strong understanding of the costs involved with the numerous potential adaptive behaviors and investments that are currently available—although we do know that costs are involved, because if these behaviors and investments were costless, we would expect them to be much more broadly employed at present (Hsiang & Narita 2012). For example, the fact that some households currently have air conditioners and some do not tells us that some families find this investment worth the cost it imposes and some do not, perhaps because the latter households do not experience extreme heat as often or because purchasing an air conditioner would require that the household forego other expenses that are more essential to their well-being, such as food or education.

Developing a full understanding of the economics of adaptation is an important question closely related to the analysis presented throughout this report. In future analyses, we hope that researchers will provide the details needed to evaluate carefully both the costs and benefits of various adaptive actions and investments, which will enable the design of policies that optimally facilitate adaptation. We note, however, that the quantity of information required to undertake such an exercise is even greater than what we use here: In addition to knowing (1) how the climate affects people, we must also know (2) how adaptation mediates this effect in quantitative terms (i.e., the *benefits* of adaptation), as well as (3) what populations sacrifice in order to undertake these adaptations (i.e., the *costs* of adaptation). Material in this report has relied heavily on (1), insights that are just now becoming available due to scientific advances, whereas both (2) and (3) require additional research innovations that build on what has already been achieved. To understand how adaptation mitigates the impact of climate on a certain dimension of the economy, we must first develop techniques to measure the effect of climate in the absence of additional adaptation, which is the focus of this report and the research underlying it, and then we must develop techniques to measure how new behavioral changes or defensive investments alter the quantitative structure of this linkage and the cost of these actions. Because the latter remains a generally unanswered question, our assessment is that the potential gains from adaptation remain unknown, but they may be understood in the near future as research advances.

Because the current body of research is insufficient to project expected patterns of adaptation and their costs and benefits, this book has been exclusively focused on the direct effects of climate change—assuming populations respond to climate similarly as they have in the recent past—and their general equilibrium effects. Nonetheless, because we are certain populations will adapt even in the absence of government actions, it is worth considering what some example adjustments might look like for illustrative purposes, even if we cannot fully quantify their impact and cannot yet evaluate their full economic costs or benefits.

AGRICULTURE

Agriculture is a sector where producers have been adapting to their climate for millennia. We expect that as the climate changes in the future, farmers will make numerous adjustments in an effort to cope with these changes. As explained earlier, it remains difficult to evaluate fully the cost, benefits, and effectiveness of each one of these adjustments individually, although we do have some sense of what various adjustments might look like based on historically observed adaptations.

For example, we expect that farmers will adjust which crops they plant, shifting toward varieties or products that are more conducive to their new local conditions, probably because they are more tolerant of extreme heat (Mendelsohn, Nordhaus, & Shaw, 1994). Current research is insufficient to evaluate the costs of this transition, so it is difficult to know how many farmers will make which crop transitions at which points in time and what their net benefits will be. It is also likely that some producers will change croplands to rangeland, that farmers will expand their use of irrigation to help mitigate the effects of rainfall loss and extreme heat, and that farmers will change their planting dates to earlier in the season so that crops will be exposed to less adverse planting conditions. It is also possible that patterns of agricultural production will migrate northward, so that land in the North and West that was not previously used for agriculture but has rising productivity due to warming is brought into production. Longer growing seasons in other parts of the country may enable double- or triple-cropping, even if individual crop yields decline. Individuals from farming communities may also simply migrate out of those communities as economic production declines, as was observed in the Dust Bowl (Hornbeck 2012) and in more recent years (Feng, Oppenheimer, & Schlenker 2012). Finally, genetic technologies and advances in breeding may produce more heat-tolerant and drought-tolerant varieties of crops in the future. In the past, such efforts have had mixed success, with some advances revolutionizing production in local areas, such as the development of varieties that enabled widespread cultivation in the American West (Olmstead & Rhode 2011), while in other cases breeding advances brought little to no benefits for decades, such as the persistent heat sensitivity of maize in the Eastern United States (Schlenker & Roberts 2011; Burke & Emerick 2012). Because genetic innovations are of a "hit or miss" nature, they are more speculative and more difficult to depend on in comparison to other adaptive measures, such as irrigation, where technologies already exist.

Given the state of research, we lack the necessary information to quantify the potential economic benefits and costs associated with these various forms of adaptation. However, we can use existing data to get a quantitative sense for the collective benefits for a subset of these adjustments. Populations have adapted to their local climates in the past, and that provides us with some information about how effectively they utilize technologies that are already within reach. For example, irrigation is used extensively in the West but less so in the East, a fact that makes maize production in the West less sensitive to extreme heat (Schlenker & Roberts 2009).

To address this question, we can do a thought experiment in which we ask what maize yields would look like if farmers in the East started adopting the farming practices of farmers in the West. This exercise is useful for helping us think about the potential heat-resistance of this particular sector, but it is only half of the story because dramatically expanding irrigation and changing varieties will have costs that we are not measuring (Schlenker, Roberts, & Lobell 2013). A benefit of this approach is that it allows us to model future adaptations using simple assumptions that are calibrated to the actual, real-world behavior of adapting individuals, while a weakness of this approach is that it does not allow us to identify separately the effects of specific adaptive actions. For example, we cannot attribute separate gains to irrigation as opposed to changes in the varieties that farmers plant.

To conduct this thought experiment, we also need to have some sense for how quickly populations in the East can adopt the practices of the West. It is possible that in the future this transition could occur quickly, if policy promotes this behavior. But without knowing whether such policies will occur, a reasonable starting place is to observe how quickly adaptations of this form have occurred in the past. In the case of maize, recent research has estimated the speed with which the relationship between temperature and yields in the East has evolved to look like the relationship in the West (Burke & Emerick 2012). The measured rate of convergence is 0.28 percent per year, a very modest value that stands in contrast to some historical experiences in which new technologies were deployed more rapidly (Olmstead & Rhode 2011).

When we calibrate our model to allow for this thought experiment in adaptation, where farmers who experience hotter weather adopt techniques that weaken their dependence on temperature at historically observed rates (see appendix B for details), we see a modest reduction in projected losses in maize production as shown in figure 22.1. In the median case, national production is estimated to fall only 49 percent by the end of the century if farmers undertake additional adaptations, rather than the estimated 60 percent that would occur if we assume maize producers do not adapt beyond their current levels.

The "residual damage" that remains in this particular adaptation scenario is not dramatically smaller than the damage estimated in the benchmark scenario, primarily because observed historical adaptations have been relatively slow. Importantly, however, we stress that (1) future technological change and adoption may proceed at a faster (or slower) pace relative to recent history, and (2) this is just one type of possible adaptation. Other adaptations, such as the northward movement of maize production, change in varietals, double- and triple-cropping, and genetic innovation, which are not captured in this simple exercise, may have greater (or smaller) effects. Furthermore, we do not know the costs associated with these changes (e.g., overhauling irrigation systems is expensive), so we cannot be sure whether it will make economic sense to follow this trajectory.

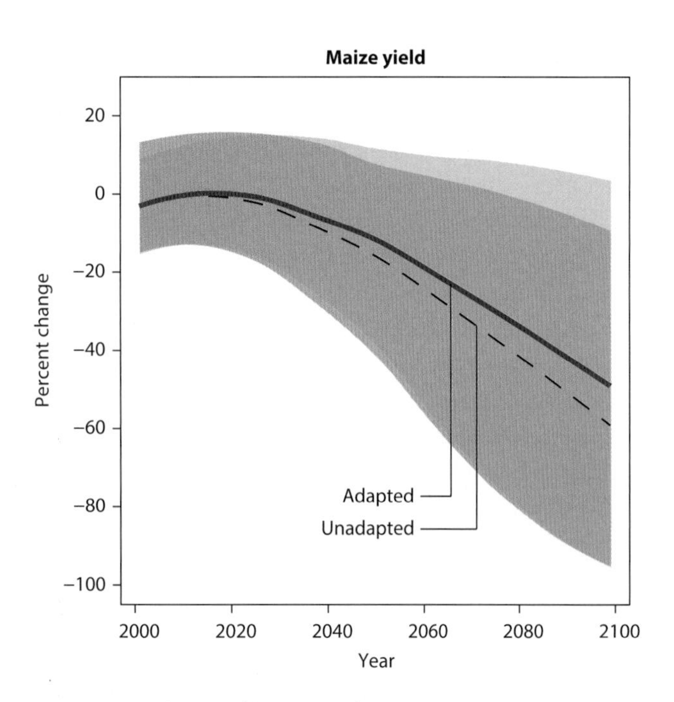

FIGURE 22.1. **National Maize Production Loss**

RCP 8.5, without adaptation and a scenario with adaptation calibrated to historical patterns; for example, expanded irrigation (median and 90 percent range)

LABOR

Behavioral change and defensive investments may also help reduce some, but not all, of the impacts on labor productivity associated with rising temperatures across the United States. The extent to which adaptation can mitigate these risks will vary by industry, in particular between high- and low-risk sectors, but also among subsectors within these categories.

For high-risk sectors, laborers exposed to heat may shift working hours to cooler times of the day or reschedule activities for a cooler day. It is unclear how much such temporal substitutions can minimize the impact of extreme temperatures, though one study by Graff Zivin and Neidell (2014) found very limited intertemporal substitution (shifting activity across days) of high-risk labor. Intratemporal substitution (shifting labor to different times of the day) has not significantly altered the observed influence of warm days on labor supply, perhaps because in many cases workers have little discretion about their activities during core business hours. Other behavioral adaptive strategies focus on improved work practices, such as periodic rest, hydration, and training to facilitate acclimatization over time, which have long been part of the Occupational Safety and Health Administration's guidelines for worker safety. Finally, behavioral adaptation may also take the form of labor migration from the warmer areas of the South to the North, where conditions may be more bearable.

Adaptation investments can also help to reduce heat exposure and the risk of heat-related impacts on labor productivity, including engineering controls, such as air-conditioning and ventilation, which are feasible when production occurs indoors. It is likely that over the course of the next century, new technologies will be developed to mitigate climate's impact on labor productivity further. While it is impossible to predict what these technologies may make possible, this assessment provides insight into the potential value of such innovations to help businesses and American workers cope with future climate change.

HEALTH

As the climate changes, Americans will likely change their behaviors and make defensive investments to protect their health. Historically, public health programs and education campaigns have been effective in encouraging individuals to adopt behaviors that are both privately beneficial and improve community-level health outcomes. Privately motivated investments, particularly investments in air-conditioning, are likely to have been important in reducing mortality due to extreme heat (Barrecca et al. 2013). Since 1950, heat-related mortality has declined in association with the rapid rise in residential air-conditioning usage, with penetration rates climbing roughly 1.5 percent per year from zero in 1950 to 80 percent in 2004. The downward historical trend in heat-related mortality, if extrapolated into the future, might suggest that we can continue to make strong gains in cutting future temperature-related mortality. However, caution is warranted, as growth in residential air-conditioning penetration has slowed substantially in the past decade, with overall penetration growth rates slowing to around 0.4 percent per year. (If this current rate is simply extrapolated into the future, air-conditioning penetration would be expected to hit 100 percent around 2055.)

In this context, it is difficult to know for sure how future temperature-related mortality will evolve. While continued air-conditioning uptake will accrue some benefits, it is unclear how much further continued air-conditioning adoption can depress heat-related mortality. Much heat-related mortality occurs outside of homes (Barrecca et al. 2013), and the heat-mortality response in heavily air-conditioned populations—such as in the South, where penetration is close to complete—remains similar in magnitude to the estimates presented in this analysis. Thus, it seems unlikely that autonomous expansion of air-conditioning usage alone will mitigate all heat-related mortality in the future, and it is likely that adaptation on other margins will be necessary.

Nonetheless, we may implement a thought experiment, similar to the one earlier for maize, where we use regional patterns of heat-related mortality to consider potential gains from currently available technologies and practices. In this case, we assume the South represents highly adapted counties and imagine that other locations begin to behave as the South does. We then use trends in sensitivity over time to calibrate the rate at which cooler populations start to behave as the South does. Results from this calibrated thought experiment are shown in figure 22.2, where we see substantial reductions in temperature-related mortality in the RCP 8.5 scenario. In the median case for this thought experiment, additional mortality is estimated to rise by only 7 deaths per 100,000 by the end of the century, compared to 13 deaths per 100,000 in our baseline projections. As mentioned earlier, it is unclear whether this relatively rapid and effective adaptation is feasible, as our estimate relies on extrapolating the downward trend in heat-related mortality that was likely driven by rapid adoption of air-conditioning, and future adoption of air-conditioning may have limited impact because penetration rates cannot exceed 100 percent.

An additional feature of these results worth noting is the substantial broadening of the 90 percent confidence interval in figure 22.2 when adaptation is modeled explicitly (light blue shading). Uncertainty increases when we account for adaptation using empirical results because the parameters governing adaptation responses are estimated using real-world data and thus exhibit statistical uncertainty (see appendix B for details). Given the current state of research, these uncertainties are large, causing projections to be more uncertain. We hope future research on adaptation may effectively reduce these uncertainties.

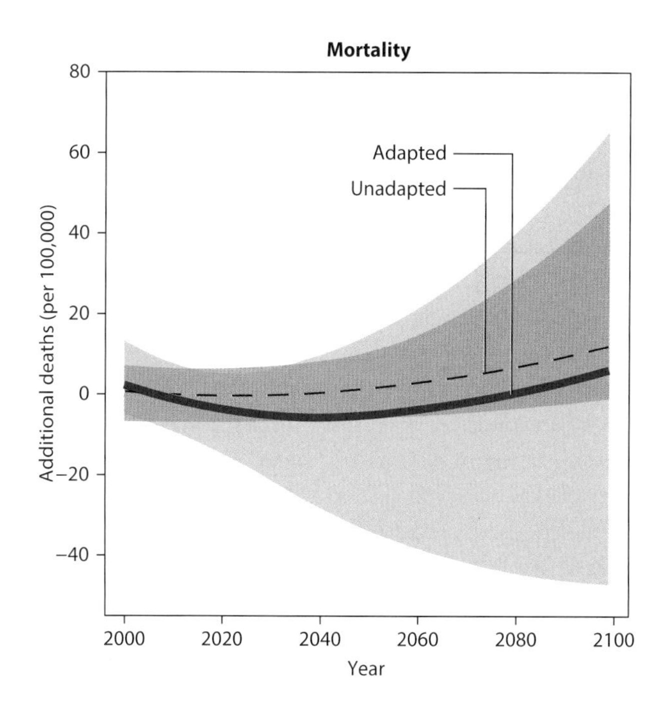

FIGURE 22.2. **National Temperature-Related Mortality Rate Changes**

RCP 8.5, without adaptation and a scenario with adaptation calibrated to historical patterns; for example, continued air-conditioning adoption (median and 90 percent range; deaths per 100,000)

CRIME

Temperature-related effects on violent crime rates is yet another case where behavior changes or defensive investments may potentially offset future climate impacts. In general, it is thought that violent crime increases in the heat because of physiologic and psychological responses (Hsiang, Burke, & Miguel 2013), and it is unclear how these responses can be mitigated directly. However, it is possible that using adaptive law-enforcement practices, such as deploying more police officers on abnormally hot days, potential criminals can be deterred enough to offset the effects of elevated temperature. The continued adoption of air-conditioning may also help, as may other small adjustments that reduce the exposure of individuals to higher temperatures. Still, existing evidence suggests that this type of adaptation is challenging: Unlike with mortality and maize yields, we do not see strong evidence that warmer populations exhibit very different violent crime responses to temperature than do colder populations (Ranson 2014), suggesting that there are not currently technologies or practices in use that substantially mitigate this impact.

Nonetheless, historically, the linkage between temperature and violent crime has weakened very slowly (Ranson 2014), and we can use this trend in a similar fashion to the examples given earlier, where we model populations as adapting by allowing their behavior to converge to the empirically observed response of modern warm counties. We display results from this thought exercise in figure 22.3, where we see that accounting for historically observed patterns of adaptation mitigates a small portion of the climate-related change in crime rates in RCP 8.5. In the median case, additional violent crime is estimated to rise 3.5 percent by the end of the century if populations undertake additional adaptations, rather than the estimated 3.7 percent rise that we project if we assume populations do not adapt beyond their current levels.

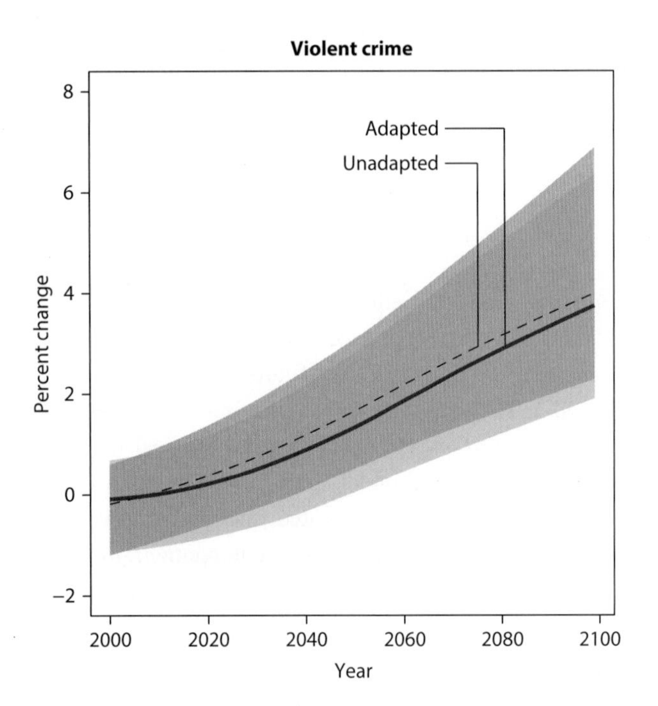

FIGURE 22.3. **National Climate-Related Violent Crime Changes**

RCP 8.5, without adaptation and a scenario with adaptation calibrated to historical patterns; for example, weather-dependent deployment of police officers (median and 90 percent range)

ENERGY

Several defensive investments and behavioral changes can reduce the impact of the projected changes in temperature discussed in chapter 4. Improvements in the efficiency of air conditioners and building shells can reduce the amount of electricity required to maintain comfortable indoor temperatures on increasingly hot days, as can changes in how building systems are operated. More expensive electricity could lead to improvements in the efficiency of electricity-consuming devices and behavioral shifts that affect electricity demand. Breakthroughs in electricity-storage technology and demand-response programs could enable wholesale power-market operators and utilities to meet climate-driven increases in daily peak demand without necessitating the construction of as much new generation capacity.

All of these effects are captured to some extent in RHG-NEMS, the energy-sector model used in quantifying the impact of projected changes in temperature on electricity demand, total energy demand, and energy costs described in chapter 10. The rate of technological learning in air-conditioning is affected by changes in electricity prices, which occur in our modeling. The same is true for consumer demand for cooling services and efficiency of end-use appliances and devices. RHG-NEMS estimates match historically observed responses as found in the econometric literature, but could be too conservative about future efficiency advances.

The supply side is the area where adaptive advances are likely to have the greatest mitigating effect on climate-driven energy cost increases. The development and deployment of grid storage technology would significantly reduce the need for the additional generation capacity described in chapter 10. The cost of such technology, however, remains prohibitively high for widespread commercialization, which is why little of it occurs in RHG-NEMS.

COASTAL COMMUNITIES

As shown in the previous chapter, the risks of climate change to coastal communities are some of the least sensitive to changes in global emissions, at least in this century.

Fortunately, households, businesses, community organizations, and local governments along the coast have considerable adaptive capacity. To explore the extent to which the construction of coastal defenses, such as seawalls, building modifications, and beach nourishment, can reduce the economic cost of inundation from mean sea-level rise and SLR-driven increases in storm surge described in chapter 11, we partnered with Industrial Economics, Inc. (IEc), the developers of the National Coastal Property Model (NCPM).

NCPM comprehensively examines the contiguous U.S. coast at a detailed 150 × 150 m (about 500 × 500 feet) grid level; incorporates site-specific elevation, storm surge, and property value data; estimates cost-effective responses to the threats of inundation and flooding; and provides economic-impact results for four categories of response: shoreline armoring, beach nourishment, structural elevation, and property abandonment (Neumann et al. 2014). The model was originally developed to address the threat of SLR and was modified to incorporate the effect of storm surge on estimates of vulnerability, impact, adaptation response, and economic damage (see appendix C).

IEc assessed the cost of inundation and greater storm surge from mean SLR, using the same local SLR projections used in the RMS North Atlantic Hurricane Model. In one scenario, IEc assumed no defensive investments are made (consistent with our baseline analysis described in chapter 11) and found costs between now and 2100 similar to those from RMS. In a second scenario, IEc assessed the extent to which defensive investments that can be made by individual property owners (i.e., structural elevation) can reduce these costs. In a third scenario, IEc adds beach nourishment to the adaptation options basket, a defensive investment that generally requires collective community action. In a fourth scenario, shoreline armoring is added, the option that likely requires the greatest degree of collective/public action.

IEc finds that more than two thirds of projected inundation damage from *likely* SLR in each decade of the century can be avoided through proactive investments in shoreline armoring and beach nourishment, though both will require substantial public coordination. Adaptation is less effective in coping with lower probability, higher SLR projections, but can still cut projected costs by more than half. IEc finds adaptation similarly effective in reducing SLR increases in costal storm flooding, with structural elevation added to shoreline armoring and beach nourishment.

A range of barriers can prevent adaptation from occurring in an economically optimal fashion, including government-backed flood insurance that shields coastal homeowners from the cost of hurricane-related flooding and local opposition to shoreline armoring or structural elevation. Indeed, these factors exacerbate coastal property risks today. IEc finds that 86 percent of expected hurricane flood damage at current sea levels could be avoided through economically efficient adaptive investments that are not occurring.

INFORMING ADAPTATION

While Americans will likely reduce at least some of the impacts of climate change on coastal property, energy systems, crime rates, public health, labor productivity, and agricultural production through behavioral change and defensive investments, such adaptive measures are unlikely to occur (at least in a relatively efficient manner) without adequate information regarding the economic risks these investments and behavioral changes are intended to address. Climate change is not, and will not, manifest through consistent year-to-year increases in temperature or changes in precipitation. Storm damage does not occur evenly every year, and neither will climate-driven changes in storm flooding. The weather will continue to be variable. If adaptive decisions are made based either on that year's weather or past experience, businesses, households, and policy makers will always be behind the curve. The goal of this assessment is to provide the best available information on what is coming down the road so that these individuals may make well-informed decisions.

A better understanding of the costs, benefits, and limitations of adaptation is also critical in informing household, business, and policy decisions. In our view, empirical work on the benefits of adaptation is currently highly limited and uncertain, while empirical research on the cost of these adaptations is almost nonexistent. Determining the economic valuations of specific adaptive investments and actions is a critical area of research because there are many unanswered questions. Reliable quantitative estimates will be key in determining the best private-sector and policy responses to climate change.

TECHNICAL APPENDIXES

APPENDIX A

PHYSICAL CLIMATE PROJECTIONS

ROBERT KOPP AND D.J. RASMUSSEN

1. TEMPERATURE, PRECIPITATION, AND HUMIDITY PROJECTIONS

In constructing the ensemble of temperature and precipitation projections used in this analysis, we are trying to address two challenges. First, and preeminently, ensembles of opportunity like the Coupled Model Intercomparison Project (CMIP) ensembles are not probability distributions because they are arbitrarily compiled; any sampling from such a distribution would thus not be random and would likely undersample "tails" of the probability distribution (Tebaldi and Knutti 2007). Yet while simple climate models like MAGICC (Meinshausen et al. 2011) project probability distributions of global mean temperature change, they lack the spatial and temporal resolution needed to estimate climate risk, which downscaled global climate model (GCM) output can provide. We seek to combine these strengths. Second, traditional pattern scaling approaches estimate forced climate change but neglect unforced variability (e.g., Mitchell 2003). The interactions between these two factors may play a significant role in the timing and amplitude of climate-change damage, so here we estimate the combination and forced and unforced variability.

We start with an estimated probability distribution of global mean temperatures over time from a simple climate model, use this distribution to weight local projections of monthly temperature and precipitation from more complex global climate models and from surrogate models employed to ensure the tails of the probability distribution are represented, then use historical relationships to translate monthly values to daily values. We also construct a probabilistic estimate of wet-bulb temperatures based on daily projections.

1.1. Global mean temperature

Projections of global mean temperature for the four Representative Concentration Pathways (RCPs) were calculated using MAGICC6 (Meinshausen et al. 2011) in probabilistic mode. MAGICC is a commonly used simple climate model

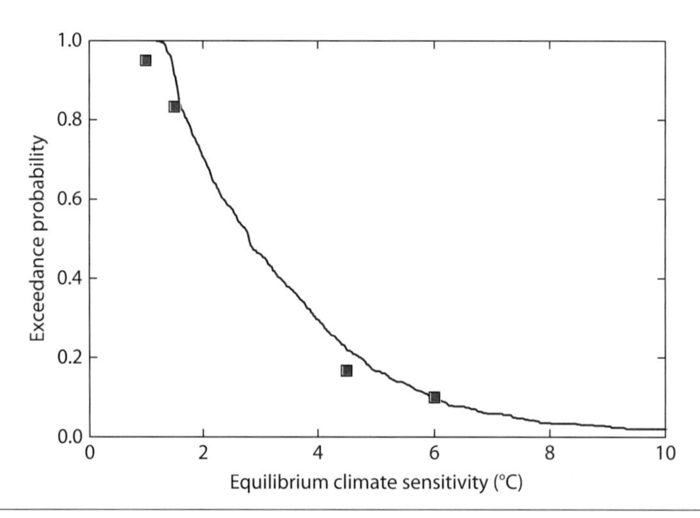

FIGURE A1. Survival function of climate sensitivities from MAGICC. Red squares indicate the statements made by AR5.

that represents the atmosphere, ocean, and carbon cycle at a hemispherically averaged level. The distribution of input parameters for MAGICC that we employ has been constructed from a Bayesian analysis based on historical observations (Meinshausen et al. 2009; Rogelj, Meinhausen & Knutti 2012) and the climate sensitivity probability distribution of the Intergovernmental Panel on Climate Change's Fifth Assessment Report (AR5) (figure A1) (Collins et al. 2013).

The climate sensitivity probability distribution from AR5 is based on several lines of information. Observational, paleoclimatic, and feedback analysis evidence indicate 5th/17th/83rd percentiles of 1.0°C/1.5°C/4.5°C. Additional evidence from climate models suggests a 90th percentile of 6.0°C (Collins et al. 2013).

For each RCP, we used 600 MAGICC model runs provided by M. Meinshausen (pers. comm.). The 5th/17th/83rd/90th percentiles of equilibrium climate sensitivity for these 600 runs are 1.5°C/1.6°C/4.9°C/5.9°C per CO_2 doubling. The differences in climate sensitivity between MAGICC and AR5 in part reflect sampling and in part the constraints needed to fit historical observations with the MAGICC model structure.

1.2. Global climate model output

Because of computational constraints, GCM results are often calculated with horizontal resolutions too coarse (e.g., around 2° × 2°) to assess climate change vulnerabilities and impacts at a spatial scale of interest to regional planners and policy makers. Additionally, GCM projections that are directly available from the CMIP5 archive contain systematic model biases that must be corrected before being employed to address climate impacts. In this study, projections of monthly average temperature, minimum daily temperature, maximum daily temperature, and precipitation were obtained from a monthly bias-corrected and spatially disaggregated (BCSD) archive derived from select CMIP5 models (Brekke et al. 2013). For the continental United States, projections are disaggregated to 1/8 × 1/8 degree (~14 km) horizontal resolution, while 1/2 × 1/2 degree (~56 km) model output with global coverage over land only is used to provide projections for Alaska and Hawaii. A detailed inventory of the models used with each RCP is shown in Table A1.

Rather than use the absolute model projections of temperature and precipitation, we assume that the models are best at projecting changes from a baseline period from their own historical climate estimates, here selected as 1981–2010. Model projections are then mapped and added to observed temperature and precipitation normals (1981–2010) at stations from the Global Historical Climatology Network (GHCN) (Arguez et al. 2012) (http://go.usa.gov/KmqH). The GHCN data set is a database of meteorological variables measured daily worldwide and is the most comprehensive set of climate

TABLE A1 CMIP5 models included in temperature and precipitation projections

Model	*RCP 8.5*	*RCP 6.0*	*RCP 4.5*	*RCP 2.6*
access1-0	x		x	
access1-3	x		x	
bcc-csm1-1	x	x	x	x
bcc-csm1-1-m	x		x	
bnu-esm	x		x	x
canesm2	x		x	x
ccsm4	x	x	x	x
cesm1-bgc	x		x	
cesm1-cam5	x	x	x	x
cmcc-cm	x		x	
cmcc-cm5	x		x	
csiro-mk3-6-0	x	x	x	x
fgoals-g2	x		x	x
fio-esm	x	x	x	
gfdl-cm3	x	x	x	x
gfdl-esm2g	x	x	x	x
gfdl-esm2m	x	x	x	
giss-e2-h-cc			x	
giss-e2-r	x	x	x	x
giss-e2-r-cc			x	
hadgem2-ao	x	x	x	x
hadgem2-cc	x		x	
hadgem2-es	x	x	x	x
inmcm4	x		x	
ipsl-cm5a-lr	x	x	x	x
ipsl-cm5a-mr	x	x	x	x
ipsl-cm5b-lr	x		x	
miroc-esm	x	x	x	x
miroc-esm-chem	x	x	x	x
miroc5	x	x	x	x
mpi-esm-lr	x		x	x
mpi-esm-mr	x		x	x
mri-cgcm3	x		x	x
noresm1-m	x	x	x	x
noresm1-me	x	x	x	x

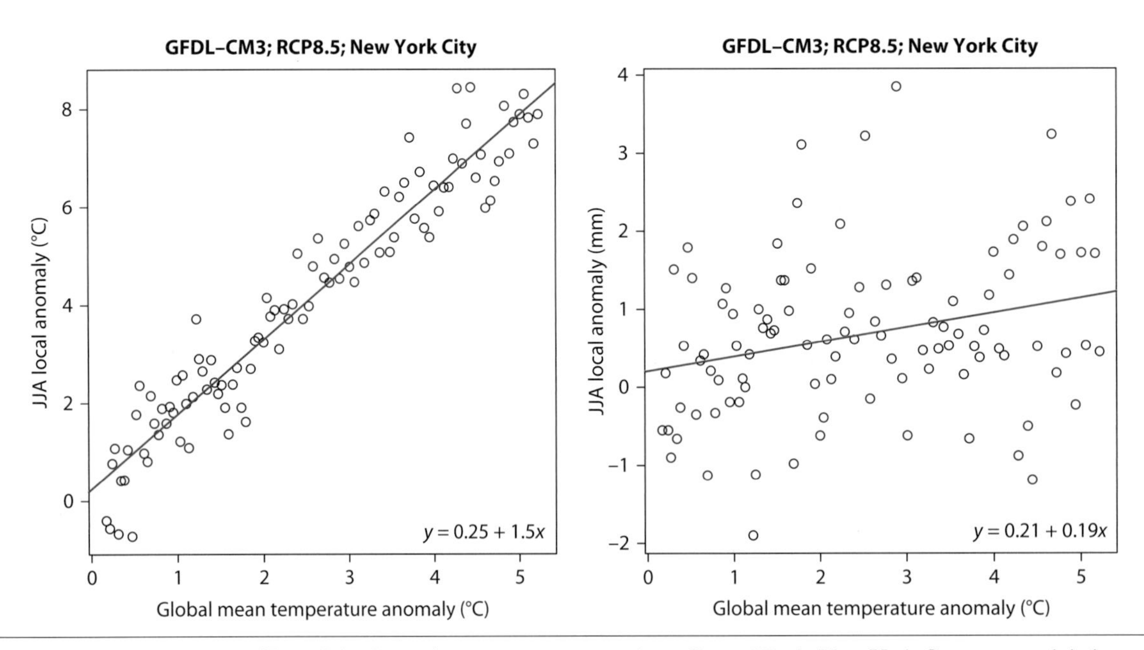

FIGURE A2. Left: Local summer (June-July-August) temperature anomaly at Central Park, New York City, versus global mean temperature anomaly for the GFDL-CM3 model under RCP 8.5. Right: as for Left, but for daily precipitation rate (mm/day).

data within the United States. The use of station-level normals accounts for local meteorological phenomena, such as the urban heat-island effect and land-sea interaction, that are not well reproduced by the gridded BCSD model output. Only GHCN stations that met the strictest of the National Climatic Data Center's data-completion requirements for the 30-year monthly climate normals definitions were used (2,688 sites for temperature, 2,722 sites for precipitation).

1.3. Pattern fitting

We regard the output of each model as the sum of forced climate change and unforced climate variability. We further assume that the forced climate change can be approximated as linear in the long-term (30-year) running average of global mean temperature. Accordingly, for each CMIP5 model and scenario i and each at station j, we fit the changes from the 1981–2010 reference levels for seasonal temperature and precipitation to the linear model

$$y_{i,j}(\Delta T, t) = \hat{k}_{i,j}\Delta T + b_{i,j} + \epsilon_{i,j}(t) \qquad \text{(A1)}$$

following *Mitchell* (2003). Here, ΔT is the running-average change in global mean temperature relative to the reference period (1981–2010), $\hat{k}$ is the estimated seasonal pattern, $\hat{k}\Delta T$ is the estimated forced climate change, $b_{i,j}$ is the y-intercept, and $\epsilon(t)$ is an estimated temporal pattern of unforced variability. In our analysis, we use a single realization of unforced variability from each CMIP5 model. Figure A2 shows an example regression for one particular model and scenario (GFDL-CM3 under RCP 8.5) at New York City for summertime monthly mean temperature and daily precipitation rate.

1.4. Probability weighting

We divide the unit interval [0, 1] into ten bins, with a somewhat higher density of bins at the tails of the interval. (Specifically, the bins are centered at the 4th, 10th, 16th, 30th, 50th, 70th, 84th, 90th, 94th, and 99th percentiles.) The quantiles of global mean temperature change corresponding to the bounds and center of each bin are taken from the MAGICC6 output. CMIP5 model output is categorized into bins based on the projected change in global mean temperature from 1981–2010 to 2081–2099.

In bins, primarily at the tail of the distribution, not represented by at least 2 CMIP5 models, we generate model surrogates sufficient to bring the number of models plus surrogates to two. To generate a model surrogate, we take the global mean temperature projection from MAGICC output corresponding to the central quantile of the bin. If there is no CMIP5 output in the bin, we pick two models with global mean temperature projections close to the bin, such that one model pattern reflects a large net increase in contiguous United States (CONUS) precipitation with temperature and one reflects a net decrease or lesser increase in CONUS precipitation with temperature. If there is a single CMIP5 model in the bin, we pick a single model with a precipitation pattern complementing the one in the bin. We then use the patterns from the selected models to scale the global mean temperature projection and add the residuals from the same models, generating a surrogate model that includes both forced change and unforced variability. A table of the models and surrogate models used, along with the spatial patterns of temperature and precipitation change, is provided in Tables A2–A5 and Figures A6–A7.

TABLE A2 Selected patterns and probability weights used for RCP 2.6

Quantile	*Model*	*Weight*	*2080–2099 Global ΔT (°C)*	*2080–2099 CONUS ΔP (%)*
0.00	gfdl-esm2g	0.04	0.25	-1.41
0.04	scaled fio-esm	0.04	0.39	-0.70
0.10	giss-e2-r	0.02	0.46	2.42
0.10	scaled giss-e2-r	0.02	0.46	2.39
0.16	scaled fgoals-g2	0.04	0.52	2.99
0.20	fgoals-g2	0.04	0.58	3.34
0.30	scaled mpi-esm-lr	0.10	0.64	3.38
0.40	mpi-esm-lr	0.10	0.78	4.13
0.40	mpi-esm-mr	0.03	0.78	2.57
0.43	bcc-csm1-1	0.03	0.82	0.33
0.46	noresm1-m	0.03	0.84	0.59
0.47	ccsm4	0.03	0.86	5.01
0.55	noresm1-me	0.03	0.94	6.42
0.55	mri-cgcm3	0.03	0.92	3.78
0.63	miroc5	0.04	1.03	4.11
0.65	ipsl-cm5a-mr	0.04	1.07	-1.39
0.66	hadgem2-ao	0.04	1.08	-0.23
0.68	bnu-esm	0.04	1.13	1.02
0.76	ipsl-cm5a-lr	0.04	1.25	6.10
0.82	hadgem2-es	0.02	1.40	9.26
0.84	cesm1-cam5	0.02	1.46	13.56
0.85	csiro-mk3-6-0	0.02	1.45	10.41
0.85	canesm2	0.02	1.47	7.79
0.90	scaled miroc-esm-chem	0.02	1.62	7.43
0.91	miroc-esm	0.02	1.65	7.25
0.93	miroc-esm-chem	0.03	1.69	7.78
0.97	gfdl-cm3	0.03	1.92	6.14
0.99	scaled hadgem2-es	0.01	2.18	14.41
0.99	scaled miroc-esm-chem	0.01	2.18	10.05

TABLE A3 Selected patterns and probability weights used for RCP 4.5

Quantile	*Model*	*Weight*	*2080–2099 Global ΔT (°C)*	*2080–2099 CONUS ΔP (%)*
0.04	scaled gfdl-esm2g	0.04	0.93	4.63
0.07	gfdl-esm2g	0.04	0.99	4.93
0.09	fio-esm	0.02	1.03	1.64
0.10	scaled gfdl-esm2m	0.02	1.03	2.37
0.17	gfdl-esm2m	0.03	1.17	2.70
0.18	giss-e2-r-cc	0.03	1.17	1.98
0.20	giss-e2-r	0.03	1.19	2.02
0.21	inmcm4	0.05	1.21	7.65
0.25	fgoals-g2	0.05	1.31	2.89
0.29	giss-e2-h-cc	0.05	1.36	6.51
0.37	bcc-csm1-1-m	0.05	1.49	-4.01
0.41	cesm1-bgc	0.02	1.56	4.52
0.42	bcc-csm1-1	0.02	1.57	4.23
0.44	mpi-esm-lr	0.02	1.62	5.52
0.45	noresm1-m	0.02	1.60	5.92
0.45	ipsl-cm5b-lr	0.02	1.62	9.24
0.45	ccsm4	0.02	1.62	6.16
0.45	mri-cgcm3	0.02	1.63	10.28
0.46	noresm1-me	0.02	1.67	1.67
0.48	mpi-esm-mr	0.02	1.68	4.53
0.48	miroc5	0.02	1.71	0.34
0.58	cnrm-cm5	0.02	1.88	6.97
0.70	access1-3	0.02	2.10	9.22
0.71	cmcc-cm	0.02	2.14	0.21
0.71	bnu-esm	0.02	2.13	1.06
0.72	ipsl-cm5a-lr	0.02	2.17	-2.60
0.74	access1-0	0.02	2.21	3.10
0.74	ipsl-cm5a-mr	0.02	2.23	-4.01
0.74	csiro-mk3-6-0	0.02	2.24	13.46
0.77	canesm2	0.02	2.30	13.32
0.78	hadgem2-cc	0.02	2.31	2.77
0.78	cesm1-cam5	0.02	2.30	6.44
0.79	hadgem2-ao	0.02	2.36	-1.42
0.82	miroc-esm	0.02	2.46	4.19
0.84	hadgem2-es	0.02	2.55	8.15
0.85	miroc-esm-chem	0.02	2.53	11.40
0.88	gfdl-cm3	0.02	2.70	10.00
0.90	scaled miroc-esm-chem	0.02	2.80	12.58
0.90	scaled hadgem2-ao	0.02	2.80	-1.68
0.95	scaled miroc-esm-chem	0.03	3.23	14.51
0.95	scaled hadgem2-ao	0.03	3.23	-1.94
0.99	scaled miroc-esm-chem	0.01	4.12	18.54
0.99	scaled hadgem2-ao	0.01	4.12	-2.47

TABLE A4 Selected patterns and probability weights used for RCP 6.0

Quantile	*Model*	*Weight*	*2080–2099 Global ΔT (°C)*	*2080–2099 CONUS ΔP (%)*
0.04	scaled gfdl-esm2m	0.04	1.31	2.22
0.04	scaled fio-esm	0.04	1.31	-0.78
0.10	scaled gfdl-esm2m	0.02	1.42	2.42
0.10	scaled fio-esm	0.02	1.42	-0.85
0.15	gfdl-esm2g	0.04	1.50	2.40
0.18	fio-esm	0.04	1.56	-0.93
0.24	giss-e2-r	0.07	1.62	3.23
0.25	gfdl-esm2m	0.07	1.66	2.82
0.36	noresm1-m	0.07	1.85	7.76
0.41	noresm1-me	0.05	1.94	4.27
0.43	bcc-csm1-1	0.05	1.96	5.09
0.44	miroc5	0.05	1.98	1.39
0.49	ccsm4	0.05	2.11	8.42
0.65	csiro-mk3-6-0	0.04	2.41	1.45
0.67	hadgem2-ao	0.04	2.49	-1.54
0.69	ipsl-cm5a-lr	0.04	2.55	-1.53
0.72	ipsl-cm5a-mr	0.04	2.61	-5.75
0.78	cesm1-cam5	0.04	2.76	17.36
0.82	miroc-esm	0.02	2.93	10.84
0.85	miroc-esm-chem	0.02	3.03	9.40
0.86	hadgem2-es	0.02	3.06	6.42
0.87	gfdl-cm3	0.02	3.10	11.80
0.90	scaled miroc-esm-chem	0.02	3.25	10.08
0.90	scaled hadgem2-es	0.02	3.25	6.83
0.95	scaled miroc-esm-chem	0.03	3.79	11.74
0.95	scaled hadgem2-es	0.03	3.79	7.95
0.99	scaled miroc-esm-chem	0.01	4.47	13.87
0.99	scaled hadgem2-es	0.01	4.47	9.40

In the final probability distribution, the models and surrogates in a bin are weighted equally such that the total weight of the bin corresponds to the target distribution for 2081–2099 temperature. For example, if there are four models in the bin centered at the 30th percentile and stretching from the 20th to the 40th percentiles, each will be assigned a probability of 20%/4 = 5%. Thus the projected distribution for global mean temperature approximates the target (Figure A3).

For models that end in 2100, we extend projections to 2200 by assuming that global mean temperature beyond 2100 follows the quantile of the MAGICC output that corresponds to that model's position in the 2081–2099 average. We apply the model's own pattern of forced change and use residuals (i.e., unforced variability) that are equal to the residuals from 2000 to 2099, run backwards to preserve continuity.

To assess the effect of the probability weighting on our results, we present in Tables A6 to A13 a comparison of our projections with model weights and surrogate models to those based simply on the unweighted distribution of results from CMIP5. Table A6 shows the sensitivity of regional average annual temperatures under RCP 8.5 in a weighted and

TABLE A5 Selected patterns and probability weights used for RCP 8.5

Quantile	*Model*	*Weight*	*2080–2099 Global ΔT (°C)*	*2080–2099 CONUS ΔP (%)*
0.04	scaled giss-e2-r	0.04	2.26	8.37
0.04	scaled inmcm4	0.04	2.26	1.58
0.10	scaled giss-e2-r	0.02	2.43	9.00
0.10	scaled inmcm4	0.02	2.43	1.70
0.12	giss-e2-r	0.03	2.50	9.26
0.14	inmcm4	0.03	2.58	1.81
0.18	gfdl-esm2m	0.03	2.64	5.29
0.22	gfdl-esm2g	0.05	2.77	7.20
0.33	fgoals-g2	0.05	3.03	1.52
0.39	noresm1-m	0.05	3.14	6.60
0.40	mri-cgcm3	0.05	3.19	13.38
0.43	bcc-csm1-1-m	0.02	3.24	4.85
0.45	miroc5	0.02	3.31	-0.33
0.46	ipsl-cm5b-lr	0.02	3.33	9.32
0.46	noresm1-me	0.02	3.32	2.98
0.47	bcc-csm1-1	0.02	3.34	2.00
0.50	fio-esm	0.02	3.42	6.50
0.50	cnrm-cm5	0.02	3.47	11.10
0.51	cesm1-bgc	0.02	3.48	6.61
0.51	mpi-esm-mr	0.02	3.52	5.63
0.53	mpi-esm-lr	0.02	3.55	5.33
0.53	ccsm4	0.02	3.59	5.39
0.64	access1-3	0.01	3.96	15.44
0.65	csiro-mk3-6-0	0.01	3.97	11.92
0.66	hadgem2-ao	0.01	4.04	2.02
0.66	access1-0	0.01	4.05	0.81
0.67	cesm1-cam5	0.01	4.04	10.92
0.69	cmcc-cm	0.01	4.14	5.39
0.71	bnu-esm	0.01	4.27	3.41
0.73	ipsl-cm5a-mr	0.01	4.36	-10.46
0.75	canesm2	0.01	4.41	22.06
0.75	ipsl-cm5a-lr	0.01	4.43	-5.31
0.78	hadgem2-cc	0.01	4.60	5.97
0.78	gfdl-cm3	0.01	4.61	12.92
0.78	miroc-esm	0.01	4.62	5.09
0.78	hadgem2-es	0.01	4.63	6.94
0.83	miroc-esm-chem	0.04	4.90	4.90
0.84	scaled miroc-esm-chem	0.04	4.93	4.93
0.90	scaled gfdl-cm3	0.02	5.45	15.26
0.90	scaled miroc-esm-chem	0.02	5.45	5.45
0.95	scaled gfdl-cm3	0.03	6.20	17.35
0.95	scaled miroc-esm-chem	0.03	6.20	6.20
0.99	scaled gfdl-cm3	0.01	8.07	22.59
0.99	scaled miroc-esm-chem	0.01	8.07	8.07

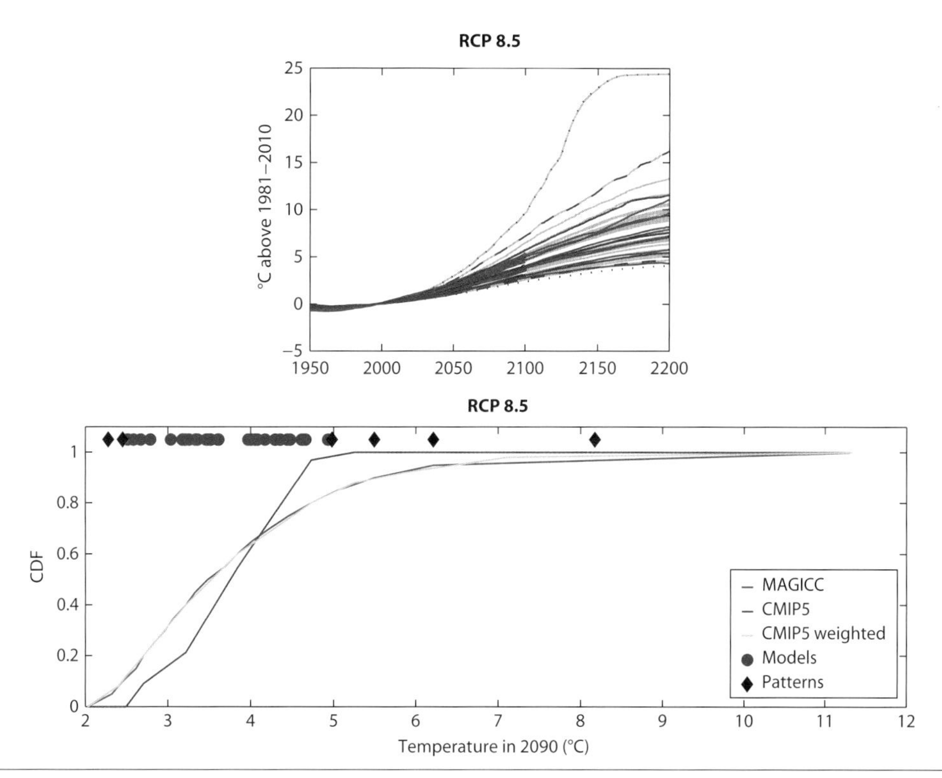

FIGURE A3. (Top) Global mean temperature trajectories for RCP 8.5 from MAGICC (blue), CMIP5 model output (red), and model surrogates (gray). Heavy blue = median, light blue = 17th/83rd percentile, dashed blue = 5th/95th percentile, dotted blue = 1st/99th percentiles. (Bottom) The distribution of 2081–2099 global mean temperature for RCP 8.5. Blue dots = CMIP5 output, black diamonds = model surrogates, blue curve = CMIP5 model output CDF, red curve = MAGICC projection, cyan curve = CMIP5 models and model surrogates weighted to align with MAGICC projections.

unweighted scheme. The largest differences between methods occur at the end of the twenty-first century in the 5th to 95th percentile interval. The differences between unweighted and weighted range from 1.3°C in Hawaii to nearly 3.0°C over Alaska. Differences in seasonal precipitation totals are presented in Tables A7 to A10, and, in general, show less of a divergence between weighting schemes compared to annual average temperature. Differences in extreme heat/humidity days (chapter 4) are presented in Tables A11 to A13. The largest differences between weighting schemes occurs in the most extreme conditions (ACP Humid Heat Stroke Index Category IV).

1.5. Daily projections

Both GCM output and surrogate output are treated at the monthly average level. To generate daily temperature and precipitation, we assume that relationship between the monthly means and the daily values come from a stationary distribution (e.g., Wood et al. 2002), which is the standard approach for BCSD downscaling. We randomly assign each future year to a historical year between 1981 and 2010. Monthly averages are mapped to daily values from the GHCN stations using the additive relationship (for temperature) or multiplicative relationship (for precipitation) from that historical year. Where daily observations are missing from the 30-year historical record, we fill in the missing days and months using relationships between daily and monthly values from gridded data sets and between the climatological 30-year normal value at the GHCN station. A gridded observational data set (Maurer et al. 2002) and the North American Regional Reanalysis (NARR) (Mesinger et al. 2006) are used to provide the daily values for the continental United States and for

TABLE A6 Average annual temperature anomaly (°C) for both weighted and unweighted distributions

RCP 8.5			*2040–2059 anomaly (Δ°C)*			*2080–2099 anomaly (Δ°C)*		
Region	1981–2010 (°C)	Method	50	17–83	5–95	50	17–83	5–95
CONUS	11.9	unweighted	4.4	3.1–5.5	2.6–6.4	9.2	6.4–10.6	5.9–12.4
		weighted	4.4	2.6–5.8	2.5–6.4	9.4	6.1–12.4	5.7–14.3
Northeast	8.8	unweighted	4.5	3.5–5.6	2.9–6.9	9.3	6.8–11.4	6.4–12.7
		weighted	4.5	3.0–6.4	2.4–7.0	9.3	6.6–12.8	5.5–15.9
Southeast	16.8	unweighted	3.9	2.9–4.6	2.3–5.3	8.0	6.4–9.4	4.9–10.7
		weighted	3.9	2.4–5.1	2.0–5.7	8.1	5.4–10.8	4.9–13.1
N. Great Plains	7.2	unweighted	4.7	3.1–5.9	2.2–7.0	9.6	6.6–11.1	5.7–13.5
		weighted	4.7	2.9–6.4	2.5–7.0	9.7	6.0–13.5	5.7–15.1
S. Great Plains	16.8	unweighted	4.5	3.0–5.1	2.6–5.7	8.8	6.9–10.4	6.0–11.4
		weighted	4.5	2.9–5.2	2.4–6.0	8.9	6.2–11.4	5.8–13.9
Midwest	9	unweighted	4.9	3.4–6.0	2.9–7.4	9.7	6.8–11.7	6.2–14.2
		weighted	5.0	3.0–6.8	2.8–7.4	10.2	6.3–14.2	6.1–15.7
Northwest	8.8	unweighted	4.1	3.0–5.3	2.3–6.5	8.1	6.1–10.7	5.8–12.2
		weighted	4.2	2.3–6.0	2.2–6.6	8.1	5.8–12.2	4.9–13.6
Southwest	12.4	unweighted	4.5	3.2–5.5	2.6–6.2	8.6	6.9–10.7	6.2–12.5
		weighted	4.5	2.7–5.7	2.4–6.2	8.6	6.4–12.5	5.9–14.4
Alaska	-1.7	unweighted	5.7	3.9–7.2	3.6–8.6	10.5	7.7–14.8	6.6–16.2
		weighted	5.8	3.9–8.0	2.4–9.1	11.4	7.7–16.3	6.6–19.1
Hawaii	23.6	unweighted	2.5	1.8–3.4	1.6–4.0	5.0	4.1–7.4	3.7–8.1
		weighted	2.6	1.6–3.6	1.2–4.4	5.2	3.7–7.7	3.3–9.4

Alaska and Hawaii, respectively. Where daily precipitation projections exceed twice the historical daily maximum and ten times the model's mean daily precipitation rate for the month, we invoke a "spill over" routine that evenly distributes two-thirds of the incident daily precipitation to the nearest adjacent two days within the month. Daily maximum and minimum temperatures for Alaska and Hawaii were calculated from 3-hourly NARR data. For the rare case when the daily downscaled $T_{min} < T_{avg} < T_{max}$ is not satisfied, T_{min} and T_{max} are approximated as T_{avg} -2.5 K and T_{avg} + 2.5 K, respectively. An example of the daily weather generation is shown in Figure A4.

1.6. Wet-bulb temperatures

At each GHCN site, we estimate a relationship between dry-bulb and wet-bulb temperature and the associated error with (1) a simple linear model and (2) a piecewise linear model with a single breakpoint. The model with the smallest Bayesian information criterion (BIC) is used. The simple linear model is of the form:

$$T_w(T) = b_0 + \beta_0 T_d + \epsilon. \qquad \text{(A2)}$$

The piecewise linear model is of the form:

$$T_w(T) = b_1 + \beta_1 min(T_d, T_0) + \beta_2 max(0, T_d - T_0) + \epsilon. \qquad \text{(A3)}$$

TABLE A7 Weighted and unweighted regional average winter precipitation change (Δ%)

	DJF RCP 8.5		*2040–2059 Change (Δ%)*			*2080–2099 Change (Δ%)*		
Region	*1981–2010 (mm)*	*Method*	*50*	*17–83*	*5–95*	*50*	*17–83*	*5–95*
CONUS	159.7	unweighted	7.0	-1.4–10.3	-3.8–15.5	11.2	1.6–24.2	-9.5–29.2
		weighted	2.5	-2.0–10.1	-2.9–15.5	7.3	-1.4–21.4	-1.7–25.6
Northeast	228.4	unweighted	12.6	3.1–17.5	-14.5–23.9	23.3	8.6–33.9	1.7–44.4
		weighted	4.4	-1.4–16.7	-2.5–23.9	15.8	-0.8–33.0	-1.8–37.8
Southeast	323.4	unweighted	4.0	-3.4–11.4	-12.9–14.5	3.8	-6.3–21.9	-22.9–32.7
		weighted	-0.3	-3.4–9.8	-9.3–14.5	-0.9	-2.3–16.2	-14.9–31.4
N. Great Plains	40.6	unweighted	19.7	8.0–27.6	-11.6–37.6	37.8	24.2–55.4	13.9–70.6
		weighted	17.5	-1.1–27.6	-2.6–37.6	33.7	-0.2–49.7	-1.6–70.6
S. Great Plains	121.6	unweighted	-0.9	-12.4–9.4	-13.4–12.6	-2.3	-18.7–15.9	-34.6–24.2
		weighted	-1.5	-11.0–7.0	-13.0–11.7	-1.4	-11.8–10.5	-25.9–22.8
Midwest	135.0	unweighted	9.9	3.2–17.3	-7.0–28.0	21.7	10.1–35.1	-11.1–39.8
		weighted	6.7	-1.5–14.3	-6.1–28.0	17.0	-0.5–32.3	-1.4–35.5
Northwest	241.6	unweighted	9.3	-1.7–14.9	-13.4–19.9	13.7	-1.4–26.5	-15.1–42.1
		weighted	4.0	-2.1–12.3	-10.7–19.9	5.5	-2.4–26.5	-14.4–42.1
Southwest	110.4	unweighted	5.5	-8.8–15.5	-14.9–30.3	13.3	-5.2–37.5	-19.3–54.8
		weighted	-0.9	-6.2–12.8	-14.9–30.3	1.2	-2.7–36.0	-19.3–51.6
Alaska	105.8	unweighted	12.1	3.2–21.2	0.4–27.7	30.6	15.0–42.1	5.6–65.2
		weighted	6.0	-1.5–20.6	-1.6–25.0	15.5	0.3–40.4	-1.4–65.2
Hawaii	601.1	unweighted	-7.1	-16.9–5.2	-21.2–10.2	-3.9	-13.8–8.4	-28.3–28.0
		weighted	-3.3	-15.6–2.3	-21.2–5.8	-3.1	-13.0–6.7	-28.3–16.4

Here, T_w is the wet-bulb temperature, T_d the dry-bulb temperature, b_i are y-intercepts, β_i are slopes, and T_o is the breakpoint. Errors are assumed to come from a stationary normal distribution, $\epsilon \sim N(0, \sigma^2)$. The model is fit to historical (1981–2010) maximum daily wet-bulb temperatures calculated using the Wobus method (Doswell et al. 1982, Marsh and Hart 2012) from 3-hourly 2-m temperature, 2-m dew point temperature and pressure from NARR. Examples are shown in Figure A5. While commonly used, the Wobus method does introduce small errors compared to more precise calculations (Davies-Jones 2008). These errors are generally <1 K and do not significantly affect the ACP Humid Heat Stroke index introduced in chapter 4.

The regression provides the distribution of T_w conditional on T_d for the baseline climatology. To account for the effects of climate change, we shift the conditional distribution upward by $\beta_o \Delta T_f$, where ΔT_f is the local forced summertime temperature change given by $\hat{k}\Delta T$ in equation A1. In particular, we use the relationship

$$T_w(T_d, \Delta T_f) = T_w(T_d - \Delta T_f) + \beta_o \Delta T_f \qquad \text{(A4)}$$

to generate estimates of future wet-bulb temperatures. Note that, for cases where the simple linear model best captures the distribution, this expression reduces to that model.

The extrapolation of the historical relationship between wet-bulb and dry-bulb temperatures effectively assumes that the distribution of relative humidities remains near constant. In fact, because the land warms faster than the ocean, relative humidity over land is expected to decrease in a warmer climate (Sherwood and Fu 2014). The failure of our historically based method to account for this shift may result in a slight upward bias in projected wet-bulb temperatures in areas not in proximity to large bodies of water such as the oceans or the Great Lakes.

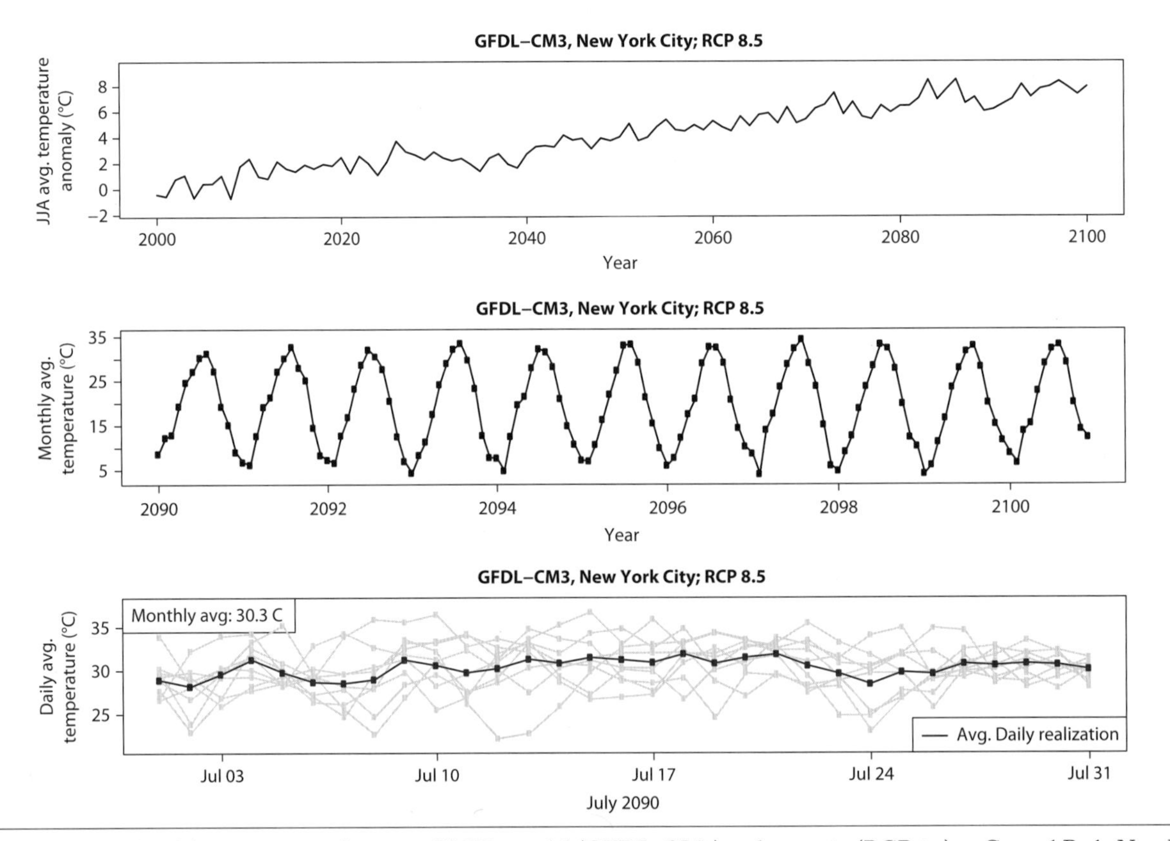

FIGURE A4. Example of data generation from one CMIP5 model (GFDL-CM3) and scenario (RCP 8.5) at Central Park, New York, showing (a) seasonal projections for the century (relative to 1981–2010), (b) monthly projections for 2091–2100, and (c) ten independent daily weather projections for average temperature in July in 2099. The blue line is the average of all ten projections and is equal to the July 2099 monthly average.

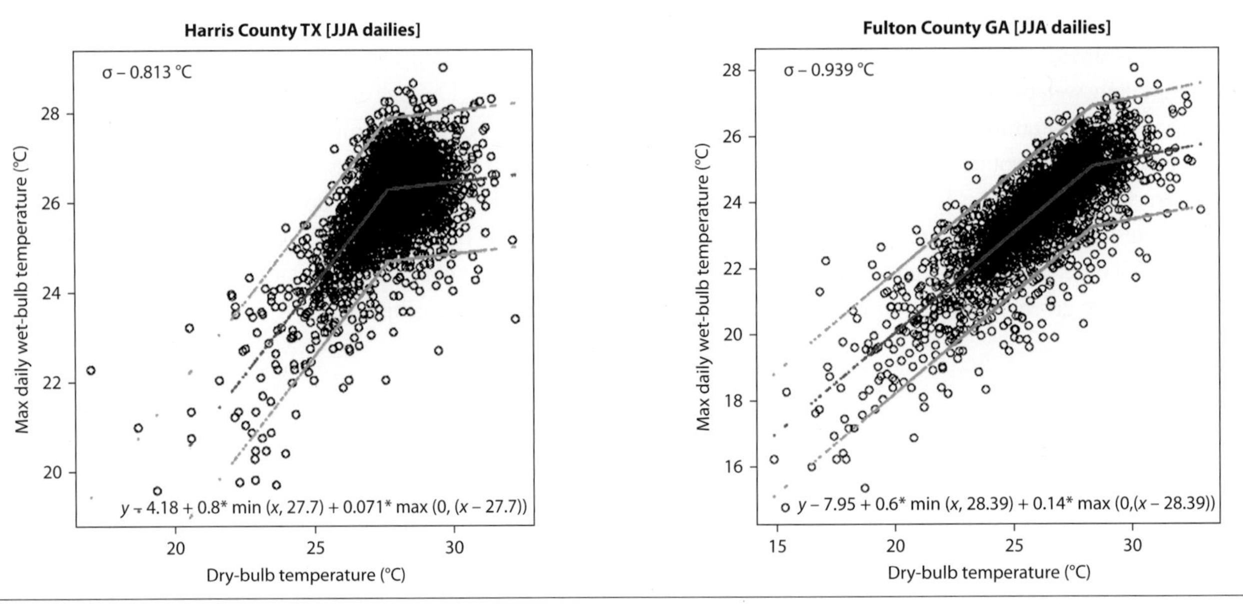

FIGURE A5. Fit of historical summertime (June-July-August) daily maximum wet-bulb temperatures to daily average temperature at Fulton County, Georgia, and Harris County, Texas. Equations are shown for the best-fit piecewise linear model. The standard deviation from the best-fit model is also shown.

TABLE A8 Weighted and unweighted regional average spring precipitation change (Δ%)

MAM RCP 8.5			*2040–2059 Change (Δ%)*			*2080–2099 Change (Δ%)*		
Region	*1981–2010 (mm)*	*Method*	*50*	*17–83*	*5–95*	*50*	*17–83*	*5–95*
CONUS	196.1	unweighted	6.6	2.6–12.5	-3.2–16.1	10.7	2.2–17.6	-5.0–22.2
		weighted	5.0	0.6–10.6	0.3–15.2	7.0	0.7–15.8	-3.3–18.7
Northeast	277.5	unweighted	10.1	3.6–16.4	-3.9–18.8	15.7	10.5–24.5	5.9–32.7
		weighted	7.2	0.2–15.2	-3.9–17.1	12.9	1.0–23.4	0.4–30.3
Southeast	327.9	unweighted	6.3	-3.5–14.6	-8.5–18.0	7.7	-8.6–16.9	-18.7–28.1
		weighted	2.6	0.3–11.7	-7.8–17.7	1.3	-2.7–16.9	-18.7–26.2
N. Great Plains	133.7	unweighted	16.4	7.9–21.6	-2.0–36.7	26.1	12.8–37.7	-1.6–63.4
		weighted	14.5	1.4–19.5	0.7–36.7	19.0	2.5–35.1	1.1–63.4
S. Great Plains	210.2	unweighted	2.5	-4.0–9.6	-16.4–15.3	-0.6	-12.9–9.8	-16.7–14.6
		weighted	0.7	-3.2–6.0	-10.5–15.3	-0.1	-9.3–8.0	-13.8–14.6
Midwest	250.7	unweighted	12.3	6.6–18.4	1.5–21.1	20.5	11.7–30.7	7.2–36.7
		weighted	8.5	0.8–16.6	0.5–21.1	16.0	1.4–28.6	0.8–36.7
Northwest	170.4	unweighted	6.6	-3.0–12.8	-7.8–20.9	7.3	0.2–16.1	-8.4–24.5
		weighted	1.4	-1.8–11.0	-5.9–16.2	4.8	0.3–13.7	-3.4–24.5
Southwest	83.4	unweighted	-1.4	-12.5–9.5	-18.1–24.1	-10.8	-19.5–2.1	-23.6–11.6
		weighted	0.1	-9.1–6.4	-18.1–12.5	-2.8	-17.9–0.5	-22.3–9.0
Alaska	80.7	unweighted	15.9	7.2–28.4	2.4–34.8	38.6	23.0–53.9	9.3–70.9
		weighted	11.1	0.5–21.6	0.3–33.1	35.6	1.6–52.0	1.1–60.7
Hawaii	586.1	unweighted	-0.3	-18.1–13.5	-25.9–19.1	-4.2	-23.8–19.8	-34.7–32.7
		weighted	-0.5	-18.1–8.0	-25.9–18.9	-0.9	-21.3–8.9	-34.7–25.5

TABLE A9 Weighted and unweighted regional average summer precipitation change (Δ%)

JJA RCP 8.5			*2040–2059 Change (Δ%)*			*2080–2099 Change (Δ%)*		
Region	*1981–2010 (mm)*	*Method*	*50*	*17–83*	*5–95*	*50*	*17–83*	*5–95*
CONUS	212.5	unweighted	0.1	-6.5–5.8	-15.5–8.3	-0.3	-13.4–5.4	-22.1–9.9
		weighted	-0.2	-3.8–4.4	-8.4–7.8	-0.3	-8.3–3.3	-17.2–9.9
Northeast	309.5	unweighted	2.2	-2.3–8.8	-4.1–12.7	2.3	-4.7–18.7	-6.8–26.2
		weighted	0.3	-0.7–8.0	-3.0–12.7	0.3	-2.1–12.3	-5.8–26.2
Southeast	365.5	unweighted	2.6	-7.7–10.5	-14.2–19.1	-2.4	-14.5–12.4	-21.5–18.1
		weighted	0.2	-5.3–5.9	-14.2–11.9	-0.5	-13.7–11.5	-17.1–16.1
N. Great Plains	167.2	unweighted	-5.2	-13.4–4.2	-20.0–19.1	-10.8	-23.3–3.3	-36.0–12.3
		weighted	-0.5	-10.2–2.8	-19.7–8.5	-3.5	-22.2–2.2	-25.8–12.3
S. Great Plains	237.4	unweighted	-1.6	-8.9–6.0	-18.3–13.7	-7.0	-20.2–5.5	-23.5–13.2
		weighted	-0.1	-8.4–4.6	-16.8–9.0	-0.5	-18.4–1.5	-21.7–12.5
Midwest	304.2	unweighted	-3.2	-9.3–7.2	-17.1–12.9	-3.9	-16.8–7.6	-29.8–12.8
		weighted	-0.3	-6.7–6.5	-11.0–8.9	-0.2	-10.3–5.9	-21.4–11.8
Northwest	73.1	unweighted	-5.4	-22.5–7.5	-28.6–18.4	-15.6	-31.5–3.9	-55.8–29.0
		weighted	-1.1	-18.3–1.9	-24.6–8.4	-5.0	-29.4–0.3	-49.1–14.4
Southwest	78.3	unweighted	3.7	-8.9–15.9	-21.4–22.3	1.4	-20.8–23.3	-28.4–34.6
		weighted	0.9	-2.1–12.4	-16.7–19.0	1.0	-10.0–20.3	-26.9–31.5
Alaska	184.7	unweighted	14.8	9.0–20.8	4.4–25.8	28.9	18.2–40.6	12.2–64.0
		weighted	12.5	0.7–18.5	0.6–22.9	22.1	1.9–38.0	0.5–57.0
Hawaii	472.8	unweighted	-0.1	-14.1–18.3	-30.8–29.7	-3.6	-15.2–22.3	-37.6–49.0
		weighted	-0.2	-8.7–12.1	-30.8–22.2	-0.5	-12.8–17.4	-37.6–45.5

TABLE A10 Weighted and unweighted regional average autumn precipitation change (Δ%)

SON RCP 8.5			*2040–2059 Change (Δ%)*			*2080–2099 change (Δ%)*		
Region	*1981–2010 (mm)*	*Method*	*50*	*17–83*	*5–95*	*50*	*17–83*	*5–95*
CONUS	182.0	unweighted	3.8	-0.4–8.7	-2.0–10.8	5.8	-1.9–10.9	-7.0–19.9
		weighted	2.2	-0.2–7.9	-2.0–10.2	3.3	-0.1–8.9	-7.0–18.8
Northeast	288.2	unweighted	3.9	-3.4–9.6	-5.4–13.7	6.9	-0.6–13.5	-9.0–15.9
		weighted	1.3	-0.2–6.5	-4.4–10.5	3.1	-0.2–12.0	-9.0–13.9
Southeast	306.0	unweighted	6.3	-1.7–16.5	-6.4–18.5	9.1	0.1–20.3	-15.6–27.7
		weighted	3.9	-0.0–12.6	-6.4–17.7	3.7	0.1–18.4	-15.6–26.6
N. Great Plains	93.1	unweighted	4.4	-3.4–11.8	-8.9–20.3	9.5	-3.1–15.6	-10.1–35.0
		weighted	0.9	-0.8–11.8	-8.8–20.3	2.0	-0.5–12.0	-8.9–23.9
S. Great Plains	201.9	unweighted	3.5	-7.9–11.0	-11.4–16.5	1.8	-15.6–13.0	-23.7–25.3
		weighted	-0.1	-7.0–8.9	-11.4–15.0	0.0	-15.6–5.7	-23.7–22.5
Midwest	233.4	unweighted	5.1	-1.2–9.3	-10.4–12.7	6.5	-1.5–12.7	-13.4–16.7
		weighted	2.2	-0.4–8.2	-4.5–12.7	3.5	-0.2–11.0	-13.4–16.0
Northwest	168.4	unweighted	2.3	-10.5–14.6	-12.1–16.7	9.2	-6.9–18.5	-12.9–27.6
		weighted	0.1	-7.4–13.8	-11.9–16.7	0.9	-3.9–15.7	-12.0–20.3
Southwest	80.3	unweighted	-1.6	-9.3–11.8	-22.4–19.6	-1.7	-9.6–13.8	-27.6–24.2
		weighted	-0.0	-6.3–7.5	-17.5–18.6	0.1	-8.9–11.6	-17.9–17.7
Alaska	170.1	unweighted	16.1	10.4–21.6	5.0–27.4	32.4	25.4–40.1	22.6–48.2
		weighted	11.7	0.7–18.5	0.2–26.5	31.2	1.4–36.8	0.9–41.3
Hawaii	627.7	unweighted	5.1	-8.2–21.0	-17.9–26.2	13.3	-2.5–43.8	-15.4–64.0
		weighted	1.0	-1.6–15.4	-17.9–26.2	5.1	-0.3–35.8	-15.4–44.3

TABLE A11 Weighted and unweighted projections of expected regional average (population-weighted) category II+ ACP Humid Heat Stroke Index days per summer

RCP 8.5			*2020–2039*	*2040–2059*	*2080–2099*	*2120–2139*	*2140–2159*	*2180–2199*
Region	*1981–2010*	*Method*						
CONUS	4.65	weighted	10.21	17.21	34.93	49.88	54.42	60.01
		unweighted	10.36	16.83	33.59	49.1	53.83	59.73
Northeast	3.25	weighted	8.11	16.15	39.47	60.73	66.81	74.18
		unweighted	8.28	15.65	37.81	60.72	67.3	75.75
Southeast	8.24	weighted	18.52	30.49	56.73	74.05	78.68	83.36
		unweighted	18.82	29.87	55.63	74.4	79.43	84.45
N. Great Plains	1.45	weighted	3.16	5.58	14.28	25.25	30.09	36.79
		unweighted	3.18	5.41	12.91	22.63	27.27	33.98
S. Great Plains	8.85	weighted	14.9	19.92	30.06	38.54	41.42	45.33
		unweighted	14.91	19.63	29.23	37.03	39.69	43.38
Midwest	4.82	weighted	11.37	19.96	42.81	62.77	68.37	75.27
		unweighted	11.57	19.53	40.69	61.94	68.26	76
Northwest	0.16	weighted	0.52	1.48	9.66	23.68	30.27	39.17
		unweighted	0.5	1.36	7.5	20.13	26.96	36.9
Southwest	0.05	weighted	0.11	0.26	1.5	4.78	6.75	10.3
		unweighted	0.1	0.23	0.96	2.83	4.05	6.53
Alaska	0	weighted	0	0	0.18	3.92	7.77	15.36
		unweighted	0	0	0.06	1.39	3.17	8.44
Hawaii	1.63	weighted	6.76	16.89	48.26	67.16	70.95	75.92
		unweighted	7.16	16.14	47.14	68.53	72.88	77.12

TABLE A12 Weighted and unweighted projections of expected regional average (population-weighted) category III+ ACP Humid Heat Stroke Index days per summer

RCP 8.5			*2020–2039*	*2040–2059*	*2080–2099*	*2120–2139*	*2140–2159*	*2180–2199*
Region	*1981–2010*	*Method*						
CONUS	0.03	weighted	0.17	0.75	7.49	21.82	28.22	37.15
		unweighted	0.17	0.65	5.35	18.04	24.8	35.13
Northeast	0.05	weighted	0.22	1.03	9.82	29.47	38.1	49.93
		unweighted	0.23	0.87	7.12	25.25	34.53	48.68
Southeast	0.01	weighted	0.08	0.48	8.8	28.22	36.92	49.49
		unweighted	0.09	0.4	5.57	22.64	32.19	47.38
N. Great Plains	0.02	weighted	0.1	0.33	2.99	10.08	14.05	20.22
		unweighted	0.1	0.3	2.09	7.16	10.6	16.34
S. Great Plains	0.01	weighted	0.04	0.15	2.48	9.22	12.43	17.29
		unweighted	0.04	0.12	1.37	6.55	9.74	15.04
Midwest	0.08	weighted	0.49	1.95	14.23	34.62	42.87	53.54
		unweighted	0.49	1.73	11.31	30.34	39.48	51.91
Northwest	0	weighted	0.01	0.07	1.73	9.3	14.56	22.52
		unweighted	0.01	0.06	0.88	5.43	9.74	17.61
Southwest	0	weighted	0.01	0.02	0.18	1.35	2.27	3.82
		unweighted	0.01	0.02	0.07	0.37	0.71	1.54
Alaska	0	weighted	0	0	0	1.38	2.55	5.93
		unweighted	0	0	0	0.09	0.3	1.39
Hawaii	0	weighted	0	0.01	3.1	18.73	27.22	36.62
		unweighted	0	0.01	1.02	11.94	20.94	33.12

TABLE A13 Weighted and unweighted projections of expected regional average (population-weighted) category IV ACP Humid Heat Stroke Index days per summer

RCP 8.5			*2020–2039*	*2040–2059*	*2080–2099*	*2120–2139*	*2140–2159*	*2180–2199*
Region	*1981–2010*	*Method*						
CONUS	0	weighted	0	0.01	0.79	6.27	10.23	17.13
		unweighted	0	0.01	0.29	2.79	5.47	11.54
Northeast	0	weighted	0	0.01	1.14	9	14.87	24.92
		unweighted	0	0.01	0.36	3.93	8.24	18.23
Southeast	0	weighted	0	0	0.36	5.54	9.88	18.26
		unweighted	0	0	0.08	1.54	3.67	9.75
N. Great Plains	0	weighted	0	0	0.24	2.82	4.89	8.89
		unweighted	0	0	0.07	0.9	2	4.76
S. Great Plains	0	weighted	0	0	0.05	1.17	2.11	4.15
		unweighted	0	0	0.01	0.14	0.38	1.25
Midwest	0	weighted	0	0.02	2.22	13.35	20.07	30.47
		unweighted	0	0.02	0.96	7.65	13.41	24.2
Northwest	0	weighted	0	0	0.18	2.9	5.51	11.29
		unweighted	0	0	0.05	0.71	1.94	5.75
Southwest	0	weighted	0	0	0.02	0.41	0.83	1.58
		unweighted	0	0	0.01	0.04	0.09	0.27
Alaska	0	weighted	0	0	0	0.73	1.51	2.3
		unweighted	0	0	0	0	0.01	0.1
Hawaii	0	weighted	0	0	0.01	2.43	4.99	12.03
		unweighted	0	0	0	0.08	0.42	3.92

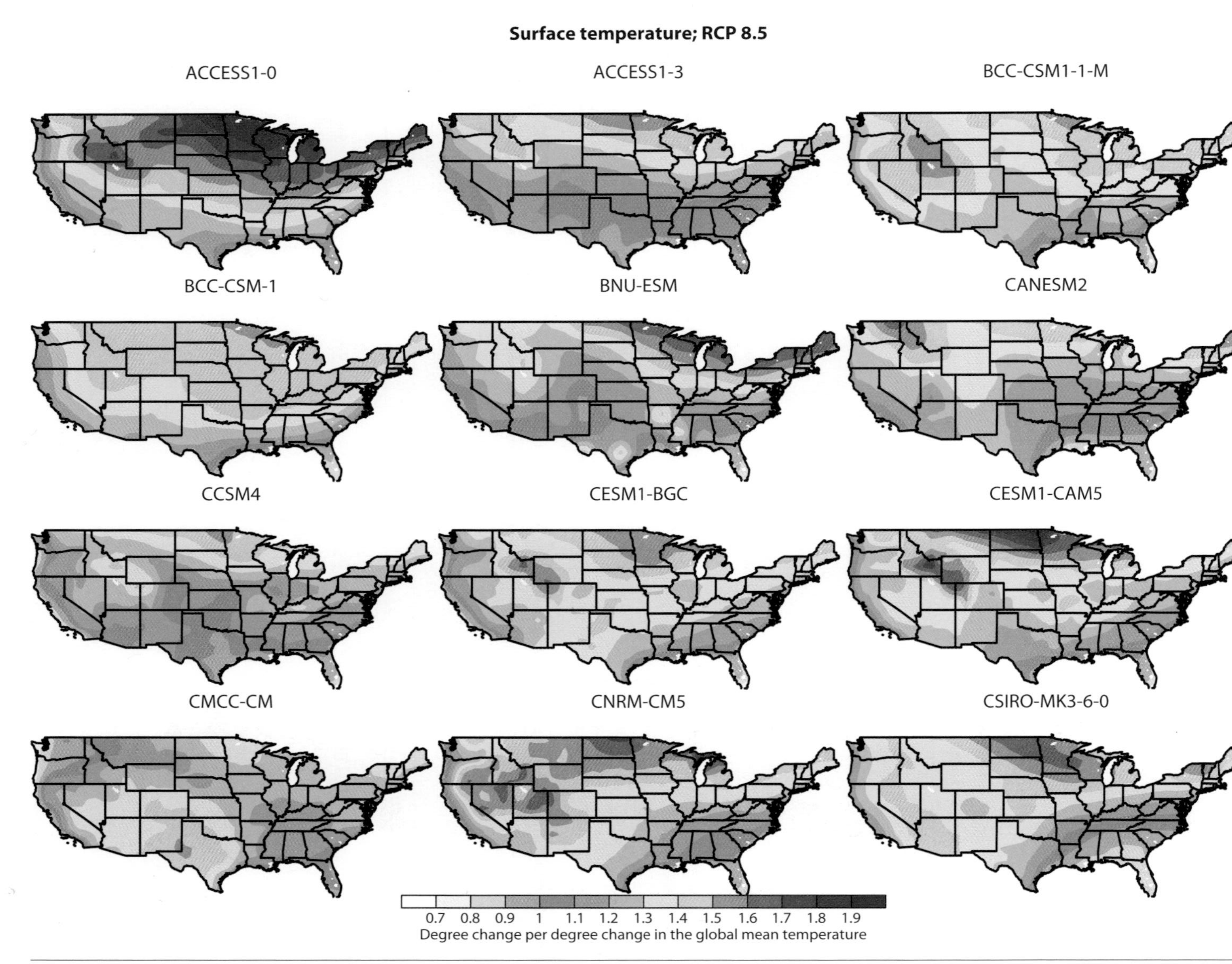

FIGURE A6. Annual surface temperature patterns for each CMIP5 model for RCP 8.5.

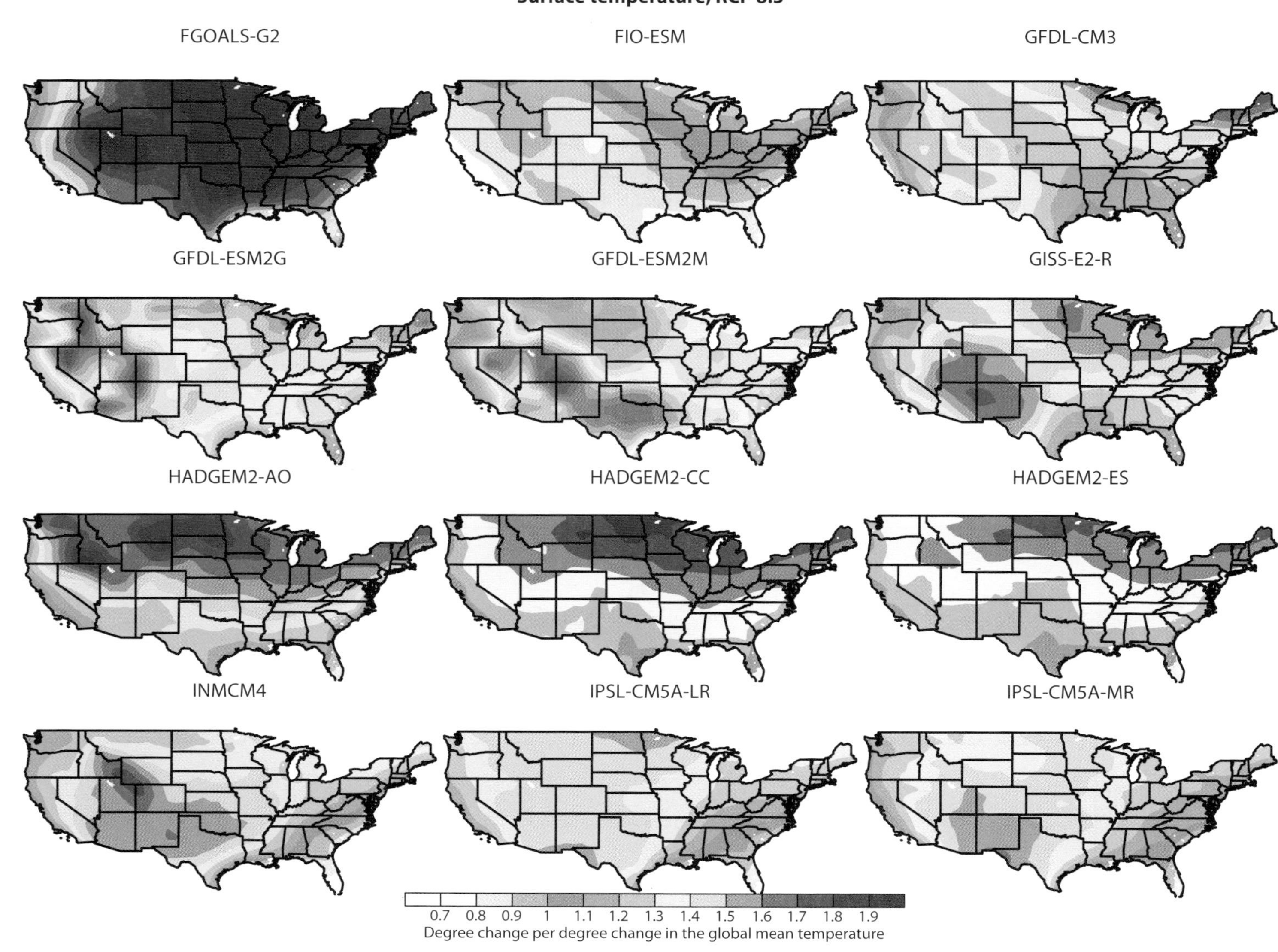

FIGURE A6. *(continued)*

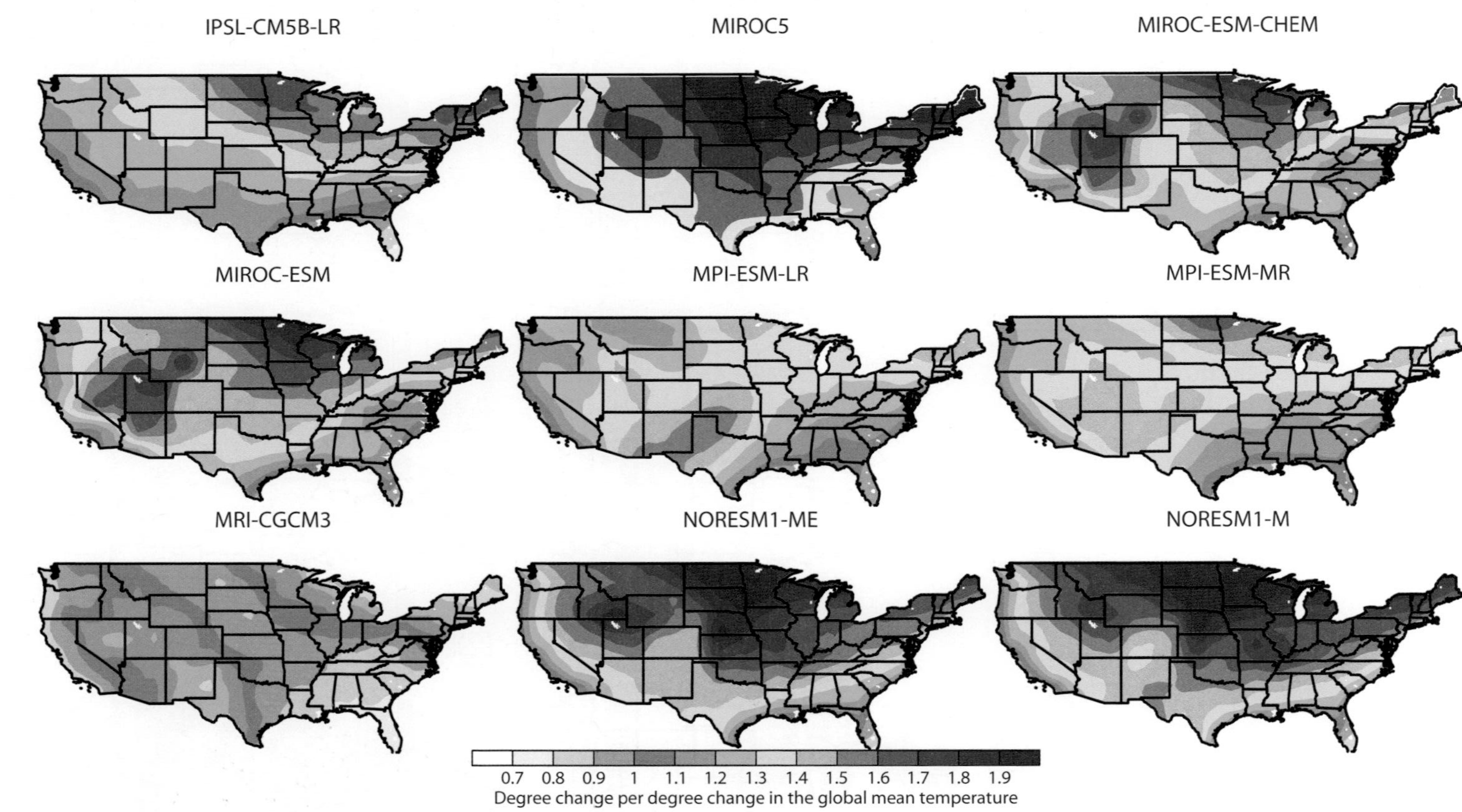

FIGURE A6. *(continued)*

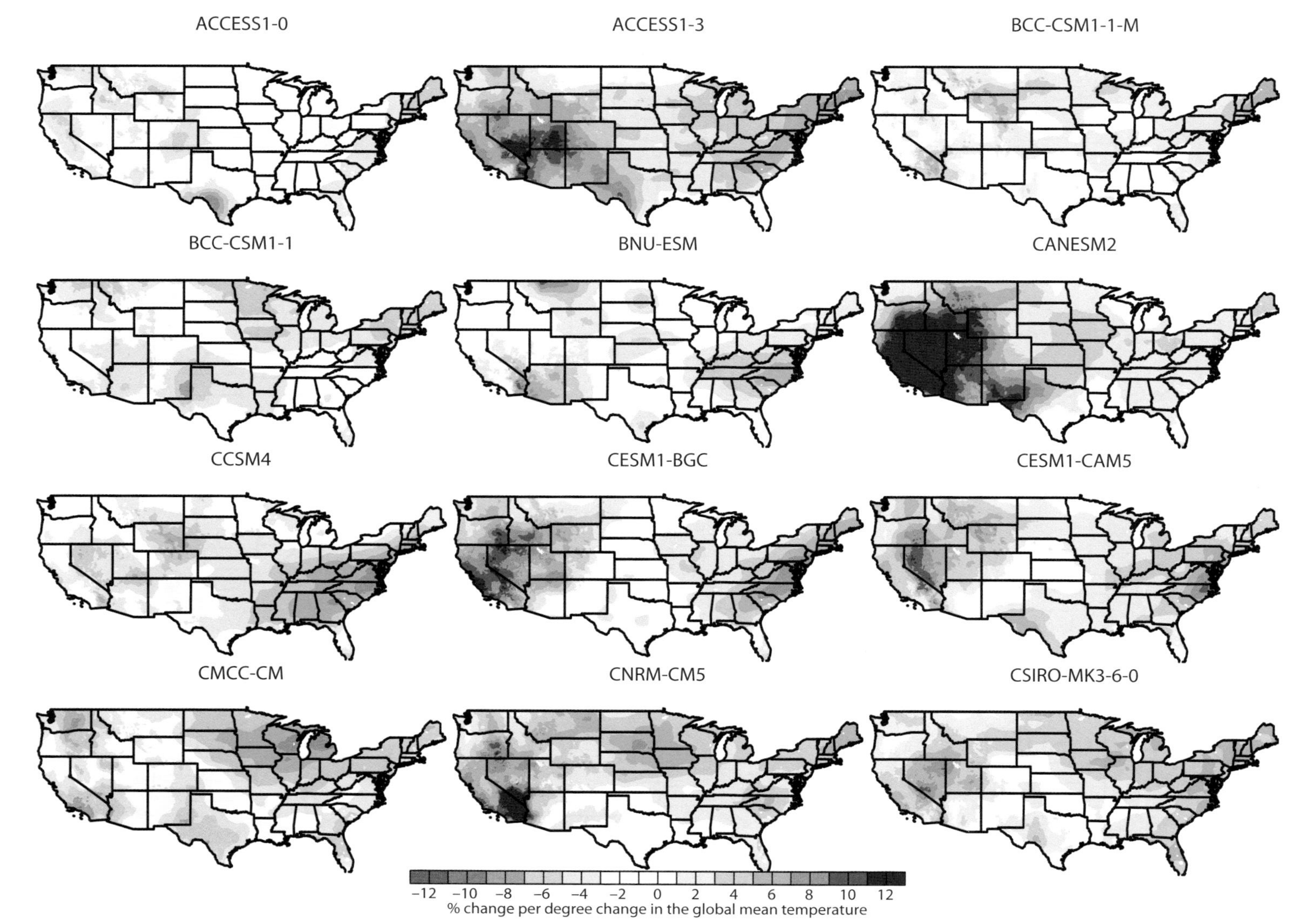

FIGURE A7. Annual daily precipitation rate patterns for each CMIP5 model.

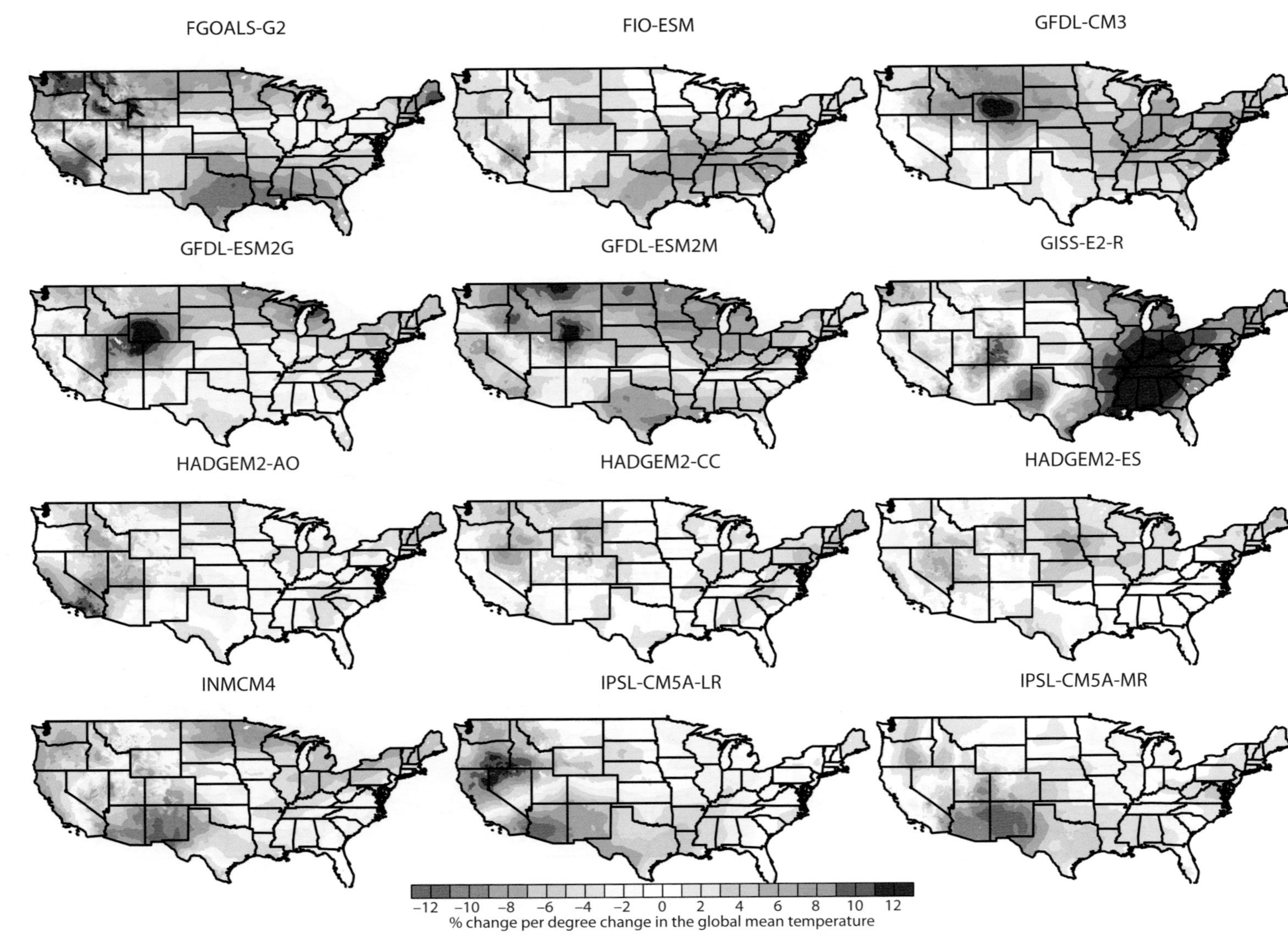

FIGURE A7. *(continued)*

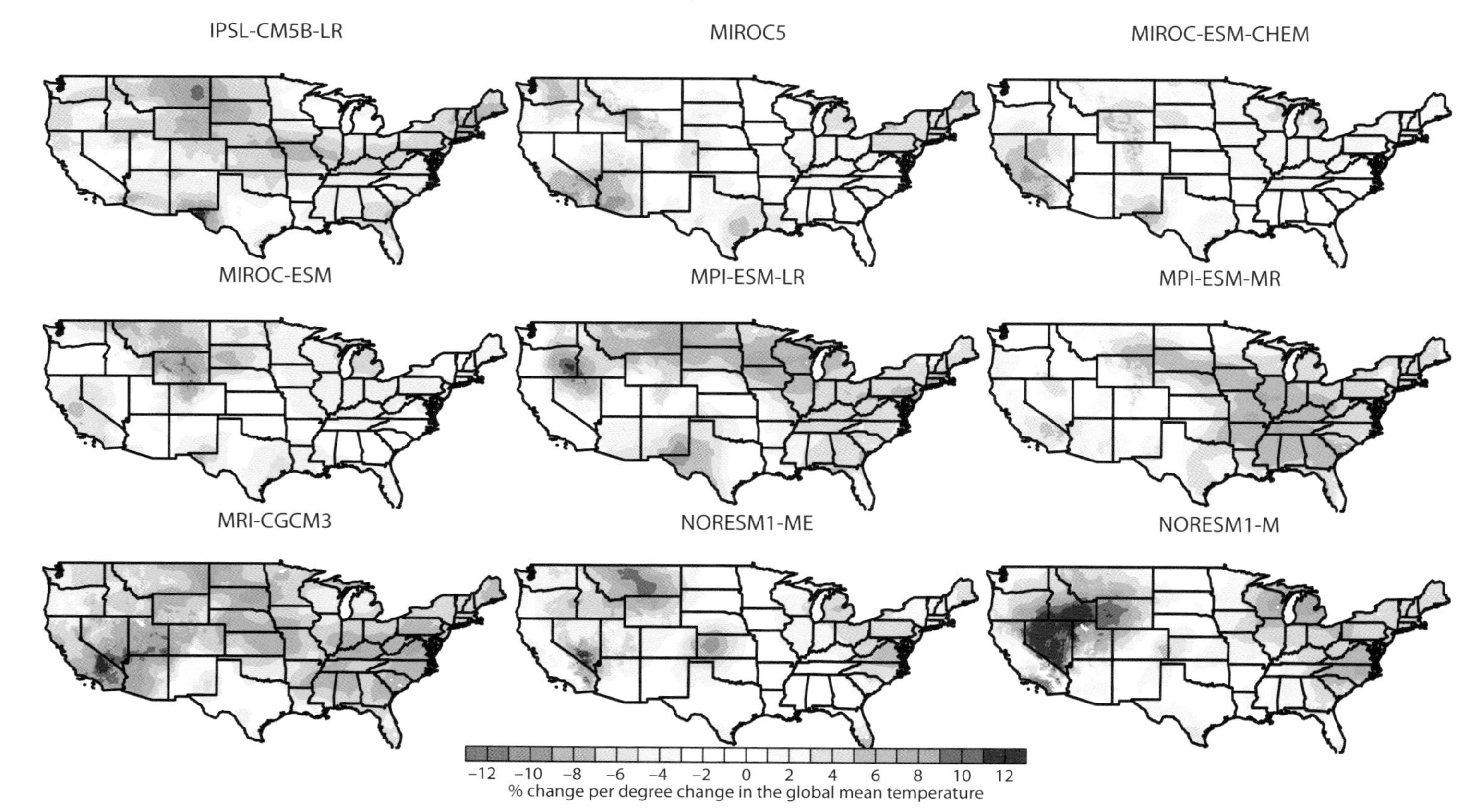

FIGURE A7. *(continued)*

2. SEA-LEVEL RISE PROJECTIONS

The sea-level rise projections used in this analysis are described in depth in Kopp et al. (2014). We briefly summarize key elements of the methodology.

At a globally averaged scale, sea-level change is driven primarily by changes in the amount of heat stored in the ocean (thermal expansion) and the amount of ice stored in glaciers and ice sheets. Smaller contributions to sea-level change are made by human-induced changes in the amount of water stored on land, through groundwater extraction, and through dam construction.

For impact analysis, it is important to estimate not just global mean sea-level (GMSL) change but also local sea-level change, as the impacts are experienced at particular localities, not the global mean. Local sea-level (LSL) change differs from GMSL change due to numerous effects, including:

- ocean dynamics and "steric" variations in the temperature and salinity of the ocean,
- changes in Earth's gravitational field and rotation and flexure of the Earth's lithosphere due to the redistribution of mass between land ice and the ocean (known as static-equilibrium effects),
- land motion and other effects associated with the ongoing response to the redistribution of mass since the end of the last ice age (known as glacial isostatic adjustment, GIA), and
- nonclimatic effects, such as tectonics and sediment compaction due to both natural processes and fluid (groundwater or hydrocarbon) withdrawal.

Figure A8 summarizes the process used in constructing the sea-level rise projections for each RCP.

Projected sea-level rise due to thermal expansion, ocean dynamics, and steric effects (known here collectively as oceanographic effects) is based on the projections of the CMIP5 models. One realization was used from each available model (Table A14), and each model was treated as an equally likely sample from an underlying normal distribution. Changes in the mass balance of glaciers were also indirectly based on the CMIP5 models, via the surface mass balance model of Marzeion, Jarosch & Hofer (2012). The projections of Marzeion, Jarosch & Hofer (2012) for each CMIP5 model were similarly treated as equally likely samples from an underlying normal distribution.

Projections of changes in the Greenland and Antarctic ice sheets were assumed to follow log-normal distributions, with likely (17th to 83rd percentile) ranges from IPCC AR5 (Church et al. 2013). As AR5 does not provide estimates beyond the 17th to 83rd percentiles, the ratio of the 95th-to-83rd and 5th-to-17th percentiles from the expert elicitation study of Bamber and Aspinall (2013) were used to set the shape of the tail projections.

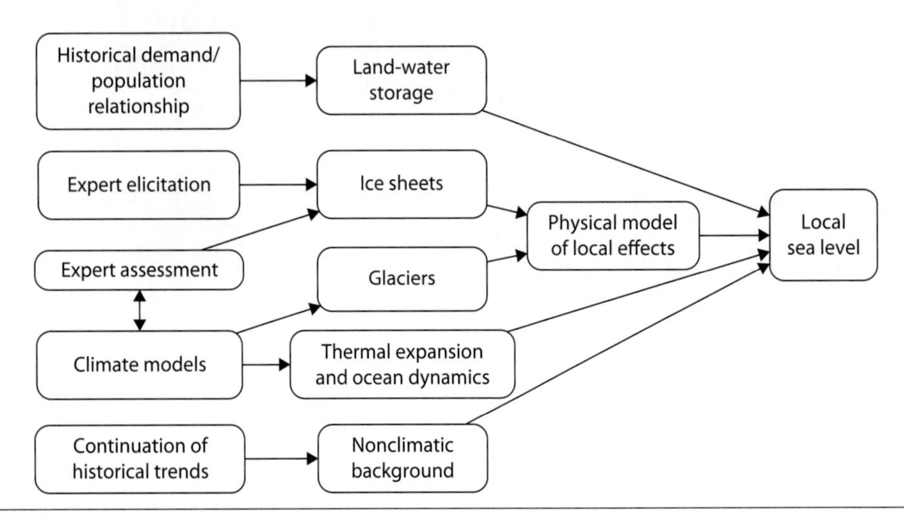

FIGURE A8. Flow of sea-level rise projection construction

TABLE A14 CMIP5 models used for sea-level projections

Model	*Oceanographic Effects*			*Glaciers*		
	RCP 8.5	*RCP 4.5*	*RCP 2.6*	*RCP 8.5*	*RCP 4.5*	*RCP 2.6*
access1-0	1	1				
access1-3	1	1				
bcc-csm1-1	2	2	2	2	2	2
bcc-csm1-1-m	1	1	1			
canesm2	1	2	2	1	2	2
ccsm4	1	1	1	1	1	1
cmcc-cesm	1					
cmcc-cm	1	1				
cmcc-cms	1	1				
cnrm-cm5	2	2	1	2	2	1
csiro-mk3-6-0	1	1	1	2	2	1
gfdl-cm3	1	2	1	1	1	1
gfdl-esm2g	1	1	1			
gfdl-esm2m	1	1	1			
giss-e2-r	2	2	2	2	2	
giss-e2-r-cc	1	1				
hadgem2-cc	1					
hadgem2-es	1		2	2	2	2
inmcm4	1	1		1	1	
ipsl-cm5a-lr	2	2	2	2	2	2
ipsl-cm5a-mr	1	2	1			
miroc-esm	1	2	1	1	1	1
miroc-esm-chem				1	1	1
miroc5	1	1	1	1		
mpi-esm-lr	2	2	2	2	2	2
mpi-esm-mr	1	1	1			
mri-cgcm3	1		1	1	1	1
noresm1-m	1	2	1	1	2	1
noresm1-me	1	1	1			

1 = to 2100, 2 = to 2200.

Estimates of GMSL change due to human-caused changes in land water storage are based on models of the relationship between total global population, impoundment of water in dams, and groundwater depletion (Chao et al. 2008; Wada et al. 2012; Konikow 2011).

Background rates of GIA and nonclimatic process were estimated from tide-gauge data using a Gaussian process model similar to that employed by Kopp (2013). Note that this approach assumes that background rates of sea-level change will continue unchanged; to the extent that these result from human activities, such as hydrocarbon extraction in western Gulf states, economic and policy changes can decrease or increase these rates in a fashion for which our projections do not account.

TABLE A15 Sea-level rise projections

Feet Above 2000	*RCP 8.5*				*RCP 4.5*				*RCP 2.6*			
	Likely	*90% Range*	*99% Range*	*1-in-1000*	*Likely*	*90% Range*	*99% Range*	*1-in-1000*	*Likely*	*90% Range*	*99% Range*	*1-in-1000*
GLOBAL												
2030	0.4–0.5	0.3–0.6	0.3–0.7	0.8	0.4–0.5	0.3–0.6	0.3–0.7	0.7	0.4–0.5	0.3–0.6	0.3–0.7	0.7
2050	0.8–1.1	0.7–1.3	0.5–1.6	2.0	0.7–1.0	0.6–1.1	0.5–1.4	1.9	0.7–0.9	0.6–1.1	0.5–1.4	1.8
2100	2.0–3.3	1.7–4.0	1.3–5.8	8.1	1.5–2.5	1.2–3.1	0.8–4.8	7.1	1.2–2.1	0.9–2.7	0.6–4.6	6.9
2150	3.2–5.9	2.6–7.6	1.8–12.1	17.8	2.0–4.3	1.4–5.7	0.7–10.0	15.6	1.5–3.5	1.1–4.9	0.7–9.6	15
2200	4.4–9.2	3.2–12.3	2.0–20.7	31.1	2.4–6.4	1.3–8.9	0.2–16.9	27.3	1.7–5.3	0.9–7.8	0.2–16.4	26.5
PORTLAND, ME												
2030	0.4–0.8	0.2–0.9	0.0–1.2	1.3	0.4–0.7	0.2–0.8	0.1–1.0	1.1	0.3–0.8	0.2–0.9	0.0–1.1	1.3
2050	0.7–1.4	0.4–1.7	0.1–2.2	2.5	0.6–1.3	0.4–1.5	0.1–2.0	2.3	0.5–1.3	0.3–1.6	-0.1–2.1	2.4
2100	1.7–3.8	1.0–4.6	0.1–6.6	9.4	1.3–2.9	0.7–3.6	0.0–5.5	7.9	0.9–2.4	0.4–3.1	-0.3–5.2	7.6
2150	3.1–5.7	2.4–7.2	1.5–12.9	18.7	1.7–4.4	0.8–5.9	-0.3–11.2	16.5	1.3–3.3	0.8–4.9	0.3–10.5	16.4
2200	4.0–8.3	2.9–11.1	1.6–21.4	32.1	1.8–6.2	0.6–8.9	-0.9–18.8	28.4	1.1–4.8	0.5–7.8	-0.3–18.2	28.4
BOSTON, MA												
2030	0.4–0.9	0.3–1.0	0.1–1.2	1.4	0.4–0.8	0.3–0.9	0.1–1.1	1.2	0.4–0.8	0.3–1.0	0.0–1.2	1.3
2050	0.8–1.6	0.6–1.8	0.2–2.3	2.6	0.8–1.4	0.5–1.7	0.2–2.1	2.4	0.7–1.4	0.4–1.7	0.0–2.2	2.6
2100	2.0–4.0	1.3–4.9	0.4–6.8	9.7	1.5–3.1	1.0–3.8	0.3–5.8	8.2	1.2–2.6	0.7–3.3	0.0–5.4	7.9
2150	3.5–6.1	2.8–7.6	2.0–13.3	19.3	2.1–4.8	1.2–6.3	0.1–11.7	17	1.6–3.7	1.2–5.3	0.7–10.9	16.8
2200	4.5–8.9	3.5–11.8	2.2–22.2	33	2.4–6.7	1.1–9.5	-0.4–19.4	29	1.7–5.4	1.0–8.3	0.3–18.8	29.0
NEWPORT, RI												
2030	0.4–0.9	0.3–1.1	0.0–1.3	1.5	0.5–0.8	0.4–0.9	0.2–1.1	1.2	0.4–0.8	0.3–1.0	0.1–1.2	1.3
2050	0.9–1.6	0.6–1.9	0.3–2.3	2.6	0.8–1.5	0.6–1.7	0.3–2.1	2.4	0.7–1.4	0.5–1.7	0.1–2.2	2.5
2100	2.1–4.1	1.4–5.0	0.5–6.9	9.9	1.6–3.2	1.1–3.9	0.3–5.9	8.3	1.2–2.7	0.8–3.4	0.1–5.5	8.0
2150	3.7–6.3	3.0–7.8	2.0–13.6	19.7	2.2–5.0	1.3–6.5	0.2–11.9	17.2	1.8–3.9	1.3–5.5	0.7–11.1	17.1
2200	4.7–9.2	3.6–12.1	2.3–22.5	33.6	2.5–7.0	1.3–9.7	-0.3–19.7	29.4	1.8–5.6	1.1–8.5	0.4–19.1	29.3
NEW YORK, NY												
2030	0.4–1.0	0.2–1.2	0.0–1.4	1.6	0.5–0.8	0.4–1.0	0.2–1.1	1.2	0.4–0.9	0.3–1.0	0.1–1.2	1.4
2050	0.9–1.6	0.6–1.9	0.3–2.4	2.7	0.9–1.5	0.6–1.7	0.3–2.2	2.5	0.8–1.5	0.5–1.7	0.2–2.2	2.5
2100	2.1–4.2	1.4–5.1	0.5–6.9	9.9	1.7–3.3	1.1–4.0	0.4–6.0	8.3	1.3–2.8	0.9–3.5	0.2–5.6	8.0
2150	3.7–6.5	2.9–8.0	1.8–13.7	20	2.3–5.1	1.4–6.7	0.3–11.9	17.1	1.9–4.0	1.4–5.6	0.8–11.1	16.9
2200	4.9–9.5	3.8–12.3	2.3–22.6	33.6	2.7–7.2	1.5–9.9	-0.1–19.7	29.4	2.0–5.7	1.3–8.7	0.6–19.0	29.1

ATLANTIC CITY, NJ												
2030	0.6–1.0	0.4–1.1	0.2–1.3	1.5	0.6–0.9	0.5–1.0	0.3–1.2	1.3	0.5–1.0	0.4–1.1	0.2–1.3	1.5
2050	1.0–1.8	0.8–2.0	0.4–2.5	2.8	1.0–1.6	0.8–1.9	0.5–2.3	2.6	0.9–1.6	0.7–1.9	0.3–2.4	2.7
2100	2.4–4.5	1.7–5.3	0.8–7.2	10.3	2.0–3.6	1.4–4.3	0.7–6.3	8.7	1.7–3.1	1.2–3.8	0.5–6.0	8.3
2150	4.2–6.9	3.4–8.4	2.5–14.1	20.6	2.8–5.6	1.9–7.2	0.7–12.5	17.8	2.4–4.5	1.9–6.1	1.3–11.7	17.5
2200	5.5–10.0	4.4–13.0	3.0–23.4	34.4	3.3–7.9	2.1–10.6	0.4–20.5	30.4	2.7–6.4	2.0–9.4	1.2–19.8	30
PHILADELPHIA, PA												
2030	0.4–0.9	0.3–1.1	0.1–1.3	1.4	0.5–0.8	0.4–0.9	0.2–1.1	1.2	0.4–0.9	0.3–1.0	0.1–1.3	1.4
2050	0.9–1.6	0.6–1.9	0.3–2.3	2.6	0.8–1.5	0.6–1.7	0.3–2.1	2.4	0.8–1.5	0.5–1.7	0.2–2.2	2.5
2100	2.1–4.1	1.4–5.0	0.5–6.8	9.8	1.7–3.3	1.1–4.0	0.4–5.9	8.3	1.3–2.8	0.9–3.5	0.2–5.6	7.9
2150	3.7–6.3	2.9–7.9	2.0–13.5	19.9	2.3–5.1	1.4–6.6	0.3–11.7	17	1.9–4.0	1.4–5.5	0.8–11.1	16.8
2200	4.9–9.3	3.8–12.2	2.4–22.5	33.4	2.7–7.1	1.5–9.8	0.0–19.5	29.3	2.0–5.7	1.3–8.7	0.6–18.9	29
LEWES, DE												
2030	0.5–0.9	0.4–1.1	0.2–1.3	1.4	0.5–0.8	0.4–1.0	0.3–1.1	1.2	0.5–0.9	0.4–1.0	0.2–1.2	1.3
2050	1.0–1.7	0.7–1.9	0.4–2.4	2.7	0.9–1.5	0.7–1.7	0.5–2.2	2.5	0.9–1.5	0.7–1.7	0.4–2.1	2.6
2100	2.3–4.2	1.7–5.0	0.8–7.0	9.9	1.9–3.4	1.4–4.1	0.7–6.0	8.4	1.5–2.9	1.1–3.6	0.5–5.8	8.1
2150	3.9–6.7	3.2–8.2	2.2–13.8	20.4	2.6–5.3	1.8–6.7	0.8–12.1	17.5	2.1–4.3	1.7–5.8	1.1–11.4	17.2
2200	5.2–9.8	4.2–12.7	2.8–23.0	34.1	3.1–7.4	1.9–10.1	0.5–20.0	30	2.4–6.1	1.7–9.1	0.9–19.4	29.6
BALTIMORE, MD												
2030	0.5–0.9	0.3–1.0	0.1–1.2	1.3	0.5–0.8	0.4–0.9	0.2–1.1	1.2	0.5–0.8	0.3–1.0	0.1–1.2	1.3
2050	0.9–1.6	0.7–1.8	0.4–2.3	2.6	0.9–1.5	0.7–1.7	0.4–2.1	2.5	0.8–1.4	0.6–1.7	0.3–2.1	2.5
2100	2.2–4.1	1.6–4.9	0.7–6.8	9.7	1.7–3.2	1.3–3.9	0.6–5.9	8.2	1.4–2.8	1.0–3.5	0.4–5.6	7.9
2150	3.7–6.4	3.0–7.9	2.1–13.4	19.9	2.4–5.0	1.6–6.5	0.6–11.8	17.1	2.0–4.1	1.5–5.6	1.0–11.1	16.8
2200	5.0–9.4	3.9–12.3	2.6–22.5	33.4	2.9–7.1	1.7–9.8	0.3–19.5	29.3	2.2–5.8	1.5–8.8	0.7–19.0	29
WASHINGTON, D.C.												
2030	0.5–0.9	0.4–1.0	0.2–1.2	1.3	0.5–0.8	0.4–0.9	0.2–1.1	1.2	0.5–0.8	0.3–1.0	0.2–1.2	1.3
2050	0.9–1.5	0.7–1.8	0.4–2.2	2.6	0.9–1.4	0.7–1.6	0.4–2.0	2.4	0.8–1.4	0.6–1.6	0.3–2.0	2.5
2100	2.2–4.0	1.6–4.8	0.7–6.7	9.6	1.7–3.2	1.2–3.9	0.6–5.8	8.1	1.4–2.8	1.0–3.4	0.4–5.6	7.9
2150	3.7–6.3	3.0–7.9	2.1–13.4	19.8	2.4–5.0	1.6–6.5	0.6–11.7	17	2.0–4.0	1.5–5.6	1.0–11.1	16.8
2200	5.0–9.4	3.9–12.2	2.6–22.4	33.3	2.8–7.1	1.7–9.7	0.3–19.4	29.3	2.1–5.8	1.4–8.7	0.7–18.9	28.9

(*continued*)

TABLE A15 Sea-level rise projections (*continued*)

Feet Above 2000	RCP 8.5				RCP 4.5				RCP 2.6			
	Likely	*90% Range*	*99% Range*	*1-in-1000*	*Likely*	*90% Range*	*99% Range*	*1-in-1000*	*Likely*	*90% Range*	*99% Range*	*1-in-1000*
NORFOLK, VA												
2030	0.6–1.0	0.5–1.1	0.3–1.3	1.4	0.6–0.9	0.5–1.0	0.3–1.2	1.3	0.6–0.9	0.5–1.1	0.3–1.2	1.3
2050	1.1–1.7	0.9–2.0	0.6–2.4	2.8	1.1–1.6	0.9–1.8	0.6–2.2	2.6	1.0–1.6	0.8–1.8	0.5–2.2	2.7
2100	2.5–4.4	1.9–5.2	1.1–7.2	10.1	2.1–3.6	1.6–4.2	1.0–6.2	8.5	1.8–3.1	1.4–3.8	0.8–6.0	8.4
2150	4.3–7.0	3.5–8.6	2.6–14.1	20.6	3.0–5.6	2.2–7.1	1.2–12.4	17.9	2.5–4.6	2.1–6.2	1.5–11.8	17.6
2200	5.7–10.3	4.7–13.2	3.4–23.4	34.6	3.6–7.9	2.5–10.5	1.0–20.3	30.6	2.9–6.7	2.2–9.6	1.4–20.0	30.1
WILMINGTON, NC												
2030	0.4–0.7	0.3–0.9	0.1–1.1	1.2	0.4–0.7	0.3–0.8	0.2–1.0	1	0.4–0.7	0.3–0.8	0.1–1.0	1.1
2050	0.8–1.4	0.6–1.6	0.3–2.0	2.4	0.7–1.2	0.5–1.4	0.3–1.8	2.3	0.7–1.2	0.5–1.4	0.3–1.8	2.3
2100	1.9–3.6	1.4–4.3	0.7–6.4	9.2	1.5–2.8	1.1–3.4	0.5–5.6	7.9	1.2–2.4	0.8–3.1	0.3–5.3	7.8
2150	3.3–5.9	2.7–7.5	1.9–13.0	19.5	2.0–4.5	1.3–6.0	0.5–11.3	17	1.6–3.7	1.1–5.3	0.5–11.0	16.8
2200	4.5–8.9	3.5–11.9	2.4–22.2	33.3	2.3–6.5	1.3–9.3	0.0–19.1	29.5	1.6–5.5	0.9–8.4	0.1–18.9	29
CHARLESTON, SC												
2030	0.5–0.8	0.3–0.9	0.2–1.1	1.2	0.5–0.8	0.4–0.9	0.2–1.0	1.1	0.5–0.8	0.4–0.9	0.2–1.0	1.1
2050	0.9–1.4	0.7–1.6	0.5–2.0	2.6	0.8–1.3	0.7–1.5	0.5–1.9	2.3	0.8–1.3	0.7–1.5	0.5–1.8	2.3
2100	2.1–3.8	1.6–4.5	0.9–6.6	9.4	1.7–3.0	1.3–3.6	0.8–5.8	8.1	1.4–2.6	1.0–3.3	0.5–5.6	7.9
2150	3.6–6.2	3.0–7.9	2.2–13.2	19.8	2.4–4.9	1.7–6.3	0.8–11.6	17.3	1.9–4.0	1.5–5.6	0.9–11.2	17
2200	4.8–9.4	3.9–12.4	2.9–22.5	33.6	2.8–7.0	1.8–9.7	0.4–19.5	30.1	2.1–5.9	1.4–8.9	0.6–19.2	29.5
FORT PULASKI, GA												
2030	0.5–0.8	0.3–0.9	0.2–1.1	1.2	0.5–0.8	0.4–0.9	0.2–1.0	1.1	0.5–0.8	0.4–0.9	0.2–1.0	1.1
2050	0.9–1.4	0.7–1.7	0.5–2.0	2.6	0.9–1.3	0.7–1.5	0.5–1.9	2.3	0.8–1.3	0.7–1.5	0.5–1.8	2.3
2100	2.2–3.8	1.7–4.6	0.9–6.6	9.4	1.7–3.0	1.3–3.7	0.8–5.8	8.1	1.4–2.7	1.1–3.3	0.6–5.6	7.9
2150	3.7–6.3	3.1–8.0	2.3–13.3	19.8	2.4–4.9	1.7–6.4	0.8–11.6	17.4	2.0–4.1	1.5–5.7	1.0–11.2	17
2200	5.0–9.5	4.0–12.5	3.0–22.7	33.8	2.9–7.1	1.8–9.8	0.5–19.5	30.2	2.2–6.0	1.4–8.9	0.7–19.2	29.5
MIAMI, FL												
2030	0.4–0.7	0.3–0.9	0.1–1.1	1.2	0.4–0.7	0.3–0.8	0.1–1.0	1.1	0.4–0.7	0.3–0.8	0.2–0.9	1.0
2050	0.8–1.3	0.6–1.5	0.4–1.9	2.5	0.7–1.2	0.6–1.4	0.4–1.7	2.2	0.7–1.1	0.6–1.3	0.4–1.7	2.2
2100	2.0–3.6	1.5–4.3	0.9–6.4	9.3	1.5–2.8	1.1–3.4	0.6–5.6	8.1	1.2–2.4	0.9–3.0	0.5–5.4	7.9
2150	3.3–6.2	2.6–7.9	1.6–13.3	20.1	2.1–4.6	1.3–6.1	0.4–11.5	17.6	1.6–3.8	1.1–5.4	0.6–11.0	17.2
2200	4.5–9.4	3.3–12.6	2.1–22.5	34.4	2.4–6.8	1.3–9.6	-0.1–19.3	30.7	1.7–5.7	0.9–8.6	0.1–19.0	30.1

PENSACOLA, FL												
2030	0.4–0.7	0.3–0.8	0.1–0.9	1.0	0.4–0.6	0.3–0.8	0.1–0.9	1	0.4–0.7	0.2–0.8	0.1–1.0	1.1
2050	0.8–1.3	0.6–1.5	0.4–1.8	2.4	0.7–1.1	0.6–1.3	0.3–1.6	2.1	0.7–1.1	0.5–1.3	0.3–1.7	2.2
2100	1.8–3.5	1.3–4.2	0.7–6.2	9.1	1.4–2.7	1.0–3.3	0.5–5.4	7.8	1.1–2.3	0.7–3.0	0.3–5.2	7.6
2150	3.1–5.9	2.4–7.6	1.5–12.8	19.4	1.9–4.5	1.2–5.9	0.4–11.0	17	1.5–3.7	1.0–5.2	0.4–10.9	16.7
2200	4.2–9.0	3.1–12.1	1.9–21.8	33.3	2.2–6.5	1.2–9.3	-0.2–18.7	29.7	1.6–5.4	0.8–8.3	0.1–18.6	29.1
GRAND ISLE, LA												
2030	1.1–1.4	1.0–1.5	0.8–1.6	1.7	1.1–1.3	0.9–1.4	0.8–1.6	1.7	1.0–1.4	0.9–1.5	0.8–1.7	1.8
2050	1.9–2.4	1.7–2.6	1.5–3.0	3.5	1.8–2.3	1.7–2.4	1.5–2.8	3.3	1.8–2.3	1.6–2.5	1.4–2.8	3.3
2100	4.1–5.8	3.6–6.5	2.9–8.6	11.5	3.7–5.0	3.3–5.6	2.7–7.7	10.1	3.4–4.6	3.0–5.3	2.6–7.5	9.9
2150	6.6–9.4	5.8–11.1	4.9–16.3	22.9	5.4–7.9	4.7–9.4	3.8–14.5	20.6	4.9–7.1	4.5–8.7	3.9–14.3	20.2
2200	8.8–13.6	7.7–16.8	6.5–26.6	38.1	6.8–11.2	5.7–13.9	4.3–23.3	34.5	6.1–10.0	5.4–13.0	4.6–23.2	33.9
GALVESTON, TX												
2030	0.8–1.1	0.7–1.2	0.6–1.4	1.5	0.8–1.1	0.7–1.2	0.6–1.3	1.4	0.8–1.1	0.7–1.2	0.5–1.4	1.5
2050	1.5–2.0	1.3–2.2	1.1–2.6	3.1	1.4–1.8	1.3–2.0	1.1–2.3	2.8	1.4–1.8	1.2–2.0	1.0–2.4	2.9
2100	3.2–4.9	2.7–5.7	2.0–7.6	10.6	2.8–4.1	2.4–4.7	1.9–6.8	9.2	2.5–3.8	2.1–4.4	1.6–6.7	9.0
2150	5.3–8.1	4.6–9.8	3.6–15.0	21.5	4.1–6.6	3.4–8.1	2.5–13.1	19.2	3.7–5.9	3.2–7.4	2.6–13.0	18.8
2200	7.2–12.0	6.0–15.1	4.8–24.8	36.1	5.1–9.4	4.0–12.1	2.7–21.5	32.5	4.4–8.3	3.7–11.2	2.9–21.4	31.9
SAN DIEGO, CA												
2030	0.4–0.6	0.3–0.6	0.2–0.7	0.8	0.3–0.6	0.3–0.6	0.1–0.8	0.8	0.4–0.6	0.3–0.6	0.2–0.8	0.8
2050	0.7–1.2	0.6–1.3	0.4–1.7	2.3	0.7–1.1	0.5–1.2	0.3–1.6	2.1	0.6–1.0	0.5–1.2	0.3–1.6	2.1
2100	1.9–3.4	1.5–4.1	1.0–6.3	9.1	1.4–2.7	1.0–3.3	0.5–5.5	8	1.2–2.3	0.9–3.0	0.5–5.3	7.8
2150	3.1–5.9	2.4–7.7	1.7–13.0	19.6	2.0–4.5	1.4–6.1	0.5–11.2	17.6	1.6–3.8	1.1–5.3	0.6–11.0	17.1
2200	4.2–9.2	3.0–12.6	1.7–22.5	34	2.4–6.8	1.3–9.6	0.0–19.2	30.4	1.7–5.6	0.9–8.6	0.2–19.0	30
SAN FRANCISCO, CA												
2030	0.3–0.5	0.3–0.6	0.2–0.7	0.8	0.3–0.5	0.2–0.6	0.1–0.8	0.8	0.3–0.5	0.3–0.6	0.2–0.7	0.8
2050	0.7–1.1	0.5–1.3	0.3–1.6	2.3	0.6–1.0	0.5–1.2	0.3–1.5	2.1	0.6–1.0	0.4–1.1	0.2–1.5	2.1
2100	1.8–3.2	1.4–4.0	0.8–6.2	9	1.3–2.6	1.0–3.2	0.5–5.4	7.9	1.1–2.3	0.8–2.9	0.4–5.2	7.7
2150	2.9–5.7	2.3–7.5	1.5–12.9	19.4	1.9–4.4	1.2–5.9	0.4–11.1	17.4	1.4–3.6	1.0–5.2	0.5–10.9	17
2200	3.9–9.0	2.8–12.3	1.5–22.2	33.8	2.2–6.6	1.1–9.3	-0.1–19.1	30.3	1.5–5.4	0.7–8.4	0.0–18.9	29.9

(*continued*)

TABLE A15 Sea-level rise projections (*continued*)

Feet Above 2000	*RCP 8.5*				*RCP 4.5*				*RCP 2.6*			
	Likely	*90% Range*	*99% Range*	*1-in-1000*	*Likely*	*90% Range*	*99% Range*	*1-in-1000*	*Likely*	*90% Range*	*99% Range*	*1-in-1000*
ASTORIA, OR												
2030	0.1–0.3	0.0–0.3	-0.1–0.4	0.5	0.1–0.3	0.0–0.3	-0.1–0.5	0.5	0.1–0.3	0.0–0.3	-0.1–0.4	0.5
2050	0.2–0.6	0.1–0.8	-0.1–1.2	1.8	0.2–0.6	0.1–0.7	-0.1–1.1	1.6	0.2–0.6	0.0–0.7	-0.1–1.1	1.6
2100	1.0–2.3	0.6–3.0	0.1–5.3	7.9	0.5–1.7	0.2–2.3	-0.2–4.5	6.9	0.3–1.5	0.0–2.1	-0.4–4.4	6.8
2150	1.6–4.2	0.9–5.9	0.2–11.1	17.8	0.7–3.1	0.0–4.6	-0.7–9.7	15.8	0.3–2.4	-0.2–4.0	-0.6–9.6	15.5
2200	2.1–7.0	1.0–10.1	-0.3–19.8	31.5	0.6–4.8	-0.4–7.5	-1.6–17.2	28.2	-0.1–3.8	-0.8–6.7	-1.5–17.1	27.7
SEATTLE, WA												
2030	0.3–0.5	0.2–0.5	0.2–0.6	0.7	0.3–0.5	0.2–0.5	0.1–0.7	0.7	0.3–0.5	0.2–0.5	0.1–0.6	0.7
2050	0.6–1.0	0.5–1.1	0.3–1.5	2.1	0.6–0.9	0.4–1.1	0.2–1.4	1.9	0.5–0.9	0.4–1.1	0.2–1.4	1.9
2100	1.6–3.0	1.3–3.7	0.8–5.9	8.5	1.3–2.4	0.9–3.0	0.5–5.2	7.5	1.1–2.2	0.7–2.8	0.3–5.1	7.4
2150	2.6–5.2	2.0–6.9	1.3–12.0	18.6	1.8–4.1	1.1–5.6	0.3–10.7	16.6	1.3–3.4	0.9–5.0	0.5–10.6	16.3
2200	3.5–8.3	2.4–11.3	1.2–21.0	32.4	2.1–6.2	1.0–8.9	-0.2–18.4	29.1	1.4–5.2	0.7–8.1	0.0–18.3	28.6
JUNEAU, AK												
2030	-1.2– -1.1	-1.3– -1.0	-1.4– -0.9	-0.8	-1.2– -1.1	-1.3– -1.0	-1.4– -0.9	-0.8	-1.2– -1.1	-1.3– -1.0	-1.4– -0.9	-0.8
2050	-1.9– -1.6	-2.1– -1.5	-2.2– -1.1	-0.5	-2.0–-1.7	-2.1– -1.5	-2.2– -1.2	-0.7	-2.0– -1.7	-2.2– -1.5	-2.3– -1.2	-0.7
2100	-3.5– -2.4	-3.9– -1.7	-4.3–0.6	3.2	-3.9–-2.8	-4.2– -2.2	-4.6– -0.1	2.2	-4.1– -3.0	-4.4– -2.4	-4.7– -0.2	2.2
2150	-5.2– -2.6	-5.9– -1.0	-6.8–4.2	10.8	-6.0–-3.8	-6.6– -2.3	-7.3–2.9	8.6	-6.4– -4.3	-6.8– -2.7	-7.2–2.9	8.6
2200	-7.0– -2.3	-8.2–0.7	-9.5–10.3	22.1	-8.3–-4.4	-9.3– -1.6	-10.4–8.1	18.4	-8.8– -5.1	-9.6– -2.2	-10.3–8.2	18.3
ANCHORAGE, AK												
2030	-0.1–0.3	-0.2–0.4	-0.4–0.6	0.8	-0.1–0.4	-0.3–0.5	-0.5–0.8	0.9	-0.1–0.2	-0.2–0.4	-0.4–0.6	0.7
2050	-0.2–0.5	-0.4–0.8	-0.7–1.2	1.7	-0.2–0.6	-0.4–0.8	-0.8–1.3	1.6	-0.2–0.4	-0.4–0.7	-0.7–1.0	1.5
2100	-0.6–1.2	-1.2–2.0	-2.0–4.0	6.8	-0.3–1.3	-0.8–2.0	-1.5–3.9	6.5	-0.2–1.2	-0.6–1.8	-1.2–3.9	6.5
2150	-1.5–2.3	-2.7–4.2	-4.5–9.2	16	-0.5–2.4	-1.4–4.0	-2.7–8.9	15.1	-0.4–2.2	-1.2–3.8	-2.5–9.3	15.2
2200	-1.3–4.5	-3.0–7.5	-5.3–17.0	29.2	-0.5–4.2	-1.8–6.9	-3.5–16.4	27.4	-0.4–4.1	-1.6–7.1	-3.6–17.1	28
HONOLULU, HI												
2030	0.4–0.6	0.3–0.7	0.2–0.8	0.9	0.4–0.6	0.3–0.7	0.2–0.8	0.9	0.4–0.6	0.3–0.7	0.2–0.8	0.9
2050	0.8–1.2	0.6–1.4	0.4–1.9	2.5	0.7–1.1	0.6–1.3	0.4–1.7	2.2	0.6–1.1	0.5–1.3	0.3–1.7	2.2
2100	2.1–3.8	1.6–4.6	0.9–6.9	10	1.5–2.8	1.1–3.5	0.6–5.8	8.7	1.2–2.5	0.9–3.2	0.4–5.6	8.5
2150	3.3–6.6	2.4–8.7	1.4–14.3	21.5	2.1–4.9	1.3–6.5	0.4–12.0	19	1.5–4.0	1.0–5.7	0.5–11.7	18.6
2200	4.4–10.4	2.9–14.2	1.3–24.8	37.7	2.4–7.3	1.2–10.4	-0.3–20.5	33.2	1.6–6.1	0.7–9.3	-0.1–20.0	32.8

We combine probability distributions for the different components contributing to sea-level rise to construct sea-level rise projections for RCP 8.5, RCP 4.5, and RCP 2.6. We do not construct projections for RCP 6.0, as sea-level rise in RCP 4.5 and RCP 6.0 is indistinguishable in the twenty-first century (Church et al. 2013), and insufficient model output is available to extend RCP 6.0 projections for oceanographic and glacial effects beyond 2100.

Our local sea-level rise projections (Table A15) can be compared with those of other sources. For example, under RCP 8.5, we project that sea level at New York City will *likely* rise over the twenty-first century by between 0.7 and 1.3 m (2.1–4.2 ft), and will *very likely* rise between 0.4 and 1.5 m (1.4–5.1 ft). By comparison, Miller et al. (2013)'s "low" projection is 70 cm (comparable to our 17th percentile for RCP 8.5), their "central" projection is 1.0 m (comparable to our 50th percentile), their "high" projection is 1.4 cm (comparable to our 90th percentile), and their "higher" projection is 1.6 m (comparable to our 97th percentile). We project a 1-in-200 probability that New York City will experience more than 2.1 m (6.9 ft) of sea-level rise, and a 1-in-1000 probability it will experience more than 3.1 m (9.9 ft).

Our 1-in-1000 probability GMSL projection for RCP 8.5, 2.5 m (8 ft) in 2100, is similar to other estimates of the maximum sea-level rise physically possible in the current century (e.g., Church et al. 2013; Miller et al. 2013). Accordingly, we interpret 1-in-1000 probability projections as the maximum physically plausible. Such high sea-level rise requires a fairly rapid collapse of the West Antarctic Ice Sheet (WAIS) (Little, Oppenheimer & Urban 2013) following its destabilization by effects such as grounding line retreat feedbacks (Schoof 2007, Gomez et al. 2012, 2013) or ice-cliff collapse feedbacks (Bassis and Walker 2012; Pollard and DeConto 2013). Observations indicate that the WAIS may already have been destabilized (Joughin, Smith & Medley 2014; Rignot et al. 2014), but do not yet suggest that the ensuing collapse will occur at the rates needed to reach the maximum-plausible twenty-first century sea-level rise. Other users may have alternative assessments of the likely rate of collapse, which may render high-end outcomes more likely than we estimate.

REFERENCES

Arguez, A., I. Durre, S. Applequist, R. S. Vose, M. F. Squires, X. Yin, R. R. Heim Jr, and T. W. Owen (2012), NOAA's 1981–2010 US climate normals: An overview, *Bulletin of the American Meteorological Society*, *93* (11), 1687–1697, doi: 10.1175/BAMS-D-11-00197.1.

Bamber, J. L., and W. P. Aspinall (2013), An expert judgement assessment of future sea level rise from the ice sheets, *Nature Climate Change*, *3*, 424–427, doi:10.1038/nclimate1778.

Bassis, J. N., and C. C. Walker (2012), Upper and lower limits on the stability of calving glaciers from the yield strength envelope of ice, *Proceedings of the Royal Society A: Mathematical, Physical and Engineering Science*, *468* (2140), 913–931, doi:10.1098/rspa.2011.0422.

Brekke, L., B. L. Thrasher, E. P. Maurer, and T. Pruitt (2013), *Downscaled CMIP3 and CMIP5 Climate Projections: Release of Downscaled CMIP5 Climate Projections, Comparison with Preceding Information, and Summary of User Needs*, 116 pp., U.S. Department of the Interior, Bureau of Reclamation, Technical Service Center, Denver, Colorado.

Chao, B. F., Y. H. Wu, and Y. S. Li (2008), Impact of artificial reservoir water impoundment on global sea level, *Science*, *320* (5873), 212–214, doi:10.1126/science.1154580.

Church, J. A., P. U. Clark, et al. (2013), Chapter 13: Sea level change, in *Climate Change 2013: The Physical Science Basis*, edited by T. F. Stocker, D. Qin, G.-K. Plattner, M. Tignor, S. K. Allen, J. Boschung, A. Nauels, Y. Xia, V. Bex, and P. Midgley, Cambridge University Press.

Collins, M., R. Knutti, et al. (2013), Chapter 12: Long-term climate change: Projections, commitments and irreversibility, in *Climate Change 2013: the Physical Science Basis*, edited by T. F. Stocker, D. Qin, G.-K. Plattner, M. Tignor, S. K. Allen, J. Boschung, A. Nauels, Y. Xia, V. Bex, and P. Midgley, Cambridge University Press.

Davies-Jones, R. (2008), An efficient and accurate method for computing the wet-bulb temperature along pseudoadiabats, *Monthly Weather Review*, *136* (7), 2764–2785, doi:10.1175/2007MWR2224.1.

Doswell, C. A., III, J. T. Schaefer, D. W. M. T. W. Schlatter, and H. B. Wobus (1982), Thermodynamic analysis procedures at the National Severe Storms Forecast Center, in *Ninth Conference on Weather Forecasting and Analysis*, pp. 304–309, American Meterological Society, Seattle, WA.

Gomez, N., D. Pollard, J. X. Mitrovica, P. Huybers, and P. U. Clark (2012), Evolution of a coupled marine ice sheet–sea level model, *Journal of Geophysical Research*, *117*, F01,013, doi:10.1029/2011JF002128.

Gomez, N., D. Pollard, and J. X. Mitrovica (2013), A 3-D coupled ice sheet – sea level model applied to Antarctica through the last 40 ky, *Earth and Planetary Science Letters*, *384*, 88–99, doi:10.1016/j.epsl.2013.09.042.

Joughin, I., B. E. Smith, and B. Medley (2014), Marine ice sheet collapse potentially underway for the Thwaites Glacier Basin, West Antarctica, *Science*, *344*, 735–738, doi:10.1126/science.1249055.

Konikow, L. F. (2011), Contribution of global groundwater depletion since 1900 to sea-level rise, *Geophysical Research Letters*, *38*, L17,401, doi:10.1029/2011GL048604.

Kopp, R. E. (2013), Does the mid-Atlantic United States sea level acceleration hot spot reflect ocean dynamic variability?, *Geophysical Research Letters*, *40*, 3981–3985, doi:10.1002/grl.50781.

Kopp, R. E., R. M. Horton, C. M. Little, J. X. Mitrovica, M. Oppenheimer, D. J. Rasmussen, B. H. Strauss, and C. Tebaldi (2014), Probabilistic 21st and 22nd century sea-level projections at a global network of tide gauge sites, *Earth's Future*, doi:10.1002/2014ER000239.

Little, C. M., M. Oppenheimer, and N. M. Urban (2013), Upper bounds on twenty-first-century Antarctic ice loss assessed using a probabilistic framework, *Nature Climate Change*, *3*, 654–659, doi:10.1038/nclimate1845.

Marsh, P. T., and J. A. Hart (2012), SHARPPY: A Python Implementation of the Skew-T/Hodograph Analysis and Research Program, in *92nd American Meteorological Society Annual Meeting*, American Meteorological Society, New Orleans, LA.

Marzeion, B., A. H. Jarosch, and M. Hofer (2012), Past and future sea-level change from the surface mass balance of glaciers, *The Cryosphere*, *6*, 1295–1322, doi:10.5194/tc-6-1295-2012.

Maurer, E. P., A. W. Wood, J. C. Adam, D. P. Lettenmaier, and B. Nijssen (2002), A long-term hydrologically based dataset of land surface fluxes and states for the conterminous United States, *Journal of Climate*, *15* (22), 3237–3251.

Meinshausen, M., N. Meinshausen, W. Hare, S. C. Raper, K. Frieler, R. Knutti, D. J. Frame, and M. R. Allen (2009), Greenhouse-gas emission targets for limiting global warming to 2°C, *Nature*, *458* (7242), 1158–1162, doi:10.1038/nature00817.

Meinshausen, M., S. C. B. Raper, and T. M. L. Wigley (2011), Emulating coupled atmosphere-ocean and carbon cycle models with a simpler model, MAGICC6 – Part 1: Model description and calibration, *Atmospheric Chistry and Physics*, *11* (4), 1417–1456, doi:10.5194/acp-11-1417-2011.

Mesinger, F., G. DiMego, E. Kalnay, K. Mitchell, P. C. Shafran, W. Ebisuzaki, D. Jović, J. Woollen, E. Rogers, E. H. Berbery, et al. (2006), North American regional reanalysis, *Bulletin of the American Meteorological Society*, *87* (3), 343–360, doi: 10.1175/BAMS-87-3-343.

Miller, K. G., R. E. Kopp, B. P. Horton, J. V. Browning, and A. C. Kemp (2013), A geological perspective on sea-level rise and impacts along the U.S. mid-Atlantic coast, *Earth's Future*, *1*, 3–18, doi:10.1002/2013EF000135.

Mitchell, T. D. (2003), Pattern scaling: an examination of the accuracy of the technique for describing future climates, *Climatic Change*, *60* (3), 217–242, doi:10.1023/A:1026035305597.

Pollard, D., and R. M. DeConto (2013), Modeling drastic ice retreat in Antarctic subglacial basins, in *AGU Fall Meeting Abstracts*, Abstract GC34A-03, San Francisco, CA.

Rignot, E., J. Mouginot, M. Morlighem, H. Seroussi, and B. Scheuchl (2014), Widespread, rapid grounding line retreat of Pine Island, Thwaites, Smith, and Kohler glaciers, West Antarctica, from 1992 to 2011, *Geophysical Research Letters*, *41*, 3502–3509, doi:10.1002/2014GL060140.

Rogelj, J., M. Meinshausen, and R. Knutti (2012), Global warming under old and new scenarios using IPCC climate sensitivity range estimates, *Nature Climate Change*, *2* (4), 248–253, doi:10.1038/nclimate1385.

Schoof, C. (2007), Ice sheet grounding line dynamics: Steady states, stability, and hysteresis, *Journal of Geophysical Research*, *112* (F3), F03S28, doi:10.1029/2006JF000664.

Sherwood, S., and Q. Fu (2014), A drier future?, *Science*, *343* (6172), 737–739, doi:10.1126/science.1247620.

Tebaldi, C., and R. Knutti (2007), The use of the multi-model ensemble in probabilistic climate projections, *Philosophical Transactions of the Royal Society A: Mathematical, Physical and Engineering Sciences*, *365* (1857), 2053–2075, doi:10.1098/rsta.2007.2076.

Wada, Y., L. P. H. van Beek, F. C. Sperna Weiland, B. F. Chao, Y.-H. Wu, and M. F. P. Bierkens (2012), Past and future contribution of global groundwater depletion to sea-level rise, *Geophysical Research Letters*, *39*, L09,402, doi:10.1029/2012GL051230.

Wood, A. W., E. P. Maurer, A. Kumar, and D. P. Lettenmaier (2002), Long-range experimental hydrologic forecasting for the eastern United States, *Journal of Geophysical Research: Atmospheres*, *107* (D20), 4429, doi:10.1029/2001JD000659.

APPENDIX B
CLIMATE IMPACTS

SOLOMON HSIANG, AMIR JINA, AND JAMES RISING

1. INTRODUCTION

Recent years have seen an explosion of empirical research leading to an unprecendented advance in our knowledge of the interaction of society and the environment. As the quantity and quality of existing studies continues to increase, the literature as a whole becomes increasingly effective at answering societally important questions, so long as this growth can be leveraged and integrated into our existing knowledge base. The approach of this report toward environmental assessment is an effort to integrate this research holistically. Our vision was to create a system to not only incorporate recent discoveries, but also to be updated with new research and new findings as they become available in the future. We present below a flexible, open-source, and adaptive system to combine our best estimates of environmental impacts, allowing us constantly to learn the broad societal effects from an evolving body of research. The current approach is not limited to climate, however, as it can be extended to project many different types of impacts (e.g., from policy changes). The tools we have designed can become a central hub enabling researchers to collaborate on a larger body of socially important research.

We identify and employ a meta-analytical approach (described in section 2) that draws on Bayesian methods commonly used in medical research and previously implemented in Hsiang et al. (2013b). Using these techniques, we design an open-source tool that can update composite dose-response functions in real time as new research becomes available (see figure B2). With the method of meta-analysis in place, we then identify a group of rigorous studies across a number of climate-impacted sectors. Combining the dose-response functions from individual studies (detailed in section 3), we generate a series of composite response functions for each sector. Finally, to understand the impact of climate upon each sector, we take the product of our response functions and the downscaled physical clmate projections described in appendix A, giving us partial equilibrium impacts out to the end of the current century (described in section 4). These impacts are then used, alongside detailed sectoral models (described in appendix C), as input to computationally model the general equilibrium effects of changes in climate (described in appendix D).

2. META-ANALYSIS APPROACH

The empirical impact functions are treated as conditional distributions, conditioned on weather variables such as mean temperature and precipitation. This representation facilitates meta-analysis and also captures the range of uncertainty in the empirical estimates. Each distribution is evaluated at a given quantile only when it is applied to data, as described in section 4.

2.1. Hierarchical Bayesian Modeling

The impact estimates that combine results from more than one study apply a Bayesian hierarchical model structure, as described by Gelman et al. (2013). This approach simultaneously estimates a distribution of possible underlying effect sizes, as well as a degree of "partial-pooling." If the individual study estimates are consistent with a single underlying effect, their estimates are pooled to estimate the effect accurately. However, if the study estimates are inconsistent with each other, the underlying effect is estimated to be only loosely informed by each study, which is considered to have its own idiosyncratic effect.

Consider a collection of impact functions, $f_i(\beta_i|T)$, for $i \in \{1, \ldots, N\}$ indexing independently published results. Here, $f_i(\beta i|T)$ is a probability distribution for β_i conditioned on a weather variable T. We wish to combine these estimates into a single conditional distribution, $g(\hat{\beta}|T)$, where $\hat{\beta}$ is called the "hyper-parameter." We treat each value of T independently, so we will write these functions as $f_i(\beta_i)$ and $g(\hat{\beta})$.

Under hierarchical merging, the conditional parameter distributions are required to be Gaussian distributions, and later, Gaussian parameter estimate errors are used in all applicable response functions. The governing equations are

$$\theta_i \sim N(\hat{\beta}, \tau^2)$$
$$\beta_i \sim N(\theta_i, \sigma_i^2)$$

where β_i is a measured parameter, corresponding to a true (unobserved) parameter θ_i that characterizes the response for study i. σ^2 is the standard error of β_i. We are interested in $\hat{\beta}$, the underlying hyperparameter, and τ^2, the variance between models. We apply noninformative priors to $\hat{\beta}$ and τ. That is, $p(\hat{\beta}) \propto 1$ and $p(\tau) \propto 1$. The values of β_i and σ^2 are provided by the published studies, and the rest of the parameters are simultaneously estimated.

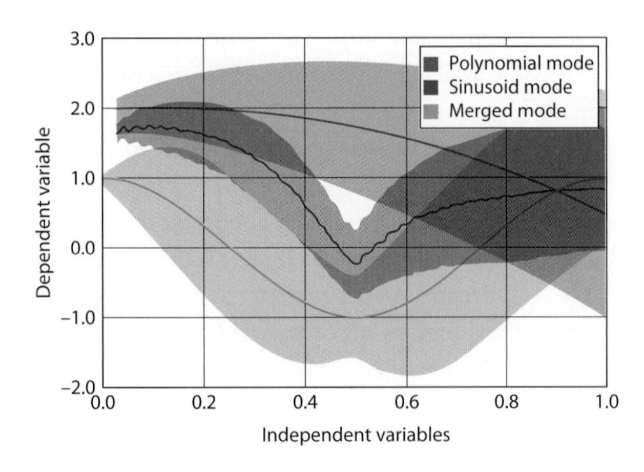

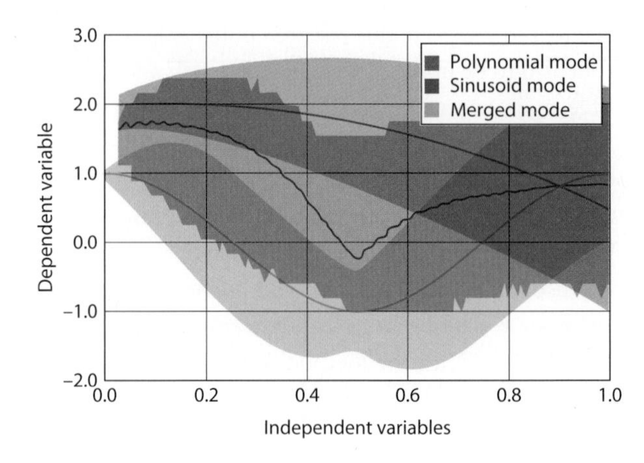

FIGURE B1. Pooled (**left**) and Bayesian hierarchical (**right**) estimates for a constructed polynomial and sinusoidal response function. The pooled distribution is calculated as $p(\hat{\beta}|x) = p(\bigcap_{i=1}^{N} \hat{\beta} = \beta_i|x) \propto \prod_{i=1}^{N} p(\beta_i|x)$, where $p(\beta_i|x)$ is the conditional distribution for either the polynomial or sinusoidal response function at a given value of x. The 95% confidence interval of the pooled result does not overlap with the individual estimates when they are far from each other. The confidence intervals on $\hat{\beta}$ are wider, reflecting the uncertainty in resolving the two estimates.

An analytic solution exists for how to generate draws from the posterior distribution of this hierarchical model and is described in chapter 5 of Gelman et al. (2013). We approximate the posterior by producing draws and constructing a histogram for each conditional distribution.

Figure B1 shows an example of how pooled and Bayesian hierarchical results differ for a combination of two imagined impact functions.

2.2. Distributed Meta-analysis System

To support the management of empirical results, the meta-analysis combination process, and the application of these results to data, we constructed a new tool called DMAS, the Distributed Meta-Analysis System.[1] The heart of DMAS is a database of results that can be easily recombined into many different meta-analyses (see Figure B2). This database is designed to be expanded in a decentralized manner, by "crowd-sourcing" from scientists working independently to detail their empirical findings.[2] By combining the efforts of many researchers throughout various academic disciplines in an Internet research community, the capacity of scientists to maintain up-to-date empirical relationships increases drastically. Below, we detail the numerous features that are available to promote academic exchange on DMAS. Unlike many crowd-sourcing projects, the necessary vetting of this information is made much easier by connecting each estimate with published literature. This connection to academic journals further supports the construction of comprehensive meta-analyses.

2.2.1. MODEL TYPES

The DMAS library results are conditional distributions, representing one or more parameter estimates typically in a dose-response curve. The following representations are used for impact functions in this report:

Discrete-Discrete Probability Models: The discrete-discrete probability model represents either a sampled approximation to a continuous probability density function, $f(y|x)$, at discrete values of $y \in \{y_i\}$ and $x \in \{x_i\}$, or a probability mass function of the same form. This is most appropriate when the collection of response outcomes is limited or categorical. Both the dependent and independent variables may be either categorical or sampled at a collection of numerical levels. For continuous functions, the sampling of the dependent variable, $\{y_i\}$, and the independent variable, $\{x_i\}$, can be uneven.[3] This model can be treated as a matrix $P = (p_{ij} = f(y_j|x_i))$. Discrete-discrete probability models are the ultimate form for any Bayesian hierarchical meta-analysis result, after draws from the posterior distribution are organized into histograms.

Spline Models: The spline model represents a continuous conditional probability function, using a spline to denote the log of its values.

$$f(y|x) = \begin{cases} e^{a_0 + b_0 y + c_0 y^2} & \text{for } y_0 \le y < y_1 \\ e^{a_1 + b_1 y + c_1 y^2} & \text{for } y_1 \le y < y_2 \\ \ldots & \ldots \end{cases}$$

Distinct splines are described at distinct values of the conditioning variable $x \in \{x_i\}$, which may be categorical or numerical and may be sampled unevenly. The lowest value of y_0 for each x may be $-\infty$, and the highest value of y_I may be ∞. Spline models are used for most impact functions, as they provide arbitrary resolution on the shape of the conditional distribution curve.

1 Available online at http://dmas.berkeley.edu.

2 A similar process of decentralized collection of results has begun for drug discovery (Lessl et al. 2011).

3 For calculating a CDF, we assume that each y_i represents a histogram-style bin, while under interpolation we use linear interpolation (see section 8).

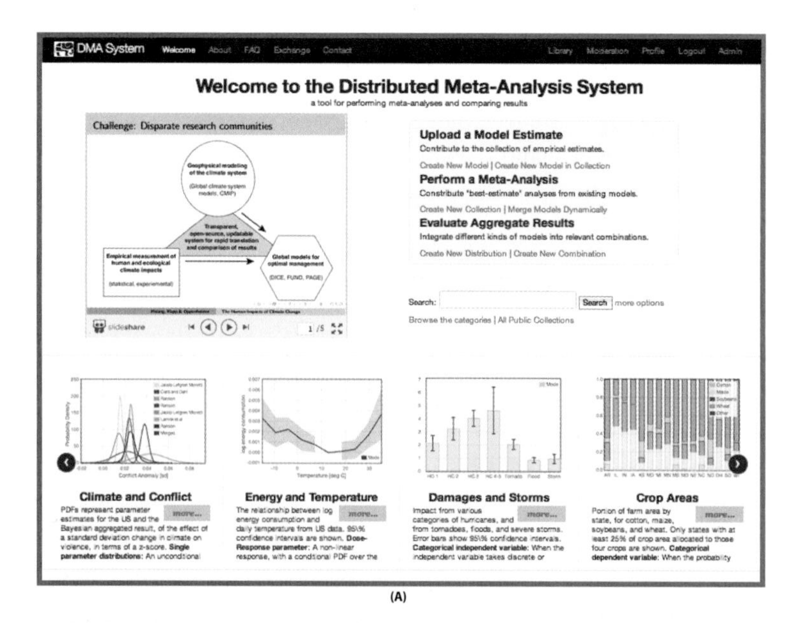

(A)

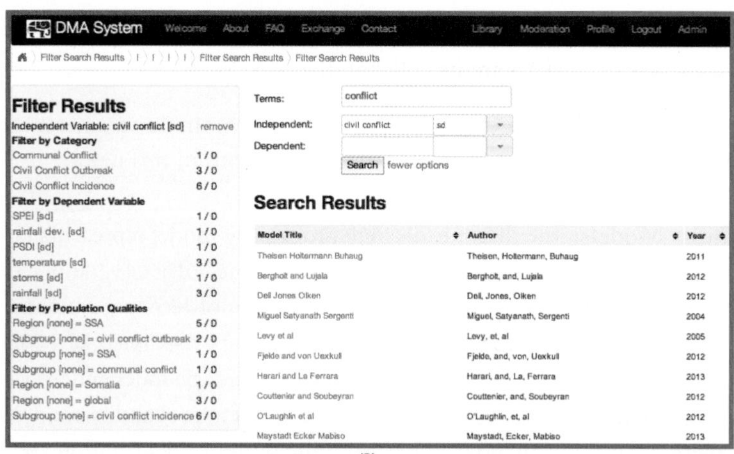

(B)

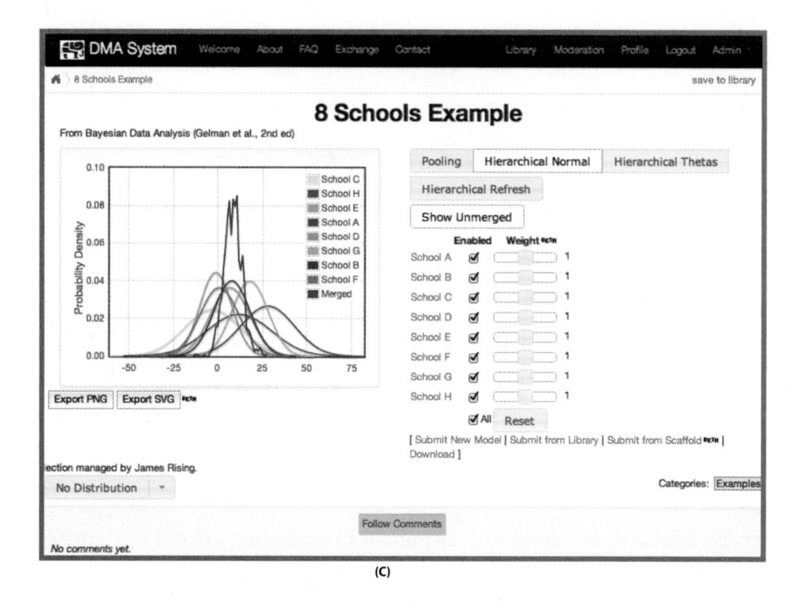

(C)

FIGURE B2. **A**: Front page of the DMAS system, showing a gallery of existing results. **B**: Search screen, allowing results to be filtered by population and study characteristics. **C**: A sample result, the 8-schools example from Gelman et al. (2013), showing options for selecting and weighting results and identifying the meta-analysis combination technique.

Bin Models: A bin model represents a model defined across continuous spans, where the distribution is constant over each span. It is a combination of information describing the width of each bin and an underlying categorical model of one of the other model types describing each bin's probability distribution. Bin models are used for degree-day impacts as in Schlenker and Roberts (2009), with an underlying spline curve representation for each bin.

2.2.2. IMPORTING MODELS

Each model type has a file format specification, and parameter estimates can be added by uploading files in these formats. We also provide a variety of simplified ways to specify models. This includes a spreadsheet-style entry for discrete-discrete probability models, a GUI model generator for uniform, Gaussian, and polynomial models, and a "feature interpreter" which allows spline models to be described in terms of any collection of their mean, variance, standard deviation, skew, mode, or arbitrary confidence intervals.

2.2.3. ADDITIONAL FEATURES

Finally, DMAS includes a wide range of features that can help support the continued evolution and use of these results. The models can be visualized, compared, and combined with weights. They can be selectively included or excluded from a meta-analysis combination, either manually or using arbitrary population or study characteristics. Arbitrary functions of results can be computed, applied to data, and output for external use. Finally, the entire database can be quickly searched using tags, parameter definitions, and study-specific meta-data.

Some of the greatest advantages of DMAS are its collaborative aspects. Scientists can curate collections of results for meta-analyses, with a moderation system for others to submit additions. They can also create crowd-sourcing templates, which ask for study results in a customized form, to collect information from many researchers quickly. In addition, each result and each collection can act as a discussion board, inviting authorized users to debate the choices used.

We hope that this tool can act as a platform to promote the meta-analysis process and make results available to both modelers and a wider audience.

3. MICRO-FOUNDING IMPACT FUNCTIONS

We develop empirical, micro-founded impact functions for a number of sectors seen to be economically important. These include agriculture, crime, health, and labor. Within each sector, we draw on statistical studies that robustly account for a number of potential confounding factors when trying to identify the impacts of climate. For the current analysis, we make no claim to having performed an exhaustive quantitative meta-analysis from the reviewed papers. Numerous high-quality and insightful studies are omitted from sectors, though many studies were used to confirm the validity of the selected papers. However, we have designed our approach to be inclusive in the long run by building an open-source system for meta-analysis and collaboration. Incorporating each study took considerable effort, often requiring new data, efforts on the part of the original authors and ourselves to rerun analyses, and extensive discussions to ensure an accurate interpretation of results. In this process, we are indebted to each of the authors listed below. Our final selection required studies to meet the following criteria:

1. **Nationally representative**. We required that studies be conducted at national level or be drawn from a representative random sample of the entire United States. This was of particular relevance to health-sector studies. For example, many studies that we considered performed detailed time-series analysis of single or multiple cities (e.g., Curriero et al. 2002; Anderson and Bell, 2009). While these were high-quality studies, inclusion would have required either a weighting scheme based on city populations or an assumption of national generalizability.
2. **Analyze recent time periods in U.S. history**. As we are concerned with potential effects of adaptation, we preferred studies that identified effects as close to the present as possible.

3. **Robust to unobserved factors that differ across spatial units** (jurisdictions, counties, or states). We placed an emphasis on studies that were able to control for unobservable differences between spatial units of analysis with the inclusion of fixed effects. This required the use of longitudinal or panel data, as cross-sectional comparisons between could suffer from omitted variable bias.
4. **Identify responses to high-frequency climatic variables** (days or weeks). The importance of using high-frequency data to estimate climate impacts is demonstrated by all papers included, building on early work by Deschênes and Greenstone (2007), and in one case finding large effects by considering subdaily temperature responses (Schlenker and Roberts, 2009).
5. **Identify responses to the full distribution of temperature and rainfall measures**. Many studies looked at single climatic events, or parts of the temperature or rainfall distribution (e.g., heat-waves in Anderson and Bell, 2011). As we are modeling annual impacts, we chose only those studies that included the full distribution of realized climate outcomes, and ensured the validity of results by comparison to numerous studies looking at single phenomena or subpopulations.
6. **Account for seasonal patterns and trends in the outcomes**. Cyclicality and seasonality of responses to climate forcings is a source of major concern, so we selected only those studies that robustly accounted for seasonal patterns and time trends in their analysis.
7. **Ecologically valid**. We required studies to be valid for real-life circumstances and levels of exposure, which led us to prefer studies that were quasi-experimental in design, using observational data. For example, in the case of labor, numerous laboratory studies exist on the intensive margin effects of temperature upon productivity (e.g., Seppanen, Fisk & Lei 2006). As these raised a question of ecological validity when applied to the labor sector, we chose to not include them.

Many of the impacts of climate change will unfold over years, but distinguishing between the role of climate change and the role of social, technological, and economic evolution is very difficult over any long time horizon. Our criteria for selecting studies requires that long-term trends are accounted for and are not reflected in the measured impact response functions. As a result, the impacts that we measure are from weather "shocks," short-term changes in temperature and precipitation that are not captured by long-term trends. This approach has both strengths and weaknesses. Its key strength is that it clearly identifies the impacts of weather as distinct from longer-term changes. However, it may miss many of the long-term impacts of climate change that do not take the form of increases in the size, frequency, and duration of weather shocks.

We identify a number of studies using panel data to isolate the variation within the relevant spatial unit, while controlling for unobservable difference between units. Estimates from each of the studies were combined, as detailed in section 2. We have been conservative in our choice of studies for the current analysis, using only studies that we think most credibly identify the impact of climate upon specific outcomes in each sector. However, our approach allows for future studies to be incorporated, introducing new findings, and modifying the current results. The following is a complete list of empirical response functions used in this study, with detailed discussion of each of the studies below (shown in figure B4):

Agriculture	Maize yields vs. temperature (East)
	Maize yields vs. temperature (West)
	Maize yields vs. precipitation (East)
	Maize yields vs. precipitation (West)
	Wheat yield vs. temperature
	Soybean yields vs. temperature (East)
	Soybean yields vs. temperature (West)
	Soybean yields vs. precipitation (East)
	Soybean yields vs. precipitation (West)

	Cotton yields vs. temperature
	Cotton yields vs. precipitation
	Maize yields vs. 100 ppm CO_2 increase
	Wheat yields vs. 100 ppm CO_2 increase
	Soybean yields vs. 100 ppm CO_2 increase
	Cotton yields vs. 100 ppm CO_2 increase
Crime	Violent crime vs. temperature
	Violent crime vs. precipitation
	Property crime vs. temperature
	Property crime vs. precipitation
Health	Mortality vs. temperature (all age)
	Mortality vs. temperature (younger than 1 year)
	Mortality vs. temperature (1–44 years)
	Mortality vs. temperature (45–64 years)
	Mortality vs. temperature (65 years and up)
Labor	Hours worked in high-risk industries vs. temperature
	Hours worked in low-risk industries vs. temperature

3.1. Agriculture

Schlenker and Roberts (2009)

Outcome data: Yields for maize, soybeans, and cotton from the U.S. Department of Agriculture National Agricultural Statistical Service.

Climate data: PRISM temperature and rainfall, temporally downscaled to daily resolution.

Sample period: 1950 to 2009

Sample unit: County-years, for counties with recorded yields of maize, soybeans, or cotton.

Methodology: Piecewise linear response of log(yield) to cumulative temperature (degree days) and polynomial response to precipitation (seasonal total), controlling for county fixed effects and state-specific quadratic trends. Piecewise linear models are specific to each crop type, with thresholds that capture the beneficial effects of temperatures below a certain point, and the deleterious effects above.

Result: Modified version of Schlenker and Roberts (2009), SI appendix, p. 9, fig. A3; and p. 20, fig. 10.

Impact function: We contacted the authors of the study to select a preferred response function from the multiple methods they had employed, selecting a piecewise linear specification using degree days for temperature and seasonal total precipitation. We obtained impact functions for each of the three crops studied, for both temperature and precipitation. The authors note the distinct difference in response between counties to the east and west of the 100th meridian for maize and soybeans, so we obtained separate response functions in for these regions. On December 19th, 2013, we were sent a complete list of response functions that were updated to span the time period up to and including 2011 (as presented in Berry, Roberts & Schlenker 2012).

Hsiang, Lobell, Roberts, and Schlenker (2013a)

Outcome data:	USDA-NASS
Climate data:	University of Delaware monthly temperature and precipitation
Sample period:	1950 to 2007
Sample unit:	County-year
Methodology:	Nonlinear response of log(yield) to crop-specific seasonal average temperature and precipitation, controlling for county and year fixed effects.
Result:	Hsiang et al. 2013a, p. 19.
Impact function:	We use the response of wheat to seasonal average temperature presented in the paper. Results were obtained from the author. Calorie weighted averages were taken between maize and wheat in order to combine results, as detailed in section 4.

McGrath and Lobell (2013)

Outcome data:	Yield from 1960 to 2004 from FAOStat.
Climate data:	Keeling CO2 concentrations and country average P/PET.
Methodology:	Process model that develops the response of different crops to carbon dioxide concentrations and growing season P/PET from empirical studies. This is then used to estimate the changes to historical yields under a 100-ppm increase in CO2.
Result:	McGrath and Lobell (2013), p. 5, fig. 4 (obtained results for the United States from the authors).
Impact function:	We contacted the authors and received estimates of the CO2 fertilization relationship with yields of different crops on January 17th, 2014, specifically for the United States. Data were for eight different crop types. We used an average of all types for cotton estimates.

3.1.1. STORAGE

In addition to the above impacts on yields, we observe that farmers store crops for sale in the future, and so the overall impact of climate on supply of crops may appear smoother than if there were no storage. For our projections, we also make use of Fisher et al. (2012, Appendix p. xi, table A4) to estimate crop consumption as a moving average process of crop production. We estimated the following equation for crop c,

$$\ln(consumption)_{c,t} = \sum_{l=0}^{L}\left[\beta_{c,l} \times \ln(production)_{c,t-1}\right] + \theta_c t + \gamma_c t^2 + \varepsilon_{c,t}$$

where $L = 2$, except for soybeans where $L = 3$, and we account for linear and quadratic time-trends. Example results of this model are shown in figure B3. We project the smoothing of future crops with a time-series structure that incorporates these empirical results on storage. Weights for each crop are constructed from the lagged coefficients, $\beta_{c,l}$, presented in section 4.1.

3.2. Crime

Jacob, Lefgren, and Moretti (2007)

Outcome data:	FBI National Incident Based Reporting System.
Climate data:	Weekly temperature and precipitation from the NCDC GHCN-Daily database.
Sample period:	1995 to 2001
Sample unit:	Jurisdiction-weeks

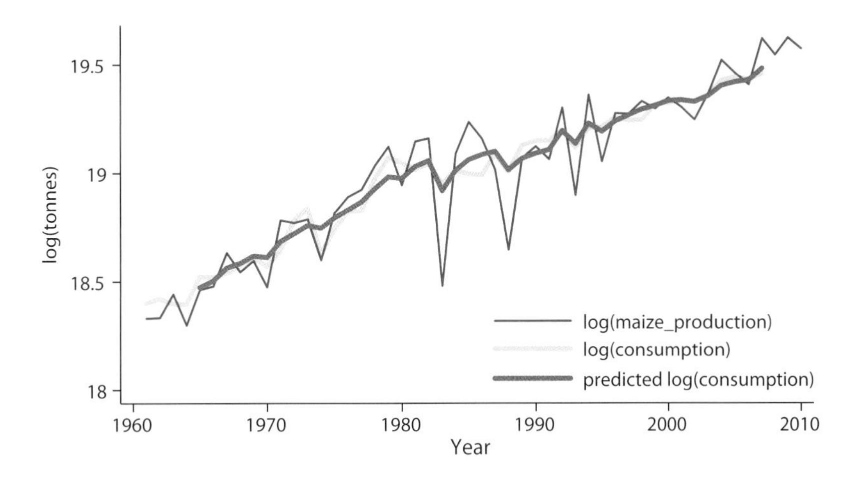

FIGURE B3. Predicted consumption of maize, modeled as a moving average of production. Predicted values compare well to observed consumption and allow us to project the smoothed consumption values out to the end of the century.

Methodology: Linear response of log(crime_rate) to average temperature and precipitation, controlling for jurisdiction-by-year and month fixed effects, as well as jurisdiction-specific fourth-order polynomials in day of year.

Result: Modified version of Jacob, Lefgren & Moretti 2007, pp. 508–509, table 2.

Impact function: We obtained data and replication files from the authors and generated coefficients for a month-long exposure window, to account for displacement of crime, as noted in the text. The climate variables are at weekly resolution, and in order to make this comparable to Ranson (2014) we reran the analysis using maximum temperatures and then scaled the coefficients in Jacob, Lefgren & Moretti (2007). We did this by first dividing the coefficient for the monthly exposure by 7 to get a daily response, and further by 4 to account for the lagged climate variables. This resulted in the marginal effect on crime of a 1°F increase in daily temperature. Taking a reference point of zero response at a temperature of 65°F (to coincide with the central point of the reference bin of (Ranson, 2014)) we derived a linear response of violent crimes and property crimes to temperature and precipitation.

Ranson (2014)

Outcome data: FBI Universal Crime Reporting Data.

Climate data: Daily temperature and precipitation from the NCDC GHCN-Daily database.

Sample period: 1960 to 2009

Sample unit: County-months

Methodology: Nonlinear response of log(crime_rate) to maximum temperature and precipitation, controlling for county-by-year and state-by-month fixed effects. Temperature is transformed into number of days within 10°F bins, with the 60–69°F bin as a reference point.

Result: Ranson (2014), p. 9, fig. 4.

Impact function: We contacted the author and received updated estimates of the percentage change for each of eight different classes of crimes on March 12th, 2014. To derive response functions, we grouped these into violent crimes (murder, rape, aggravated assault, and simple assault) and property crimes (robbery, burglary, larceny, and vehicle theft), and combined results within each class of crimes.

3.3. Health

Deschênes and Greenstone (2011)

Outcome data: National Center for Health Statistics Compressed Mortality Files.
Climate data: Daily temperature and precipitation from NCDC.
Sample period: 1968 to 2002
Sample unit: County-years
Methodology: Nonlinear response of mortality to temperature, controlling for county-by-age-group and state-by-year-by-age-group fixed effects. Temperature is transformed into number of days in a year-long window within 10°F bins, with the 50–59°F bin as a reference point.
Result: Modifed version of Deschênes and Greenstone (2011), p. 9, fig. 2.
Impact function: We contacted the authors and received estimates on November 5th, 2013. To make the study comparable to Barreca et al. (2013), the main analysis was rerun with log(mortality) as an outcome.

Barreca, Clay, Deschênes, Greenstone, and Shapiro (2013)

Outcome data: Mortality from the Mortality Statistics of the United States (pre-1959) and the Multiple Cause of Death files (post-1959).
Climate data: Daily temperature and precipitation from the NCDC GHCN-Daily database.
Sample period: 1929 to 2004
Sample unit: State-months
Methodology: Nonlinear response of log(mortality) to temperature, controlling for state-by-month and year-month fixed effects, and state-by-month-specific quadratic time trends. Temperature is transformed into number of days in a 2-month window within 10°F bins, with the 60–69°F bin as a reference point.
Result: Modified version of Barreca et al. (2013), p. 37, table 3, panel B.
Impact function: We contacted the authors and received estimates on November 5th, 2013. The preferred specification, to account for forward displacement, was to use monthly mortality with a 2-month exposure window to temperature. We used the estimated response from 1960 to 2004. To make this response comparable to the response of Deschênes and Greenstone (2011), the analysis was rerun with the reference point changed to the 50–59°F bin. To scale the coefficients, we divided each coefficient value by a factor of 6. We also obtained age-specific response functions for ages 0–1, 1–44, 45–64, and 65+.

3.4. Labor

Graff Zivin and Neidell (2014)

Outcome data: Hours worked from the American Time Use Survey.
Climate data: Daily temperature, precipitation, and humidity from NCDC.
Sample period: 2003 to 2006
Sample unit: Person-days
Methodology: Seemingly unrelated regression allowing for correlated errors between time spent working, or indoor and outdoor leisure. Nonlinear response to maximum temperatures controlling for

	county, year-by-month, and day of week fixed effects, as well as individual level controls. Temperature is transformed into number of days within 5°F bins, with the 76–80°F bin as a reference point.
Result:	High-risk: Graff Zivin and Neidell (2014), p. 15, fig. 3; Low-risk: Graff Zivin and Neidell (2014), p. 16, fig. 4.
Impact function:	We contacted the authors prior to publication and received full estimates for high-risk and low-risk labor responses to temperature on December 18th, 2013.

4. APPLICATION OF IMPACT FUNCTIONS

We apply two approaches for sampling the conditional distributions for each impact function: a Monte Carlo approach and a constant quantile approach. The Monte Carlo approach captures the full range of uncertainty in impact function estimates, under the assumption that each impact function is independent. We randomly select quantiles for each of the 26 empirical distributions. The constant quantile approach applies the same quantile across all distributions. In particular, we use a low quantile ($p = 0.33333$), median quantile ($p = 0.5$), and high quantile ($p = 0.66667$). The ordinality of the quantiles is chosen so that these describe, in essence, low-, median-, and high-impact scenarios (see Figure B5). High quantiles correspond to greater losses in yield and labor productivity, and greater increases in crime and mortality, within the range of statistical uncertainty.

Under either approach, the same quantile is used across the entire range of the conditioning variable. By evaluating each impact function at a quantile, we generate a single-dimensional, deterministic function that is used in the evaluation of the impact for each Monte Carlo or constant quantile run.

The impact results for crime, labor productivity, and mortality are all estimated by binning weather values. In these cases, we construct a continuous impact curve by linearly interpolating between the midpoints of these bins.

Impact results were produced in "result sets" that consisted of an impact value for each year in each county in the United States. Result sets are grouped into "batches," with a result set for each RCP (2.6, 4.5, 6.0, and 8.5), each GCM model or model surrogate (from 28 to 44 models in each RCP), and each of ten weather realizations, for a total of 1440 result sets per batch. Example values for a single county drawn from a single result set are shown in Figure B6. We produced 25 batches of Monte Carlo results. These results are later combined and weighted as described in section 7. Twenty-six conditional distributions were used to generate 15 impact results. These calculations are described later. Throughout, impacts are ultimately reported relative to 2012, the baseline year for the CGE model. The definitions of each impact described later do not reflect this.

In the following, we use the notation T_{AVG} for mean daily temperature; T_{MIN} and T_{MAX} for minimum and maximum daily temperature, respectively; and P for for precipitation. All weather variables are provided at the county level and on a daily basis. Below, where counties are indexed by j, weather variables are implicitly also indexed by j. We indexed days in the year by d and years by t.

4.1. Agricultural Yields and Production

Percent changes in agriculture production, relative to 2012, were generated using fixed, county-specific growing seasons. The growing season, denoted $S(j)$ for county j, is determined using the centroid of the county applied to the planting and harvesting dates in Sacks et al. (2010). For maize and wheat, for which Sacks et al. (2010) provide two calendars (two croppings for maize, and summer and winter wheat), the calendar that represented the greatest portion of land area in each county was used.

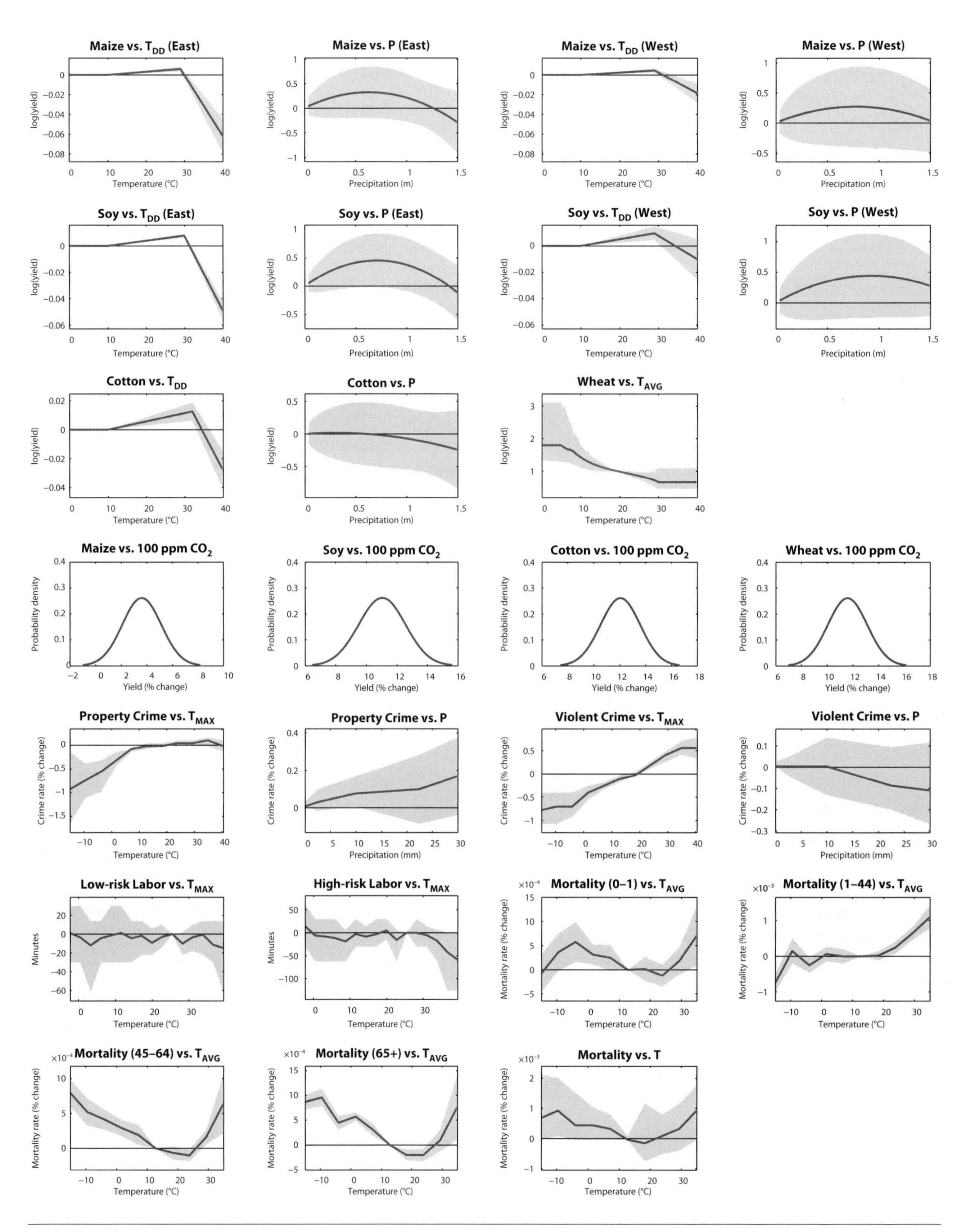

FIGURE B4. All 26 dose-response functions used in our analysis. Ninety-five percent confidence intervals are shaded.

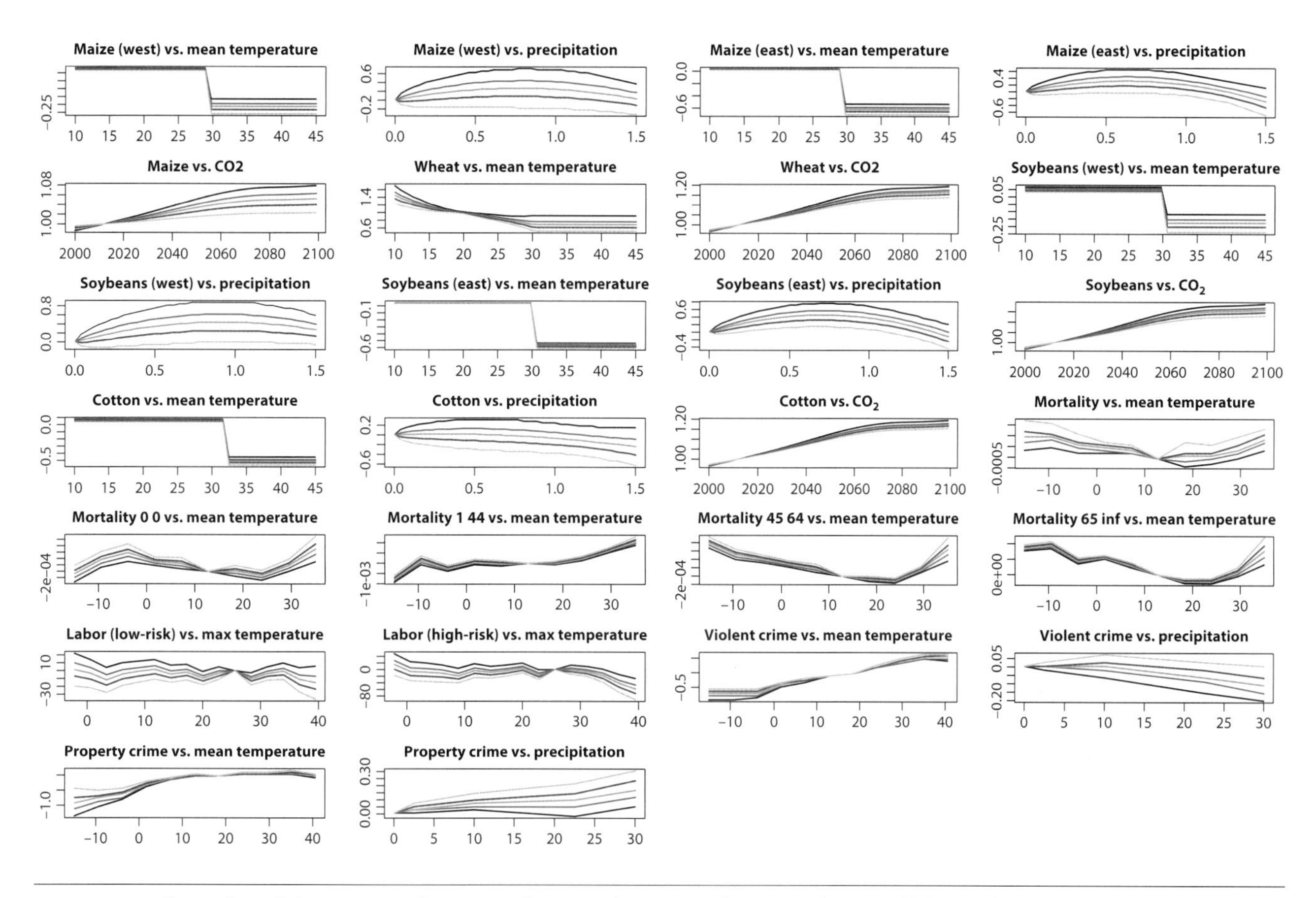

FIGURE B5. Quantiles of the response functions, showing the impact functions that would be used in evaluation for quantiles of $p \in \{0.1, 0.3, 0.5, 0.7, 0.9\}$. CO2 fertilization graphs show evolution for RCP 4.5.

Relative changes in yield are calculated based on seasonal temperatures and precipitation as follows:

Wheat Wheat uses a seasonal average temperature response function:

$$Y_{jt} = f\left(\frac{1}{N(S(j))}\sum_{d \in S(j)} T_{AVG,\, d}\right)$$

where $f(\cdot)$ is calculated by Hsiang et al. (2013a) as a function of average mean daily temperature over the growing season, and $N(S(j))$ is the number of days in the growing season for county j. This functional form was only used for wheat, as a degree-day representation was unavailable.

Cotton Cotton uses a single degree-day function:

$$Y_{jt} = e^{f}(0.01\, \Sigma_{d \in s(j)} DD_{low}(T_{MAX,d}, T_{MIN,d}), 0.01\, \Sigma_{d \in s(j)} DD_{high}(T_{MAX,d}, T_{MIN,d})) + g(1 \times 10^{-3}\, \Sigma_{d \in s(j)} Pd)$$

where DD_{low} and DD_{high} are growing degree days below and above the crop-specific breakpoint specifiedin Schlenker and Roberts (2009) and calculated as specified there using the minimum and maximum daily temperatures. The functions $f(\cdot)$ and $g(\cdot)$ translate degree days and precipitation, respectively, into yield effects.

Maize and Soybeans Maize and soybeans have two degree-day responses:

$$Y_{jt} = \begin{cases} e^{f_{east}}(0.01\sum_{d\in s(j)} DD_{low}(T_{MAX,d}T_{MIN,d}), 0.01\sum_{d\in s(j)} DD_{high}(T_{MAX,d}T_{MIN,d})) + g_{east}(1\times 10^{-3}\sum_{d\in s(j)} P_d) \\ e^{f_{west}}(0.01\sum_{d\in s(j)} DD_{low}(T_{MAX,d}T_{MIN,d}), 0.01\sum_{d\in s(j)} DD_{high}(T_{MAX,d}T_{MIN,d})) + g_{west}(1\times 10^{-3}\sum_{d\in s(j)} P_d) \end{cases}$$

Here, $f_{east}(\cdot)$ and $g_{east}(\cdot)$ are used to the east of the 100th meridian, excluding Florida. In this area, irrigation is less common and the response to increased temperatures is more extreme.

Figure B7 shows distributions of degree days for RCP 6.0, under the MIROC-ESM-CHEM model. CO_2 fertilization is modeled as a multiplicative factor applied to yields and estimated as a linear increase for each additional 100 ppm of CO_2:

$$Y'_{jt} = Y_{jt}\left(1 + \frac{C_t - C_{2012}}{100} X\right)$$

where C_t is the CO_2 concentration in year t under a given RCP, and X is the estimated CO_2 fertilization effect, which varies from 3% to 12% depending on the crop, from McGrath and Lobell (2013). McGrath and Lobell (2013) do not provide a value for cotton, so the Bayesian combination of all provided crop effects is used for it.

Economic output from the agricultural sector is not synonymous with yield, due to strategic storage. We model output as an autoregressive process of yields, as estimated from USDA data. For grains and cotton, the expression is

$$I_{jt} = 0.51Y_{jt} + 0.28Y_{j,t-1} + 0.21Y_{j,t-2}$$

A 4-year moving average is used for soybeans:

$$I_{jt} = 0.2Y_{jt} + 0.52Y_{j,t-1} + 0.19Y_{j,t-2} + 0.1Y_{j,t-3}$$

4.2. Crime

Both violent and property crime are calculated as

$$I_{jt} = \left(1 + 0.01\frac{1}{12}\sum_{d\in y(t)} f\left(T_{MAX,d}\right)\right)\left(1 + 0.01\frac{1}{12}\sum_{d\in y(t)} g(P_d)\right)$$

where $y(t)$ is the set of days in year t, $f(\cdot)$ is a function of daily maximum temperature, and $g(\cdot)$ is a function of daily precipitation. Both $f(\cdot)$ and $g(\cdot)$ are calculated as a Bayesian combination of the effect for each subcategory of crime estimated by Ranson (2014) and the average effect for violent or property crime from Jacob et al. (2007).

4.3. Mortality

Both average and age-specific mortalities are calculated as

$$I_{jt} = \sum_{d\in y(t)} f\left(T_{AVG,d}\right)$$

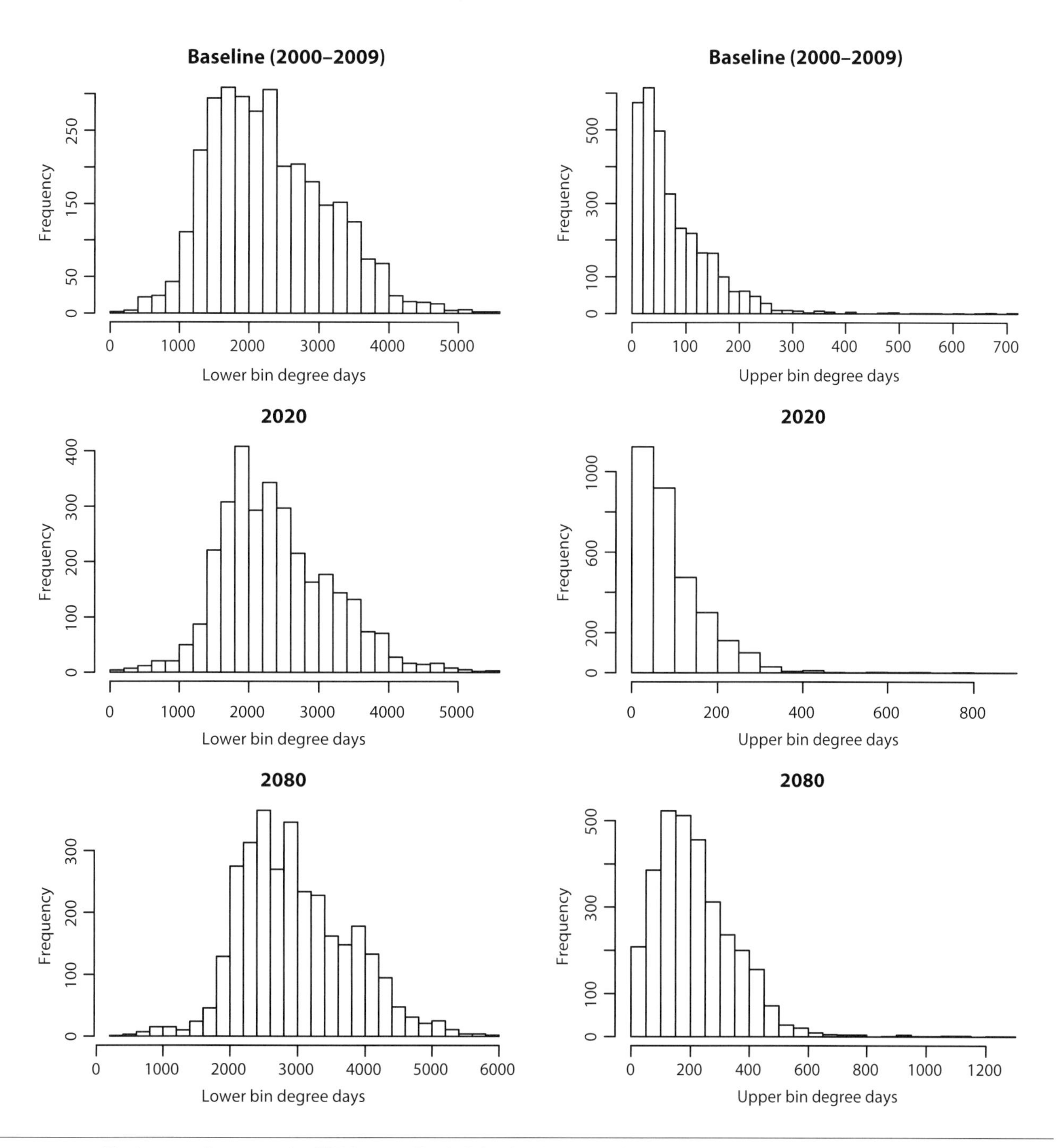

FIGURE B6. Evolution of impacts for FIPS 17161, Rock Island, Illinois, under RCP 6.0 as modeled by MIROC MIROCESM-CHEM. In the top panel, the left axis measures linear changes in temperature (°C) and CO_2 (100 ppmv). The right axis measures changes in rainfall relative to the 2000–2009 annual mean. In the lower panel, both axes show fractional changes in each of the sectors relative to the 2000–2009 annual mean.

The parameters of $f(\cdot)$ are calculated as a Bayesian combination of the results from Deschênes and Greenstone (2011) and a corrected form of Barreca et al. (2013). Age-specific mortalities, for newborns, ages 1 to 44, ages 45 to 64, and ages 65 and up, are provided by Barreca et al. (2013).

Mortality is reported both as percentage changes and as differences in the mortality rate. In either case, the pooled and age-specific mortality rates per county are from the Centers for Disease Control and Prevention (2013).

4.4. Labor Productivity

The structure of the labor productivity calculation is identical for high-risk and low-risk sectors:

$$I_{jt} = \frac{H + \frac{1}{60} \sum_{d \in y(t)} f\left(T_{MAX,d}\right)}{H}$$

where H is the average number of hours worked per year in the baseline. For high-risk labor, $H = 7.67 \times 365$, and for low-risk labor, $H = 6.92 \times 365$. The parameters of $f(\cdot)$ are provided by Graff Zivin and Neidell (2014).

5. ADAPTATION

The empirical approach used to estimate adaptation applies observed rates of adaptation from recent history to regional differences in response functions to capture the observed capacity for adaptation. Here, we describe the general structure for estimating adaptation. Additional details are added in later sections specific to each adapted response.

Each impact response curve starts from the same baseline as the unadapted case in the year 2000. Within each county, it then transitions asymptotically toward a future shape. The evolution of each parameter of the distribution follows,

$$\beta(t) = \beta(\infty) + (\beta(t_{before}) - \beta(\infty))e^{-(t-t_{before})/\tau}$$

where $\beta(\infty)$ is the maximum possible adaptation, which in the cases considered here represents no response to temperature; $\beta(t_{before})$ is the parameter from a historical period, as estimated for a period centered on t_{before}; and τ determines the rate of evolution of the parameter.

Using parameters estimated for a second period closer to the present day than t_{before}, we calculate the rate of adaptation, τ, as

$$\tau = -\frac{t_{after} - t_{before}}{\log(\beta(t_{after}) - \beta(\infty)) / (\beta(t_{before}) - \beta(\infty))}$$

where $\beta(t_{after})$ is the parameter value during a period centered on t_{after}. This process is represented on the left of figure B8.

We further manipulate τ into an incremental product, $\gamma = e^{-1/\tau}$, so that

$$\beta(t+1) = \gamma\beta(t) + (1 - \gamma)\beta(\infty) \tag{B1}$$

With the exception of maize, a different $\beta(\infty)$ is used for determining the rate of evolution (τ and γ) as is used for estimating the final evolution of the parameter values. The $\beta(\infty)$ used to estimate the actual evolution of the parameters

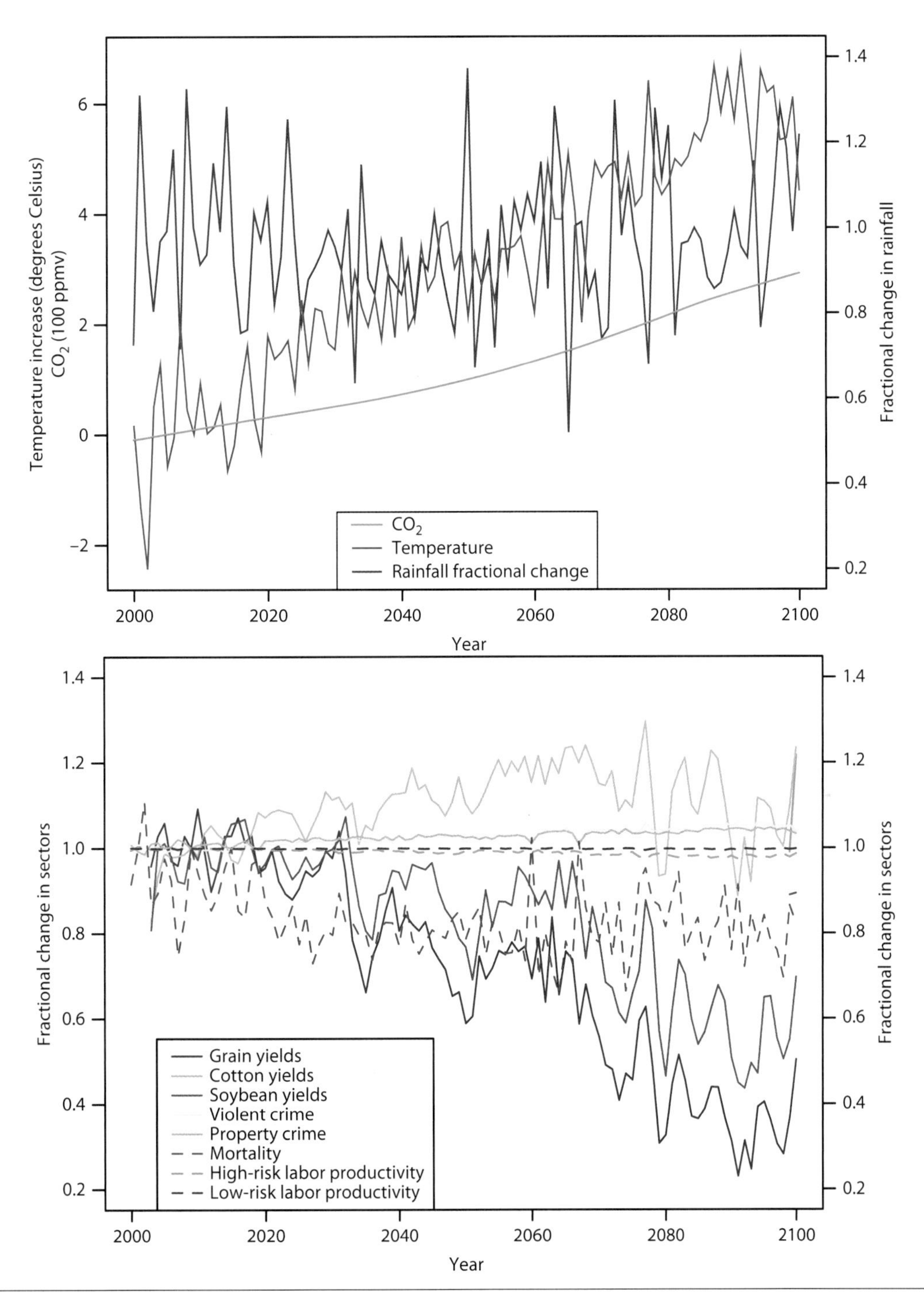

FIGURE B7. Distributions of the degree days across all U.S. counties in the lower (left) and upper (right) bins of maize, which has a 29°C bin threshold. The rows represent the average of degree days between 2000 and 2009, and for the years 2020 and 2080.

is thought of as the parameter value for a fully adapted region of a particular temperature. As a result, it is a function of average temperature: the response curve for New England, under the current climate, is considered to be fully adapted to a colder average temperature than the response curve for the Southwest. To estimate how $\beta(\infty)$ changes with temperature, we use values from several regions in the United States, treated each as fully adapted to its current climate.

Specifically, we construct a linear approximation to the evolution of $\beta(\infty)$ across temperatures, using the parameters and average temperature $\bar{T}$ of various regions. Let this function be denoted as $f(\bar{T})$. For the dynamic evolution of the adapted response curves, we take $\beta(\infty) = f(\bar{T}_t)$. At any point in time, $\bar{T}$ is calculated as the average temperature over the previous 15 years. This allows the fully adapted coefficients both to be predicted between the temperatures of observed regions and extrapolated beyond them. The process is shown on the right of figure B8.

Each section that follows describes the evidence used to estimate the temporal evolution of impact responses (the rate of adaptation, τ) and the spatial estimation (how $\beta(\infty)$ varies with $\bar{T}$).

5.1. Maize in the Eastern United States

Increasing the availability of irrigation is a key adaptation to climate change, as additional water for evapotranspiration allows crops to continue to grow at higher temperatures. Fields to the east of the 100th meridian are much less consistently irrigated than those to the west. This helps explain the lower response to killing degree-days in the west ($\beta = -0.21$, compared to −0.62) (Schlenker and Roberts, 2009).

5.1.1. TEMPORAL EVOLUTION

The killing degree-day coefficient has evolved over the course of the past half century, as shown on the left of figure B9. Using Monte Carlo draws from each of these regional parameter distributions independently and fitting exponentials, γ is calculated to be 0.99722 ± 0.00573. We approximate the distribution of values for γ with a Gaussian, as shown on the right of figure B9. This corresponds to a very slow rate of adaptation, with a time constant of about 360 years (Burke and Emerick, 2013).

We hold constant the growing degree-days coefficient; only the killing degree-day coefficient adapts. Because the growing degree-day coefficient is higher in the east than in the west, this may result in an optimistic consideration of the trade-offs that result from adaptation.

5.1.2. SPATIAL ESTIMATION

No extrapolation is performed for maize. Instead, each year the killing degree-day coefficient for all eastern counties is updated according to equation B1.

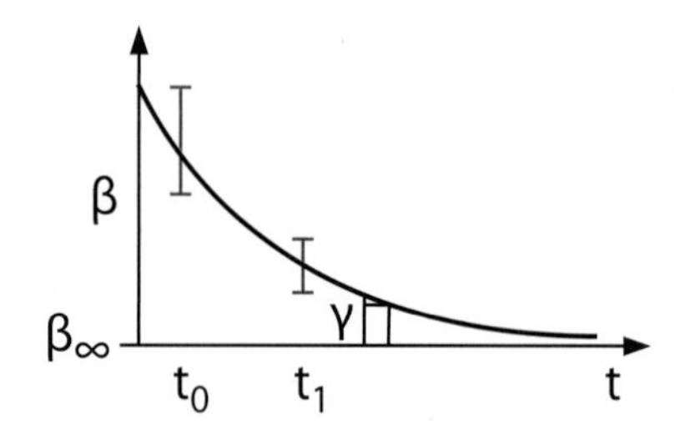

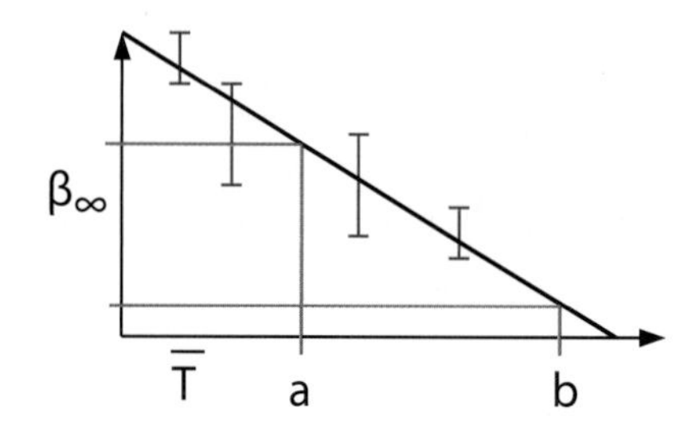

FIGURE B8. **Left:** Estimation of temporal evolution of variable values. A historical estimate (at t_0) and a current estimate (at t_1) are used to fit an exponential, characterized by γ. **Right:** Estimation of potential adaptation as a function of average temperature. Regional estimates (in red) are used to produce a linear approximation for how each coefficient varies with average temperature ($\bar{T}$). This approximation is then evaluated (green) to determine coefficients at future times.

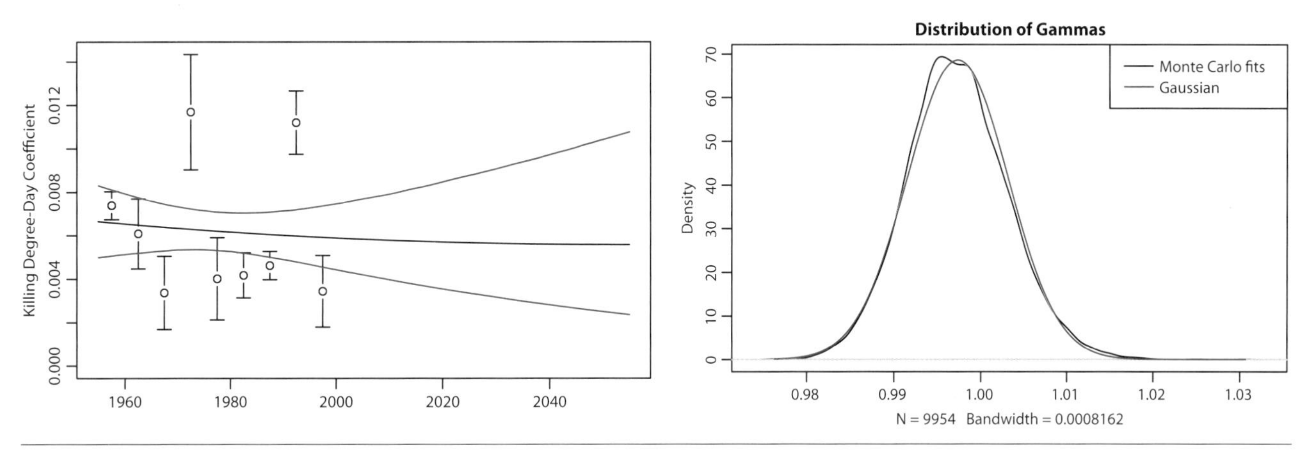

FIGURE B9. **Left:** Killing degree-day coefficients computed for 5-year periods between 1955 and 2000. Red lines show the 90% confidence interval on the exponential evolution of the killing degree-day coefficient. **Right:** Values of γ computed in the Monte Carlo, and the corresponding Gaussian approximation in red.

5.2. Temperature-Related Mortality

5.2.1. TEMPORAL EVOLUTION

For temperature-related mortality, we assume that there are two parallel adaptations occuring: one for cold temperatures and one for hot temperatures. The rate of temporal evolution for these processes is estimated as the mean of the values of γ for each bin for which $T < 65°F$, and for each bin for $T > 65°F$, respectively. Figure B10 shows the coefficient estimates for each bin and the calculation of γ for the median of these coefficient distributions. In the evaluation of actual impacts, Monte Carlo draws are taken from the various distributions, and the value of γ is estimated separately for each Monte Carlo run. The time constant for the cold-related mortality is 31.3 years and for heat-related deaths is 84.0 years.

5.2.2. SPATIAL ESTIMATION

Estimates were given for four regions: the Northeast, Midwest, South, and West. These, along with the pooled estimate, were used to construct a linear approximation (see figure B11).

5.3. Violent and Property Crime

The process for estimating adaptation for crime is similar to the process for temperature-related mortality. Pooled results for all forms of property crime and all forms of violent crime were used, as segregated by regions with different maximum temperatures, from Ranson (2014). Because the maximum temperature for the national estimate was very similar to the segregated 65°F region, and because the national estimate for the highest temperature bin is outside of the coefficient range of the segregated regions, we used the 65°F response function as the baseline for estimating adaptive capacity.

5.3.1. TEMPORAL EVOLUTION

Estimates for periods centered at 1960 and 2000 were used to estimate the temporal evolution of crime adaptation. Two rates were calculated: one for bins below 65°F and one for bins above that value (which also corresponds to the dropped bin in Ranson (2014)). These calculations are shown in figures B12–B13.

For violent crime, the time-constant is 71.2 years for the lower range of temperatures and 139 years for the upper range. For property crime, the time-constants are 793 years and 16.4 years, respectively.

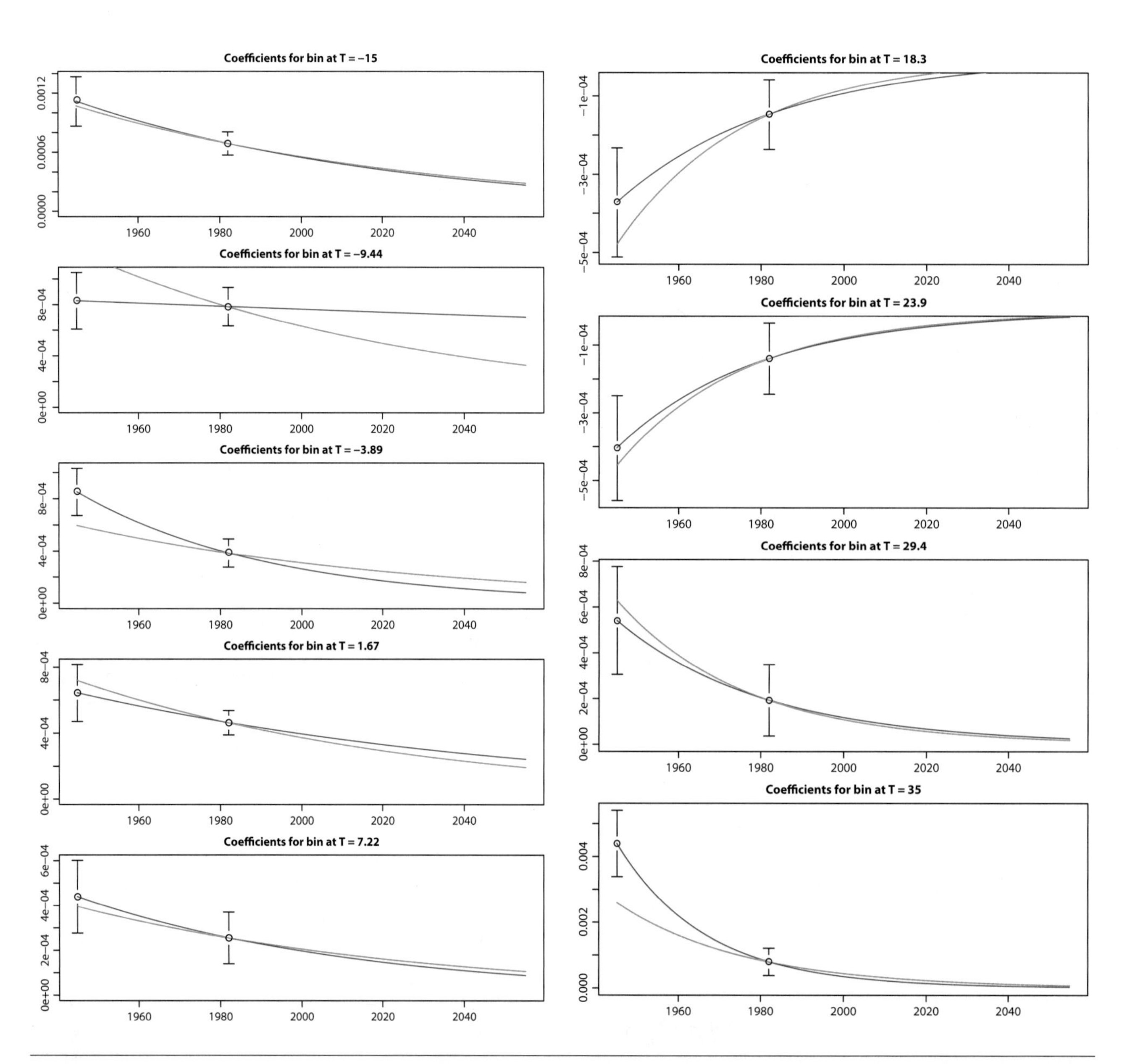

FIGURE B10. Historical coefficient estimates for cold-related (left) and heat-related (right) mortality. Red shows the time evolution based only on the coefficient described in each panel, while green shows the pooled time evolution for all coefficients in that set.

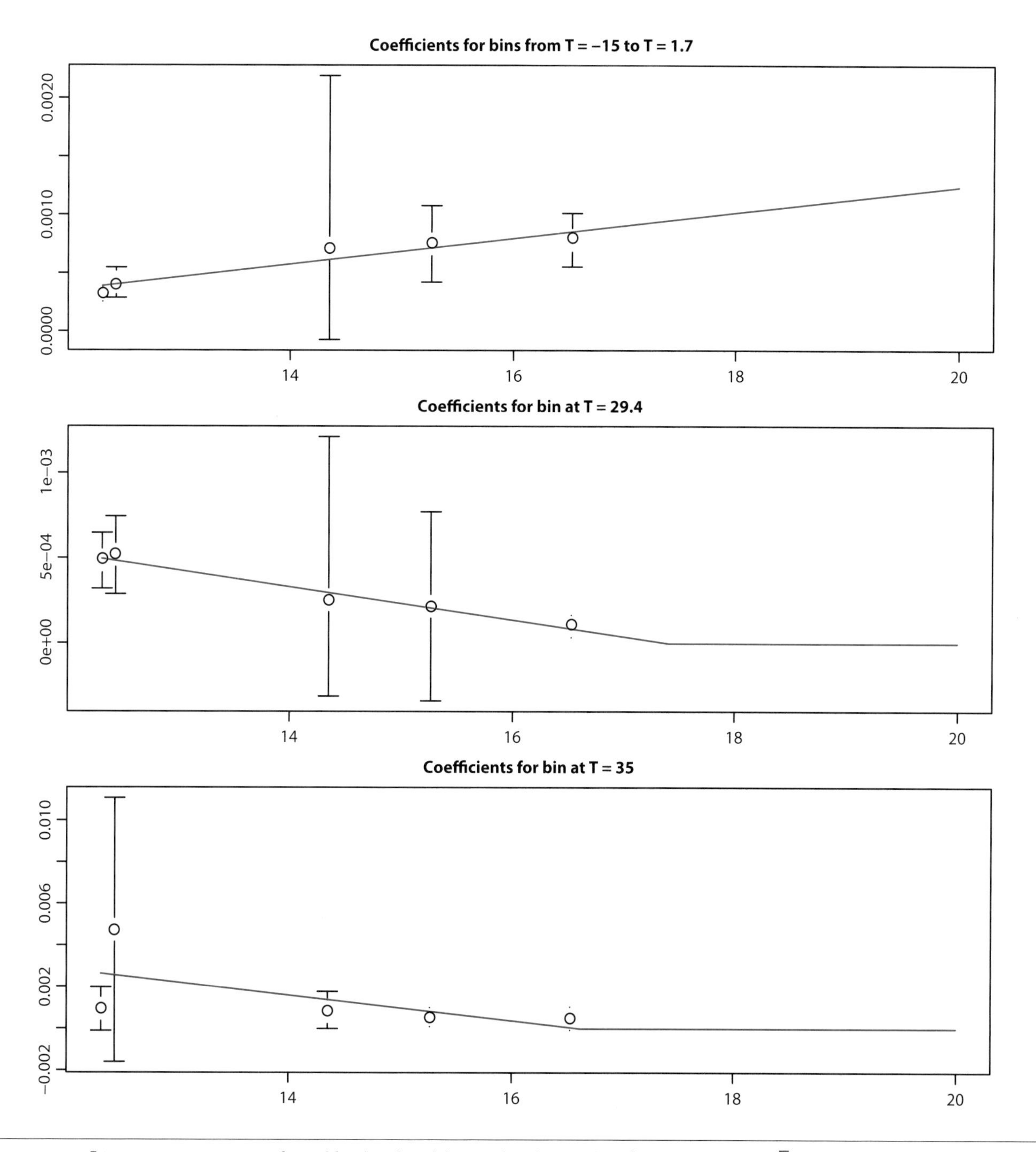

FIGURE B11. Linear approximations for cold-related and heat-related mortality β as a function of $\overline{T}$. Excess mortality for days of a given temperature is clipped at 0. A single coefficient was provided for cold-related mortality ($T < 65°F$).

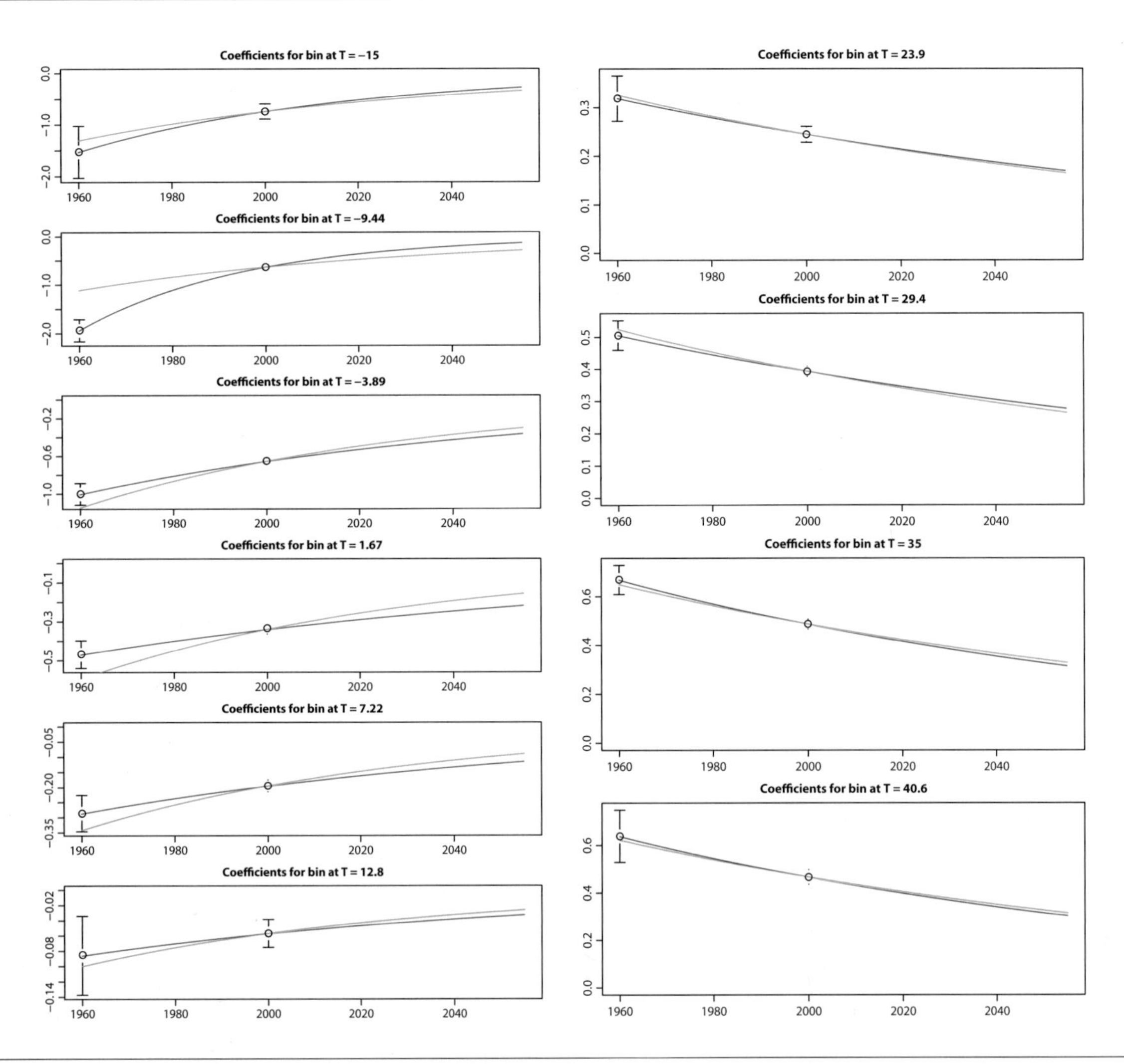

FIGURE B12. Historical coefficient estimates for cold-related (left) and heat-related (right) mortality. Red shows the time evolution based only on the coefficient described in each panel, while green shows the pooled time evolution for all coefficients in that set.

5.3.2. SPATIAL ESTIMATION

Estimates were analyzed from regions with an average maximum temperature of $T = 45°F$, $T = 55°F$, $T = 65°F$, and $T = 75°F$. Figures B14 and B15 show these estimates by bin.

6. RETURN INTERVALS

The number of events exceeding a given threshold is used to communicate return periods of extreme events. We identify the threshold for the 1-in-20 event and count the number of events that exceed this threshold for each 20-year window over the course of the twenty-first century.

The expected number of events across models for an investment in year t is calculated as a weighted sum

$$\sum_i w_i e_{it}$$

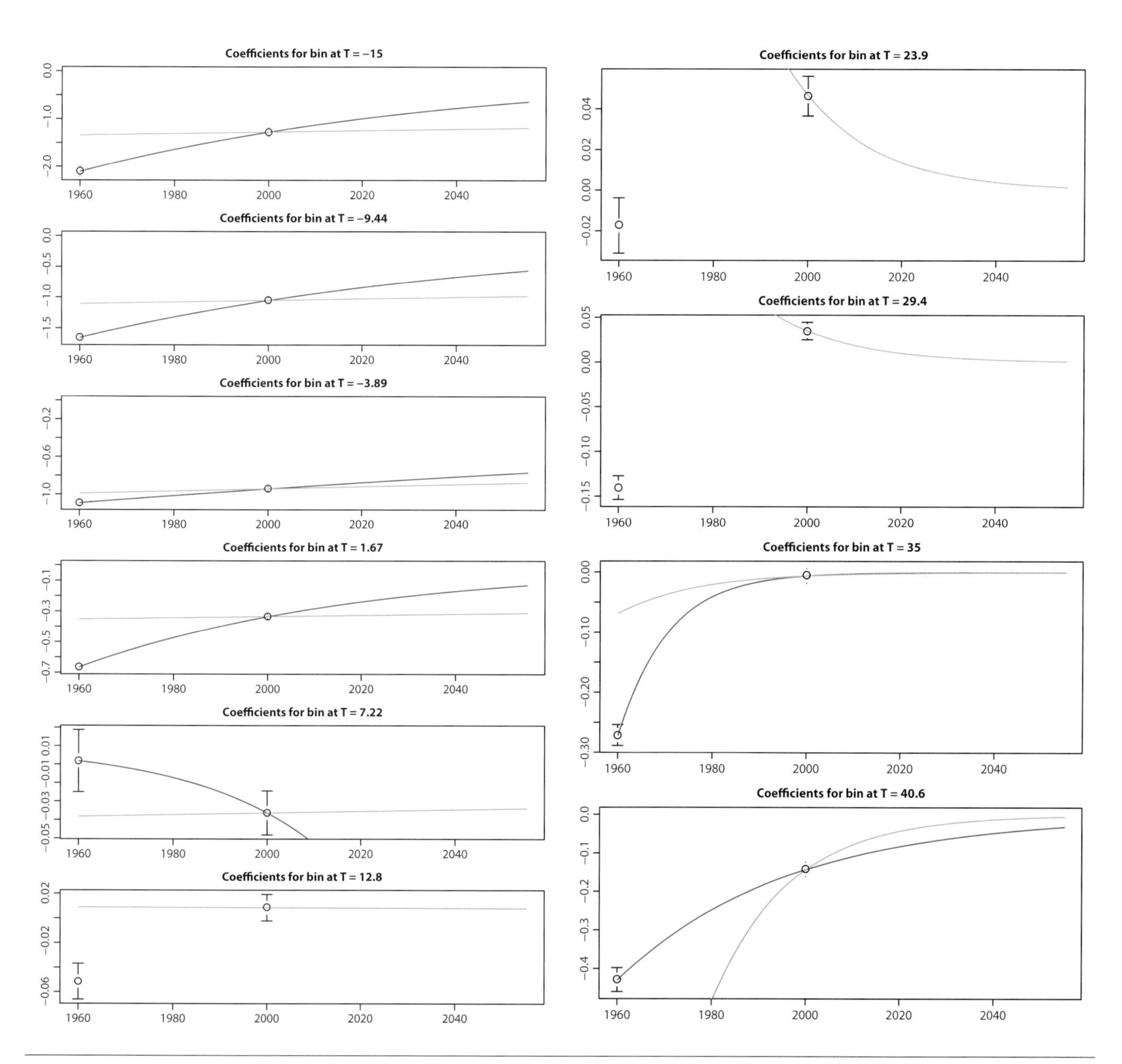

FIGURE B13. Historical coefficient estimates for property crime driven by maximum temperatures below (left) and above (right) 65°F. Red shows the time evolution based only on the single coefficient, while green shows the pooled time evolution for all coefficients in that set. The value of $T = 23.8889$, and $T = 29.4444$ cannot be estimated because the two estimates are on opposite sides of the zero-line.

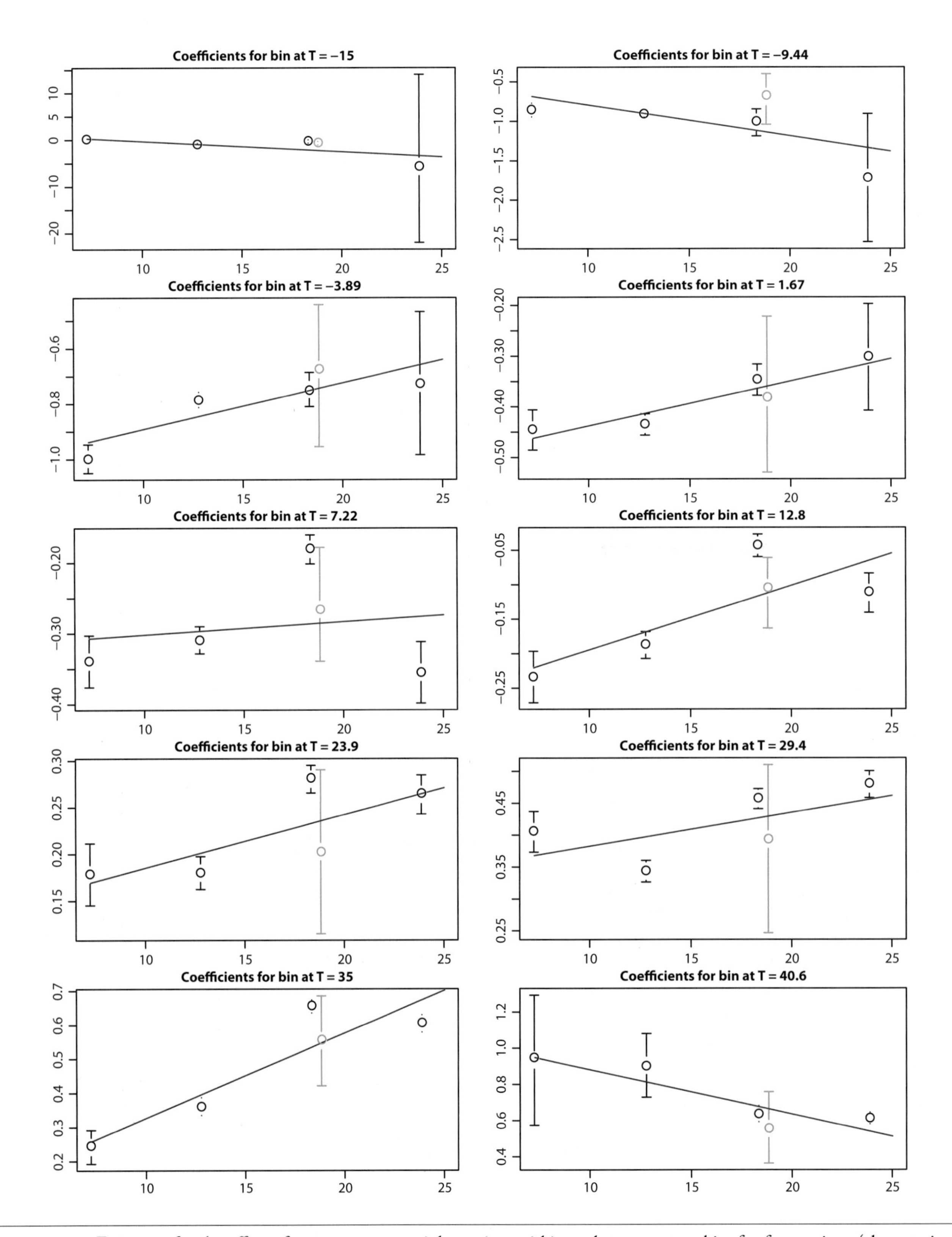

FIGURE B14. Estimates for the effect of temperatures on violent crime within each temperature bin, for four regions (characterized by $\bar{T}_{max} = 45$, $\bar{T}_{max} = 55$, $\bar{T}_{max} = 65$, and $\bar{T}_{max} = 75$). The green estimate is the national average estimate, at a national average maximum temperature of 65.88603°F, but was not used for the estimation. The red lines were used for interpolating and extrapolating coefficient values, based on temperature.

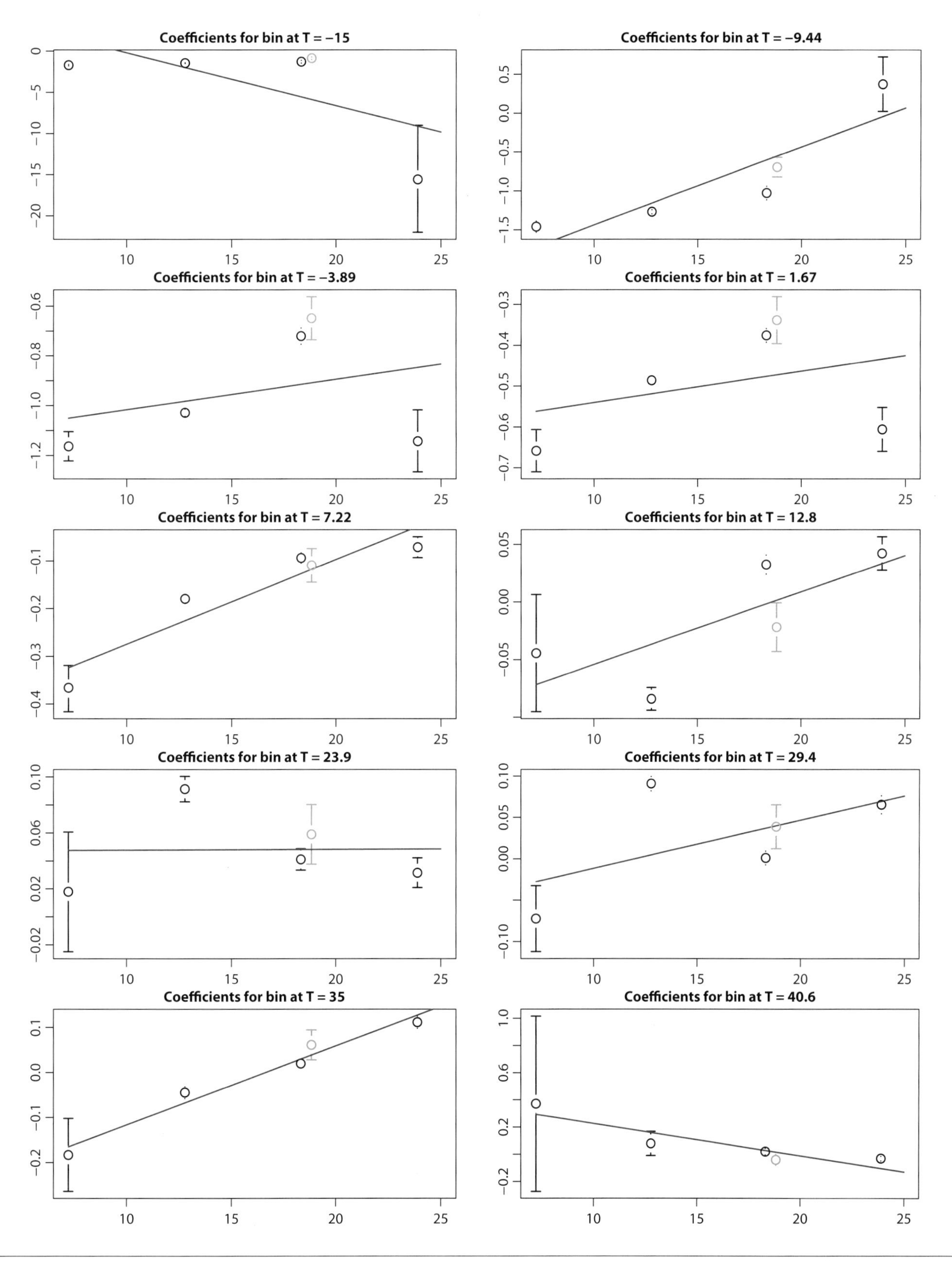

FIGURE B15. Same as figure B14 for property crime.

where w_i is the weight given to model i, and e_{it} is the number of events exceeding a threshold experienced in the 20 years following year t. That is, for a threshold x, e_{it} is the number of impacts $y_{i\tau} \geq x$ for $\tau \in t, \ldots, t+19$. The threshold is chosen such that the expected number of events for the period 1990–2009 is approximately 1. These thresholds vary slightly between RCPs, due to the different collections of models used to estimate each RCP. The curves are shifted slightly ($\Delta y < 0.1$) for display purposes.

7. IMPACT AGGREGATION

Of the 36,000 results generated across all RCPs, models, weather realizations, and Monte Carlo draws of the impact functions, 0.3% were dropped because of outliers in the estimates of additional mortality in the age 1–44 cohort. These outliers were unrealistically high estimates of national averaged mortality, in excess of 100 additional deaths per 100,000. The results from runs that had outlier mortality rates were dropped for all impacts.

County-level results are aggregated to the state, NCA region, and national levels, as weighted sums (see figure B16). This section describes the weighting of results for these aggregations.

7.1. Grain Aggregation for CGE Analysis

Grain yields (maize and wheat) are both combined within the same county and aggregated to higher scales by calorie totals. Within each county, the average impact is

$$I_i = \frac{I_i^{wheat} A_i^{wheat} C^{wheat} + I_i^{maize} A_i^{maize} C^{maize}}{A_i^{wheat} C^{wheat} + A_i^{maize} C^{maize}}$$

where A_i^{wheat} is the average acres of wheat planted between 2000 and 2005, and A_i^{maize} is the average acres for maize. C^{wheat} and C^{maize}, the calorie density of wheat and maize, are taken to be 1690 calories/kg and 1615 calories/kg, respectively. The weighting of each county result for aggregation to higher scales is

$$W_i = A_i^{wheat} C^{wheat} + A_i^{maize} C^{maize}$$

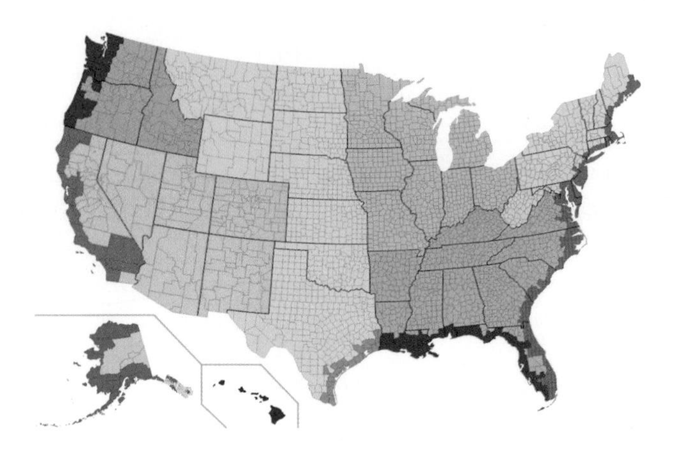

FIGURE B16. Results were aggregated for passing to the CGE model at both state and NCA region (colors).

7.2. Cotton and Soybean Aggregation for CGE Analysis

Cotton and soybean results are aggregated by weighting by the total area planted, as averaged over the period 2000–2005. Since impacts are proportional to changes in yields, and yields are calculated relative to area planted, this is equivalent to weighting counties by production.

7.3. Agricultural Aggregation for Distributions and Maps

All crops are weighted by production (in 1000 MT) for constructing total agriculture impact distributions and maps. Production is calculated from the USDA reported bushels measurement using 56 lbs/bushel and 60 lbs/bushel for soy and wheat, and averaged over the period 2000–2005. The data used here and in the CGE aggregations above come from reproduction data for Schlenker and Roberts (2009).

7.4. Crime Aggregation

Counties are weighted by the number of reported property and violent crimes from the Uniform Crime Statistics, averaged over the period 2000–2005, and provided for reproduction of Ranson (2014).

Counties that are not explicitly identified at the county level (of which there are 172) are aggregated using the mean country rates of property and violent crime. Furthermore, because we use the 2010 census for county populations, these rates are scaled by 0.9339, the ratio of the average national population in the period 2000–2005 to the population in 2010, before being scaled by the individual county populations in 2010, to maintain comparability.

7.5. Labor Aggregation

Labor employment by county is averaged over the period 2000–2005, as reported by the Bureau of Labor Statistics (Bureau of Labor Statistics, 2014). Following Graff, Zivin, and Neidell (2014), high-risk sectors consist of agriculture, forestry, fishing and hunting; mining, quarrying, and oil and gas extraction; utilities; construction; manufacturing; and transportation and warehousing. All others are considered low-risk.

The Bureau of Labor Statistics data exclude the counties represented by FIPS codes 02105, 02195, 02198, 02230, and 02275, as these were created after 2005.

7.6. Mortality Aggregation

Counties are weighted by 2010 census populations for aggregating mortality. All estimates except for the NCA region estimates used in the text use total populations, irrespective of age cohort. For the aggregation from sub-NCA regions (in figure B16) to NCA regions used in the text, the census totals for each age are summed into impact cohorts and used for weighting.

7.7. Distributions Across Result Sets

Within each RCP, models are weighted to capture a desired distribution of temperatures, as described in section 1.4 of appendix A. The weighting process for results for a given weather realization of a given RCP involves constructing an

ECDF as follows. Let the calculated value of an impact in a given county and year for model $m \in \{1, \ldots, M\}$ be I_m. The CDF of this impact across all models is

$$F(I) = \frac{1}{\sum_m w_m} \sum_{m \text{ for } I_m \leq I} w_m$$

where w_m is the weight given to model m provided in tables A3 to A6. Extending this process to multiple weather realizations and multiple batches is done by simply including all available results in the weighted ECDF. Unweighted results are reported in table B1.

8. LINEAR EXTRAPOLATION ASSUMPTION

In this section, we consider the consequences of extending impact functions linearly beyond the support of the data that were used to estimate their form. In our main analysis, we do not linearly extrapolate response functions in an effort to provide conservative estimates, however we examine here the effect of this decision on our impact estimates. The standard implementation of the impact response functions assumes a flat extrapolation, where the coefficients used in the edge bins remain constant when extrapolated beyond the response function support. In several of the empirical impacts, the response as a function of temperature appears to grow linearly with temperature at the extreme. This is particularly true for mortality and high-risk labor productivity. Agricultural yields are implicitly linearly extrapolated (and estimated as such) because they are in terms of degree-days. Figure B17 displays how quantiles of impact responses change under a linear extrapolation assumption.

The effect of the linear extrapolation assumption is shown in the tails of the distributions and at extreme temperatures (end of century for RCP 8.5). Most impact distributions become wider, although property crime shows narrowing of impacts under linear extrapolation. This is because, while crime generally increases with temperature, it decreases at the very high end of both temperature and precipitation. Violent crime shows very little change because of the flat response at high temperatures.

The effects on the tails of the health and labor distributions is large: at the 90th percentile in high-risk labor, the impact increases from −0.5% to −0.7% (a 40% increase), and in mortality, the additional mortality increases from 20 per 100,000 to 24 per 100,000 (a 20% increase).

TABLE B1 Unweighted impact result percentiles

	RCP 8.5					RCP 4.5					RCP 2.6				
Percentiles	5	17	50	83	95	5	17	50	83	95	5	17	50	83	95
Agricultural Yields															
2080–2099	−42.44	−34.14	−16.49	0.42	13.26	−26.74	−19.37	−4.26	5.09	9.54	−16.64	−12.1	−2.99	1.34	4.57
2040–2059	−18.27	−11.31	−3.96	4.69	8.61	−14.11	−10.76	−1.59	4.83	8.28	−12.9	−9.58	−3.43	1.3	4.28
2020–2039	−9.55	−6.16	−1.92	2.58	7.98	−9.32	−6.86	−1.16	3.71	9.27	−13.61	−8.33	−2.11	1.32	2.77
Agricultural Yields (w/o CO_2 fertilization)															
2080–2099	−59.76	−54.04	−40.86	−28.97	−19.24	−35.23	−28.84	−15.04	−6.71	−2.91	−19.12	−14.76	−5.99	−1.84	1.33
2040–2059	−27.49	−21.36	−14.89	−7.26	−3.91	−20.7	−17.56	−8.94	−3.01	0.15	−16.72	−13.49	−7.59	−3.09	−0.29
2020–2039	−13.59	−10.39	−6.36	−2.07	3.09	−12.45	−10.05	−4.54	0.12	5.44	−16.29	−11.1	−5.11	−1.72	−0.42
High-Risk Labor															
2080–2099	−2.53	−2.06	−1.44	−0.98	−0.76	−1.18	−0.89	−0.59	−0.3	−0.13	−0.54	−0.42	−0.28	−0.11	0.18
2040–2059	−1.03	−0.77	−0.49	−0.26	−0.15	−0.74	−0.63	−0.38	−0.17	−0.02	−0.47	−0.4	−0.31	−0.13	0.15
2020–2039	−0.5	−0.38	−0.16	−0.02	0.11	−0.44	−0.37	−0.2	−0.03	0.12	−0.44	−0.33	−0.21	−0.03	0.15
Low-Risk Labor															
2080–2099	−0.68	−0.46	−0.24	−0.1	0	−0.24	−0.16	−0.08	−0.03	0	−0.12	−0.08	−0.04	0.01	0.05
2040–2059	−0.22	−0.15	−0.08	−0.02	0.02	−0.15	−0.1	−0.06	−0.01	0.03	−0.11	−0.07	−0.04	0.01	0.05
2020–2039	−0.11	−0.07	−0.03	0.01	0.05	−0.1	−0.06	−0.03	0.01	0.05	−0.1	−0.06	−0.02	0.02	0.06
Mortality															
2080–2099	0.75	3.85	9.63	18.04	24.86	−2.53	−2.06	−1.44	−0.98	−0.76	−0.68	−0.46	−0.24	−0.1	0
2040–2059	−3.39	−1.27	2.48	5.89	8.75	−1.03	−0.77	−0.49	−0.26	−0.15	−0.22	−0.15	−0.08	−0.02	0.02
2020–2039	−3.82	−1.83	1.15	4.27	6.13	−0.5	−0.38	−0.16	−0.02	0.11	−0.11	−0.07	−0.03	0.01	0.05
Property Crime															
2080–2099	0.34	0.51	0.74	0.97	1.07	−0.03	0.14	0.51	0.83	1.01	−0.18	−0.01	0.23	0.46	0.61
2040–2059	−0.01	0.13	0.33	0.59	0.73	−0.13	0.03	0.36	0.67	0.85	−0.14	−0.03	0.22	0.44	0.59
2020–2039	−0.29	−0.07	0.1	0.31	0.46	−0.25	−0.09	0.19	0.48	0.63	−0.2	−0.06	0.11	0.32	0.42
Violent Crime															
2080–2099	1.84	2.2	3.05	3.76	4.41	0.21	0.74	1.54	2.31	2.54	−0.27	0.05	0.69	1.27	1.48
2040–2059	0.34	0.7	1.19	1.75	2.41	−0.02	0.42	1.09	1.65	1.84	−0.25	0.26	0.79	1.02	1.52
2020–2039	−0.29	0.04	0.45	0.9	1.34	−0.47	0.07	0.56	0.97	1.3	−0.25	0.01	0.48	0.81	1.06

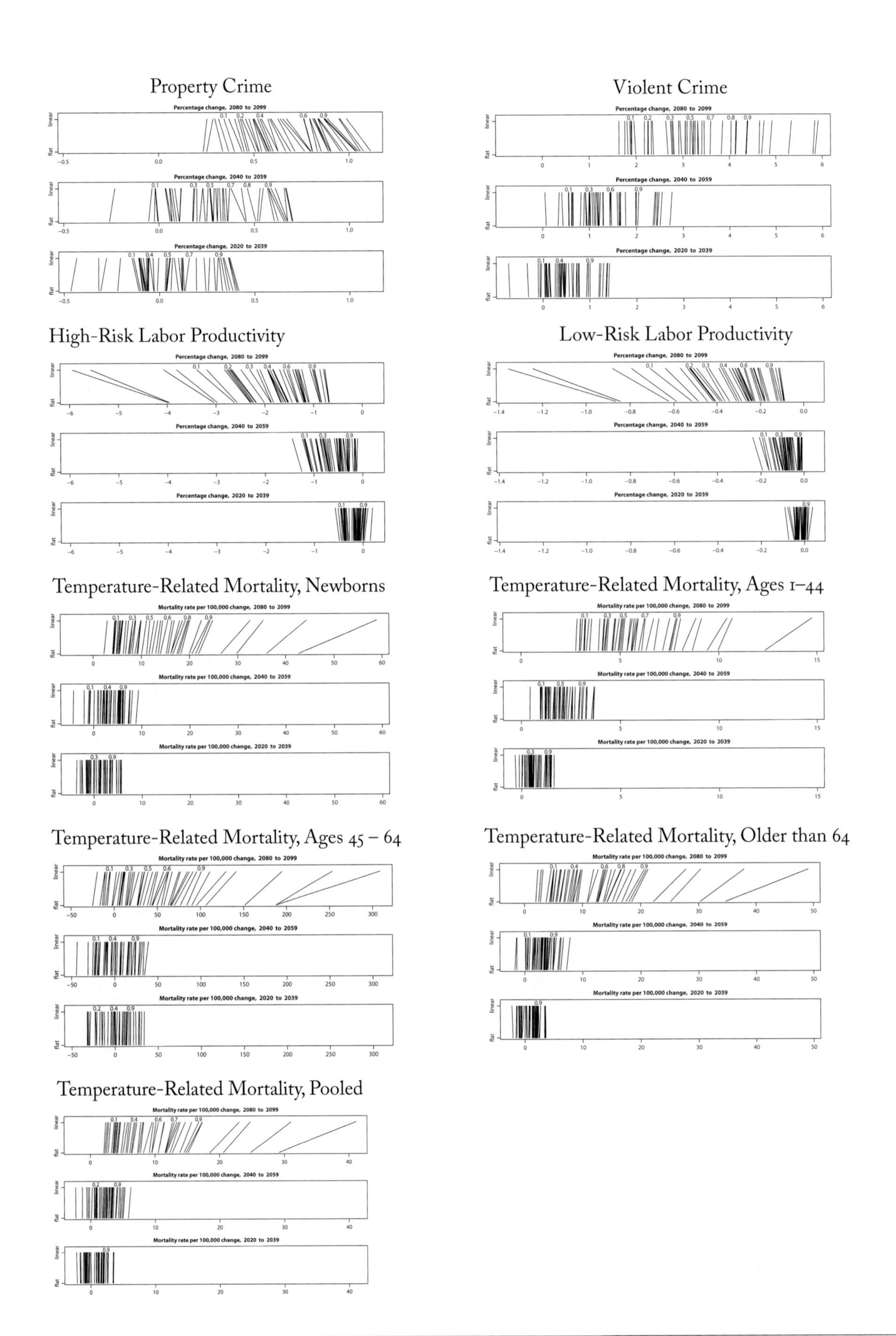

FIGURE B17. The evolution of impact response quantiles, under flat and linear extrapolation, under RCP 8.5 at a national level. The lines are quantiles (labeled at the top), showing how these shift between flat extrapolation (bottom) and linear extrapolation (top).

REFERENCES

Anderson, B. G., and M. L. Bell (2009), Weather-related mortality: How heat, cold, and heat waves affect mortality in the United States, *Epidemiology (Cambridge, Mass.)*, *20*(2), 205–13, doi:10.1097/EDE.0b013e318190ee08.

Anderson, G. B., and M. L. Bell (2011), Heat waves in the United States: mortality risk during heat waves and effect modification by heat wave characteristics in 43 U.S. communities, *Environmental Health Perspectives*, *119*(2), 210–18, doi:10.1289/ehp. 1002313.

Barreca, A., K. Clay, O. Deschênes, M. Greenstone, and J. S. Shapiro (2013), Adapting to climate change: The remarkable decline in the US temperature-mortality relationship over the 20th century, *NBER Working Paper*.

Berry, S. T., M. J. Roberts, and W. Schlenker (2012), Corn Production Shocks in 2012 and Beyond: Implications for Food Price Volatility Model (September).

Bureau of Labor Statistics, U.S. Department of Labor (2014), Occupational employment statistics, http://data.bls.gov/oes/, accessed on March 20, 2014.

Burke, M., and K. Emerick (2013), Adaptation to climate change: Evidence from US Agriculture.

Centers for Disease Control and Prevention (2013), Compressed Mortality File 1999-2010 on CDC WONDER Online Database, released January 2013, Data are compiled from Compressed Mortality File 1999-2010 Series 20 No. 2P, accessed at http: //wonder.cdc.gov/cmf-icd10.html on March 22, 2014.

Curriero, F. C., K. S. Heiner, J. M. Samet, S. L. Zeger, L. Strug, and J. A. Patz (2002), Temperature and mortality in 11 cities of the eastern United States, *American Journal of Epidemiology*, *155*(1), 80–87.

Deschênes, O., and M. Greenstone (2007), The economic impacts of climate change: Evidence from agricultural output and random fluctuations in weather, *The American Economic Review 91*(1), 354–85.

Deschênes, O., and M. Greenstone (2011), Climate change, mortality, and adaptation: Evidence from annual fluctuations in weather in the US, *American Economic Journal: Applied Economics*, *3*(October), 152–85.

Fisher, A., W. Hanemann, M. J. Roberts, and W. Schlenker (2012), The economic impacts of climate change: Evidence from agricultural output and random fluctuations in weather: Comment, *American Economic Review*, *102*(7), 3749–60.

Gelman, A., J. B. Carlin, H. S. Stern, D. B. Dunson, A. Vehtari, and D. B. Rubin (2013), *Bayesian data analysis*, CRC Press.

Graff Zivin, J., and M. Neidell (2014), Temperature and the allocation of time: Implications for climate change, *Journal of Labor Economics*, *32*(1), 1–26, doi:10.1086/671766.

Hsiang, S., D. Lobell, M. Roberts, and W. Schlenker (2013a), Climate change and crop choice: Evidence from Australia, Brazil, China, Europe, and the United States, *Working paper*.

Hsiang, S. M., M. Burke, and E. Miguel (2013b), Quantifying the influence of climate on human conflict, *Science*, *341*(6151), 1235,367, doi:10.1126/science.1235367.

Jacob, B., L. Lefgren, and E. Moretti (2007), The dynamics of criminal behavior evidence from weather shocks, *Journal of Human Resources*, *42*(3).

Lessl, M., J. S. Bryans, D. Richards, and K. Asadullah (2011), Crowdsourcing in drug discovery, *Nature Reviews Drug Discovery*, *10*(4), 241–42.

McGrath, J. M., and D. B. Lobell (2013), Regional disparities in the CO_2 fertilization effect and implications for crop yields, *Environmental Research Letters*, *8*(1), 014,054, doi:10.1088/1748-9326/8/1/014054.

Ranson, M. (2014), Crime, weather and climate change, *Journal of Environmental Economics and Management*, 67(3): 274–302.

Sacks, W. J., D. Deryng, J. A. Foley, and N. Ramankutty (2010), Crop planting dates: An analysis of global patterns, *Global Ecology and Biogeography*, *19*(5), 607–20.

Schlenker, W., and M. J. Roberts (2009), Nonlinear temperature effects indicate severe damages to U.S. crop yields under climate change, *Proceedings of the National Academy of Sciences of the United States of America*, *106*(37), 15,594–8, doi:10.1073/pnas.0906865106.

Seppanen, O., W. Fisk, and Q. Lei (2006), Effect of temperature on task performance in office environment, *LBNL-60946*.

APPENDIX C
DETAILED SECTORAL MODELS

MICHAEL DELGADO, SHASHANK MOHAN, ROBERT MUIR-WOOD, AND PAUL WILSON

1. INTRODUCTION

Most of the climate impacts quantified in the *American Climate Prospectus* rely on econometrically derived impact functions. To assess the impact of changes in temperature on energy demand and expenditures and changes in sea level and storm activity on coastal property and infrastructure, we use more detailed sectoral models. For energy-sector impacts, we use RHG-NEMS, a version of the National Energy Modeling System of the Energy Information Administration (EIA, 2013). For coastal impacts, we use the North Atlantic Hurricane Model of Risk Management Solutions (RMS).

2. ENERGY-IMPACT MODELING

As discussed in chapter 10, we limited our analysis of climate-related energy-sector impacts to changes in electricity demand, energy expenditures, and electricity-generation capacity because of changes in heating and cooling requirements. While a handful of studies have attempted to quantify these relationships, Deschênes and Greenstone (2011) provide the only nationally applicable, recent, empirical estimate of changes in energy demand resulting from changes in heating and cooling requirements. However, this paper provides information on changes in residential demand only.

To supplement the findings in Deschênes and Greenstone, we use RHG-NEMS, a version of the EIA's National Energy Modeling System (NEMS) (EIA 2013) maintained by the Rhodium Group. RHG-NEMS is a detailed, multisector, bottom-up model of U.S. energy supply and demand with detailed residential, commercial, and industrial demand modules, electricity and primary energy supply modules, and energy market and macroeconomic modules, among others. Consumer demand curves are derived from detailed consumer choice algorithms that allow consumers to substitute, subject to limited foresight, from a wide array of available end-use technologies in order to satisfy their energy-service demands (see section 2.2).

2.1. Physical Data

We model changes in energy demand by varying the number of heating degree days (HDDs) and cooling degree days (CDDs), which in turn drive heating and cooling service requirements in NEMS. HDDs and CDDs are a measure of the approximate number of degrees that a living space must be heated or cooled for each day summed over all the days in the year. Daily minimum and maximum temperature projections were drawn from the outputs of the climate-modeling process (see appendix A), meaning that the projections are both consistent with the other impact categories in this study and that they covary in the macroeconomic modeling process described in appendix D. The method for calculating HDDs and CDDs is described in tables C1 and C2 (see for example UK Met Office, 2012).

2.2. RHG-NEMS

Our analysis of the impact on residential and commercial heating and cooling demand due to climate change relies heavily on RHG-NEMS. EIA uses NEMS to produce its Annual Energy Outlook (AEO), which projects the production, conversion, consumption, trade, and price of energy in United States through 2040. NEMS is an energy-economic model that combines a detailed representation of the U.S. energy sector with a macroeconomic model provided by IHS Global Insight. The version of RHG-NEMS used for this analysis is keyed to the 2013 version of the AEO. The NEMS model is documented in full in U.S. Energy Information Administration (2009). Documentation of the macroeconomic and energy-sector assumptions used in the AEO 2013 version of NEMS is available in U.S. Energy Information Administration (2013a).

RHG-NEMS is designed as a modular system with a module for each major source of energy supply, conversion activity, and demand sector, as well as the international energy market and the U.S. economy (figure C1). The integrating module acts as a control panel, executing other RHG-NEMS modules to ensure energy-market equilibrium in each projection year. The solution methodology of the modeling system is based on the Gauss-Seidel algorithm. Under this approach, the model starts with an initial solution, energy quantities and prices, and then iteratively goes through each of the activated modules to arrive at a new solution. That solution becomes the new starting point and the above process

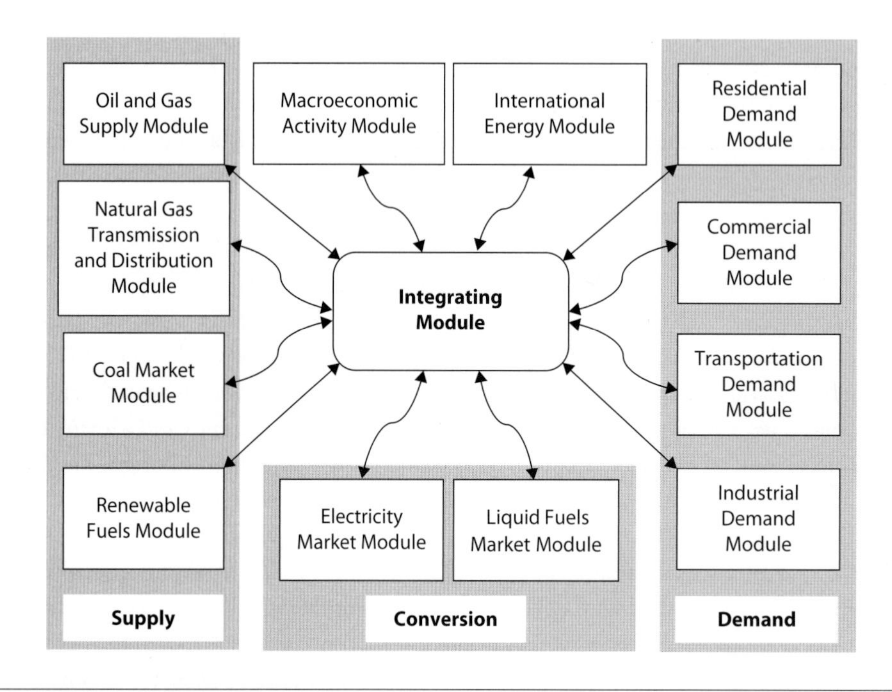

FIGURE C1. Basic NEMS structure and information flow.

TABLE C1 Formula for calculating daily HDD-65

Condition	Formula Used
$T_{min} > 65F$	$D_h = 0$
$\frac{(T_{max} + T_{min})}{2} > 65F$	$D_h = \frac{(65F - T_{min})}{4}$
$T_{max} >= 65F$	$D_h = \frac{(65F - T_{min})}{2} - \frac{(T_{max} - 65F)}{4}$
$T_{max} < 65F$	$D_h = 65F - \frac{(T_{max} + T_{min})}{2}$

TABLE C2 Formula for calculating daily CDD-65

Condition	Formula Used
$T_{max} < 65F$	$D_c = 0$
$\frac{(T_{max} + T_{min})}{2} < 65F$	$D_c = \frac{(T_{max} - 65F)}{4}$
$T_{min} <= 65F$	$D_c = \frac{(T_{max} - 65F)}{2} - \frac{(65F - T_{min})}{4}$
$T_{min} > 65F$	$D_c = \frac{(T_{max} + T_{min})}{2} - 65F$

repeats itself. The cycle repeats until the new solution is within the user-defined tolerance of the previous solution. Then the model has "converged," producing the final output.

2.2.1. RESIDENTIAL AND COMMERCIAL DEMAND MODULES

In RHG-NEMS, energy-consumption estimates of residential and commercial sectors are modeled using the residential and commercial demand modules. The residential demand module projects energy demand by end-use service, fuel type, and Census division (table C3). Similarly, the commercial demand module projects energy demand by end-use service and fuel type for 11 different categories of buildings in each Census division (table C4). Both modules use energy prices and macroeconomic projections from the RHG-NEMS system to estimate energy demand based on extensive exogenous inputs including consumer behavior, appliance efficiency and choices, and government policies.

One of the factors affecting energy demand in RHG-NEMS is climate. Future climate is represented as annual HDDs and CDDs by census region in both the residential and commercial demand modules. The modules incorporate the future change in heating demand (due to change in HDDs) and cooling demand (due to change in CDDs) to inform decisions about appliance purchases as well as total energy consumption. For further details see residential demand module (U.S. Energy Information Administration, 2013c) and commercial demand module (U.S. Energy Information Administration, 2013b) documentations.

2.3. Impact Estimation

Because of the complexity and run time of RHG-NEMS, running the model for each of our weather draws was infeasible. Instead, we constructed representative scenarios that varied the HDDs and CDDs from our baseline and recorded

TABLE C3 Residential module: key variables

Index Value	Census Division	Housing Type	Fuel	Technology Choice
1	New England	Single family	Distillate	Space heating
2	Middle Atlantic	Multifamily	LPG	Space cooling
3	East North Central	Mobile home	Natural gas	Clothes washing
4	West North Central		Electricity	Dishwashing
5	South Atlantic		Kerosene	Water heating
6	East South Central		Wood	Cooking
7	West South Central		Geothermal	Clothes drying
8	Mountain		Coal	Refrigeration
9	Pacific		Solar	Freezing

TABLE C4 Commercial module: key variables

Index Value	Census Division	Building Type	Fuel	End-Use Service
1	New England	Assembly	Electricity	Space heating
2	Middle Atlantic	Education	Natural gas	Space cooling
3	East North Central	Food sales	Distillate oil	Water heating
4	West North Central	Food service	Residual oil	Ventilation
5	South Atlantic	Health care	LPG	Cooking
6	East South Central	Lodging	Steam coal	Lighting
7	West South Central	Office - large	Motor gasoline	Refrigeration
8	Mountain	Office - small	Kerosene	Office equipment - PCs
9	Pacific	Mercantile & service	Wood	Office equipment - other
10		Warehouse	Municipal solid waste	Misc. end-use loads
11	U.S. Total	Other	Hydro	
12			Waste hear	
13			Other gaseous fuels	

the response of the model in the variables of interest: changes in energy and electricity demand, energy and electricity expenditures, and electricity-generation capacity. Using the results of these runs, we built the response functions to estimate, for each census region, the percent change in a particular variable due to absolute changes in the number of HDDs and CDDs. Table C5 presents the percent change in electricity demand resulting from an increase in 1 HDD or CDD. Tables C6 and C7 present the percent change in residential and commercial energy expenditures, respectively, resulting from an increase in 1 HDD or CDD. Table C8 presents the change in national electricity generation-capacity resulting from an increase in 1 HDD or CDD.

The final distributions were estimated using a Monte Carlo simulation of the changes in county, state, regional, and national energy expenditures, electricity demand, and changes in national electricity generation capacity given the set of all modeled changes in county daily temperature extremes. The HDD-CDD response functions given earlier are applied to all counties within each census region. This process yields a set of covarying distributions for percent changes in electricity and energy demand at the county level for each year in the analysis. Changes in state, regional, and national quantities use population-weighted average changes in HDDs and CDDs at the state level. Base values are EIA estimates of

TABLE C5 Predicted change in residential and commercial electricity demand from a change in 1 HDD/CDD (percent)

Census Region	HDD (%)	CDD (%)
New England	0.00056	0.00329
Middle Atlantic	0.00150	0.00895
East North Central	0.00101	0.00924
West North Central	0.00023	0.00755
South Atlantic	0.00040	0.00718
East South Central	−0.00138	0.00620
West South Central	−0.00047	0.00742
Mountain	0.00061	0.00881
Pacific	0.00130	0.00673

TABLE C6 Predicted change in residential energy demand from a change in 1 HDD/CDD (percent)

Census Region	HDD (%)	CDD (%)
New England	0.00102	0.00511
Middle Atlantic	0.00116	0.00012
East North Central	0.00040	0.00011
West North Central	0.00010	0.00012
South Atlantic	0.00102	0.00907
East South Central	−0.00193	0.00759
West South Central	0.00494	0.00010
Mountain	−0.00010	0.00014
Pacific	0.00239	0.00769

TABLE C7 Predicted change in commercial energy demand from a change in 1 HDD/CDD (percent)

Census Region	HDD (%)	CDD (%)
New England	0.00013	0.00160
Middle Atlantic	0.00176	0.00621
East North Central	0.00157	0.00776
West North Central	0.00037	0.00317
South Atlantic	−0.00030	0.00505
East South Central	−0.00056	0.00423
West South Central	−0.00681	0.00410
Mountain	0.00138	0.00322
Pacific	0.00048	0.00600

TABLE C8 Predicted change in national electricity capacity required from a change in 1 HDD/CDD (GW)

	HDD (GW)	CDD (GW)
U.S. total	0.08184	0.2856

state-level electricity sales (U.S. Energy Information Administration 2014a) and energy expenditures (U.S. Energy Information Administration 2014b). For maps displaying county-level absolute changes in electricity and energy demand, we multiply each county's share of the state population by the state's base energy or electricity demand to find a base county-level value. In some cases where the statewide energy or electricity-demand distribution departs significantly from the population distribution, this will misrepresent the magnitude of changes at the county level. However, the data portrayed in these maps are intended for illustrative purposes and do not factor into any state or national calculations.

In calculating total direct costs and benefits, including all impact categories, we used changes in residential and commercial energy expenditures, derived from the percent changes in this quantity provided by RHG-NEMS and the base energy expenditures by state from EIA (U.S. Energy Information Administration,2014b).

3. COASTAL IMPACTS

3.1. Introduction to RMS

RMS is a leading provider of products and services for the quantification, management, and transfer of catastrophe risk. RMS offers risk-assessment models for more than 50 territories worldwide, encompassing major perils such as earthquakes, tropical cyclones, other windstorms, floods, terrorist attacks, and infectious disease. Founded at Stanford University in 1988, RMS operates today from ten offices in the United States, Europe, India, China, and Japan, with headquarters in Newark, California. RMS is a subsidiary of the UK-based Daily Mail and General Trust plc media enterprise.

RMS employs professionals with backgrounds in actuarial and statistical sciences, meteorology, physics, geology, seismology, structural and civil engineering, management consulting, economics, and finance. RMS also utilizes a global network of academic contacts and consulting engineers who are retained for periodic review of RMS technology or for specific projects.

3.2. RMS Modeling Overview

This section describes in general terms the RMS modeling process for a number of catastrophe risk perils.

Analytical and statistical models for quantifying catastrophe risk have become a key component in how insurance and reinsurance companies manage the risk of natural disasters. These models are founded on a range of analytical, engineering, and empirical techniques. Given the scientific understanding of natural catastrophes, the methodology behind these models incorporates established principles of meteorology, seismology, wind and earthquake engineering, and other related fields to model the frequency, severity, and physical characteristics of natural catastrophes. To estimate the losses resulting from such catastrophes, RMS has developed a computer program and certain related utilities that provide a mathematical representation of catastrophes and that simulate the ultimate damage and insured losses to property for each simulated eveng.

Figure C2 summarizes the data and processes used to model the financial loss from catastrophe events within RMS's commercially available models.

The RMS property-loss models consist of three main elements:

1. **Precompiled databases** defining the events, site hazards (for example, distance to coast for hurricane, topography for surge), and details about the building-stock vulnerability.
2. **Analytical modules** including stochastic, hazard, vulnerability, and financial analysis components, as described later.
3. **Exposure data** describing the properties of the exposure analyzed (typically for insuring the property exposure) including its location, building information, and details of insurance coverages including covered perils and financial terms.

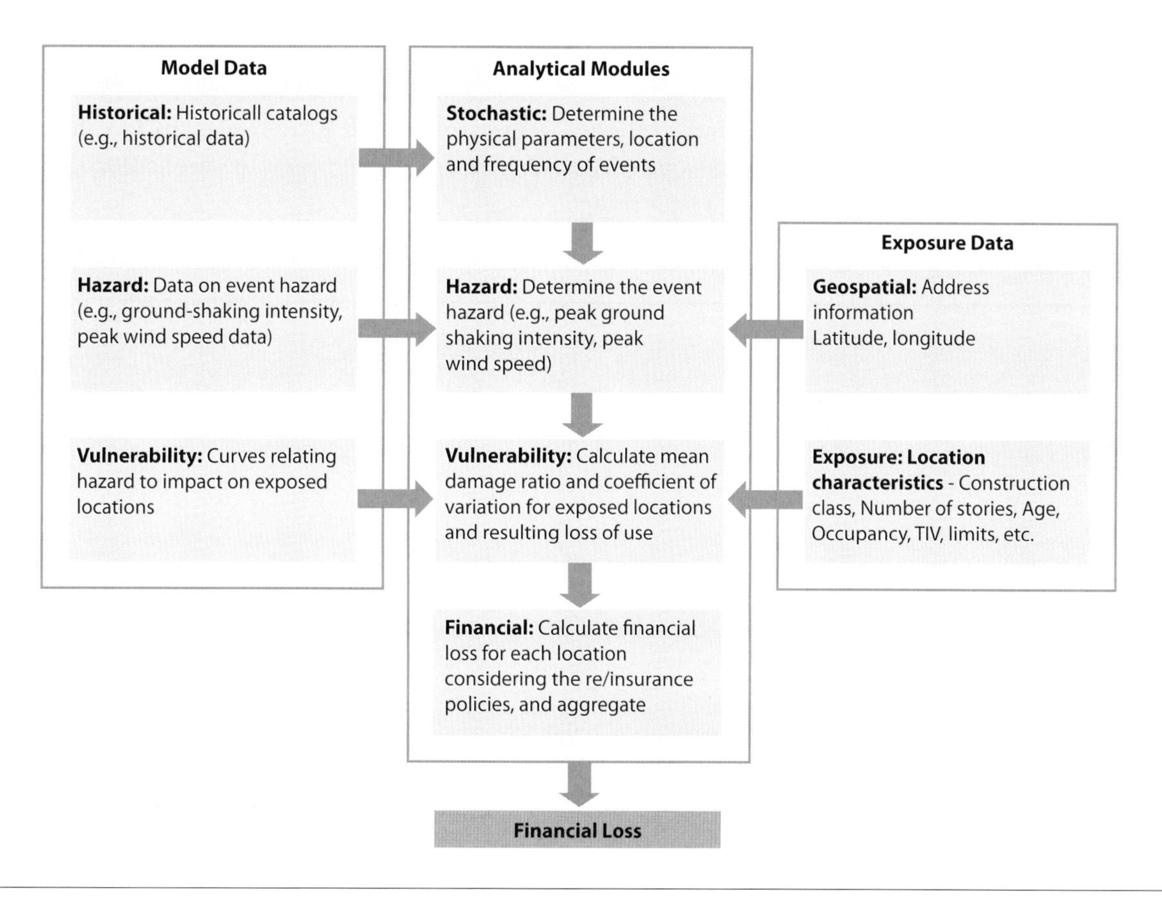

FIGURE C2. The RMS modeling process.

Regardless of the peril, RMS models follow a consistent structure comprising four principal modules.

1. **Stochastic module**: simulates tens of thousands of hypothetical events, each uniquely identified by its location, physical characteristics, and rate of occurrence, to provide a representation of the universe of physically plausible combinations of key event parameters more comprehensive than could be obtained from the sparse catalog of historical events.
2. **Hazard module**: determines an event's intensity of effect (such as windspeed or flood depth) at a given location given the parameters of the event and their interaction with the site conditions.
3. **Vulnerability module**: generates an estimate of the mean damage ratio (MDR) to a specific exposure as a function of the hazard, as well as a coefficient of variation (CV) around the mean, reflecting uncertainty both in the estimation of the hazard and the resulting damage. The MDR is the repair cost to the exposure represented as a percentage of the value.
4. **Financial analysis module**: applies the MDR and the related CV generated by the vulnerability module for each event to estimate the distribution of losses for each event at each location and in the aggregate across all the events in the simulation.

The RMS models can be used to perform both "deterministic" and "probabilistic" analyses. For a deterministic analysis, the RMS models produce a loss estimate for a single historical or scenario event. For a probabilistic analysis, the model generates a loss for each of the tens of thousands of stochastic events, and the results are displayed from highest to lowest in the form of a cumulative probability distribution, also called an occurrence exceedance probability (EP) curve. The EP curve provides the full spectrum of possible losses and their related annual probabilities of exceedance for a given set of insurance exposures or contracts. By selecting a particular annual probability (e.g., the 1% or "100-year average return period"), it becomes possible to identify the magnitude of the associated loss at this probability. The integral under the EP curve is the amount that would need to be set aside each year to fund all future losses — known as the average annualized loss. The EP curve loss distribution is not a prediction of future losses, but is solely intended to be illustrative of the range of possible losses and the likelihood of occurrence of these losses thereof. In each RMS model, the first three modules comprise a peril- and region-specific implementation, discussed in further detail herein. The financial analysis module consists of elements common to all models, as described later.

3.2.1. FINANCIAL ANALYSIS MODULE

The financial analysis module applies the MDR and the related CV generated by the vulnerability module for each event to estimate the distribution of losses both by location and in the aggregate for a given set of exposure data. To calculate the loss, the MDR for each coverage (i.e., building, contents, and business interruption/alternative living expenses) at each location is applied to the insured values, resulting in an estimate of the ground-up economic loss to the value at risk at such location before applying the effects of insurance.

Secondary Uncertainty

Secondary uncertainty represents the uncertainty in the amount of loss, given that a specific event has occurred, and is represented in the model output by the event standard deviations. The financial analysis module incorporates uncertainty from the following four sources: hazard uncertainty, vulnerability uncertainty, specification uncertainty, and portfolio data uncertainty.

- Each peril has its own characteristic type of hazard uncertainty. Uncertainties related to windstorm perils include windfield and terrain effects.
- Vulnerability uncertainty refers to the uncertainty in the amount of damage a building may sustain given a specific local hazard condition.
- The uncertainty in the magnitude of loss can also be affected by the level of specificity of the relevant model parameters. Where it is possible to differentiate risk, a less detailed model produces a loss with a higher level of uncertainty. Specification uncertainty can arise from several sources: for example, the level of geographic resolution, construction characteristics, and local conditions can all be modeled at varying levels of detail. Refinements in a model's capacity accurately to reflect higher levels of detail in loss generation reduce the uncertainty in the model's loss estimates.
- Portfolio data uncertainty reflects the level of detail in the exposure data entered into the model, such as around the construction type, year of construction, and number of stories. More detailed data should mean reduced levels of uncertainty in the model. For example, using portfolio data with location information at the street address level instead of at the ZIP code level generally decreases the secondary uncertainty.

Loss Amplification

After a major catastrophic event, claims costs can exceed the normal cost of settlement due to a unique set of economic, social, and operational factors. Claims analysis studies from both the 2004 and 2005 hurricane season, as well as in research into other major catastrophes such as Hurricane Andrew in 1992 and the Northridge earthquake in 1994, show that loss amplification can be a significant source of loss.

One component of loss amplification is economic demand surge, the increase in the cost of labor and materials needed to make repairs as a result of shortages after a large event. However, in the aftermath of C-8 Hurricane Katrina, RMS launched a research initiative to evaluate and quantify factors beyond economic demand-surge that could further amplify losses in severe catastrophes. As a result of this research, RMS has updated and expanded the demand surge module by quantifying three major factors under a broader loss amplification methodology: economic demand surge, claims inflation, and coverage expansion in super-catastrophe (Super-Cat) scenarios.

- **Economic demand surge**: Economic demand surge (EDS) is defined as the rapid increase in the cost of building materials and labor cost as demand for repair exceeds supply or capacity of the construction sector after a major hurricane. Demand-surge factors have been developed independently for each coverage, with building and business interruption coverage contributing more than contents. EDS varies by region in the United States and in Canada.

- **Claims inflation**: The claims inflation factor considers the increased cost associated with an insurer's inability to adjust claims fully when the number of claims is too high to work in accordance with standard claim adjustment procedures. Claims inflation can only be triggered if an event has activated economic demand surge. The impact of this factor varies with the estimated number of claims occurring for an event. Overall, claims inflation has a minor impact compared to the other two post-event loss amplification (PLA) components.
- **Super-Cat**: This component of loss amplification encompasses expanding losses from a range of factors including containment failures (flooding from levee or dam failure, pipeline or tank rupture, release of toxic chemicals), long-term evacuation (due to breakdown of local infrastructure, such as water or sewage systems and resulting contamination), or systemic economic consequences of a severe catastrophe (absence of labor force to make repairs; hotels and retail facilities incapable of opening due to lack of tourism). Primary escalation for Super-Cat events occurs with respect to time element (such as business interruption and adjusted living expenses) losses. This effect was modeled for several major urban areas in the United States.

Claims inflation and Super-Cat loss amplification effects occur with more extreme events than those that would cause EDS. The effects of these three components of loss amplification compound multiplicatively.

3.3. The RMS North Atlantic Hurricane Model

The RMS North Atlantic Hurricane Model explicitly includes direct damage from wind and storm surge to property, contents, and interrupted businesses in the Gulf states, Florida, and the East Coast of the United States.

By virtue of using empirical claims calibration, the model implicitly includes related hurricane property damage from tree-fall, wind missiles, rain infiltration, and so forth. The RMS North Atlantic Hurricane Model does not include the following potential sources of nonmodeled loss:

- Inland flooding from hurricane-related rainfall, though some wind-driven rain losses are implicitly included through the empirical claims calibration.
- Chaotic weather systems such as tornadoes and hailstorms that sometimes accompany hurricanes, though damage caused by such weather systems may be implicitly included through the empirical claims calibration to the extent of their prevalence in past events to which the claims data pertain.
- Losses from any source occurring in areas that experience wind speeds below a modeling threshold of 50 mph.
- Claims adjustment expenses.
- Off-premises power losses.
- Losses not covered in insurance policies.
- Policy coverage expansion, for example due to legal mandates.
- Contingent and other indirect time-element coverages (i.e., contingent business interruption).

Furthermore the RMS North Atlantic Hurricane Model does not include modeled storms that at no point over their life cycles affect land as a hurricane (i.e., category 1 or greater).

The RMS North Atlantic Hurricane Model's geographic coverage extends across 21 states of the continental United States and the District of Columbia, Hawaii, the Gulf of Mexico, the islands of the Caribbean, and Canada. The 23 states/municipalities of the United States included within the RMS North Atlantic Hurricane Model are Alabama, Connecticut, Delaware, District of Columbia, Florida, Georgia, Hawaii, Louisiana, Maine, Maryland, Massachusetts, Mississippi, New Hampshire, New Jersey, New York, North Carolina, Pennsylvania, Rhode Island, South Carolina, Texas, Vermont, Virginia, and West Virginia. The geographic extent of the model coverage is a reflection of the significant decay of wind hazard away from the coast. Tropical cyclones are fueled by hot, moisture-laden air from the ocean. Because hurricanes are dependent on ocean evaporation to sustain energy, the relatively lower moisture content over land contributes to the dissipation of winds after hurricane landfall.

While inland states may incur hurricane-related losses from nonmodeled sources such as inland flooding, their contribution to direct wind-related losses is very minor. During the period 1950 to 2013, approximately 98%[1] of insured losses from Atlantic hurricanes occurred within the RMS modeled hurricane states and is principally related to inland flood.

3.3.1. HURRICANE MODEL

Stochastic Event Module

The stochastic module consists of a set of tens of thousands of stochastic events that represent more than 100,000 years of hurricane activity. RMS scientists have used state-of-the-art modeling technologies to develop a stochastic event set made of events that are physically realistic and span the range of all possible storms that could occur in the coming years.

At the heart of the stochastic module is a statistical track model that relies on advanced statistical techniques (Hall and Jewson 2007) to extrapolate the HURDAT catalog (Jarvinen et al. 1984) and generate a set of stochastic tracks having similar statistical characteristics to the HURDAT historical tracks. Stochastic tracks are simulated from genesis (starting point) to lysis (last point) using a semi-parametric statistical track model that is based on historical data. Simulated hurricane tracks provide the key drivers of risk, including landfall intensity, landfall frequency, and landfall correlation.

The stochastic event module is calibrated to ensure that simulated landfall frequencies are in agreement with the historical record. Target landfall rates are computed on a set of linear coastal segments (or RMS gates) by smoothing the historical landfall rates. The RMS smoothing technique uses long coastal segments, obtained by extending each RMS gate in both directions and keeping the orientation constant. Historical storms that cross an extended gate contribute to the landfall rate at the corresponding original segment.

Importance sampling of the simulated tracks is performed to create a computationally efficient event set of 30,000 events used for loss cost determinations. Each of these events has a frequency of occurrence given by its mean Poisson rate. Because event frequencies were calibrated against history (HURDAT 1900–2011), this set of Poisson rates represents the RMS baseline model, and this rate set is called the "RMS historical rate set."

Windfield (or Wind Hazard) Module

The wind hazard module calculates windfields from both landfalling hurricanes and bypassing storms. Once tracks and intensities have been simulated by the stochastic module, the windfield module simulates 10-meter, 3-second peak gusts on a variable resolution grid. Size and shape of the time-stepping windfields are generated using an analytical wind profile derived from Willoughby (2006). Peak gusts are the driver of building damage and are used as the basis of many building codes worldwide, including the United States.

Wind-Vulnerability Module

The wind-vulnerability module of the RMS U.S. Hurricane Model relates the expected proportion of physical damage for buildings and contents to the modeled peak 3-second gust wind speed at that location. The severity of the damage is expressed in terms of a mean damage ratio (MDR), which is defined as the ratio of the expected cost to repair the damaged property relative to its replacement cost. The wind-vulnerability module also estimates the variability in the expected damage ratio and models this with a beta distribution. Time-element losses due to business interruption and additional living expenses are estimated using occupancy-dependent facility restoration functions that relate the expected duration of the loss of use to the modeled building damage.

1 Derived from Property Claim Services Data, 1950–2013.

3.3.2. STORM-SURGE MODEL

The RMS North Atlantic Hurricane storm-surge model utilizes the latest technology to quantify the risk of storm surge for the U.S. Atlantic and Gulf coastlines (from Texas to Maine). The model system uses wind and pressure fields, over the lifetime of the storm, from the stochastic event set of the RMS North Atlantic Hurricane Model as forcing (i.e., as input) for the state-of-the-art MIKE 21 hydrodynamic model system to estimate still-water (i.e., without wave effects) storm surges and wave impacts. The impact of rainfall accumulation during an event is not considered in the modeling of storm-surge levels.

Storm-Surge Hazard Module

Storm surge in the numerical model is driven by the wind stress from the time-stepping windfield and the cyclones' low-pressure field, together with the changes in sea level that accompany the tides. The storm surge is dependent on the storm's characteristics in the lead up to landfall, with a large storm surge resulting from the accumulated impact of low pressures and wind stress acting for many days before the storm nears the coast. Storms may often weaken just at landfall but still retain much of the surge associated with their offshore intensity. For example, Hurricane Katrina, which reached category 5 intensity in the Gulf of Mexico but weakened to category 3 at landfall, produced a powerful 27-foot surge on the Mississippi coast.

In urban coastal areas, the resolution of the surge footprint is 100 meters. This resolution decreases to 500-meters and greater in non-urban areas. To calculate the surge hazard at a specified location, the elevation of the terrain is subtracted from the surge height at the location to determine the depth of flooding. Once the elevation is subtracted from the surge height, the wave height is calculated based on the depth of flooding and the velocity of the flow. The velocity of the flow is dependent on the proximity to the open sea and the effects of local topography on the flow. For buildings that meet the construction class and year built criteria and are within a FEMA zone, the base flood elevation is applied to the lower floor of the property to calculate an effective flood depth. RMS has implemented improved elevation data based on the USGS National Elevation Dataset (NED), enhanced with airborne LIDAR data where available, and stored on an underlying 100-meter uniform resolution grid.

Storm-Surge Vulnerability Module

Storm-Surge Vulnerability

The U.S. storm-surge vulnerability module relates the expected amount of physical damage to buildings and contents to the modeled storm-surge flood depth and the effects of wave action, if present. Like the wind vulnerability module, the severity of the damage is expressed in terms of a mean damage ratio (MDR), while the variability in the damage is characterized using a beta probability distribution. The U.S. storm-surge vulnerability functions for buildings and contents are derived from a database of flood-related damage observations compiled by the United States Army Corps of Engineers (USACE) from major flood events in the United States. This database consists of flood-depth-damage-ratio relationships that relate the incurred flood-related loss to the observed flood depth for different occupancy types. The storm-surge vulnerability functions implemented in RiskLink also incorporate structural-engineering-based adjustments to account for the effects of the building height and construction class on the expected damage and are calibrated and validated using claims data from multiple storms.

Low-Velocity Flooding and Wave Action

There are two sets of vulnerability functions in the storm-surge model—one representing the damage caused by low-velocity flooding and a second for modeling the additional damage caused by wave action. The low-velocity flooding curves, which are referred to as surge-vulnerability functions, relate the MDR of a building or its contents to the estimated flood depth according to its occupancy type, construction class, and number of stories. The wave-vulnerability functions are used for loss calculations when a location has been characterized as unsheltered and the water depth exceeds 7 feet. For a given building type and surge depth, the damage ratio predicted by the wave-vulnerability function will be higher than that simply from low-velocity flow.

3.4. Modifying the RMS North Atlantic Hurricane Model to Account for Climate Change

The RMS North Atlantic Hurricane Model can be adjusted to account for changed climate conditions by modifying various components of the detailed model described earlier.

- Local sea-level rise projections (Kopp et al. accepted, 2014) can be incorporated via adjusting the storm-surge hazard module.
- Changes in hurricane frequency and intensity (Emmanuel 2013; Knutson et al. 2013) can be incorporated via adjustments to the stochastic event module and specifically the frequency of occurrence of each event in the event set that is, by creation of alternative rate sets.

Each adjustment can be applied independently or in combination to create multiple alternative versions of the RMS North Atlantic Hurricane Model conditioned on a specified future climate state. For the purposes of this report, many hundred alternative versions of the model were created to span the range of time periods, RCPs, sea-level rise projections, as well as alternative models / inputs for the changes in hurricane frequency and intensity.

3.4.1. LOCAL SEA-LEVEL RISE PROJECTIONS

The storm-surge hazard model calculates the local inundation of each stochastic event on 18 local grids nested within a lower-resolution regional gird covering the domain of the model. The local sea-level rise projections from Kopp et al. (accepted, 2014) were mapped to the individual local grids to assign a grid-specific increase in sea level. The surge height of the footprint of each event was then adjusted at this local level before being recombined into a coherent "event" footprint (which can cover multiple local grids).

3.4.2. CHANGE IN HURRICANE FREQUENCY AND INTENSITY

The projections of future hurricane activity in Knutson et al. (2013) provide the percentage change in activity in the Atlantic basin by category of storm for the multimodel ensemble mean from phase 5 of the Coupled Model Intercomparison Project using Representative Concentration Pathway 4.5. These changes were mapped directly to the RMS historical rate set according to the maximum lifetime intensity category of each storm in the event set. Knutson et al. provide the relevant changes by storm intensity category for early- and late-twenty-first-century time slices, from which linear interpolation / extrapolation was used to infer the changes at the relevant time periods required for this report.

The projections of future hurricane activity in Emmanuel (2013) (supplemented through personal communications) were obtained as absolute values for basin activity and for activity at landfall (across the entire North Atlantic domain) by maximum storm intensity (as measured by the maximum 1-minute sustained wind speed). Projections were provided for six individual General Circulation Models from phase 5 of the Coupled Model Intercomparison Project using Representative Concentration Pathway 8.5. Projections were provided for current, mid, and late-twenty-first-century time slices. The difference between the future and current activity is used to define a percentage change in activity, and linear interpolation / extrapolation is used to infer the percentage changes at the relevant time periods consistent with this report. The changes are mapped directly the RMS historical rate set according to maximum lifetime category or maximum landfall category of each storm in the event set to create multiple alternative rate sets.

REFERENCES

Deschênes, O., Greenstone, M., 2011. Climate change, mortality, and adaptation: Evidence from annual fluctuations in weather in the US. *American Economic Journal: Applied Economics* 3, 152–85.

EIA, 2013. 2013 Annual Energy Outlook.

Emmanuel, K., 2013. Downscaling CMIP5 climate models shows increased tropical cyclone activity over the 21st century. *Proceedings of the National Academy of Sciences of the United States of America* 110, 12219–24.

Hall, T. M. & Jewson, S. (2007). Statistical modelling of North Atlantic tropical cyclone tracks. *Tellus* A 59.4: 486–498.

Jarvinen, B. R., Neumann, JC. J. and Davis, M. A. S. (1984). A tropical cyclone data tape for the North Atlantic Basin, 1886–1983, contents, limitations, and uses. *NOAA Tech. Memo.*, NWS NHC 22.

Knutson, T. R., Sirutis, J. J., A. Vecchi, G., Garner, S., Zhao, M., Kim, H.-S., Bender, M., Tuleya, R. E., Held, I. M., Villarini, G., 2013. Dynamical downscaling projections of twenty-first-century Atlantic hurricane activity: CMIP3 and CMIP5 model-based scenarios. *Journal of Climate* 26, 6591–17.

Kopp, R. E., Horton, R. M., Little, C. M., Mitrovica, J. X., Oppenheimer, M., Rasmussen, D. J., Strauss, B. H., Tebaldi, C., accepted, 2014. Probabilistic 21st and 22nd century sea-level projections at a global network of tide gauge sites. *Earth's Future* 10.1111/eft2.2014ef000239.

UK Met Office, 2012. How have cooling degree days (CDD) and heating degree days (HDD) been calculated in UKCP09? URL http://ukclimateprojections.metoffice.gov.uk/22715

U.S. Energy Information Administration, 2009. The national energy modeling system: An overview. Tech. Rep. DOE/EIA0581(2009). URL http://www.eia.gov/oiaf/aeo/overview/index.html

U.S. Energy Information Administration, 2013a. *Annual Energy Outlook* 2013. US DOE/EIA. URL http://www.eia.gov/oiaf/aeo/overview /index.html

U.S. Energy Information Administration, 2013b. Commercial demand module of the National Energy Modeling System: Model documentation 2013. Tech. rep. URL http://www.eia.gov/forecasts/aeo/nems/documentation/commercial/pdf/m066(2013).pdf

U.S. Energy Information Administration, 2013c. Residential demand module of the National Energy Modeling System: Model documentation 2013. Tech. rep. URL http://www.eia.gov/forecasts/aeo/nems/documentation/residential/pdf/m067(2013).pdf

U.S. Energy Information Administration, 2014a. State electricity profiles. URL http://www.eia.gov/electricity/state/

U.S. Energy Information Administration, 2014b. Table F30: Total energy consumption, price, and expenditure estimates, 2012. URL http://www.eia.gov/state/seds/data.cfm?incfile=/state/seds/sep_fuel/html/fuel_te.html&sid=US

Willoughby, H. E., Darling, R. W. R., & Rahn, M. E. Parametric representation of the primary hurricane vortex. Part II: A new family of sectionally continuous profiles. *Monthly weather review* 134.4 (2006): 1102-1120.

APPENDIX D

INTEGRATED ECONOMIC ANALYSIS

MICHAEL DELGADO AND SHASHANK MOHAN

1. INTRODUCTION

This report attempts to provide insight into the potential impacts of climate change in a range of economic sectors and at fine temporal and geographic scales. This includes both the immediate physical impact of climate change, as well as its broader economic ramifications. For example, farmers in the Midwest may experience changes in crop and labor productivity (the physical impacts) due to changes in temperature and precipitation. Changes in the value of the agriculture sector output (the direct costs and benefits) will impact a greater number of people as that value is removed from (or added to) the local economy. Finally, macroeconomic effects such as changes in prices, in cross-regional and cross-sectoral investment, and in long-run growth stemming from lost productivity and stranded capital assets may lessen or amplify these direct costs and benefits.

Appendices A, B, and C describe the methodologies used in this report to estimate physical impacts. This appendix describes the methodology used to estimate direct costs and benefits and their broader macroeconomic effects.

2. DIRECT COSTS AND BENEFITS

The physical impacts of climate change assessed in the *American Climate Prospectus* are reported as quantity changes relative to current levels. For example, climate-driven changes in labor productivity are measured in terms of full-time equivalent (FTE) employees, assuming current labor market conditions. In translating these changes in quantity to changes in value, we measure them against current economic structure and prices. Unless otherwise noted, all direct cost and benefit values are presented in real 2011 U.S. dollars, either total, per capita, or as a share of national, regional, or state economic output in 2012. Economic data are from U.S. Bureau of Economic Analysis (BEA; 2014a, 2014b). In areas where greater sectoral resolution than available in the BEA data was required, we used the more detailed social accounts provided in IMPLAN Group (2011) to distribute BEA totals.

2.1. Agriculture

As described in appendix B, changes in agricultural yield are converted to changes in agricultural production quantities after incorporating the interannual effects of storage and speculation. Output is modeled as an autoregressive process of yields, as estimated from U.S. Department of Agriculture data.

To calculate the direct costs and benefits from changes in production quantity, we assume that a given change in agricultural output results in a proportional change in the value of that sector's gross output in the reference year. The value of impact-sector gross output in a given state and year was calculated as the portion of all agriculture output made up by the affected crop in a given region in the IMPLAN data set (IMPLAN Group 2011) times national gross output given by BEA (U.S. Bureau of Economic Analysis 2014a). The potential for crop switching or land-use change was not accounted for in calculating direct costs and benefits; instead, areas that experienced changes in productivity, either positive or negative, had that productivity applied to the existing production in that area. The total cost for all agriculture was found by summing the impacts from the component crops.

2.2. Labor

Changes in labor productivity are weighted by employee hours worked from the county to the NCA region level. To calculate the direct costs and benefits, we assume that the direct effect of a climate-driven labor productivity change is equal to state value added (U.S. Bureau of Economic Analysis 2014b), distributed among high- and low-risk sectors using IMPLAN data (IMPLAN Group 2011) times the change in labor productivity. This is an assumption of zero elasticity of substitution between labor and other inputs, but does not assume the loss of value of other inputs to production. These dynamics are explored in our analysis of possible macroeconomic effects (see section 3).

2.3. Mortality

Changes in mortality in appendix B are reported as temperature-driven changes in all-cause mortality by age cohort (less than 1, 1 to 44, 45 to 64, and 65+, from Deschênes and Greenstone [2011] and Barreca et al. [2013]).

We use two approaches to assess the direct costs and benefits of these mortality impacts. First, we apply the value of a statistical life (VSL) used by the EPA of $7.9 million as a benchmark estimate of Americans' "willingness to pay" to reduce mortality risk. This includes both market and nonmarket costs. We also use an alternative valuation technique, in which we estimate the value of full-time-equivalent (FTE) employee-years lost and/or gained due to climate-driven changes in mortality. We estimate this by extrapolating cohort mortality rates, extrapolated to single years of age using national death rates by 5-year cohort (listed in table D4) and assuming a uniform distribution within each 5-year cohort. To this we apply current labor-force participation rates by age cohort. BEA (U.S. Bureau of Economic Analysis 2014b) state-level value added was divided by current state-specific FTE employment to arrive at a value per employee, which was multiplied by state cohort-specific population and the labor participation data to arrive at a time series of expected value lost per mortality in each cohort bin by state. This was discounted at an annual rate of 3% to arrive at a net present value expected loss per death in each cohort and state. This was multiplied by the change in mortality to arrive at an expected value lost for each state and cohort, which was combined to form probabilistic regional and national labor income mortality cost and benefit estimates.

2.4. Crime

Crime impacts are valued using the method described in Heaton (2010). Changes in property and violent crime are multiplied by current crime levels using the FBI's "Crime in the United States" data set (U.S. Federal Bureau of Investigation 2012), and the cost estimates, given in 2007 U.S. dollars, are adjusted to 2011 U.S. dollars using the BEA national real GDP deflator.

2.5. Coastal impacts

Coastal impacts were derived directly from the process-modeling method described in appendix C. The total costs of climate damage include the average marginal (annual) costs of property falling below sea level, as well as climate-driven changes in average annual loss from hurricanes and other coastal storms, averaged over 20-year periods.

2.6. Energy

The direct costs and benefits of climate-driven changes in energy demand were assessed using RHGNEMS, a version of the National Energy Modeling System (NEMS), developed by the U.S. Energy Information Administration (2009) for use in the *Annual Energy Outlook* (see, for example, U.S. Energy Information Administration 2013) and maintained by the Rhodium Group. A detailed description of those methods is given in appendix C.

3. MACROECONOMIC EFFECTS

The direct costs and benefits described above have broader economic ramifications by changing relative prices, diverting investment, and altering trade flows, among other effects. As is discussed in chapter 14, many approaches have been used to capture these economic dynamics at regional, national, and global scales. A number of researchers have begun employing computable general equilibrium (CGE) economic models using a mixed complementarity problem (MCP) equilibrium (e.g., Yang et al. 1996, Jorgenson et al. 2004, Abler et al. 2009, Backus et al. 2010). First proposed by Arrow and Debreu (1954), this type of general equilibrium analysis is a branch of economics that represents the modeled system as completely self-contained, allowing feedbacks of changes in technology, factor (labor, capital, or resource) supplies, or any other parameter to spread throughout the economy through changes in prices and quantities in a theoretically consistent manner. Many modern CGE models explicitly represent multiple individually optimizing agents (both consumers and producers) in a unified framework that ensures price and goods equilibria across all regions and sectors in the modeled economy.

We have elected to use a CGE model in this analysis because of the ability to examine the effect of fine spatial, temporal, and sectoral resolution impacts in a way that tracks impacts, and their interaction, through time.

3.1. CGE modeling

There are many varieties of CGE models; one of the most important design distinctions is the method for representing time or change. The simplest type of model is a comparative statics model. This approach represents the current economy as a single-year equilibrium that is shocked by some experiment. This type of analysis was frequently used in early climate-impact assessments (for an overview, see Tol [2008]). It is useful for understanding how today's economy would perform in an alternative state of the world, such as one with a much hotter climate, but is unable to estimate the impacts on non-steady-state changes such as impacts to growth from insufficient capital availability after a storm or overcapitalization of certain industries in periods of declining demand.

A second alternative is an intertemporal optimization CGE model, originally developed in Ramsey (1928), Cass (1965), and Koopmans (1965), in which a single CGE optimization framework allows agents to substitute between activities in time as well as across sectors and regions. While many variants on this model have since been developed, the central feature of this model is that agents are able to anticipate changes in future prices and may adjust their behavior in each model period to adapt to the future. This type of model is also useful for policy analysis, but with a very different purpose: Because agents in this model are perfectly optimizing over the entire model horizon, this model portrays the best-case outcome and is thus useful for finding optimal policies in response to a specific change.

Finally, CGE models may be recursive-dynamic. In this variant, consumers optimize their behavior in each period with knowledge of conditions in that period only, and dynamic equations link decisions in one period to the constraints on behavior in the next period. For example, if in one period it is optimal to spend more and save less, this will result in a lower level of savings to draw on in the next period.

The goal of the *American Climate Prospectus* is to assess the potential economic impacts of climate change in the United States under current economic and business practices, in order to provide information on how those practices may need to change in the future to reduce climate risk. As such,we chose a recursive-dynamic CGE model to explore the macroeconomic effects of the direct costs and benefits described above.

4. RHG MODEL OF THE U.S. ECONOMY (RHG-MUSE)

RHG-MUSE is a dynamic recursive CGE model of the U.S. economy. It is written in the GAMS mathematical programming language, with the static core defined in the MCP-specific sublanguage MPSGE, developed by Rutherford (1987). MUSE is solved annually from 2011 to 2099 using the PATH solver (Dirkse and Ferris 1995) and simulates the growth of the U.S. economy with changes in labor, capital, and productivity. The model is calibrated using the 2011 IMPLAN social accounting matrices (SAMs) at the state level. The model has 50 regions (for computational reasons, Washington, D.C., is considered part of Maryland in this model) and nine sectors.

4.1. Economic data sources

The IMPLAN social accounts used in this analysis are a detailed data set describing the inputs and outputs of 440 sectors, 4 factors, 9 households, 6 types of government, and an account for corporations, capital additions and deletions, and each of foreign and domestic trade for all 50 U.S. states and Washington, D.C. IMPLAN SAMs are similar to traditional input-output tables, with the exception that they include flows between institutions such as households and government. This enables the tracking of factors (such as labor use by industry and compensation to households) and other flows not included in a traditional I/O table, enabling a comprehensive accounting of all flows in the economy.

The IMPLAN data set is typically paired with the IMPLAN model, which models the propagation of certain economic changes as they flow through the accounts tables. This methodology is roughly similar to our calculation of direct effects and does not endogenously represent changes in prices, technology, or growth.

We have incorporated this data set into our model using an adapted version of Thomas Rutherford's analysis tools implan98 (Rutherford 2004), IMPLAN2006inGAMS (Rausch 2008), and IMPLAN2010inGAMS. Our build process reads each state's IMPLAN SAM as an individual state model, validates the data to ensure that all flows balance (each agent, sector, and transfer type must have inputs and outputs sum to zero), and then aggregates the data into a multiregional national data set. The resulting aggregated model includes the 9 sectors shown in Table D1 (see table D5 for the

TABLE D1 Sectors used in the MUSE model

Agriculture
Indoor services
Outdoor services
Infrastructure
Manufacturing
Mining
Energy
Real estate
Transportation

full sectoral aggregation scheme) and can be flexibly aggregated from 50 states to as few as 2 regions. Additional adjustments are made to ensure that domestic trade is balanced across all regions and that the data set is internally consistent. At the conclusion of this process, the data set is in equilibrium; that is, the three criteria for an MCP solution (described in section 4.3.5) are met without a change in prices.

4.2. RHG-MUSE model structure

As described above, RHG-MUSE uses a hybrid "recursive-dynamic" framework to examine the long-run equilibrium response to climate impacts. Recursive-dynamic models find an equilibrium solution that optimizes agent behavior (subject to the objectives and constraints defined in the model) within a period, which sets prices and quantities that are linked to subsequent periods through dynamic *updating equations*, which govern the calibration data and constraints applied to the optimization problems in subsequent periods. In other words, the simulation progresses from one year to the next, pausing each year to optimize each agent's welfare, then updating the values for the next year with new information and continuing on.

The optimization that takes place each year is a function only of the previous year's events and is only able to change variables pertaining to that year. Therefore, we refer to the set of annual optimization equations as the RHG-MUSE *static core*, to set it apart from the *updating equations*, which apply changes derived in the *static core* to subsequent periods.

The structure used to run RHG-MUSE is written in Python and Javascript; the data preparation, run management, and *updating equations* are written in the mathematical programming language GAMS; and the *static core* is written in the GAMS sublanguage MPSGE and solved using the PATH GAMS solver (Dirkse and Ferris, 1995). Throughout the following sections, we refer to the mathematical symbols given in the leftmost column of table D2.

TABLE D2 Symbols used in RHG-MUSE

Indexing sets:	
r	50 states - Washington, D.C., combined with Maryland
s	9 sectors, given in Table D1
g	9 goods, 1 for each sector, interchangeable with s
m	Local (*loc*), domestic (*dtrd*) and international (*ftrd*)

TABLE D2 Symbols used in RHG-MUSE (*continued*)

Symbol	Description
t	Domestic (*dtrd*) and international (*ftrd*)
y	89 time periods, running from 2011 to 2099
a	Single years of age, from 0 to 65
Static parameters:	
Base-year data	
k^{E0}	2011 capital returns endowment
e_r^{Tot}	2011 RMS projected value of exposed property
μ_r^S	2011 regional marginal propensity to save
$\mu_{r,s,g}^0$	2011 use of Armington goods by sector
$k_{r,s}^0$	2011 capital use by sector
$l_{r,s}^0$	2011 labor use by sector
$o_{r,s}^{Y0}$	2011 output to local markets
$o_{r,s}^{N0}$	2011 output to domestic markets
$o_{r,s}^{F0}$	2011 output to foreign markets
$y_{r,s}^0$	2011 total output
$d_{r,g}^0$	2011 consumer goods demand
Calibration data	
r^0	Reference growth rate
r^L	Reference labor productivity growth rate
$p_{r,a,y}$	Reference population projection
θ^x	Initial extant capital share
δ	Annual depreciation rate
Positive optimization variables:	
Sector activity levels	
C_r	Consumption activity level index
$Y_{r,s}$	New production activity level index
$Y_{r,s}^X$	Extant production activity level index
$A_{r,s}$	Regional Armington aggregate activity level index
G_r	Government activity level index
K	Extant capital exchange activity level index
Commodity prices	
P_r^C	Consumption good price index
$P_{r,s}^A$	Armington aggregate good price index
$P_{r,s}^Y$	Local output good price index
P_s^N	Nationally traded good price index
P_r^F	International trade price index
P_r^G	Public sector output good price index
P_r^L	Regional wage rate index
R_r^N	Return index on new vintage capital
$R_{r,s}^X$	Return index on extant capital in production
R^E	Return index in extant capital exchange
P^S	Savings good price index

Consumer agent budget indices

R_r^A	Regional agent endowment index

Free optimization variables:

Auxiliary/constraint variables

I_r	Regional investment level
$\hat{I}_r$	Adjustment to investment endowments to balance savings price differences
$\hat{R}^k$	Realized rate of return after savings price adjustments
Θ_r^S	Regional share of national savings
$\theta_{r,s}^{RS}$	Share of regional investment by good

Non-optimization variables:

Growth variables

$k_{r,y,y}^{EX}$	Real extant capital earnings endowment
$k_{r,s,y}^{KN}$	Reference new capital use
$k_{r,s,y}^{KX}$	Reference extant capital use
$U_{r,s,y}^{LN}$	Labor used in new production
$U_{r,s,y}^{KN}$	Capital used in new production
γ_y^L	Labor productivity factor

Climate-impact variables

$I_{r,y}^A$	Agricultural productivity climate factor (2011 = 1, $I_{r,y}^A > 1$ means yield increase)
θ_r^A	Share of RHG-MUSE agriculture made up of affected crops (grains, oilseeds, cotton)
$I_{r,s,y}^L$	Labor productivity climate factor (2011 = 1, $I_{r,s,y}^L > 1$ means productivity increase)
$I_{r,y}^{In}$	RMS projection of inundation damage (2011 = 0, $I_{r,y}^{In} \geq 0$ by definition)
$I_{r,y}^{In}$	Inundation damage, adjusted for previous property loss due to sea-level rise
$I_{r,y}^{St}$	RMS projection of storm damage (2011 = 0, $I_{r,y}^{St} \geq 0$ by definition)
$I_{r,y}^{St'}$	Storm damage, adjusted for property losses due to sea-level rise
$I_{r,y}^{BI}$	RMS projection of business interruption (2011 = 0, $I_{r,y}^{BI} \geq 0$ by definition)
$I_{r,y}^{BI'}$	Business interruption, adjusted for property losses due to sea-level rise
$I_{r,g,y}^E$	Energy expenditure climate factor (2011 = 1, $I_{r,g,y}^E > 1$ means exependiture increase)
$I_{r,a,y}^M$	Climate-related change in per-capita mortality rate (2011 = 0, $I_{r,a,y}^M > 0$ means rate increase)
$M_{r,a,y}$	Cumulative climate-related mortality change, with mortality by age advancing annually
$l_{r,y}^M$	Labor force, adjusted for climate-related mortality

4.3. Annual "static core" optimization module

The RHG-MUSE *static core* is written in the MPSGE modeling language, developed by Rutherford (1987) and documented in *MPSGE: A User's Guide* (Markusen & Rutherford 2004). MPSGE models are characterized by collections of individually optimizing agents, both producers and consumers, interacting in a coherent general-equilibrium framework (see section 4.3.5). Each agent consumes goods with a constant elasticity of substitution (CES) utility or consumption functions (for consumers and producers, respectively), and each producer converts its inputs into outputs with a constant elasticity of transformation (CET) production function. For the sake of brevity, we will describe the quantities and variables used to create the CES "nests" defining each agent's behavior, but will not provide the full set of equations defining RHG-MUSE. Böhringer and Wiegard (2003) provide a comprehensive overview of the equations governing the MCP equilibrium described by an MPSGE model.

The RHG-MUSE model core includes one aggregate consumer per region and two firms per region and sector (one for new vintage production and one for existing, or extant production). Government is represented by a single producer for each region. The consumer, whose budget is equal to the regional GDP (calculated by income, which is the sum of labor and capital earnings, tax revenue, net exports, and borrowing), divides its endowment in constant shares between consumption, savings, and government services. The producer receives the revenue from consumer, government, and investment spending and purchases labor, capital, and intermediate goods in order to produce regionally differentiated local products, national tradable products, or foreign exports. While additional details define the specific behavior of producers and consumers (described later), this dynamic forms the core relationship of the RHG-MUSE static core.

4.3.1. PRODUCTION

Each sector produces three products, subject to a transformation nest allowing substitution between local, domestic, and international goods. Producers make use of intermediate inputs with zero substitution (i.e., any given sector cannot decide to change the mix of aggregated goods from which it makes its product) and factors. RHG-MUSE uses a putty-clay formulation that allows substitution of capital and labor in the first year of production but fixes their ratio in subsequent periods.

Figures D1 and D2 offer a graphical representation of this structure, showing the substitutability of capital and labor in new production in contrast with the fixed input shares of capital, labor, and intermediate goods in extant production.

4.3.2. CONSUMPTION

All income from the two factors of production — labor and capital — in each region accumulates to a representative consumer. In addition to the factor income, the consumers are endowed with the net foreign borrowing available each year due to the net trade imbalance. The consumers allocate this combined endowment into consumption by households and government and savings (see figure D3). See section 4.3.4 for additional detail on the representation of trade. Government consumption is modeled as a Leontief production function with Armington good inputs and a region-specific government service as the product (see figure D5); household consumption is represented as a Cobb-Douglass production function, with Armington goods as inputs and a region-specific consumption good as the product (see figure D6).

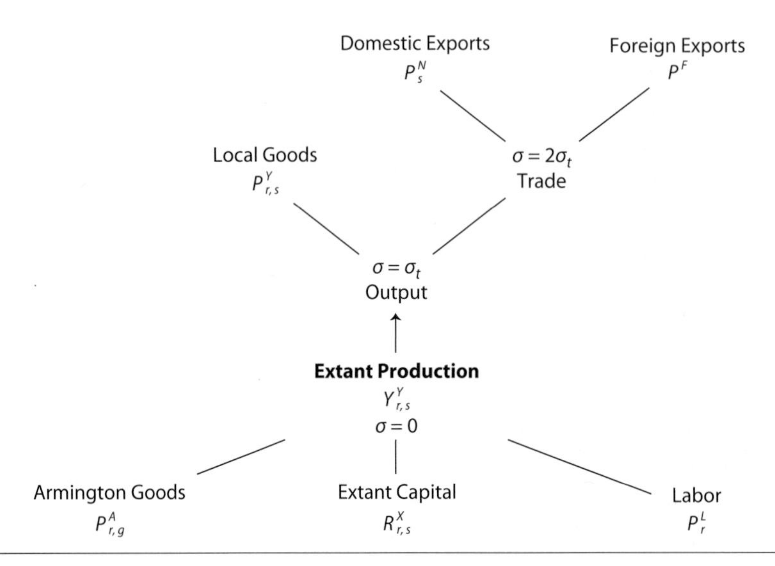

FIGURE D1. Extant production restricts the use of capital, labor, and intermediate goods to remain in fixed proportion. The relationship between capital and labor in extant production, which is fixed within each year, is updated every year with the new capital-labor mix as new production ages.

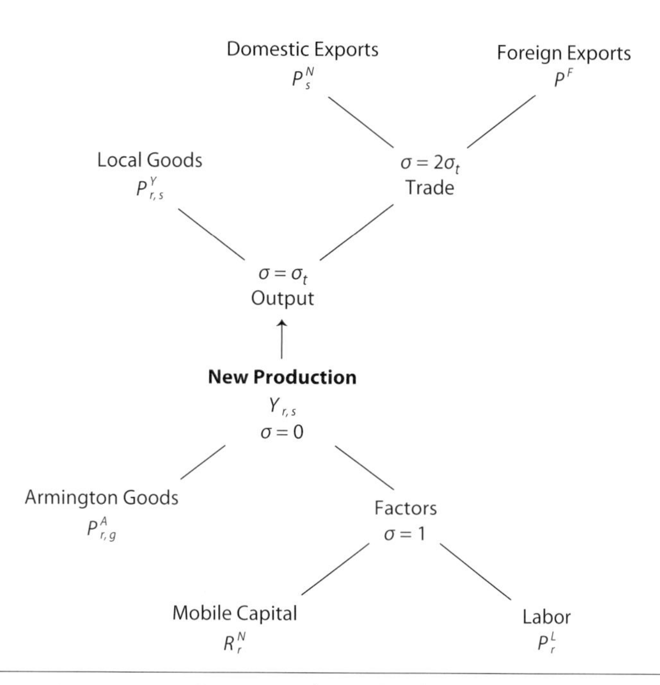

FIGURE D2. New production allows price-sensitive substitution of capital and labor, while maintaining the balance between factors and intermediate inputs seen in the social accounts.

4.3.3. CAPITAL, SAVINGS, AND INVESTMENT

RHG-MUSE distinguishes between malleable and non-malleable capital. All extant capital stock is assumed to be non-malleable and is fixed to a particular region and sector, but new capital stock is mobile across regions and sectors (see figure D7).

Consumers have a constant marginal propensity to save, meaning they allocate a fixed portion of their income to saving (see figure D3). Investment generated through current-year savings is immediately transformed into new capital, driving an equalization of the rate of return across regions. The real value of savings is distributed as investment according to the distribution of investment expenditures across regions and sectors in the base-year SAM, but it is spent at a price equal to the regional price of investment goods, weighted by the regional distribution of new capital formation. This modification allows the model to capture changes in regional investment goods prices while maintaining the regional distribution of savings. This composite savings price is achieved through the use of four constraints.

Equation D1 sets the return on new capital, such that the returns paid to owners of stock in the national pool of new capital reflect changes in the price of investment goods in the regions and sectors in which new capital is being applied. The quantity $\Sigma_s[PA_{r,s}\theta_{r,s}^{RSAV}]$ is the regional average Armington good price weighted by the share of each good in the investment good bundle, determined by 2011 shares. Because all prices are indexed to 1 in 2011, this sets the effective return, $\hat{R}^K$, to be directly proportional to the rate of return in each region and inversely proportional to the change in the weighted average goods price in the regions and sectors where investments are made.

Equation D2 provides a similar function for the savings market, such that the price of savings *PS* is proportional to the regional average cost of investment goods weighted by the goods used in investment, $\Sigma_s[PA_{r,s}\theta_{r,s}^{RSAV}]$, averaged over all regions weighted by the regional distribution of investment, I_r.

Equation D3 assigns the value of Θ_r^S, an accounting variable that measures a region's share of national savings in a given year.

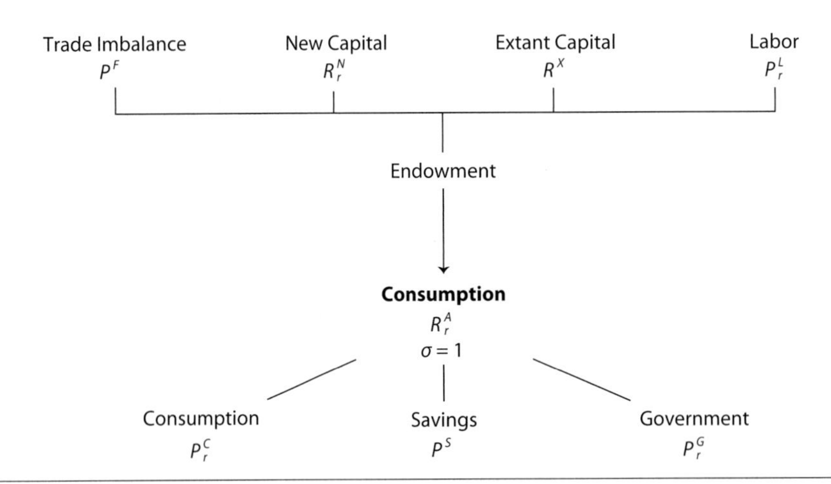

FIGURE D3. Regional demand is split between household consumption, savings, and government services. Endowments are the sum of labor and capital earnings (which are adjusted to include tax earnings), and regional borrowing. Note the variable sign of the trade imbalance quantities; negative endowments represent a forced expenditure rather than income.

Finally, equation D4 makes a bookkeeping adjustment that balances the effects on regional GDP of having national savings and capital prices but local and sectoral differences in goods prices for investment or capital outlays, ensuring a closed economic system.

Capital earnings from new capital formation are returned to the owner of that capital through the savings price constraints and savings goods. This allows RHG-MUSE to track regional capital ownership separately from capital use and to allow changes in investment good prices in the region of capital use to affect returns in the region of ownership and to allocate returns accordingly. Earnings from extant capital are also distributed to the owner, but through an exchange market, such that ownership and use rates are differentiated by region but rates of return equalize.

4.3.4. TRADE

Commodities produced by the nine sectors are modeled as nested Armington goods, in which local, domestic, and international goods are assumed to be imperfect substitutes for one another (Armington, 1969). Each region's international trade deficit is assumed to be fixed at base year (2011) nominal levels. All regional consumption of goods is a composite of local production, domestic trade from other regions, and foreign goods (see figure D4) and all regional production is output to a mix of local, domestic, and international markets (see figures D1 and D2).

Net foreign borrowing is fixed nominally in RHG-MUSE, meaning that as the economy grows, the deficit will shrink as a share of GDP. This is consistent with the representation of national deficits in many recursive-dynamic CGE models and satisfies the theoretical principle that a large deficit is unsustainable in the long run. However, the United States, in addition to many other countries, has maintained a deficit representing a relatively stable share of GDP over multiple decades. This would indicate that modeling net foreign borrowing as a constant share of GDP might be more realistic. In the version of RHG-MUSE used in the *American Climate Prospectus,* we chose to use the former; that is, to fix regional borrowing at 2011 nominal levels regardless of GDP changes for two reasons. First, RHG-MUSE is not forward-looking, and currently there is no penalty for increased borrowing; therefore, representing foreign borrowing as a fixed share of GDP enables an unrealistic positive feedback to borrowing. Second, when the foreign borrowing rate is fixed to GDP, damage to GDP from climate change (or elsewhere) is magnified through reduced borrowing. This may have a realistic interpretation, as global damage from climate change could tighten international lending markets, leading to a reduction in U.S. borrowing from the counterfactual "no-impact" baseline. This effect is, however, highly uncertain, and we have chosen to model fixed nominal borrowing as a more conservative estimate of macroeconomic damage. We encourage further research in this area so that future work may attempt to quantify the impact of climate damage on currency and lending markets.

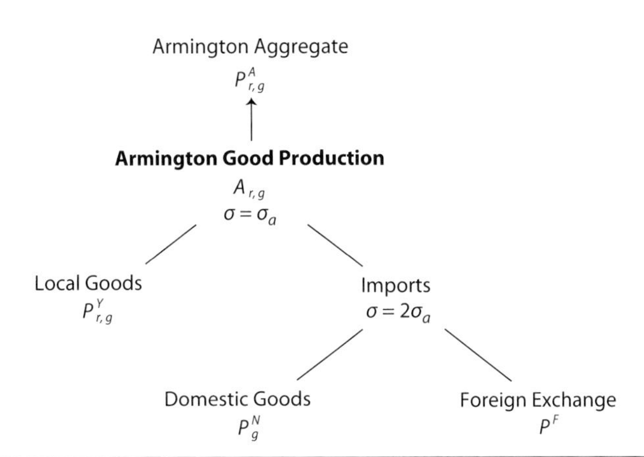

FIGURE D4. Armington trade assumes that imports are imperfect substitutes for local goods, and vice versa. Similar to the transformation nests that allow producers to substitute local sales for exports, the Armington aggregate nest allows consumers of goods (all of which consume the Armington good $p^A_{r,s}$) to substitute between local and imported goods. Because these are treated as imperfect substitutes, a loss of productivity or utility will accrue to the producer or consumer, respectively, that substitutes away from their 2011 mix.

4.3.5. YEARLY STATIC CORE OPTIMIZATION

Every year, the *static core's* calibration data are updated using the *updating equations* (described in section 4.4) and is then resolved for the state of the economy and any impacts occurring in that year. This recursive-dynamic framework mimics an agent behaving optimally for current conditions but not preparing for events occurring in future model periods directly. Forward-looking behavior is simulated, to the extent that it exists today, in the preferences expressed in the current IMPLAN data. For example, while savings would be an irrational behavior in a purely myopic world, agents in RHG-MUSE do exhibit a constant marginal propensity to save, as current regional savings rates are preserved throughout the model horizon.

While the *updating equations* are a set of assignment statements using data from one year to determine values in the next, the *static core* is an optimization problem in which each region's representative agent maximizes its own utility subject to a budget of endowed goods (in RHG-MUSE, this budget stems from labor earnings, returns on owned capital, and foreign borrowing). Mathematically, the single-year optimization problem takes the form of a "mixed complementarity-problem" (MCP), which is defined by a set of inequalities that, when solved, provide by definition the solution to the set of optimization problems for all regions simultaneously. The preferences and technologies in the RHG-MUSE *static core*

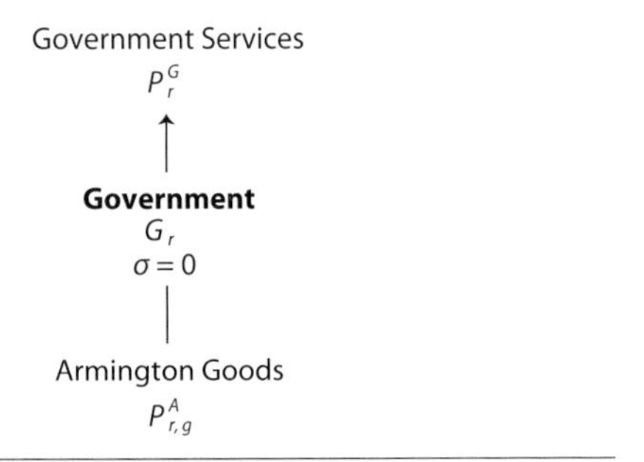

FIGURE D5. Government is modeled as a single regional producer that converts Armington aggregate goods into government consumption estimates. The public services produced are a private good making up a fixed share of GDP.

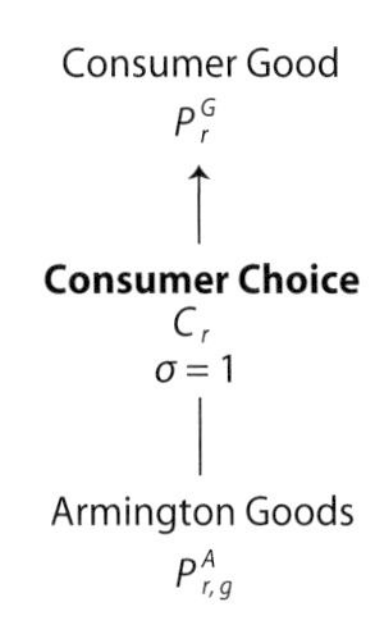

FIGURE D6. Real consumption is fixed as a share of regional GDP. Consumers have Cobb-Douglass utility functions, such that the share of consumption expenditures on any good is fixed to the 2011 consumption expenditure share.

Regional Extant Capital

$R^X_{r,s}$

↑

Extant Capital Market

K

$\sigma = 0$

|

Capital Exchange Volume

R^E

$$\hat{R}^K \sum_{k \in s} \left[P^A_{i,k} \theta^{RS}_{i,k} \right] = R^N_i \qquad \forall (i \in r) \qquad \text{(D1)}$$

$$\sum_{i \in r} \left[\sum_{k \in s} \left[P^A_{i,k} \theta^{RS}_{i,k} \right] I_i \right] = P^S \sum_{i \in r} [I_i] \qquad \text{(D2)}$$

$$\Theta^S_i * \sum_{j \in r} \left[\mu^S_i * RA_i \right] = \mu^S_i * RA_i \qquad \forall (i \in r) \qquad \text{(D3)}$$

$$\hat{I}_i P^S = r^0 \left[\Theta^S_i \sum_{j \in r} \left[R^N_j I_j \right] - R^N_i I_i \right] + I_i \sum_{k \in s} \left[P^A_{i,k} \theta^{RS}_{i,k} - P^S \right] \qquad \forall (i \in r) \qquad \text{(D4)}$$

FIGURE D7. Savings, investment, and capital flows are governed by the behavior of a number of agents as well as by four constraints. Savings is a fixed share of regional GDP, and the money saved is distributed across regions as investment, which subsequently becomes new capital, such that the rate of return is equalized across regions. Existing capital held by regional agents is used in fixed quantities by region and sector. Allocations are determined by the usage of new capital in the year of investment, with a fixed proportion of capital depreciating each year, and returns are generated through a homogeneous stock market.

are expressed through nested CES utility/consumption functions and CET production functions, which are related by three fundamental equilibrium conditions:

- Zero profit condition
  Producers must have zero economic profits after all payments to intermediate producers, owners of factors such as capital and labor, outlays for investment and savings, and any other expenses. In effect, this constraint means that producers cannot waste revenue.
- Market clearance condition
  The total endowment by consumers plus the total production by producers of all goods and services must equal the total consumption of each good by both consumers and producers. In effect, this constraint means that all goods must have an equal number of sources and sinks.
- Consumer budget condition
  Consumers may not spend more than that with which they are endowed. Endowments may include labor income, capital earnings, and borrowing, and expenditures include private and government consumption and savings. In effect, this constraint means that consumers cannot spend money to which they do not have access.

In RHG-MUSE, we make no theoretical exceptions to the above statements in the form of imperfect competition or intertemporal constraints. We do, however, allow some markets to have demand fall short of supply. In this case, the price for the associated good will be zero, and the market will still clear.

It is important to note that this is not a single-agent model. Single-agent models are frequently employed in energy and climate policy analysis in order to determine a utility-maximizing policy for the entire country or planet. Such models have a single objective function that optimizes utility for the entire system. For example, the Nordhaus (1994) models DICE and RICE use population-weighted discounted utility of regional per capita consumption. Measures are usually taken to ensure that inequality is not too greatly exacerbated (the DICE/RICE maximand uses a diminishing marginal utility of consumption), but in principle increases in inequality could be found to be "optimal" if the net global utility payoff were positive, regardless of the distributional consequences.

Instead, the optimization carried out in the RHG-MUSE *static core* is structured such that each of the regional consumer and producer agents are *individually optimizing*. This is a property of the MCP equilibrium uses by the MPSGE language — the solution to the set of conditions detailed in the model by definition maximizes the utility of each agent described in the nests in sections 4.3.1 to 4.3.4. Therefore, any increases in inequality observed in the model outputs may only have come from changes in the calibrating data (e.g., population, capital ownership rates), from economic forces beyond the agent's control (prices), or from climate impacts, and the optimal solution each year will be at least as good for each agent, known as a Pareto improvement, relative to the initial conditions before the optimization took place.

Of course, while RHG-MUSE in its current form does enable the examination of changes in interregional inequality, it cannot be used to study changes in intraregional inequality, such as differential impacts on various income or age groups. This topic is explored conceptually in chapter 15, and further study in this area would be useful in elucidating the distributional consequences that impacts from climate change may have.

4.4. RHG-MUSE dynamics

The static core of RHG-MUSE is based on 2011 data and could be run as a single-year model. If climate impacts were to be applied directly to the static core, this would enable a "comparative statics" study, in which the economy of 2011 is tested in a counterfactual climate setting. The dynamic *updating equations* presented in this section modify the base data each year as a function of the outputs of the previous year's optimization as well as a small number of external inputs. This structure defines a no-impact "baseline," to which climate-change scenarios are compared. Note that the baseline scenario is not truly without the influence of climate, but simply with the same influence that the climate had on the economy in 2011.

4.4.1. POPULATION GROWTH AND THE LABOR SUPPLY

The population in each region is assumed to grow at the United States average growth rate as projected by the United Nations (Raftery and Heilig 2012, UN Population Division 2012). Age cohorts maintain their share of the state population, and the change in the labor supply is equal to the change in the number of people between the ages of 15 and 64, inclusive. Migration of labor between regions is not allowed. This population model was used not because of its likelihood but because it preserves the regional and sectoral balance used to calibrate the model, enabling a faithful comparison to the economy of today. In reality, the population will likely shift toward urban centers and coastal areas. Migration will likely also play a role in the way that Americans respond to climate change, but the empirical work quantifying these changes was deemed not yet sufficient to be relied on by this report; furthermore, costs associated with large-scale migration are even more difficult to quantify. It is unclear whether the adaptive benefit from increased domestic and international migration would be larger than the increased costs.

4.4.2. CAPITAL STOCK MODEL

Capital stock changes are driven by a vintaged capital growth model. In 2011, the earnings from extant capital equal the IMPLAN capital earnings times the initial extant capitalshare, θ^X. In subsequent years, the endowment of earnings from the extant capital exchange (in real terms), k_r^{EX}, equals the value of investment in that year, plus the remaining earnings from the year before, depreciated by the annual depreciation rate δ. Each year, y, the existing capital stock and new capital stock use in each sector and region is depreciated and becomes the extant capital stock for that sector and region for the next year. Each region's share of national earnings from new capital stock is returned to the regional agents according to their share of national investment, Θ_r^S:

$$k_{i,t}^{EX} = \begin{cases} k_i^{E0}\theta^X & t = 2011 \\ (1-\delta)(k_{i,t}^{EX} + \Theta_i^S \sum_{j \in r}\left[\sum_{k \in s}\left[U_{j,k}^{KN}\right]\right] & t > 2011 \end{cases} \quad \forall i \in r, t \in y \tag{D5}$$

Each year, after the *static core* optimization, capital endowments are updated to equal the total previous year's capital stock, depreciated by the annual depreciation rate δ.

Capital and labor cannot be substituted for one another in the extant production block (see section 4.3.1), but the ratio of their use in that block is updated over time. In 2011, the ratio of capital to labor in each region and sector is determined by the 2011 IMPLAN data. In subsequent periods, the optimal capital to labor ratio used in the new production block updates this ratio in proportion to the relative size of the new and extant capital stocks:

$$l_{i,k,t}^{X} = \frac{l_{i,k,t-1}^{X} * Y_{i,k,t-1}^{X} + \gamma^{L} U_{r,s}^{LN}}{Y_{r,s}^{X} + Y_{r,s}^{N}} \qquad \forall i \in r,\ k \in s,\ t \in y \qquad \text{(D6)}$$

$$k_{i,k,t}^{X} = \frac{k_{i,k,t-1}^{X} * Y_{i,k,t-1}^{X} + U_{r,s}^{KN}}{Y_{r,s}^{X} + Y_{r,s}^{N}} \qquad \forall i \in r,\ k \in s,\ t \in y \qquad \text{(D7)}$$

4.4.3. TECHNOLOGICAL CHANGE AND ECONOMIC ADAPTATION

Productivity changes through annual increases in labor productivity:

$$\gamma_{t}^{L} = \begin{cases} 1 & y = 2011 \\ \left(1+r^{L}\right)\gamma_{t-1}^{L} & y > 2011 \end{cases} \qquad \forall t \in y \qquad \text{(D8)}$$

Changes in technology are applied to the production block as a decrease in the labor required to produce a given amount of output. To account for the rebound effect, the inverse of the productivity change is applied to the initial observed labor price for the new producer (reference prices have no effect for Leontief producers in MPSGE). Both of these then influence the production possibility curve, and thus the optimal behavior, of the producer in the optimization (see figure D8).

4.5. Integrating Climate Impacts in RHG-MUSE

Of the impacts quantified in the macroeconomic model, agriculture, labor, and energy costs and benefits as well as coastal storm–related business interruption are applied directly to the *static core*. Mortality and sea-level–related coastal damages affect the stocks of labor and capital, respectively, and are applied in the *updating equations*.

4.5.1. AGRICULTURE

Changes in agricultural productivity are implemented through changes in output productivity of the agriculture sector. Specifically, the reference quantity produced is changed by the agricultural productivity impact I_{r}^{A} times the regional share of the agriculture sector made up by maize, wheat, oilseeds, and cotton, θ_{r}^{A}, and the rebound effect is accounted for through inverse changes in the observed producer price (see figure D9).

4.5.2. LABOR

Labor productivity impacts are implemented as temporary reductions in labor productivity. The sectors affected by high- and low-risk labor impacts are given in table D3. The "risk" of an industry corresponds to the portion of the industry's labor that is exposed to outdoor temperatures. The econometric impact functions described in appendix B have the same two-tiered structure, developed using data from comparable sector structures. The mechanism for affecting labor productivity

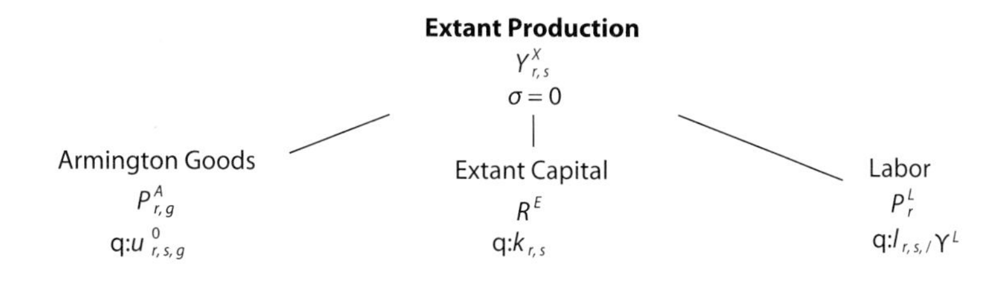

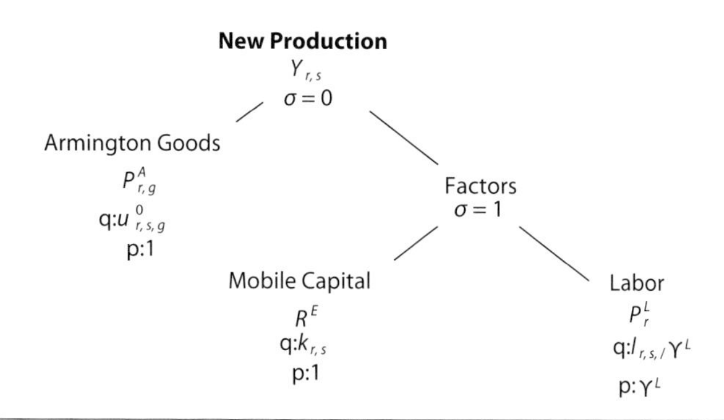

FIGURE D8. The labor-capital share used in new production each year updates extant production in the next year according to the relative size of each sector. This mechanism captures both exogenous technology change (labor productivity growth) and endogenous economic adaptation (labor-capital rebalancing). q: values signify the index used to scale the variable's quantity; p: values give the reference price used to scale the price observed by that producer.

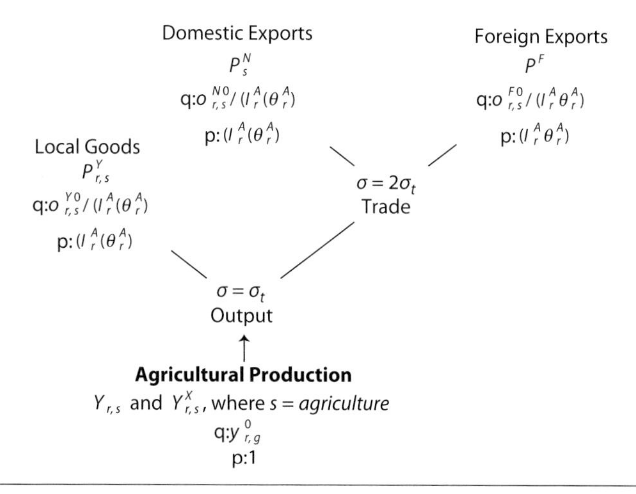

FIGURE D9. Agricultural productivity impacts affect both new and extant production identically. Sector output may be sold in local, domestic, and foreign exchange markets, and the productivity of this transformation nest is moderated by the agricultural productivity factor I^A_r times the share of total agriculture made up by affected sectors, θ^A_r. The rebound effect is accounted for by inverse changes in price. q: values signify the index used to scale the variable's quantity; p: values give the reference price used to scale the price observed by that producer.

TABLE D3 High- and low-risk labor productivity impact sectors

High Risk	Low Risk
Agriculture	Indoor services
Transportation	Real estate
Outdoor services	
Infrastructure	
Manufacturing	
Mining	
Energy	

Labor productivity impacts affect only the efficiency of labor use by these sectors, so more capital-intensive industries, such as agriculture and real estate, are less vulnerable than labor-intensive industries, such as indoor and outdoor services.

is identical to that used in baseline labor productivity growth and accounts for the rebound effect in the same way (see section 4.4.3), such that the final labor productivity adjusted for climate impacts, $\gamma^{L'}_{r,s}$ is as

$$\gamma^{L'}_{i,k} = \gamma^{L} I^{L}_{i,k} \qquad \forall i \in r, k \in s \tag{D9}$$

where the impact $I^{L}_{r,s}$ is the climate impact for the risk category corresponding to that sector.

4.5.3. ENERGY

Changes in energy demand are effectively a change in the ability to make profitable use of energy (such as for heat in industrial processes or lighting in commercial buildings) or to derive utility from energy (such as in home heating). Consumers and businesses may substitute other goods for energy, but at some cost. As a result, in RHG-MUSE, changes in energy expenditures are implemented similarly to changes in labor productivity — they affect the ability of producers to create output or consumers to derive utility from a given amount of energy goods. Because demands for intermediate goods for producers are Leontief (CES functions with an elasticity of 0), the index price is irrelevant in MPSGE; however, the rebound in consumer purchases is accounted for by an inverse change in prices (figure D10).

4.5.4. COASTAL IMPACTS

The RMS North Atlantic Hurricane Model (see appendix C) simulates damage that would occur due to sea-level rise and the hurricane patterns for each year and scenario given current property values. It is not a multiyear simulation. To avoid over-counting damages, we make two assumptions: first, in any given scenario, inundation damage is incremental, determined by the difference between projected inundation damage in the current year and the maximum previous inundation:

$$I^{In'}_{i,t} = \begin{cases} 0 & t = 2011 \\ \max\left\{I^{In}_{i,t} - \max\left\{I^{In}_{i,2011},\dots,I^{In}_{i,t-1}\right\},0\right\} & t > 2011 \end{cases} \qquad \forall t \in y \tag{D10}$$

Second, we assume that property and business activity already inundated can no longer be damaged by coastal storms, effectively assuming that reinvestment in the region takes place away from the coast. Furthermore, we do not permanently

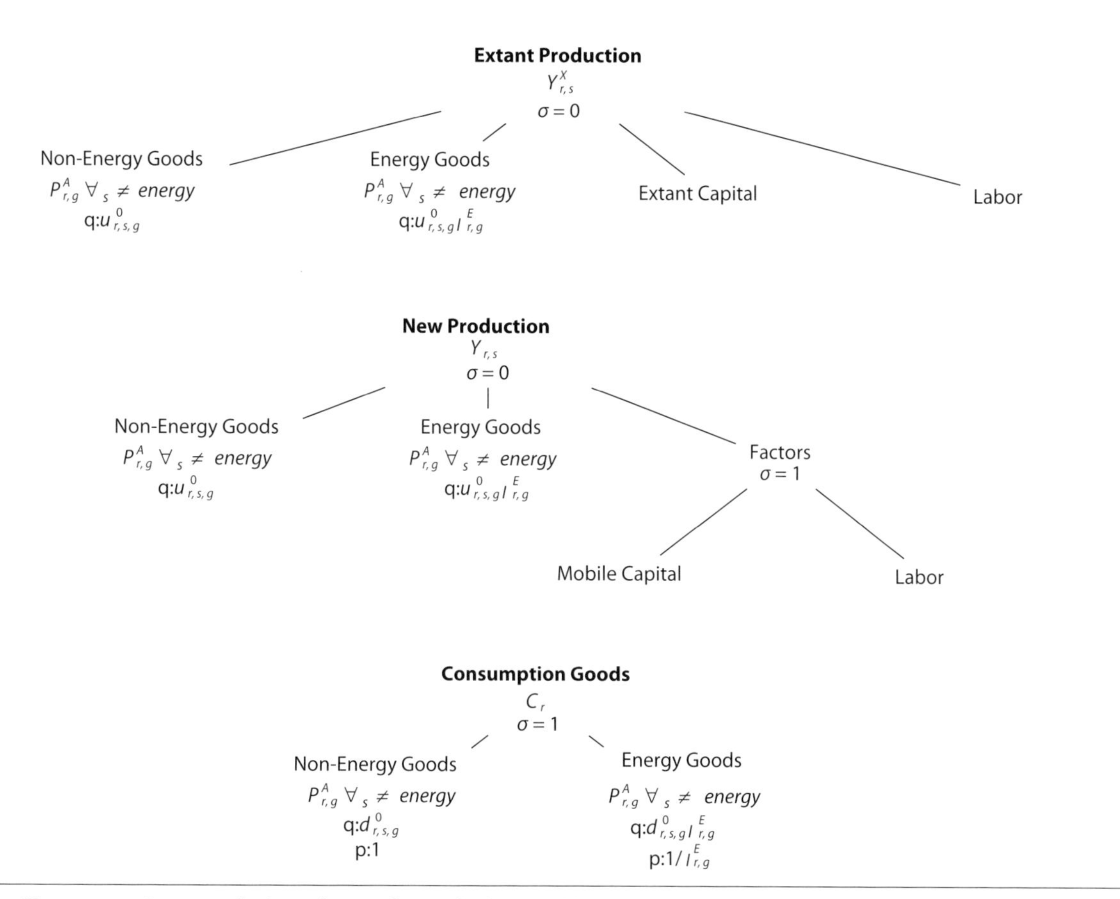

FIGURE D10. Energy. q: values signify the index used to scale the variable's quantity; p: values give the reference price used to scale the price observed by that producer.

reduce the productivity of reinvestments. This is conservative: historically, large shares of damaged property are rebuilt in areas still exposed to coastal storm damage; additionally, many businesses and investments rely on proximity to coastlines and would incur costs or lose value if moved inland. We accomplish this reduction in storm damage, $I^{St'}_{r,y}$, and business interruption, $I^{BI'}_{r,y}$, by decreasing the value of exposed property, e^{Tot}_r, by the share of total state property that has been inundated:

$$I^{St'}_{i,t} = \left[1 - \frac{I^{In'}_{i,t}}{e^{Tot}_i}\right] I^{St}_{i,t} \qquad \forall i \in r,\ t \in y \qquad \text{(D11)}$$

$$I^{BI'}_{i,t} = \left[1 - \frac{I^{In'}_{i,t}}{e^{Tot}_i}\right] I^{BI}_{i,t} \qquad \forall i \in r,\ t \in y \qquad \text{(D12)}$$

Unlike agriculture, labor, and energy impacts, which solely affect the *static core* of RHG-MUSE, coastal impacts have an effect both on the *static core* and on the *updating equations*. Property lost due to local sea-level (LSL) rise–driven inundation and damage due to tropical storms and nor'easters affect the capital stock directly, while business interruption during and immediately after storms affects output productivity.

The mechanism for damaging output is identical to that used in agricultural impacts, with the exception that business interruption affects all sectors equally (see figure D9). When a portion of the business activity is reduced in a given year and region, the total output productivity for all sectors in that region is reduced, regardless of the destination market. This will change the balance of trade in the region, as local consumers demand more imported goods to replace the lost local output; similar compensating changes will also occur in other regions as they substitute away from goods imported from the damaged region. However, this effect will be offset somewhat by the rise in operating expenses for local firms and the resulting "rebound," which will drive a reduction in demand.

Capital damage from inundation and storms is represented as a premature depreciation of extant capital (see equation D13), which reduces capital endowments by regional consumers, according to their share of ownership of capital stock, in the year in which the damage occurs (see figure D11).

$$k_{i,t}^{EX} = \begin{cases} k_i^{E0}\theta^X & t = 2011 \\ (1-\delta)(k_{i,t-1}^{EX} + \Theta_i^S \sum_{j\in r}\left[\sum_{k\in s}\left[U_{j,k}^{KN}\right]\right] & t > 2011 \end{cases} \qquad \forall i \in r, t \in y \tag{D13}$$

4.5.5. MORTALITY

Unlike the direct cost and benefit calculations described in section 2.3, which apply changes in the mortality rate to the current population, RHG-MUSE uses a population model to track mortality through time. The baseline projection is described in section 4.4.1. Changes in mortality (in persons, by age) accumulate over time, age every year, and are subtracted from the base population each year. The change in the labor force is calculated as the change in the age 15-to-64 population.

$$M_{i,q,t} = \begin{cases} 0 & t = 2011 \\ M_{i,q-1,t} + \left(p_{i,q,t} - M_{i,q-1,t}\right) * (I_{i,q,t}^M) & t > 2011 \end{cases} \qquad \forall i \in r, q \in a, t \in y \tag{D14}$$

$$l_{i,t}^M = l_{i,t}^0 \left[\frac{\sum_{q\in a} p_{i,q,t} - M_{i,q,t}}{\sum_{q\in a} p_{i,q,t}} \right] \tag{D15}$$

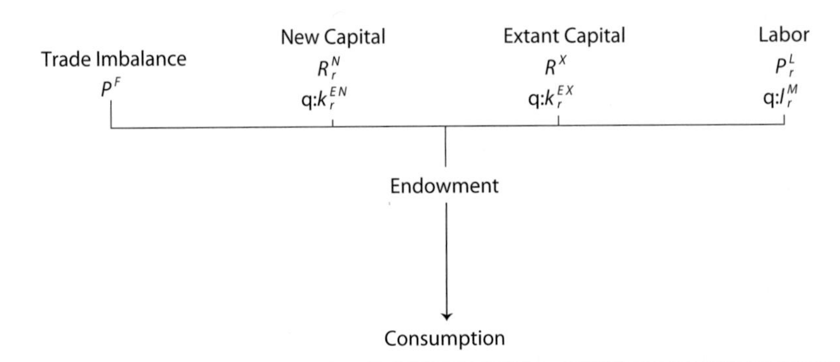

FIGURE D11. Consumption. q: values signify the index used to scale the variable's quantity; p: values give the reference price used to scale the price observed by that producer.

REFERENCES

Abler, D., Fisher-Vanden, K., McDill, M., Ready, R., Shortle, J., Wing, I. S., Wilson, T., 2009. *Economic Impacts of Projected Climate Change in Pennsylvania: Report to the Department of Environmental Protection.* Environment & Natural Resources Institute.

Armington, P. S., 1969. A theory of demand for products distinguished by place of production. Tech. Rep. 1, International Monetary Fund. URL http://www.jstor.org/stable/3866403

Arrow, K., Debreu, G., 1954. Existence of an equilibrium for a competitive economy. *Econometrica* 22, 265–29.

Backus, G., Lowry, T., Warren, D., Ehlen, M., Klise, G., Malczynski, L., Reinert, R., Stamber, K., Tidwell, V., Zagonel, A., 2010. Assessing the near-term risk of climate uncertainty: Interdependencies among U.S. states.

Barreca, A., Clay, K., Deschênes, O., Greenstone, M., Shapiro, J. S., 2013. *Adapting to climate change: The remarkable decline in the US temperature-mortality relationship over the 20th century.* Tech. Rep. NBER Working Paper No. 18692, National Bureau of Economic Research. URL http://www.nber.org/papers/w18692

Böhringer, C. T. R., Wiegard, W., 2003. Computable general equilibrium analysis: Opening a black box. Tech. Rep. ZEW Discussion Paper No. 03-56, Zentrum für Europäische Wirtschaftsforschung GmbH. URL ftp://ftp.zew.de/pub/zew-docs/dp/dp0356.pdf

Cass, D., 1965. Optimum growth in an aggregative model of capital accumulation. *Review of Economic Studies* 32, 233–40.

Deschênes, O., Greenstone, M., 2011. Climate change, mortality, and adaptation: Evidence from annual fluctuations in weather in the US. *American Economic Journal: Applied Economics* 3, 152–85.

Dirkse, S. P., Ferris, M. C., 1995. The path solver: A non–monotone stabilization scheme for mixed complementarity problems. *Optimization Methods and Software* 5, 123–56.

Heaton, P., 2010. Hidden in plain sight: What cost-of-crime research can tell us about investing in police. Tech. Rep., RAND. URL http://www.rand.org/content/dam/rand/pubs/occasional_papers/2010/RAND_OP279.pdf

IMPLAN Group, 2011. 51 states totals package. URL https://implan.com/

Jorgenson, D. W., Goettle, R. J., Hurd, B. H., Smith, J. B., 2004. *U.S. Market Consequences of Global Climate Change.* Pew Center.

Koopmans, T. C., 1965. *On the Concept of Optimal Economic Growth.* Rand McNally, Chicago, pp. 225–287.

Markusen, J., Rutherford, T., 2004. *MPSGE: A user's guide.*

Nordhaus, W. D., 1994. *Managing the Global Commons: The Economics of Climate Change.* The MIT Press.

Raftery, A.E., Li, N., Ševčíková, H., Gerland, P., Heilig., G., 2012. Bayesian probabilistic population projections for all countries. *Proceedings of the National Academy of Sciences USA* 109, 13915–21, 10.1073/pnas.1211452109.

Ramsey, F. P., 1928. A mathematical theory of saving. *Economic Journal* 38, 543–59.

Rausch, S. T. F. R., 2008. Tools for building national economic models using state-level IMPLAN social accounts. URL http://www.mpsge.org/IMPLAN2006inGAMS/IMPLAN2006inGAMS.pdf

Rutherford, T., 1987. A modeling system for applied general equilibrium analysis. Tech. Rep. Cowles Foundation Discussion Paper 836, Cowles Foundation for Research in Economics at Yale University. URL http://dido.econ.yale.edu/P/cd/d08a/d0836.pdf

Rutherford, T., 2004. Tools for building national economic models using state-level IMPLAN social accounts. URL http://www.mpsge.org/implan98.htm

Tol, R. S. J., 2008. The economic impact of climate change. URL http://www.econstor.eu/bitstream/10419/50039/1/584378270.pdf

UN Population Division, 2012. World population prospects: The 2012 revision. http://esa.un.org/unpd/wpp/Excel-Data/EXCEL_FILES/1_Population/WPP2012_POP_F01_1_TOTAL_POPULATION_BOTH_SEXES.XLS.

U.S. Bureau of Economic Analysis, 2014a. GDP by industry / VA, GO, II, EMP (1997–2013, 69 industries). URL http://www.bea.gov/industry/xls/GDPbyInd_VA_NAICS_1997-2013.xlsx

U.S. Bureau of Economic Analysis, 2014b. State GDP for all industries and regions, 2008–2013. URL http://www.bea.gov/iTable/iTableHtml.cfm?reqid=70&step=10&isuri=1&7003=200&7035=-1&7004=NAICS&7005=1%2C2%2C3&7006=XX&7036=-1&7001=1200&7002=1&7090=70&7007=2012%2C2011&7093=Levels#.U-VlmZPLS64.email

U.S. Energy Information Administration, 2009. The national energy modeling system: An overview. Tech. Rep. DOE/EIA0581(2009). URL http://www.eia.gov/oiaf/aeo/overview/index.html

U.S. Energy Information Administration, 2013. Annual energy outlook 2013. U.S. DOE/EIA. URL http://www.eia.gov/oiaf/aeo/overview/index.html

U.S. Federal Bureau of Investigation, 2012. Crime in the United States. URL http://m.fbi.gov/#http://www.fbi.gov/about-us/cjis/ucr/ucr-publications#Crime

Yang, Z., Eckaus, R. S., Ellerman, A. D., Jacoby, H. D., 1996. The MIT emissions prediction and policy analysis (EPPA) model. Tech. Rep. 6, MIT Joint Program on the Science and Policy of Global Change.

5. SUPPLEMENTAL TABLES

TABLE D4 Data used in extrapolating from "physical impact" bins to single years of age used in the CGE model

Physical Impact	Age Cohort	Cohort Deaths	Cohort Population
0 to 1	0 to 1	332,697	47,830,261
1 to 44	1 to 4	57,529	189,286,073
	5 to 9	34,262	239,385,956
	10 to 14	44,258	250,939,740
	15 to 19	157,564	255,076,636
	20 to 24	231,954	246,218,551
	25 to 34	501,208	477,763,701
	35 to 44	1,005,193	522,168,558
45 to 64	45 to 54	2,118,807	500,140,370
	55 to 64	3,258,625	358,921,022
65+	65 to 74	4,948,370	231,986,715
	75 to 84	8,109,495	154,248,308
	85+	8,372,695	56,742,313

TABLE D5 Sectoral aggregation scheme used in the CGE model

Sector	*IMPLAN Code*	*IMPLAN Name*
Agriculture	1	Oilseed farming
	2	Grain farming
	8	Cotton farming
	3	Vegetable and melon farming
	4	Fruit farming
	5	Tree nut farming
	6	Greenhouse nursery and floriculture production
	7	Tobacco farming
	9	Sugarcane and sugar beet farming
	10	All other crop farming (except algae seaweed and other plant aquaculture)
	15	Forest nurseries forest products and timber tracts
	16	Logging
	11	Cattle ranching and farming
	12	Dairy cattle and milk production
	13	Poultry and egg production
	14	Animal production except cattle and poultry and eggs (algae seaweed and other plant aquaculture)
	17	Fishing
	18	Hunting and trapping
	19	Support activities for agriculture and forestry

Sector	*IMPLAN Code*	*IMPLAN Name*
Energy	32	Natural gas distribution
	115	Petroleum refineries
	119	All other petroleum and coal products manufacturing
	31	Electric power generation transmission and distribution
	428	Federal electric utilities
	431	State and local government electric utilities
Infrastructure	33	Water sewage and other systems
	390	Waste management and remediation services
	351	Telecommunications (broadband ISP; telephone ISP)
Transportation	332	Air transportation
	333	Rail transportation
	334	Water transportation
	335	Truck transportation
	336	Transit and ground passenger transportation
	337	Pipeline transportation
	338	Scenic and sightseeing transportation and support activities for transportation
	430	State and local government passenger transit
Mining	20	Oil and gas extraction
	28	Drilling oil and gas wells
	29	Support activities for oil and gas operations
	21	Coal mining
	22	Iron ore mining
	23	Copper nickel lead and zinc mining
	24	Gold silver and other metal ore mining
	25	Stone mining and quarrying
	26	Sand gravel clay and ceramic and refractory minerals mining and quarrying
	27	Other nonmetallic mineral mining and quarrying
	30	Support activities for other mining
Manufacturing	41	Dog and cat food manufacturing
	42	Other animal food manufacturing
	43	Flour milling and malt manufacturing
	44	Wet corn milling
	45	Soybean and other oilseed processing
	46	Fats and oils refining and blending
	47	Breakfast cereal manufacturing
	48	Sugarcane mills and refining
	49	Beet sugar manufacturing
	50	Chocolate and confectionery manufacturing from cacao beans
	51	Confectionery manufacturing from purchased chocolate
	52	Nonchocolate confectionery manufacturing

(*continued*)

TABLE D5 Sectoral aggregation scheme used in the CGE model (*continued*)

Sector	*IMPLAN Code*	*IMPLAN Name*
	53	Frozen food manufacturing
	54	Fruit and vegetable canning pickling and drying
	55	Fluid milk and butter manufacturing
	56	Cheese manufacturing
	57	Dry condensed and evaporated dairy product manufacturing
	58	Ice cream and frozen dessert manufacturing
	59	Animal (except poultry) slaughtering rendering and processing
	60	Poultry processing
	61	Seafood product preparation and packaging
	62	Bread and bakery product manufacturing
	63	Cookie cracker and pasta manufacturing
	64	Tortilla manufacturing
	65	Snack food manufacturing
	66	Coffee and tea manufacturing
	67	Flavoring syrup and concentrate manufacturing
	68	Seasoning and dressing manufacturing
	69	All other food manufacturing
	70	Soft drink and ice manufacturing
	71	Breweries
	72	Wineries
	73	Distilleries
	74	Tobacco product manufacturing
	75	Fiber yarn and thread mills
	76	Broadwoven fabric mills
	77	Narrow fabric mills and schiffli machine embroidery
	78	Nonwoven fabric mills
	79	Knit fabric mills
	80	Textile and fabric finishing mills
	81	Fabric coating mills
	82	Carpet and rug mills
	83	Curtain and linen mills
	84	Textile bag and canvas mills
	85	All other textile product mills (embroidery contractors)
	86	Apparel knitting mills
	87	Cut and sew apparel contractors (exc. embroidery contractors)
	88	Men's and boy's cut and sew apparel manufacturing
	89	Women's and girl's cut and sew apparel manufacturing
	90	Other cut and sew apparel manufacturing
	91	Apparel accessories and other apparel manufacturing
	92	Leather and hide tanning and finishing

Sector	IMPLAN Code	IMPLAN Name
	93	Footwear manufacturing
	94	Other leather and allied product manufacturing
	95	Sawmills and wood preservation
	96	Veneer and plywood manufacturing
	97	Engineered wood member and truss manufacturing
	98	Reconstituted wood product manufacturing
	99	Wood windows and doors and millwork
	100	Wood container and pallet manufacturing
	101	Manufactured home (mobile home) manufacturing
	102	Prefabricated wood building manufacturing
	103	All other miscellaneous wood product manufacturing
	104	Pulp mills
	105	Paper mills
	106	Paperboard mills
	107	Paperboard container manufacturing
	108	Coated and laminated paper packaging paper and plastics film manufacturing
	109	All other paper bag and coated and treated paper manufacturing
	110	Stationery product manufacturing
	111	Sanitary paper product manufacturing
	112	All other converted paper product manufacturing
	113	Printing
	114	Support activities for printing
	116	Asphalt paving mixture and block manufacturing
	117	Asphalt shingle and coating materials manufacturing
	118	Petroleum lubricating oil and grease manufacturing
	120	Petrochemical manufacturing
	121	Industrial gas manufacturing
	122	Synthetic dye and pigment manufacturing
	123	Alkalies and chlorine manufacturing
	124	Carbon black manufacturing
	125	All other basic inorganic chemical manufacturing
	126	Other basic organic chemical manufacturing
	127	Plastics material and resin manufacturing
	128	Synthetic rubber manufacturing
	129	Artificial and synthetic fibers and filaments manufacturing
	130	Fertilizer manufacturing
	131	Pesticide and other agricultural chemical manufacturing
	132	Medicinal and botanical manufacturing
	133	Pharmaceutical preparation manufacturing

(*continued*)

TABLE D5 Sectoral aggregation scheme used in the CGE model (*continued*)

Sector	*IMPLAN Code*	*IMPLAN Name*
	134	In-vitro diagnostic substance manufacturing
	135	Biological product (except diagnostic) manufacturing
	136	Paint and coating manufacturing
	137	Adhesive manufacturing
	138	Soap and cleaning compound manufacturing
	139	Toilet preparation manufacturing
	140	Printing ink manufacturing
	141	All other chemical product and preparation manufacturing
	142	Plastics packaging materials and unlaminated film and sheet manufacturing
	143	Unlaminated plastics profile shape manufacturing
	144	Plastics pipe and pipe fitting manufacturing
	145	Laminated plastics plate sheet (except packaging) and shape manufacturing
	146	Polystyrene foam product manufacturing
	147	Urethane and other foam product (except polystyrene) manufacturing
	148	Plastics bottle manufacturing
	149	Other plastics product manufacturing (exc. inflatable plastic boats)
	150	Tire manufacturing
	151	Rubber and plastics hoses and belting manufacturing
	152	Other rubber product manufacturing (exc. inflatable rubber boats)
	153	Pottery ceramics and plumbing fixture manufacturing
	154	Brick tile and other structural clay product manufacturing
	155	Clay and nonclay refractory manufacturing
	156	Flat glass manufacturing
	157	Other pressed and blown glass and glassware manufacturing
	158	Glass container manufacturing
	159	Glass product manufacturing made of purchased glass
	160	Cement manufacturing
	161	Ready-mix concrete manufacturing
	162	Concrete pipe brick and block manufacturing
	163	Other concrete product manufacturing
	164	Lime and gypsum product manufacturing
	165	Abrasive product manufacturing
	166	Cut stone and stone product manufacturing
	167	Ground or treated mineral and earth manufacturing
	168	Mineral wool manufacturing
	169	Miscellaneous nonmetallic mineral products
	170	Iron and steel mills and ferroalloy manufacturing
	171	Steel product manufacturing from purchased steel
	172	Alumina refining and primary aluminum production
	173	Secondary smelting and alloying of aluminum

Sector	IMPLAN Code	IMPLAN Name
	174	Aluminum product manufacturing from purchased aluminum
	175	Primary smelting and refining of copper
	176	Primary smelting and refining of nonferrous metal (except copper and aluminum)
	177	Copper rolling drawing extruding and alloying
	178	Nonferrous metal (except copper and aluminum) rolling drawing extruding and alloying
	179	Ferrous metal foundries
	180	Nonferrous metal foundries
	181	All other forging stamping and sintering
	182	Custom roll forming
	183	Crown and closure manufacturing and metal stamping
	184	Cutlery utensil pot and pan manufacturing
	185	Handtool manufacturing
	186	Plate work and fabricated structural product manufacturing
	187	Ornamental and architectural metal products manufacturing
	188	Power boiler and heat exchanger manufacturing
	189	Metal tank (heavy gauge) manufacturing
	190	Metal can box and other metal container (light gauge) manufacturing
	191	Ammunition manufacturing
	192	Arms ordnance and accessories manufacturing
	193	Hardware manufacturing
	194	Spring and wire product manufacturing
	195	Machine shops
	196	Turned product and screw nut and bolt manufacturing
	197	Coating engraving heat treating and allied activities
	198	Valve and fittings other than plumbing
	199	Plumbing fixture fitting and trim manufacturing
	200	Ball and roller bearing manufacturing
	201	Fabricated pipe and pipe fitting manufacturing
	202	Other fabricated metal manufacturing
	203	Farm machinery and equipment manufacturing
	204	Lawn and garden equipment manufacturing
	205	Construction machinery manufacturing
	206	Mining and oil and gas field machinery manufacturing
	214	Air purification and ventilation equipment manufacturing
	215	Heating equipment (except warm air furnaces) manufacturing
	216	Air-conditioning refrigeration and warm air heating equipment manufacturing (laboratory freezers)
	207	Other industrial machinery manufacturing (laboratory distilling equipment)
	208	Plastics and rubber industry machinery manufacturing
	209	Semiconductor machinery manufacturing

(*continued*)

TABLE D5 Sectoral aggregation scheme used in the CGE model (*continued*)

Sector	*IMPLAN Code*	*IMPLAN Name*
	210	Vending commercial industrial and office machinery manufacturing
	211	Optical instrument and lens manufacturing
	212	Photographic and photocopying equipment manufacturing
	213	Other commercial and service industry machinery manufacturing
	217	Industrial mold manufacturing
	218	Metal cutting and forming machine tool manufacturing
	219	Special tool die jig and fixture manufacturing
	220	Cutting tool and machine tool accessory manufacturing
	221	Rolling mill and other metalworking machinery manufacturing
	222	Turbine and turbine generator set units manufacturing
	223	Speed changer industrial high-speed drive and gear manufacturing
	224	Mechanical power transmission equipment manufacturing
	225	Other engine equipment manufacturing
	226	Pump and pumping equipment manufacturing
	227	Air and gas compressor manufacturing
	228	Material handling equipment manufacturing
	229	Power-driven handtool manufacturing
	230	Other general purpose machinery manufacturing (laboratory scales and balances laboratory centrifuges)
	231	Packaging machinery manufacturing
	232	Industrial process furnace and oven manufacturing (laboratory furnaces and ovens)
	233	Fluid power process machinery
	234	Electronic computer manufacturing
	235	Computer storage device manufacturing
	236	Computer terminals and other computer peripheral equipment manufacturing
	237	Telephone apparatus manufacturing
	238	Broadcast and wireless communications equipment
	239	Other communications equipment manufacturing
	240	Audio and video equipment manufacturing
	241	Electron tube manufacturing
	242	Bare printed circuit board manufacturing
	243	Semiconductor and related device manufacturing
	244	Electronic capacitor resistor coil transformer and other inductor manufacturing
	245	Electronic connector manufacturing
	246	Printed circuit assembly (electronic assembly) manufacturing
	247	Other electronic component manufacturing
	248	Electromedical and electrotherapeutic apparatus manufacturing
	249	Search detection and navigation instruments manufacturing
	250	Automatic environmental control manufacturing
	251	Industrial process variable instruments manufacturing

Sector	*IMPLAN Code*	*IMPLAN Name*
	252	Totalizing fluid meters and counting devices manufacturing
	253	Electricity and signal testing instruments manufacturing
	254	Analytical laboratory instrument manufacturing
	255	Irradiation apparatus manufacturing
	256	Watch clock and other measuring and controlling device manufacturing
	257	Software audio and video media reproducing
	258	Magnetic and optical recording media manufacturing
	259	Electric lamp bulb and part manufacturing
	260	Lighting fixture manufacturing
	261	Small electrical appliance manufacturing
	262	Household cooking appliance manufacturing
	263	Household refrigerator and home freezer manufacturing
	264	Household laundry equipment manufacturing
	265	Other major household appliance manufacturing
	266	Power distribution and specialty transformer manufacturing
	267	Motor and generator manufacturing
	268	Switchgear and switchboard apparatus manufacturing
	269	Relay and industrial control manufacturing
	270	Storage battery manufacturing
	271	Primary battery manufacturing
	272	Communication and energy wire and cable manufacturing
	273	Wiring device manufacturing
	274	Carbon and graphite product manufacturing
	275	All other miscellaneous electrical equipment and component manufacturing
	276	Automobile manufacturing
	277	Light truck and utility vehicle manufacturing
	278	Heavy duty truck manufacturing
	279	Motor vehicle body manufacturing
	280	Truck trailer manufacturing
	281	Motor home manufacturing
	282	Travel trailer and camper manufacturing
	283	Motor vehicle parts manufacturing
	284	Aircraft manufacturing
	285	Aircraft engine and engine parts manufacturing
	286	Other aircraft parts and auxiliary equipment manufacturing
	287	Guided missile and space vehicle manufacturing
	288	Propulsion units and parts for space vehicles and guided missiles
	289	Railroad rolling stock manufacturing
	290	Ship building and repairing
	291	Boat building

(*continued*)

TABLE D5 Sectoral aggregation scheme used in the CGE model (*continued*)

Sector	*IMPLAN Code*	*IMPLAN Name*
	292	Motorcycle bicycle and parts manufacturing
	293	Military armored vehicle tank and tank component manufacturing
	294	All other transportation equipment manufacturing
	295	Wood kitchen cabinet and countertop manufacturing
	296	Upholstered household furniture manufacturing
	297	Nonupholstered wood household furniture manufacturing
	298	Metal and other household furniture manufacturing
	299	Institutional furniture manufacturing
	300	Office furniture manufacturing
	301	Custom architectural woodwork and millwork manufacturing
	302	Showcase partition shelving and locker manufacturing
	303	Mattress manufacturing
	304	Blind and shade manufacturing
	305	Surgical and medical instrument manufacturing
	306	Surgical appliance and supplies manufacturing
	307	Dental equipment and supplies manufacturing
	308	Ophthalmic goods manufacturing
	309	Dental laboratories
	310	Jewelry and silverware manufacturing
	311	Sporting and athletic goods manufacturing
	312	Doll toy and game manufacturing
	313	Office supplies (except paper) manufacturing
	314	Sign manufacturing
	315	Gasket packing and sealing device manufacturing
	316	Musical instrument manufacturing
	317	All other miscellaneous manufacturing
	318	Broom brush and mop manufacturing
Outdoor services	34	Construction of new nonresidential commercial and health care structures
	35	Construction of new nonresidential manufacturing structures
	36	Construction of other new nonresidential structures
	37	Construction of new residential permanent site single- and multi-family structures
	38	Construction of other new residential structures
	39	Maintenance and repair construction of nonresidential maintenance and repair
	40	Maintenance and repair construction of residential structures
Real estate	360	Real estate
	361	Imputed rental value for owner-occupied dwellings
Indoor services	394	Offices of physicians dentists and other health practitioners
	395	Home health care services
	396	Medical and diagnostic labs and outpatient and other ambulatory care services
	397	Hospitals

Sector	IMPLAN Code	IMPLAN Name
	398	Nursing and residential care facilities
	399	Child day care services
	400	Individual and family services
	401	Community food housing and other relief services including rehabilitation services
	319	Wholesale trade
	320	Retail: Motor vehicle and parts
	321	Retail: Furniture and home furnishings
	322	Retail: Electronics and appliances
	323	Retail: Building material and garden supply
	324	Retail: Food and beverage
	325	Retail: Health and personal care
	326	Retail: Gasoline stations
	327	Retail: Clothing and clothing accessories
	328	Retail: Sporting goods hobby book and music
	329	Retail: General merchandise
	330	Retail: Miscellaneous
	331	Retail: Nonstore
	362	Automotive equipment rental and leasing
	363	General and consumer goods rental except video tapes and discs
	364	Video tape and disc rental
	378	Photographic services
	379	Veterinary services
	365	Commercial and industrial machinery and equipment rental and leasing
	366	Lessors of nonfinancial intangible assets
	339	Couriers and messengers
	340	Warehousing and storage
	341	Newspaper publishers
	342	Periodical publishers
	343	Book publishers
	344	Directory mailing list and other publishers
	345	Software publishers
	346	Motion picture and video industries
	347	Sound recording industries
	348	Radio and television broadcasting
	349	Cable and other subscription programming
	350	Internet publishing and broadcasting
	352	Data processing hosting and related services
	353	Other information services
	354	Monetary authorities and depository credit intermediation
	355	Nondepository credit intermediation and related activities

(*continued*)

TABLE D5 Sectoral aggregation scheme used in the CGE model (*continued*)

Sector	*IMPLAN Code*	*IMPLAN Name*
	356	Securities commodity contracts investments and related activities
	357	Insurance carriers
	358	Insurance agencies brokerages and related activities
	359	Funds trusts and other financial vehicles
	367	Legal services
	368	Accounting tax preparation bookkeeping and payroll services
	369	Architectural engineering and related services
	370	Specialized design services
	371	Custom computer programming services
	372	Computer systems design services
	373	Other computer-related services including facilities management
	374	Management scientific and technical consulting services
	375	Environmental and other technical consulting services
	376	Scientific research and development services
	377	Advertising and related services
	380	All other miscellaneous professional scientific and technical services
	381	Management of companies and enterprises
	382	Employment services
	384	Office administrative services
	385	Facilities support services
	386	Business support services
	387	Investigation and security services
	388	Services to buildings and dwellings
	389	Other support services
	391	Elementary and secondary schools
	392	Junior colleges colleges universities and professional schools
	393	Other educational services
	414	Automotive repair and maintenance except car washes
	415	Car washes
	416	Electronic and precision equipment repair and maintenance
	417	Commercial and industrial machinery and equipment repair and maintenance
	418	Personal and household goods repair and maintenance
	419	Personal care services
	420	Death care services
	421	Dry-cleaning and laundry services
	422	Other personal services
	423	Religious organizations
	424	Grantmaking giving and social advocacy organizations
	425	Civic social professional and similar organizations
	426	Private households

Sector	*IMPLAN Code*	*IMPLAN Name*
	427	Postal service
	429	Other Federal Government enterprises
	432	Other state and local government enterprises
	433	Not an industry (Used and secondhand goods)
	434	Not an industry (Scrap)
	435	Not an industry (Rest of the world adjustment)
	436	Not an industry (Noncomparable imports)
	437	Employment and payroll for SL Government Non-Education
	438	Employment and payroll for SL Government Education
	439	Employment and payroll for Federal Non-Military
	440	Employment and payroll for Federal Military
	402	Performing arts companies
	403	Spectator sports
	404	Promoters of performing arts and sports and agents for public figures
	405	Independent artists writers and performers
	406	Museums historical sites zoos and parks
	407	Fitness and recreational sports centers
	408	Bowling centers
	409	Amusement parks arcades and gambling industries
	410	Other amusement and recreation industries
	383	Travel arrangement and reservation services
	411	Hotels and motels including casino hotels
	412	Other accommodations
	413	Food services and drinking places

APPENDIX E
VALUING RISK AND UNEQUAL IMPACTS

ROBERT KOPP AND SOLOMON HSIANG

To calculate the risk and inequality premiums on climate-change damage, we assume that each state of the Union is composed of homogenous agents with an income equal to that state's per capita income. For each state i, we calculate the certainty-equivalent income per capita c^*, where

$$u(c_i^*) = \sum_j p_j u(c_{i,j}). \tag{E1}$$

In this expression, the indices j denote states of the world and p_j their associated probabilities. We employ an isoelastic utility function with relative risk aversion η,

$$u(c) = \frac{c^{1-\eta}}{1-\eta} \tag{E2}$$

which implies that

$$c_i^* = \left(\sum_j p_j c_{i,j}^{1-\eta} \right)^{1/(1-\eta)}. \tag{E3}$$

Note that when $\eta = 0$, ci^* is simply the expected income of i.

We then calculate the inequality-neutral certainty-equivalent income loss. To do this, we find the equal-percentage loss that, if forfeited by all agents with certainty, yields the same welfare as the actual cross-sectional distribution of certainty-equivalent losses c_i^*. Denoting the inequality-neutral certainty-equivalent income loss as f and state populations as N_i, we compute welfare:

$$\sum_i N_i v(c_i^*) = \sum_i N_i v(c_i^0(1-f)), \qquad \text{(E4)}$$

where c_i^o is the per capita income of state i in the absence of climate impacts and the welfare function

$$\sum_i N_i v(c_i) = \sum_i N_i \frac{c_i^{1-\gamma}}{1-\gamma} \qquad \text{(E5)}$$

implies that society has a coefficient of inequality aversion γ among contemporaries.

Note that when $\gamma = 0$, this reduces to

$$\sum_i N_i c_i^* = (1-f) \sum_i N_i c_i^0 \qquad \text{(E6)}$$

$$f = \sum_i N_i (c_i^0 - c_i^*) / \sum_i N_i c_i^0 \qquad \text{(E7)}$$

so f is simply the ratio of total certainty-equivalent damage as a fraction of total income in the absence of climate impacts. For a given value of η, the inequality premium can be defined as the difference between f computed at a selected value of γ and the same variables computed at $\gamma = 0$. Similarly, for a given value of γ, the risk premium can be defined as the difference between f computed at a selected value of η and the same variables computed at $\eta = 0$.

Note also that we have intentionally chosen to compute the premium among contemporaries for unequal risk, rather than a risk premium for inequality among contemporaries. We have chosen to focus on the former because we think it more accurately captures how society understands and practically values the impacts of climate change. Relying on the latter implies that decision makers imagine each potential future realization of the national income distribution, weigh their aversion to each scenario independently, and then assign probabilities to each of these potential realizations. While feasible, we think the latter approach is less intuitive as well as less desirable, as it ignores the private risk borne by individuals and only considers risk related to the taste for socioeconomic structure in future populations.

REFERENCES

Abler, D., Fisher-Vanden, K., McDill, M., Ready, R., Shortle, J., Wing, I. S. & Wilson, T. (2009). *Economic Impacts of Projected Climate Change in Pennsylvania: Report to the Department of Environmental Protection.* Environment & Natural Resources Institute.

Adams, H. D., Guardiola-Claramonte, M., Barron-Gafford, G. A, Villegas, J. C., Breshears, D. D., Zou, C. B., . . . Huxman, T. E. (2009). Temperature sensitivity of drought-induced tree mortality portends increased regional die-off under global-change-type drought. *Proceedings of the National Academy of Sciences USA*, 106(17), 7063–66. doi:10.1073/pnas.0901438106.

Adams, R. M., McCarl, B.A., Segersen, K., Rosenzweig, C., Bryant, K. J., Dixon, B. L., Connor, R., . . . Ojima, D. (1999). The economic effects of climate change on US agriculture. In J. E. Neumann & R. O. Mendelsohn (Eds.), *The Impact of Climate Change on the United States Economy* (pp. 19–54). Cambridge, Mass.: Cambridge University Press.

Adger, W., Pulhin, J., Barnett, J., Dabelko, G., Hovelsrud, G., Levy, M., . . . Vogel, C. (2014). Human security. In C. B. Field, V. Barros, D. J. Dokken, K. J. Mach, M. D. Mastrandrea, T. E. Bilir, . . . L. L. White (Eds.), *Climate Change 2014: Impacts, Adaptation, and Vulnerability, Contribution of Working Group II to the IPCC Fifth Assessment Report* (pp. 755–791). Cambridge: Cambridge University Press.

Alaska State Legislature. (2008). *Final Report of the Alaska Climate Impact Assessment Commission.*

Aldy, J. E. & Viscusi, W. K. (2007). Age differences in the value of statistical life: Revealed preference evidence (No. 07–05) (p. 28).

Alexander, P., Brekke, L., Davis, G., Gangopadhyay, S., Grantz, K., Hennig, C., . . . Turner, T. (2011). Secure Water Act Section 9503(c) – Reclamation Climate Change and Water (Vol. 9503, p. 206). Denver, CO.

Alexandratos, N. & Bruinsma, J. (2012). World Agriculture Towards 2030/2050: The 2012 Revision (No. 12–03) (p. 160). Rome.

Allen, C. D., Macalady, A. K., Chenchouni, H., Bachelet, D., McDowell, N., Vennetier, M., . . . Cobb, N. (2010). A global overview of drought and heat-induced tree mortality reveals emerging climate change risks for forests. *Forest Ecology and Management*, 259(4), 660–84. doi:10.1016/j.foreco.2009.09.001.

American Academy of Pediatrics. (2007). Global climate change and children's health. *Pediatrics*, 120(5), 1149–52. doi:10.1542/peds.2007-2645.

American Society of Civil Engineers. (2013). *2013 Report Card for America's Infrastructure* (p. 74). Retrieved from www.infrastructurereportcard.org.

Amthor, J. S. (2001). Effects of atmospheric CO_2 concentration on wheat yield: Review of results from experiments using various approaches to control CO_2 concentration. *Field Crops Research*, 73(1), 1–34. doi:10.1016/S0378-4290(01)00179-4.

Anderegg, W., Kane, J. & Anderegg, L. (2013). Consequences of widespread tree mortality triggered by drought and temperature stress. *Nature Climate Change*, 3. doi:10.1038/nclimate1635.

Anderson, B. G. & Bell, M. L. (2009). Weather-related mortality: How heat, cold, and heat waves affect mortality in the United States. *Epidemiology*, 20(2), 205–13. doi:10.1097/EDE.0b013e318190ee08.

Anderson, G. B. & Bell, M. L. (2011). Heat waves in the United States: Mortality risk during heat waves and effect modification by heat wave characteristics in 43 US communities. *Environmental Health Perspectives*, 119(2), 210–18. doi:10.1289/ehp.1002313.

Anderson, C. A., Anderson, K. B., Deneve, K. M. & Flanagan, M. (2000). Temperature and aggression. *Advances in Experimental Social Psychology*, 32, 63–133. doi:10.1016/S0065-2601(00)80004-0.

Anderson, C. A., Bushman, B. J. & Groom, R. W. (1997). Hot years and serious and deadly assault: Empirical tests of the heat hypothesis. *Journal of Personality and Social Psychology*, 73(6), 1213–23. doi:10.1037/0022-3514.73.6.1213.

André, G., Engel, B., Berentsen, P. B. M., Vellinga, T. V & Oude Lansink, A. G. J. M. (2011). Quantifying the effect of heat stress on daily milk yield and monitoring dynamic changes using an adaptive dynamic model. *Journal of Dairy Science*, 94, 4502–13. doi:10.3168/jds.2010-4139.

Antle, J. M. & Capalbo, S. M. (2010). Adaptation of agricultural and food systems to climate change: An economic and policy perspective. *Applied Economic Perspectives and Policy*, 32, 386–416. doi:10.1093/aepp/ppq015.

Archambault, D. (2007). Efficacy of herbicides under elevated temperature and CO2. In P. C. D. Newton, R. A. Carran, G. R. Edwards & P. A. Niklaus (Eds.), *Agroecosystems in a Changing Climate* (pp. 262–79). Boston: CRC Press.

Arrhenius, S. (1896). On the influence of carbonic acid in the air upon the temperature of the ground. *Philosophical Magazine Series 5*, 41(251), 237–76. doi:10.1080/14786449608620846.

Arrow, K. J., Cropper, M. L., Gollier, C., Groom, B., Heal, G. M., Newell, R. G., . . . Weitzman, M. L. (2014). Should government use a declining discount rate in project analysis? *Review of Environmental Economics and Policy*, 8(2), 145–63. doi: 10.1093/reep/reu008.

Ashley, S. T. & Ashley, W. S. (2008). Flood fatalities in the United States. *Journal of Applied Meteorology and Climatology*, 47(3), 805–18. doi:10.1175/2007JAMC1611.1.

Atkinson, G., Dietz, S., Helgeson, J., Hepburn, C. & Sælen, H. (2009). Siblings, not triplets: Social preferences for risk, inequality and time in discounting climate change. *Economics*, 3(26), 28. doi:10.5018/economics-ejournal.ja.2009-26.

Auffhammer, M. & Aroonruengsawat, A. (2011). Simulating the impacts of climate change, prices and population on California's residential electricity consumption. *Climatic Change*, 109(S1), 191–210. doi:10.1007/s10584-011-0299-y.

Averyt, K., Fisher, J., Huber-Lee, A., Lewis, A., Macknick, J., Madden, N., . . . EW3 Scientific Advisory Committee. (2011). Freshwater use by U.S. power plants: Electricity's thirst for a precious resource. A report of the Energy and Water in a Warming World initiative. Cambridge, MA: Union of Concerned Scientists. November.

Backus, G., Lowry, T., Warren, D., Ehlen, M., Klise, G., Malczynski, L., . . . Zagonel, A. (2010). *Assessing the Near-Term Risk of Climate Uncertainty: Interdependencies Among the US States* (p. 258). Albuquerque, N.M., and Livermore, Calif.

Balbus, J. M. & Malina, C. (2009). Identifying vulnerable subpopulations for climate change health effects in the United States. *Journal of Occupational and Environmental Medicine*, 51, 33–37. doi:10.1097/JOM.0b013e318193e12e.

Barnes, E. A. (2013). Revisiting the evidence linking Arctic amplification to extreme weather in midlatitudes. *Geophysical Research Letters*, 40(17), 4734–39. doi:10.1002/grl.50880.

Barnett, A. G. (2007). Temperature and cardiovascular deaths in the US elderly: Changes over time. *Epidemiology*, 18(3), 369–72. doi:10.1097/01.ede.0000257515.34445.a0.

Barnett, T. P. & Pierce, D. W. (2009). Sustainable water deliveries from the Colorado River in a changing climate. *Proceedings of the National Academy of Sciences USA*, 106(18), 7334–38. doi:10.1073/pnas.0812762106.

Barnosky, A. D., Hadly, E. A., Bascompte, J., Berlow, E. L., Brown, J. H., Fortelius, M., . . . Smith, A. B. (2012). Approaching a state shift in Earth's biosphere. *Nature*, 486(7401), 52–58. doi:10.1038/nature11018.

Barnosky, A. D., Matzke, N., Tomiya, S., Wogan, G. O. U., Swartz, B., Quental, T. B., . . . Ferrer, E. A. (2011). Has the Earth's sixth mass extinction already arrived? *Nature*, 471, 51–57. doi:10.1038/nature09678.

Barreca, A., Clay, K., Deschênes, O., Greenstone, M., Shapiro, J. S. & Deschênes, O. (2013). Adapting to climate change: The remarkable decline in the US temperature-mortality relationship over the 20th century (January). NBER Working Paper No. 18692.

Basu, R. & Samet, J. M. (2002). Relation between elevated ambient temperature and mortality: A review of the epidemiologic evidence. *Epidemiologic Reviews*, 24(2), 190–202. doi:10.1093/epirev/mxf007.

Baylis, M. & Githeko, A. K. (2006). *The Effects of Climate Change on Infectious Diseases of Animals*. Report for the Foresight Project on Detection of Infectious Diseases, Department of Trade and Industry, UK Government.

Belasen, A. R. & Polachek, S. W. (2009). How disasters affect local labor markets: The effects of hurricanes in Florida. *Journal of Human Resources*, 44(1), 251–76. doi:10.1353/jhr.2009.0014.

Bell, M. L., Goldberg, R., Hogrefe, C., Kinney, P. L., Knowlton, K., Lynn, B., . . . Patz, J. A. (2007). Climate change, ambient ozone, and health in 50 US cities. *Climatic Change*, 82(1–2), 61–76. doi:10.1007/s10584-006-9166-7.

Bell, M., McDermott, A. & Zeger, S. (2004). Ozone and short-term mortality in 95 US urban communities, 1987–2000. *Journal of the American Medical Association*, 292(19), 2372–78. doi:10.1001/jama.292.19.2372.

Bellemare, C., Kröger, S. & Soest, A. van. (2008). Measuring inequity aversion in a heterogeneous population using experimental decisions and subjective probabilities. *Econometrica*, 76(4), 815–39. doi:10.1111/j.1468-0262.2008.00860.x.

Bender, M. A., Knutson, T. R., Tuleya, R. E., Sirutis, J. J., Vecchi, G. A., Garner, S. T. & Held, I. M. (2010). Modeled impact of anthropogenic warming on the frequency of intense Atlantic hurricanes. *Science*, 327, 454–58. doi:10.1126/science.1180568.

Bennett, C. M. & McMichael, A. J. (2010). Non-heat-related impacts of climate change on working populations. *Global Health Action*, 3: 5640. doi:10.3402/gha.v3i0.5640.

Bentz, B. J., Régnière, J., Fettig, C. J., Hansen, E. M., Hayes, J. L., Hicke, J. A., . . . Seybold, S. J. (2010). Climate change and bark beetles of the western United States and Canada: Direct and indirect effects. *BioScience*, 60(8), 602–613. doi:10.1525/bio.2010.60.8.6.

Bernard, S. M., Samet, J. M., Grambsch, A., Ebi, K. L. & Romieu, I. (2001). The potential impacts of climate variability and change on air pollution-related health effects in the United States. *Environmental Health Perspectives*, 109(Suppl 2), 199–209.

Berrittella, M., Bigano, A., Roson, R. & Tol, R. S. J. (2006). A general equilibrium analysis of climate change impacts on tourism. *Tourism Management*, 27, 913–24. doi:10.1016/j.tourman.2005.05.002.

Bigano, A., Hamilton, J. M. & Tol, R. S. J. (2005). The impact of climate change on domestic and international tourism: A simulation study. *Policy*, 15, 253–66. Retrieved from http://papers.ssrn.com/sol3/papers.cfm?abstract_id=907454.

Blois, J. L., Zarnetske, P. L., Fitzpatrick, M. C. & Finnegan, S. (2013). Climate change and the past, present, and future of biotic interactions. *Science*, 341, 499–504. doi:10.1126/science.1237184.

Bloomer, B. J., Stehr, J. W., Piety, C. A., Salawitch, R. J. & Dickerson, R. R. (2009). Observed relationships of ozone air pollution with temperature and emissions. *Geophysical Research Letters*, 36(9), L09803. doi:10.1029/2009GL037308.

Blunden, J. & Arndt, D. S. (2013). State of the climate in 2012. *Bulletin of the American Meteorological Society*, 94(8), S1–S238. doi:*10.1175/2013BAMSStateoftheClimate.1.*

Bohlken, A. T. & Sergenti, E. J. (2010). Economic growth and ethnic violence: An empirical investigation of Hindu—Muslim riots in India. *Journal of Peace Research*, 47, 589–600. doi:10.1177/0022343310373032.

Boisvenue, C. & Running, S. W. (2006). Impacts of climate change on natural forest productivity—Evidence since the middle of the 20th century. *Global Change Biology*, 12, 862–82. doi:10.1111/j.1365-2486.2006.01134.x.

Borgerson, S. G. (2008). Arctic meltdown: The economic and security implications of global warming. *Foreign Affairs*, 87, 63–77.

Bouchama, A. & Knochel, J. P. (2002). Heat stroke. *New England Journal of Medicine*, 346, 1978–88. doi:10.1056/NEJMra011089.

Bracmort, K. (2013). Wildfire Management: Federal Funding and Related Statistics (p. 15). Congressional Research Service Report 43077.

Bradfield, R., Wright, G., Burt, G., Cairns, G. & Van Der Heijden, K. (2005). The origins and evolution of scenario techniques in long range business planning. *Futures*, 37(8), 795–812. doi:10.1016/j.futures.2005.01.003.

Bradley, B. A., Wilcove, D. S. & Oppenheimer, M. (2010). Climate change increases risk of plant invasion in the Eastern United States. *Biological Invasions*, 12, 1855–72. doi:10.1007/s10530-009-9597-y.

Brekke, L., Thrasher, B. L., Maurer, E. P. & Pruitt, T. (2013). *Downscaled CMIP3 and CMIP5 Climate Projections: Release of Downscaled CMIP5 Climate Projections, Comparison with Preceding Information, and Summary of User Needs*. Denver, Colo.: U.S. Department of the Interior, Bureau of Reclamation, Technical Service Center. Retrieved from dcp.ucllnl.org/downscaled_cmip_projections/techmemo/downscaled_climate.pdf.

Brewer, M., Brown, T. C., McNutt, C. & Raff, D. (2014). *Water Resources Sector Technical Input—Interim Report in Support of the US Global Change Research Program 2014 National Climate Assessment* (vol. 2012). Alexandria, Va.

Brown, T. C., Foti, R. & Ramirez, J. A. (2013). Projected freshwater withdrawals in the United States under a changing climate. *Water Resources Research*, 49(March 2012), 1259–76. doi:10.1002/wrcr.20076.

Buckley, L. B. & Foushee, M. S. (2012). Footprints of climate change in US national park visitation. *International Journal of Biometeorology*. doi:10.1007/s00484-011-0508-4.

Buckley, B. M., Anchukaitis, K. J., Penny, D., Fletcher, R., Cook, E. R., Sano, M., . . . Hong, T. M. (2010). Climate as a contributing factor in the demise of Angkor, Cambodia. *Proceedings of the National Academy of Sciences USA*, 107(15), 6748–52. doi:10.1073/pnas.0910827107.

Buddemeier, R. W., Kleypas, J. A. & Aronson, R. B. (2004). *Coral Reefs & Global Climate Change: Potential Contributions of Climate Change to Stresses on Coral Reef Ecosystems*. Arlington, Va.

Burakowski, E. & Magnusson, M. (2012). *Climate Impacts on the Winter Tourism Economy in the United States*. New York.

Bureau of Justice Statistics. (2010). *Justice Employment and Expenditure Extracts*. Retrieved from www.bjs.gov/index.cfm?ty=tp&tid=5.

Bürger, G., Murdock, T. Q., Werner, A. T., Sobie, S. R. & Cannon, A. J. (2012). Downscaling extremes—an intercomparison of multiple statistical methods for present climate. *Journal of Climate*, 25(12), 4366–88. doi:10.1175/JCLI-D-11-00408.1.

Burke, P. J. (2012). Economic growth and political survival. *B.E. Journal of Macroeconomics*, 12(1), 43. doi:10.1515/1935-1690.2398.

Burke, M., & Emerick, K. (2012). Adaptation to climate change: Evidence from US agriculture. SSRN Working Paper, available at SSRN 2144928.

Burke, M. B., Miguel, E., Satyanath, S., Dykema, J. A. & Lobell, D. B. (2009). Warming increases the risk of civil war in Africa. *Proceedings of the National Academy of Sciences USA*, 106, 20670–74. doi:10.1073/pnas.0907998106.

Burt, C. C. (2011). Record dew point temperatures. Weather Underground: Weather extremes. Retrieved from www.wunderground.com/blog/weatherhistorian/record-dew-point-temperatures.

Butler, A. (2002). Burned: Visits to Parks Down Drastically, Even Away from Flames. *Rocky Mountain News*, July 15, 2002.

Butler, E. E. & Huybers, P. (2013). Adaptation of US maize to temperature variations. *Nature Climate Change*, 3(1), 68–72. doi:10.1038/nclimate1585.

Buzan, J. R. (2013). Implementation and model to model intercomparison of 12 heat stress metrics. M.S. thesis, University of Purdue. Retrieved from http://gradworks.umi.com/15/53/1553498.html.

CAL FIRE. (2013). CAL Fire incident information. Retrieved from http://cdfdata.fire.ca.gov/incidents/incidents_statsevents#2013.

Callendar, G. S. (1938). The artificial production of carbon dioxide and its influence on temperature. *Quarterly Journal of the Royal Meteorological Society*, 64(275), 223–40. doi:10.1002/qj.49706427503.

Calvin, K., Wise, M., Clarke, L., Edmonds, J., Kyle, P., Luckow, P. & Thomson, A. (2013). Implications of simultaneously mitigating and adapting to climate change: Initial experiments using GCAM. *Climatic Change*, 117(3), 545–60. doi:10.1007/s10584-012-0650-y.

Camargo, S. J. & Hsiang, S. M. (2012). Tropical cyclones: From the influence of climate to their socio-economic impacts. In *Extreme Events: Observations, Modeling and Economics* (p. 90).

Card, D. & Dahl, G. B. (2011). Family violence and football: The effect of unexpected emotional cues on violent behavior. *Quarterly Journal of Economics*, 126, 103–143. doi:10.1093/qje/qjr001.

Catarious Jr., D. M., Filadelfo, R., Gaffney, H., Maybee, S. & Morehouse, T. (2007). *National Security and the Threat of Climate Change* (p. 63). Alexandria, Va.

Cayan, D. R., Das, T., Pierce, D. W., Barnett, T. P., Tyree, M. & Gershunov, A. (2010). Future dryness in the southwest US and the hydrology of the early 21st century drought. *Proceedings of the National Academy of Sciences USA*, 107(50). doi:10.1073/pnas.0912391107

Climate Change Science Program. (2009). *The First State of the Coastal Cycle Sensitivity to Carbon Report Sea-Level Rise: Focus on Mid-Atlantic Region. Program* (p. 320).

Cerra, V., & Saxena, S. C. (2008). Growth dynamics: The myth of economic recovery. *American Economic Review*, 98(1), 439–57.

CH2M Hill. (2009). *Confronting Climate Change: An Early Analysis of Water and Wastewater Adaptation Costs* (p. 103).

Chalfin, A., & McCrary, J. (2013). The effect of police on crime: New evidence from US cities, 1960–2010 (No. w18815). *National Bureau of Economic Research*.

Changnon, D., Sandstrom, M. & Schaffer, C. (2003). Relating changes in agricultural practices to increasing dew points in extreme Chicago heat waves. *Climate Research*, 24(3), 243–54. doi:10.3354/cr024243.

Chassang, S. (2009). Economic shocks and civil war. *Quarterly Journal of Political Science*, 4(3), 211–28. doi:10.1561/100.00008072.

Christensen, J. H., Kumar, K. K., Aldrian, E., An, S.-I., Cavalcanti, I. F. A., Castro, M. de, . . . Zhou, T. (2013). Climate phenomena and their relevance for future regional climate change. In T. F. Stocker, D. Qin, G.-K. Plattner, M. Tignor, S. K. Allen, J. Boschung, . . . P. M. Midgley (Eds.), *Climate Change 2013: The Physical Science Basis. Contribution of Working Group I to the Fifth Assessment Report of the Intergovernmental Panel on Climate Change* (pp. 1217–1308). Cambridge: Cambridge University Press.

Church, J. A., Clark, P. U., Cazenave, A., Gregory, J. M., Jevrejeva, S., Levermann, A., . . . Unnikrishnan, A. S. (2013). Sea level change. In T. F. Stocker, D. Qin, G.-K. Plattner, M. Tignor, S. K. Allen, J. Boschung, . . . P. M. Midgley (Eds.), *Climate Change 2013: The Physical Science Basis. Contribution of Working Group I to the Fifth Assessment Report of the Intergovernmental Panel on Climate Change* (pp. 1137–1216). Cambridge: Cambridge University Press.

Clarke, L., Jiang, K., Akimoto, K., Babiker, M., Blanford, G., Fisher-Vanden, K., . . . van Vuuren, D. (2014). Assessing transformation pathways. In O. Edenhofer, R. Pichs-Madruga, Y. Sokona, E. Farahani, S. Kadner, K. Seyboth, . . . J. C. Minx (Eds.), *Climate Change 2014: Mitigation of Climate Change. Contribution of Working Group III to the IPCC Fifth Assessment Report* (pp. 413–510). Cambridge: Cambridge University Press.

Cline, W. R. (1992). *The Economics of Global Warming* (p. 399). Washington, D.C.: Institute for International Economics.

Cloern, J. E., Knowles, N., Brown, L. R., Cayan, D., Dettinger, M. D., Morgan, T. L., . . . Jassby, A. D. (2011). Projected evolution of California's San Francisco Bay-Delta-River system in a century of climate change. *PLOS ONE*, 6(9), 13. doi:10.1371/journal.pone.0024465.

Clow, D. W. (2010). Changes in the timing of snowmelt and streamflow in Colorado: A response to recent warming. *Journal of Climate*, 23(9), 2293–06. doi:10.1175/2009JCLI2951.1.

Cohn, E. G. & Rotton, J. (1997). Assault as a function of time and temperature: A moderator-variable time-series analysis. *Journal of Personality and Social Psychology*, 72(6), 1322–34. doi:*10.1037/0022-3514.72.6.1322*.

Collins, M., Knutti, R., Arblaster, J., Dufresne, J.-L., Fichefet, T., Friedlingstein, P., . . . Wehner, M. (2013). Long-term climate change: Projections, commitments and irreversibility. In T. F. Stocker, D. Qin, G.-K. Plattner, M. Tignor, S. K. Allen, J. Boschung, . . . P. M. Midgley (Eds.), *Climate Change 2013: The Physical Science Basis. Contribution of Working Group I to the Fifth Assessment Report of the Intergovernmental Panel on Climate Change* (pp. 1029–1136). Cambridge: Cambridge University Press.

Costanza, R., de Groot, R., Sutton, P., van der Ploeg, S., Anderson, S. J., Kubiszewski, I., . . . Turner, R. K. (2014). Changes in the global value of ecosystem services. *Global Environmental Change*, 26, 152–58. doi:10.1016/j.gloenvcha.2014.04.002.

Craine, J. M., Elmore, A. J., Olson, K. C. & Tolleson, D. (2010). Climate change and cattle nutritional stress. *Global Change Biology*, 16, 2901–11. doi:10.1111/j.1365-2486.2009.02060.x.

Crost, B. & Traeger, C. P. (2014). Optimal CO2 mitigation under damage risk valuation. *Nature Climate Change*, 4, 631–36. doi:10.1038/nclimate2249.

Cubasch, U., Wuebbles, D., Chen, D., Facchini, M. C., Frame, D., Mahowald, N. & Winther, J.-G. (2013). Introduction. In T. F. Stocker, D. Qin, G.-K. Plattner, M. Tignor, S. K. Allen, J. Boschung, . . . P. M. Midgley (Eds.), *Climate Change 2013: The Physical Science Basis. Contribution of Working Group I to the Fifth Assessment Report of the Intergovernmental Panel on Climate Change* (pp. 119–58). Cambridge: Cambridge University Press.

Cullen, H. M., DeMenocal, P. B., Hemming, S., Hemming, G., Brown, F. H., Guilderson, T. & Sirocko, F. (2000). Climate change and the collapse of the Akkadian empire: Evidence from the deep sea. *Geology*, 28(4), 379–82. doi:10.1130/0091-7613(2000)28<379:CCATCO>2.0.CO;2.

Curriero, F. C., Heiner, K. S., Samet, J. M., Zeger, S. L., Strug, L. & Patz, J. A. (2002). Temperature and mortality in 11 cities of the eastern United States. *American Journal of Epidemiology*, 155, 80–87. doi:10.1093/aje/155.1.80.

Curriero, F. C., Patz, J. A, Rose, J. B. & Lele, S. (2001). The association between extreme precipitation and waterborne disease outbreaks in the United States, 1948–1994. *American Journal of Public Health*, 91(8), 1194–99. doi:10.2105/AJPH.91.8.1194.

D'Anjou, R. M., Bradley, R. S., Balascio, N. L. & Finkelstein, D. B. (2012). Climate impacts on human settlement and agricultural activities in northern Norway revealed through sediment biogeochemistry. *Proceedings of the National Academy of Sciences USA*, 109(50), 20332–37. doi:10.1073/pnas.1212730109.

Dai, A. (2012). Increasing drought under global warming in observations and models. *Nature Climate Change*, 3, 52–58. doi:10.1038/nclimate1633.

Dal Bó, E. & Dal Bó, P. (2011). Workers, warriors, and criminals: Social conflict in general equilibrium. *Journal of the European Economic Association*, 9(4), 646–77. doi:10.1111/j.1542–4774.2011.01025.x.

Dale, V. H., Joyce, L. A., Mcnulty, S., Ronald, P. & Matthew, P. (2001). Climate change and forest disturbances. *BioScience*, 51(9), 723–34. doi:10.1641/0006-3568(2001)051[0723:CCAFD]2.0.CO;2.

Davcock, C., DesJardins, R., & Fennel, S. (2004). Generation cost forecasting using on-line thermodynamic models. In *Proceedings of Electric Power*. Baltimore, MD.

De Châtel, F. (2014). The role of drought and climate change in the Syrian Uprising: Untangling the triggers of the revolution. *Middle Eastern Studies*, 50(4), 521–35. doi:10.1080/00263206.2013.850076.

Dennekamp, M. & Abramson, M. J. (2011). The effects of bushfire smoke on respiratory health. *Respirology*, 16(2), 198–209. doi:10.1111/j.1440–1843.2010.01868.x.

Deryugina, T. (2011). *The Dynamic Effects of Hurricanes in the US: The Role of Non-Disaster Transfer Payments (No. 6166)* (pp. 1–56). Cambridge, Mass. Retrieved from http://econ-www.mit.edu/files/6166.

Deschênes, O. & Greenstone, M. (2011). Climate change, mortality, and adaptation: Evidence from annual fluctuations in weather in the US. *American Economic Journal: Applied Economics*, 3(4), 152–85. doi:10.1257/app.3.4.152.

Deschênes, O. & Moretti, E. (2009). Extreme weather events, mortality, and migration. *Review of Economics and Statistics*, 91(4), 659–81. doi:10.1162/rest.91.4.659.

Dietz, S.& Stern, N. (2014). Endogenous growth, convexity of damages and climate risk: How Nordhaus' framework supports deep cuts in carbon emissions." *Centre for Climate Change Economics and Policy Working Paper* 180.

Diffenbaugh, N. S., Scherer, M. & Ashfaq, M. (2012). Response of snow-dependent hydrologic extremes to continued global warming. *Nature Climate Change*, 3, 379–84. doi:10.1038/nclimate1732.

Ding, P., Gerst, M. D., Bernstein, A., Howarth, R. B. & Borsuk, M. E. (2012). Rare disasters and risk attitudes: International differences and implications for integrated assessment modeling. *Risk Analysis*, 32(11), 1846–55. doi:10.1111/j.1539–6924.2012.01872.x.

Dittrick, P. (2012). Drought raising water costs, scarcity conerns for shale plays. *Oil & Gas Journal*. Retrieved from www.ogj.com/articles/print/vol-110/issue-7d/general-interest/drought-raising-water-costs.html.

Doocy, S., Daniels, A., Murray, S. & Kirsch, T. D. (2013). The human impact of floods: A historical review of events 1980–2009 and systematic literature review. *PLOS Currents*. doi:10.1371/currents.dis.f4deb457904936b07c09daa98ee8171a.

Dukes, J. S., Pontius, J., Orwig, D., Garnas, J. R., Rodgers, V. L., Brazee, N., . . . Ayres, M. (2009). Responses of insect pests, pathogens, and invasive plant species to climate change in the forests of northeastern North America: What can we predict? *Canadian Journal of Forest Research*, 39(2), 231–48. doi:10.1139/X08–171.

Dunne, J. P., Stouffer, R. J. & John, J. G. (2013). Reductions in labour capacity from heat stress under climate warming. *Nature Climate Change*, 3(6), 563–66. doi:10.1038/nclimate1827.

Dutton, A. & Lambeck, K. (2012). Ice volume and sea level during the last interglacial. *Science*, 337, 216–19. doi:10.1126/science.1205749.

Earman, S. & Dettinger, M. (2011). Potential impacts of climate change on groundwater resources—a global review. *Journal of Water and Climate Change*, 2, 213–29. doi:10.2166/wcc.2011.034.

Earman, S., Campbell, A. R., Phillips, F. M. & Newman, B. D. (2006). Isotopic exchange between snow and atmospheric water vapor: Estimation of the snowmelt component of groundwater recharge in the southwestern United States. *Journal of Geophysical Research*, 111, 18. doi:10.1029/2005JD006470.

Ebinger, C. K. & Zambetakis, E. (2009). The geopolitics of Arctic melt. *International Affairs*, 85(6), 1215–32. doi:10.1111/j.1468–2346.2009.00858.x.

Edwards, P. N. (2011). History of climate modeling. *Wiley Interdisciplinary Reviews: Climate Change*, 2(1), 128–39. doi:10.1002/wcc.95.

Egan, T. (2006). *The Worst Hard Time: The Untold Story of Those Who Survived the Great American Dust Bowl*. Boston: Houghton Mifflin Harcourt.

Ehmer, P. & Heymann, E. (2008). *Climate Change and Tourism: Where Will the Journey Lead?* (p. 27). Frankfurt, Germany.

Elliott, J., Deryng, D., Muller, C., Frieler, K., Konzmann, M., Gerten, D., . . . Wisser, D. (2013). Constraints and potentials of future irrigation water availability on agricultural production under climate change. *Proceedings of the National Academy of Sciences USA*, 111(9), 3239–44. doi:10.1073/pnas.1222474110.

Ellis, B. (2012, August 10). Oil companies desperately seek water amid Kansas drought. CNN Money. Retrieved from www.eia.gov/electricity/monthly/current_year/february2013.pdf.

Ellsberg, D. (1961). Risk, ambiguity, and the savage axioms. *Quarterly Journal of Economics*, 75(4), 643–69.

Emanuel, K. (2007). Environmental factors affecting tropical cyclone power dissipation. *Journal of Climate*, 20(22), 5497–5509. doi:10.1175/2007JCLI1571.1.

Emanuel, K. (2013). Downscaling CMIP5 climate models shows increased tropical cyclone activity over the 21st century. *Proceedings of the National Academy of Sciences USA*, 110(30), 12219–24. doi:10.1073/pnas.1301293110.

Emelko, M. B., Silins, U., Bladon, K. D. & Stone, M. (2010). Implications of land disturbance on drinking water treatability in a changing climate: Demonstrating the need for "source water supply and protection" strategies. *Water Research*, 45(2), 461–72. doi:10.1016/j.watres.2010.08.051.

Energy Information Administration. (2011). Over half the cooling systems at US electric power plants reuse water. Retrieved from www.eia.gov/todayinenergy/detail.cfm?id=3950#.

Energy Information Administration. (2012a). How much electricity is lost in transmission and distribution in the United States? Retrieved from www.eia.gov/tools/faqs/faq.cfm?id=105&t=3.

Energy Information Administration. (2012b). State Electricity Profiles 2010, 0348 (January).

Energy Information Administration. (2013). *Electric Power Monthly*.

Engelmann, D. & Strobel, M. (2004). Inequality aversion, efficiency, and maximin preferences in simple distribution experiments. *American Economic Review*, 94(4), 857–69. doi:10.1257/0002828042002741.

Entergy. (2010). *Building a Resilient Energy Gulf Coast: Executive Report*. Washington, D.C.: Entergy. Retrieved from www.entergy.com/content/our_community/environment/GulfCoastAdaptation/Building_a_Resilient_Gulf_Coast.pdf.

EPRI. (2011). *Water Use for Electricity Generation and Other Sectors: Recent Changes (1985–2005) and Future Projections (2005–2030)*.

Fan, Q., Klaiber, H. A. & Fisher-Vanden, K. (2012). Climate change impacts on US migration and household location choice. In *Agricultural & Applied Economics Association's 2012 AAEA Annual Meeting* (pp. 1–31). Seattle, Wash.

Fankhauser, S. (1993). The economic costs of global warming: Some monetary estimates. In Y. Kaya, N. Nakicenovic', W. D. Nordhaus & F. L. Toth (Eds.), *Costs, Impacts, and Benefits of CO2 Mitigation*. Austria: International Institute for Systems Analysis.

Federal Bureau of Investigation. (2012). *Crime in the United States 2012*. Uniform Crime Reports. Retrieved from www.fbi.gov/about-us/cjis/ucr/crime-in-the-u.s/2012/crime-in-the-u.s.-2012/tables/1tabledatadecoverviewpdf/table_1_crime_in_the_united_states_by_volume_and_rate_per_100000_inhabitants_1993–2012.xls.

Fehr, E. & Schmidt, K. M. (1999). A theory of fairness, competition, and cooperation. *Quarterly Journal of Economics*, 114(4), 817–68.

Feng, S., Krueger, A. B. & Oppenheimer, M. (2010). Linkages among climate change, crop yields and Mexico-US cross-border migration. *Proceedings of the National Academy of Sciences USA*, 107(32), 14257–62. doi:10.1073/pnas.1002632107.

Feng, S., Oppenheimer, M. & Schlenker, W. (2012). *Climate Change, Crop Yields, and Internal Migration in the United States*. NBER.

Feng, S., Oppenheimer, M. & Schlenker, W. (2013). Weather Anomalies, Crop Yields, and Migration in the US Corn Belt, (September).

Fingar, T. (2008). *National Intelligence Assessment on the National Security Implications of Global Climate Change to 2030*. Washington, D.C.: Office of the Director of National Intelligence.

Fisher-Vanden, K., Wing, I. S., Lanzi, E., & Popp, D. (2013). Modeling climate change feedbacks and adaptation responses: Recent approaches and shortcomings. *Climatic Change*, 117(3), 481–95. doi:10.1007/s10584-012-0644-9.

Fishman, J., Creilson, J. K., Parker, P. A., Ainsworth, E. A., Vining, G. G., Szarka, J., . . . Xu, X. (2010). An investigation of widespread ozone damage to the soybean crop in the upper Midwest determined from ground-based and satellite measurements. *Atmospheric Environment*, 44, 2248–56. doi:10.1016/j.atmosenv.2010.01.015.

Flannigan, M. D., Krawchuk, M. A., de Groot, W. J., Wotton, B. M. & Gowman, L. M. (2009). Implications of changing climate for global wildland fire. *International Journal of Wildland Fire*, 18(5), 483. doi:10.1071/WF08187.

Flannigan, M. D., Stocks, B. J. & Wotton, B. M. (2000). Climate change and forest fires. *Science of the Total Environment*, 262(3), 221–9. doi:10.1016/S0048-9697(00)00524-6.

Flato, G., Marotzke, J., Abiodun, B., Braconnot, P., Chou, S. C., Collins, W., . . . Rummukainen, M. (2013). Evaluation of climate models. In T. F. Stocker, D. Qin, G.-K. Plattner, M. Tignor, S. K. Allen, J. Boschung, . . . P. M. Midgley (Eds.), *Climate Change 2013: The Physical Science Basis. Contribution of Working Group I to the Fifth Assessment Report of the Intergovernmental Panel on Climate Change* (pp. 741–866). Cambridge: Cambridge University Press.

Foley, C. & Holland, A. (2012). *Climate Security Part Three: Climate Change & the Homeland.* Retrieved from http://americansecurityproject.org/reports/2012/csr-part-three-climate-change-the-homeland.

Food and Agriculture Organization of the United Nations. (2011). *The State of Food Insecurity in the World: How Does International Price Volatility Affect Domestic Economies and Food Security?* Food and Agriculture Organization of the United Nations. Retrieved from www.fao.org/docrep/014/i2330e/i2330e00.htm.

Foti, R., Ramirez, J. A. & Brown, T. C. (2012). *Vulnerability of US Water Supply to Shortage: A Technical Document Supporting the Forest Service 2010 RPA Assessment* (p. 147). Fort Collins, Colo.

Francis, J. A. & Vavrus, S. J. (2012). Evidence linking Arctic amplification to extreme weather in mid-latitudes. *Geophysical Research Letters*, 39(6), L06801. doi:10.1029/2012GL051000.

Franco, G., Cayan, D. R., Moser, S., Hanemann, M. & Jones, M.A. (2011). Second California assessment: Integrated climate change impacts assessment of natural and managed systems. *Climatic Change*, 109(S1), 1–19. doi:10.1007/s10584-011-0318-z.

Frank, K. L., Mader, T. L., Harrington, Jr., J. A. & Hahn, G. L. (2001). Potential climate change effects on warm-season livestock production in the Great Plains. In *6th International Livestock Environment Symposium, American Society of Agricultural Engineers.* St. Joseph, Mich.

Frazier, T. G., Wood, N., Yarnal, B. & Bauer, D. H. (2010). Influence of potential sea level rise on societal vulnerability to hurricane storm-surge hazards, Sarasota County, Florida. *Applied Geography*, 30, 490–505. doi:10.1016/j.apgeog.2010.05.005.

Frederick, K. D. & Schwarz, G. E. (1999). *Socioeconomic Impacts of Climate Variability and Change on US Water Resources* (No. 00–21). Retrieved from www.rff.org/documents/RFF-DP-00-21.pdf.

Fritze, H., Stewart, I. T. & Pebesma, E. (2011). Shifts in Western North American snowmelt runoff regimes for the recent warm decades. *Journal of Hydrometeorology*, 12, 989–1006. doi:10.1175/2011JHM1360.1.

Frost, H. C. (2009). Statement of Dr. Herbert C. Frost, Associate Director, Natural Resource Stewardship and Science, National Park Service, US Department of the Interior, Before the Senate Subcommittee on National Parks, on the Impacts of Climate Change on National Parks in Colorado and Related Management Activities. Washington, D.C.: U.S. Senate.

Frumhoff, P., McCarthy, J., Melillo, J., Moser, S. & Wuebbles, D. (2007). *Confronting Climate Change in the US Northeast: Science, Impacts, and Solutions.* Synthesis Report of the Northeast Climate Impacts Assessment. Retrieved from www.cabdirect.org/abstracts/20083177506.html.

Garrett, K. A., Dendy, S. P., Frank, E. E., Rouse, M. N. & Travers, S. E. (2006). Climate change effects on plant disease: Genomes to ecosystems. *Annual Review of Phytopathology*, 44, 489–509. doi:10.1146/annurev.phyto.44.070505.143420.

Garrett, K. A., Forbes, G. A., Savary, S., Skelsey, P., Sparks, A. H., Valdivia, C., . . . Yuen, J. (2011). Complexity in climate-change impacts: An analytical framework for effects mediated by plant disease. *Plant Pathology*, 60, 15–30. doi:10.1111/j.1365-3059.2010.02409.x.

Gaughan, J. B., Mader, T. L., Holt, S. M., Hahn, G. L. & Young, B. (2002). Review of current assessment of cattle and microclimate during periods of high heat load. In *24th Biennial Conference of the Australian Society of Animal Production. American Society of Animal Science—Annual Meeting* (vol. 24, pp. 77–80). Adeleide, Australia: Australian Society of Animal Production. Retrieved from http://espace.library.uq.edu.au/view/UQ:98193#.Uva-SMxIqus.mendeley.

Gaughan, J., Lacetera, N., Valtorta, S. E., Khalifa, H. H., Hahn, L. & Mader, T. (2009). Response of domestic animals to climate challenges. In *Biometeorology for Adaptation to Climate Variability and Change* (pp. 131–170). doi:10.1007/978-1-4020-8921-3_7.

Gensler, J. (2014). *National Security Impacts of Climate Change.* Washington, D.C.: Bicameral Task Force on Climate Change.

Georgakakos, A., Fleming, P., Dettinger, M., Peters-Lidard, C., Richmond, T. C., Reckhow, K., . . . Yates, D. (2014). Water resources. In J. M. Melillo, T. C. Richmond & G. W. Yohe (Eds.), *Climate Change Impacts in the United States: The Third National Climate Assessment* (pp. 69–112). Washington, D.C.: U.S. Global Change Research Program. Retrieved from http://nca2014.globalchange.gov/report/sectors/water.

Gingerich, P. D. (2006). Environment and evolution through the Paleocene–Eocene thermal maximum. *Trends in Ecology & Evolution*, 21(5), 246–53. doi:10.1016/j.tree.2006.03.006.

Global Commission on the Economy and Climate. (2014). Better growth, better climate: *The new climate economy report.* Washington, D.C.: New Climate Economy.

Godfray, H. C. J., Crute, I. R., Haddad, L., Lawrence, D., Muir, J. F., Nisbett, N., . . . Whiteley, R. (2010). The future of the global food system. *Philosophical Transactions of the Royal Society of London. Series B, Biological Sciences*, 365, 2769–77. doi:10.1098/rstb.2010.0180.

Gollier, C. (2013). *Pricing the Planet's Future* (p. 248). Princeton, N.J.: Princeton University Press.

Graff Zivin, J. & Neidell, M. (2014). Temperature and the allocation of time: Implications for climate change. *Journal of Labor Economics*, 32(1), 1–26. doi:10.1086/671766.

Hamlet, A. F., Lee, S.-Y., Mickelson, K. E. B. & Elsner, M. M. (2010). Effects of projected climate change on energy supply and demand in the Pacific Northwest and Washington State. *Climatic Change*, 102, 103–28. doi:10.1007/s10584-010-9857-y.

Hansen, J., Johnson, D., Lacis, A., Lebedeff, S., Lee, P., Rind, D. & Russell, G. (1981). Climate impact of increasing atmospheric carbon dioxide. *Science*, 213(4511), 957–66. doi:10.1126/science.213.4511.957.

Hargreaves, S. (2012, July 31). Drought strains US oil production. CNN Money. Retrieved from http://money.cnn.com/2012/07/31/news/economy/drought-oil-us/index.htm.

Harrington, J. & Walton, T. (2008). Climate change in coastal areas in Florida: Sea level rise estimation and economic analysis to year 2080. Florida State University. Available at http://www.cefa.fsu.edu/content/download/47234/327898.

Hartmann, D. L., Tank, A. M. G. K., Rusticucci, M., Alexander, L. V., Brönnimann, S., Charabi, Y., . . . Zhai, P. M. (2013). Observations: Atmosphere and surface. In T. F. Stocker, D. Qin, G.-K. Plattner, M. Tignor, S. K. Allen, J. Boschung, . . . P. M. Midgley (Eds.), *Climate Change 2013: The Physical Science Basis. Contribution of Working Group I to the Fifth Assessment Report of the Intergovernmental Panel on Climate Change* (pp. 159–254). Cambridge: Cambridge University Press.

Hatfield, J. L., Cruse, R. M. & Tomer, M. D. (2013). Convergence of agricultural intensification and climate change in the Midwestern United States: Implications for soil and water conservation. *Marine and Freshwater Research*, 64(5), 423. doi:10.1071/MF12164.

Hatfield, J., Takle, G., Authors, L., Grotjahn, R., Holden, P., Environmental, W., . . . Messages, K. (2013). Agriculture. In *U.S. National Climate Assessment* (pp. 227–61).

Hatfield, J., Takle, G., Grotjahn, R., Holden, P., Izaurralde, R. C., Mader, T., . . . Liverman, D. (2014). Agriculture. In J. M. Melillo, T. C. Richmond & G. W. Yohe (Eds.), *Climate Change Impacts in the United States: The Third National Climate Assessment* (pp. 150–174). Washington, D.C.: US Global Change Research Program. Retrieved from http://nca2014.globalchange.gov/report/sectors/agriculture.

Haug, G. H., Günther, D., Peterson, L. C., Sigman, D. M., Hughen, K. A. & Aeschlimann, B. (2003). Climate and the collapse of Maya civilization. *Science*, 299(5613), 1731–35. doi:10.1126/science.1080444.

Hawkins, E. & R. Sutton (2009). The potential to narrow uncertainty in regional climate predictions. *Bulletin of the American Meteorological Society*, 90(8), 1095–1107. doi:10.1175/2009BAMS2607.1.

Hayhoe, K., Sheridan, S., Kalkstein, L. & Greene, S. (2010). Climate change, heat waves, and mortality projections for Chicago. *Journal of Great Lakes Research*, 36, 65–73. doi:10.1016/j.jglr.2009.12.009.

Heal, G. & Millner, A. (2014). Uncertainty and decision in climate change economics. *Review of Environmental Economics and Policy*, 8(1), 120–37. doi:10.1093/reep/ret023.

Heaton, P. (2010). *Hidden in Plain Sight*. Santa Monica, Calif.: RAND Corporation.

Heberger, M., Cooley, H., Herrera, P., Gleick, P. H. & Moore, E. (2012). Erratum to: Potential impacts of increased coastal flooding in California due to sea-level rise. *Climatic Change* 112(2), 523. doi:10.1007/s10584-012-0397-5

Heim, R. R. J. (2002). A review of twentieth-century drought indices used in the United States. *American Meteorological Society*, 83, 1149–65. doi:10.1175/1520-0477(2002)083<1149:AROTDI>2.3.CO;2.

Hejazi, M. I. & Markus, M. (2009). Impacts of urbanization and climate variability on floods in Northeastern Illinois. *Journal of Hydrologic Engineering*, 14, 606–16. doi;10.1061/(ASCE)HE.1943-5584.0000020.

Henderson, J., Rodgers, C., Jones, R., Smith, J. & Martinich, J. (2013). Economic impacts of climate change on water resources in the coterminous United States. *Mitigation and Adaptation Strategies for Global Change*, 18, 25. doi:10.1007/s11027-013-9483-x.

Hidalgo, F. D., Naidu, S., Nichter, S. & Richardson, N. (2010). Economic determinants of land invasions. *Review of Economics and Statistics*. doi:10.1162/REST_a_00007.

Hill, D. J., Haywood, A. M., Lunt, D. J., Hunter, S. J., Bragg, F. J., Contoux, C., . . . H. Ueda. (2014). Evaluating the dominant components of warming in Pliocene climate simulations. *Climate of the Past*, 10(1), 79–90. doi:10.5194/cp-10-79-2014.

Hoegh-Guldberg, O. (1999). Climate change, coral bleaching and the future of the world's coral reefs. *Marine Freshwater Research*, 50, 839–66. doi:10.1071/MF99078.

Hoerling, M. P., Dettinger, M., Wolter, K., Lukas, J., Eischeid, J., Nemani, R., . . . Kunkel, K. E. (2012a). Present weather and climate: Evolving conditions. In G. Garfin, A. Jardine, R. Merideth, B. M & S. LeRoy (Eds.), *Assessment of Climate Change in the Southwest United States: A Report Prepared for the National Climate Assessment* (pp. 74–100). Washington, D.C.: Island Press.

Hoerling, M. P., Eischeid, J. K., Quan, X.-W., Diaz, H. F., Webb, R. S., Dole, R. M. & Easterling, D. (2012b). Is a transition to semi-permanent drought conditions imminent in the US Great Plains? *Journal of Climate* 25, 8380–86. doi:10.1175/JCLI-D-12-00449.1.

Hofferth, S., Flood, S. M. & Sobek, M. (2013). American Time Use Survey Data Extract System: Version 2.4 [machine-readable database]. Maryland Population Research Center, University of Maryland, College Park, Maryland, and Minnesota Population Center, University of Minnesota, Minneapolis, Minnesota.

Hope, C., Anderson, J. & Wenman, P. (1993). Policy analysis of the greenhouse effect: An application of the PAGE model. *Energy Policy*, 21(3), 327–38. doi:10.1016/0301-4215(93)90253-C.

Hornbeck, R. (2012). The enduring impact of the American Dust Bowl: Short- and long-run adjustments to environmental catastrophe. *American Economic Review*, 102(4), 1477–1507. doi:10.1257/aer.102.4.1477.

Hornbeck, R. & Keskin, P. (2012). *Does Agriculture Generate Local Economic Spillovers? Short-run and Long-run Evidence from the Ogallala Aquifer.* NBER Working Paper No. 18416.

Howard, P. (2014). *Omitted Damages: What's Mission from the Social Cost of Carbon.* Retrieved from http://costofcarbon.org/files/Omitted_Damages_Whats_Missing_From_the_Social_Cost_of_Carbon.pdf.

Hoxie, N. J., Davis, J. P., Vergeront, J. M., Nashold, R. D. & Blair, K. A. (1997). Cryptosporidiosis-associated mortality following a massive waterborne outbreak in Milwaukee, Wisconsin. *American Journal of Public Health*, 87(12), 2032–35. Retrieved from www.pubmedcentral.nih.gov/articlerender.fcgi?artid=1381251&tool=pmcentrez&rendertype=abstract.

Hsiang, S. M. (2010). Temperatures and cyclones strongly associated with economic production in the Caribbean and Central America. *Proceedings of the National Academy of Sciences USA*, 107(35), 15367–72. doi:10.1073/pnas.1009510107.

Hsiang, S. M. (2011). *Essays on the social impacts of climate* (Doctoral dissertation, Columbia University).

Hsiang, S. M. & Burke, M. (2013). Climate, conflict, and social stability: What does the evidence say? *Climatic Change.* doi:10.1007/s10584-013-0868-3.

Hsiang, S. M., Burke, M. & Miguel, E. (2013). Quantifying the influence of climate on human conflict. *Science*, 341, 1235367. doi:10.1126/science.1235367.

Hsiang, S. M., Meng, K. C. & Cane, M. A. (2011). Civil conflicts are associated with the global climate. *Nature*, 476(7361), 438–41. doi:10.1038/nature10311.

Hsiang, S. M., & Narita, D. (2012). Adaptation to cyclone risk: Evidence from the global cross-section. *Climate Change Economics*, 3(2): 1–28.

Hurd, B. H. & Coonrod, J. (2012). Hydro-economic consequences of climate change in the upper Rio Grande. *Climate Research*, 53, 103–18. doi:10.3354/cr01092.

Hurd, B., Callaway, M. A. C., Smith, J. B. & Kirshen, P. (1999). Economic effects of climate change on US water resources. In R. Mendelsohn & J. E. Neumann (Eds.), *The Impact of Climate Change on the United States Economy.* Cambridge: Cambridge University Press.

Hurteau, M. D., Westerling, A. L., Wiedinmyer, C. & Bryant, B. P. (2014). Projected effects of climate and development on California wildfire emissions through 2100. *Environmental Science & Technology.* doi:10.1021/es4050133.

Izaurralde, R. C., Thomson, A. M., Morgan, J. A., Fay, P. A., Polley, H. W. & Hatfield, J. L. (2011). Climate impacts on agriculture: Implications for forage and rangeland production. *Agronomy Journal*, 103, 371–81. doi:10.2134/agronj2010.0304.

Jacob, B., Lefgren, L. & Moretti, E. (2007). The dynamics of criminal behavior: Evidence from weather shocks. *Journal of Human Resources*, 42(3), 489–527. doi:10.3368/jhr.XLII.3.489.

Jerrett, M., Burnett, R. T., Pope, C. A., Ito, K., Thurston, G., Krewski, D., . . . Thun, M. (2009). Long-term ozone exposure and mortality. *New England Journal of Medicine*, 360(11), 1085–95. doi:10.1056/NEJMoa0803894.

Jorgenson, D. W., Goettle, R. J., Hurd, B. H. & Smith, J. B. (2004). *US Market Consequences of Global Climate Change* (p. 44). Washington, D.C.

Joughin, I., Smith, B. E. & Medley, B. (2014). Marine ice sheet collapse potentially underway for the Thwaites Glacier Basin, West Antarctica. *Science* 344(6185), 735–38. doi:10.1126/science.1249055.

Joyce, L. A., Running, S. W., Breshears, D. D., Dale, V. H., Malmsheimer, R. W., Sampson, R. N., . . . Woodal, C. W. (2013). Forests. In *Climate Change Impacts in the United States: The Third National Climate Assessment.* Washington, D.C.: US Global Change Research Program.

Kampa, M. & Castanas, E. (2008). Human health effects of air pollution. *Environmental Pollution* (Barking, Essex: 1987), 151(2), 362–67. doi:10.1016/j.envpol.2007.06.012.

Kaushal, S. S., Likens, G. E., Jaworski, N. A., Pace, M. L., Sides, A. M., Seekell, D., . . . Wingate, R. L. (2010). Rising stream and river temperatures in the United States. *Frontiers in Ecology and the Environment*, 8, 461–66. doi:10.1890/090037.

Keane, R. E., Agee, J. K., Fulé, P., Keeley, J. E., Key, C., Kitchen, S. G., . . . Schulte, L. A. (2008). Ecological effects of large fires on US landscapes: Benefit or catastrophe? *International Journal of Wildland Fire*, 17(6), 696–712. doi:10.1071/WF07148.

Kelly, R. L., Surovell, T. A., Shuman, B. N. & Smith, G. M. (2013). A continuous climatic impact on Holocene human population in the Rocky Mountains. *Proceedings of the National Academy of Sciences USA*, 110(2), 443–47. doi:10.1073/pnas.1201341110.

Kemp, A. & Horton, B. (2013). Contribution of relative sea-level rise to historical hurricane flooding in New York City. *Journal of Quaternary Science*, 28, 537–41. doi:10.1002/jqs.2653.

Kennett, D. J., Breitenbach, S. F. M., Aquino, V. V., Asmerom, Y., Awe, J., Baldini, J. U. L., . . . Haug, G. H. (2012). Development and disintegration of Maya political systems in response to climate change. *Science*, 338(6108), 788–91. doi:10.1126/science.1226299.

Kenny, J. F., Barber, N. L., Hutson, S. S., Linsey, K. S., Lovelace, J. K. & Maupin, M. A. (2009). *Estimated Use of Water in the United States in 2005.* U.S. Geological Survey Circular, 1344, 52.

Kenrick, D. T. & MacFarlane, S. W. (1986). Ambient temperature and horn honking: A field study of the heat/aggression relationship. *Environment and Behavior*, 18(2), 179–91. doi:10.1177/0013916586182002.

Kim, N. K. (2014). Revisiting economic shocks and coups. *Journal of Conflict Resolution*, 29. doi:10.1177/0022002713520531.

Kimmell, T. A. & Veil, J. A. (2009). *Impact of Drought on US Steam Electric Power Plant Cooling Water Intakes and Related Water Resource Management Issues. Technology.* Retrieved from http://www.osti.gov/bridge/product.biblio.jsp?osti_id=951252.

Kjellstrom, T. & Crowe, J. (2011). Climate change, workplace heat exposure, and occupational health and productivity in central America. *International Journal of Occupational and Environmental Health*, 17(3), 270–81. doi:10.1179/107735211799041931.

Knowlton, K., Rotkin-Ellman, M., Geballe, L., Max, W. & Solomon, G. M. (2011). Six climate change-related events in the United States accounted for about $14 billion in lost lives and health costs. *Health Affairs*, 30(11), 2167–76. doi:10.1377/hlthaff.2011.0229.

Knowlton, K., Rotkin-Ellman, M., King, G., Margolis, H. G., Smith, D., Solomon, G., . . . English, P. (2009). The 2006 California heat wave: Impacts on hospitalizations and emergency department visits. *Environmental Health Perspectives*, 117(1), 61–67. doi:10.1289/ehp.11594.

Knutson, T. R., McBride, J. L., Chan, J., Emanuel, K., Holland, G., Landsea, C., . . . Sugi, M. (2010). Tropical cyclones and climate change. *Nature Geoscience* 3, 157–63. doi:10.1038/ngeo779.

Knutson, T. R., Sirutis, J. J., Vecchi, G. A., Garner, S., Zhao, M., Kim, H.-S., . . . Villarini, G. (2013). Dynamical downscaling projections of twenty-first-century Atlantic hurricane activity: CMIP3 and CMIP5 model-based scenarios. *Journal of Climate*, 26(17), 6591–6617. doi:10.1175/JCLI-D-12–00539.1.

Kollat, J. B., Kasprzyk, J. R., Thomas Jr., W. O., Miller, A. C. & Divoky, D. (2012). Estimating the impacts of climate change and population growth on flood discharges in the United States. *Journal of Water Resources Planning and Management*, 138(5), 442–52. doi:10.1061/(ASCE)WR.1943-5452.0000233.

Kopp, R. E., Horton, R. M., Little, C. M., Jerry, X., Oppenheimer, M., Rasmussen, D. J., . . . Tebaldi, C. (2014). Probabilistic 21st and 22nd century sea-level projections at a global network of tide gauge sites. *Earth's Future* 2, 287–306. doi:10.1002/2014EF000239.

Kopp, R. E., Hsiang, S. M. & Oppenheimer, M. (2013). Empirically calibrating damage functions and considering stochasticity when integrated assessment models are used as decision tools. International Conference on Climate Change Effects, Potsdam.

Kopp, R. E. & Mignone, B. K. (2012). The US government's social cost of carbon estimates after their first two years: Pathways for improvement. *Economics*, 6(15), 41. doi:10.5018/economics-ejournal.ja.2012-15.

Kopp, R. E., Simons, F. J., Mitrovica, J. X., Maloof, A. C. & Oppenheimer, M. (2009). Probabilistic assessment of sea level during the last interglacial stage. *Nature*, 462, 863–67. doi:10.1038/nature08686.

Kousky, C., Kopp, R. E., & Cooke, R. M. (2011). Risk premia and the social cost of carbon: a review. *Economics: The Open-Access, Open-Assessment E-Journal*, 5: 1–24.

Kovats, R. S. & Ebi, K. L. (2006). Heatwaves and public health in Europe. *European Journal of Public Health*, 16(6), 592–99. doi:10.1093/eurpub/ckl049.

Krist, F. J., Ellenwood, J. R., Woods, M. E., McMahan, A. J., Cowardin, J. P., Ryerson, D. E., . . . Romero, S. A. (2014). *2013–2027 National Insect and Disease Forest Risk Assessment* (p. 199).

Kumar, S., Kinter, J., Dirmeyer, P. A., Pan, Z. & Adams, J. (2013). Multidecadal climate variability and the "warming hole" in North America: Results from CMIP5 twentieth- and twenty-first-century climate simulations. *Journal of Climate*, 26, 3511–27. doi:10.1175/JCLI-D-12–00535.1.

Kunkel, K. E., Karl, T. R., Brooks, H., Kossin, J., Lawrimore, J. H., Arndt, D., . . . Wuebbles, D. (2013a). Monitoring and understanding trends in extreme storms: State of knowledge. *Bulletin of the American Meteorological Society*, 94, 499–514. doi:10.1175/BAMS-D-11–00262.1.

Kunkel, K.E., Stevens, L. E., Stevens, S. E., Sun, L., Anssen, E., Wuebbles, D., Redmond, K. T., & Dobson, J. G. (2013b). *Regional Climate Trends and Scenarios for the US National Climate Assessment: Part 6. Climate of the Northwest U.S.* Washingon, D.C.: US Department of Commerce; National Oceanic and Atmospheric Administration; and National Environmental Satellite, Data, and Information Service.

Kunreuther, H., Heal, G., Allen, M., Edenhofer, O., Field, C. B. & Yohe, G. (2012). *Risk Management and Climate Change* (No. 2012–16). Philadelphia, Pa.: The Wharton School.

Kuper, R. & Kröpelin, S. (2006). Climate-controlled Holocene occupation in the Sahara: Motor of Africa's evolution. *Science*, 313(5788), 803–807. doi:10.1126/science.1130989.

Landon-Lane, J., Rockoff, H. & Steckel, R. (2011). Droughts, floods and financial distress in the United States. In G. D. Libecap & R. H. Steckel (Eds.), *The Economics of Climate Change: Adaptations Past and Present* (pp. 73–98).

Lane, D., Jones, R., Mills, D., Wobus, C., Ready, R. C., Buddemeier, R. W., . . . Hosterman, H. (2014). Climate change impacts on freshwater fish, coral reefs, and related ecosystem services in the United States. *Climatic Change*. doi:10.1007/s10584-014-1107-2.

Larrick, R. P., Timmerman, T. A., Carton, A. M. & Abrevaya, J. (2011). Temper, temperature, and temptation: Heat-related retaliation in baseball. *Psychological Science*, 22(4), 423–28. doi:10.1177/0956797611399292.

Le Quéré, C., Peters, G. P., Andres, R. J., Andrew, R. M., Boden, T., Ciais, P., . . . Yue, C. (2014). Global carbon budget 2013. *Earth System Science Data*, 6, 235–63. doi:10.5194/essd-6-235-2014.

Leakey, A. D. B. (2009). Rising atmospheric carbon dioxide concentration and the future of C4 crops for food and fuel. *Proceedings of the Royal Society*, 276(1666), 2333–43. doi:10.1098/rspb.2008.1517.

Lenton, T. M., Held, H., Kriegler, E., Hall, J. W., Lucht, W., Rahmstorf, S. & Schellnhuber, H. J. (2008). Tipping elements in the Earth's climate system. *Proceedings of the National Academy of Sciences USA*, 105, 1786–93. doi:10.1073/pnas.0705414105.

Liang, C., Zheng, G., Zhu, N., Tian, Z., Lu, S. & Chen, Y. (2011). A new environmental heat stress index for indoor hot and humid environments based on Cox regression. *Building and Environment*, 46(12), 2472–79. doi:10.1016/j.buildenv.2011.06.013.

Linnerud, K., Mideksa, T. K. & Eskeland, G. S. (2011). The impact of climate change on nuclear power supply. *Energy Jounral*, 32(1), 149–68. doi:10.5547/ISSN0195-6574-EJ-Vol32-No1-6.

Lise, W. & Tol, R. S. J. (2002). Impact of climate on tourist demand. *Climatic Change*, 55, 429–49. doi:10.2139/ssrn.278516.

Littell, J., McKenzie, D., Peterson, D. & Westerling, A. (2009). Climate and wildfire area burned in western US ecoprovinces, 1916–2003. *Ecological Applications*, 19(4), 1003–1021. doi:10.1890/07-1183.1.

Liu, Y., Stanturf, J. & Goodrick, S. (2010). Trends in global wildfire potential in a changing climate. *Forest Ecology and Management*, 259(4), 685–97. doi:10.1016/j.foreco.2009.09.002.

Lloyds. (2013). Wildfire: A burning issue for insurers? *PsycCRITIQUES*, 28, 36. doi:10.1037/021722.

Long, S. P., Ainsworth, E. A., Leakey, A. D. B., Nösberger, J. & Ort, D. R. (2006). Food for thought: Lower-than-expected crop yield stimulation with rising CO2 concentrations. *Science*, 312(5782), 1918–21. doi:10.1126/science.1114722.

Louisiana Geographic Information Center. (2005). *Louisiana Hurricane Impact Atlas*.

Love, G., Soares, A. & Püempel, H. (2010). Climate change, climate variability and transportation. *Procedia Environmental Sciences*, 1, 130–45. doi:10.1016/j.proenv.2010.09.010.

Luce, C. H. & Holden, Z. A. (2009). Declining annual streamflow distributions in the Pacific Northwest United States, 1948–2006. *Geophysical Research Letters*, 36, 6. doi:10.1029/2009GL039407.

Luedeling, E., Zhang, M. & Girvetz, E. H. (2009). Climatic changes lead to declining winter chill for fruit and nut trees in California during 1950–2099. *PLOS ONE*, 4(7), 9. doi:10.1371/journal.pone.0006166.

Luginbuhl, R., Jackson, L., Castillo, D., & Loringer, K. (2008). Heat-Related Deaths Among Crop Workers—United States, 1992–2006. *Journal of the American Medical Association*, 300(9), 1017–18. doi:10.1001/jama.300.9.1017.

Lunt, D. J., Haywood, A. M., Schmidt, G. A., Salzmann, U., Valdes, P. J. & Dowsett, H. J. (2010). Earth system sensitivity inferred from Pliocene modelling and data. *Nature Geoscience*, 3(1), 60–64. doi:10.1038/ngeo706.

Luthi, D., Le Floch, M., Bereiter, B., Blunier, T., Barnola, J.-M., Siegenthaler, U., . . . Stocker, T. F. (2008). High-resolution carbon dioxide concentration record 650,000–800,000 years before present. *Nature*, 453(7193), 379–82. doi:10.1038/nature06949.

Mackworth, N. (1947). High incentives versus hot and humid atmospheres in a physical effort task. *British Journal of Psychology. General Section*, 38(2), 90–102.

Mackworth, N. H. (1948). The breakdown of vigilance durning prolonged visual search. *Quarterly Journal of Experimental Psychology*, 1(1), 6–21. doi:10.1080/17470214808416738.

Mader, T. L. (2003). Environmental stress in confined beef cattle. *Journal of Animal Science*, 81, 110–19.

Manabe, S. & Wetherald, R. T. (1967). Thermal equilibrium of the atmosphere with a given distribution of relative humidity. *Journal of the Atmospheric Sciences*, 24(3), 241–259.

Mares, D. (2013). Climate change and levels of violence in socially disadvantaged neighborhood groups. *Journal of Urban Health*, 90(4), 768–83. doi:10.1007/s11524-013-9791-1.

Marlon, J. R., Bartlein, P. J., Gavin, D. G., Long, C. J., Anderson, R. S., Briles, C. E., . . . Walsh, M. K. (2012). Long-term perspective on wildfires in the western USA. *Proceedings of the National Academy of Sciences USA*, 109(9), 535–43. doi:10.1073/pnas.1112839109.

Masson-Delmotte, V., Schulz, M., Abe-Ouchi, A., Beer, J., Ganopolski, A., González Rouco, J.F., Jansen, E., . . . A. Timmerman (2013). Information from paleoclimate archives. In T. F. Stocker, D. Qin, G.-K. Plattner, M. Tignor, S. K. Allen, J. Boschung, . . . P. M. Midgley (Eds.), *Climate Change 2013: The Physical Science Basis. Contribution of Working Group I to the Fifth Assessment Report of the Intergovernmental Panel on Climate Change* (pp. 383-464). Cambridge: Cambridge University Press.

Mastrandrea, M. D. (2010). Representation of climate impacts in integrated assessment models. In J. Gulledge, L. J. Richardson, L. Adkins & S. Seidel (Eds.), *Assessing the Benefits of Avoided Climate Change: Cost-Benefit Analysis and Beyond* (pp. 85–99). Arlington, Va.: Pew Center on Global Climate Change. Retrieved from www.pewclimate.org/events/2009/benefitsworkshop.

Maulbetsch, J. S. & Difilippo, M. N. (2006). *Cost and Value of Water Use at Combined-Cycle Power Plants*. California Energy Commission, PIER Energy-Related Environmental Research. CEC-500-2006-034.

McCabe, G. J. & Wolock, D. M. (2011). Independent effects of temperature and precipitation on modeled runoff in the conterminous United States. *Water Resources Research*, 47, W11522. doi:10.1029/2011WR010630.

McCrary, J. & Chalfin, A. (2012). *The Effect of Police on Crime: New Evidence from US Cities, 1960–2010*. NBER Working Paper No. 18815.

McGrath, J. M. & Lobell, D. B. (2013). Regional disparities in the CO2 fertilization effect and implications for crop yields. *Environmental Research Letters*, 8(1), 014054. doi:10.1088/1748–9326/8/1/014054.

McInerney, F. A. & Wing, S. L. (2011). The Paleocene-Eocene Thermal Maximum: A perturbation of carbon cycle, climate, and biosphere with implications for the future. *Annual Review of Earth and Planetary Sciences*, 39(1), 489–516. doi:10.1146/annurev-earth-040610-133431.

McLeman, R. A. & Hunter, L. M. (2010). Migration in the context of vulnerability and adaptation to climate change: Insights from analogues. *Wiley Interdisciplinary Reviews: Climate Change*, 1, 450–61. doi:10.1002/wcc.51.

McMichael, A. J., Woodruff, R. E. & Hales, S. (2006). Climate change and human health: Present and future risks. *Lancet*, 367(9513), 859–69. doi:10.1016/S0140-6736(06)68079-3.

Medina-Ramón, M. & Schwartz, J. (2007). Temperature, temperature extremes, and mortality: A study of acclimatisation and effect modification in 50 US cities. *Occupational and Environmental Medicine*, 64(12), 827–33. doi:10.1136/oem.2007.033175.

Meinshausen, M., Meinshausen, N., Hare, W., Raper, S. C., Frieler, K., Knutti, R., . . . Allen, M. R. (2009). Greenhouse-gas emission targets for limiting global warming to 2 C. *Nature*, 458(7242), 1158–62. doi:10.1038/nature08017.

Meinshausen, M., Raper, S. C. B. & Wigley, T. M. L. (2011). Emulating coupled atmosphere-ocean and carbon cycle models with a simpler model, MAGICC6 – Part 1: Model description and calibration. *Atmospheric Chemistry and Physics*, 11(4), 1417–56. doi:10.5194/acp-11-1417-2011.

Mendelsohn, R., Nordhaus, W. D., & Shaw, D. (1994). The impact of global warming on agriculture: A Ricardian analysis. *American Economic Review*, 84(4),753–71.

Menne, M. J., Williams, C. N. & Palecki, M. A. (2010). On the reliability of the US surface temperature record. *Journal of Geophysical Research*, 115, 1–9. doi:10.1029/2009JD013094.

Miguel, E., Satyanath, S. & Sergenti, E. (2004). Economic shocks and civil conflict: An instrumental variables approach. *Journal of Political Economy*, 112(4), 725–53. doi:10.1086/421174.

Millennium Ecosystem Assessment. (2005). *Ecosystems and Human Well-being: Synthesis. World Health* (p. 137). Washington, D.C.: Island Press. doi:10.1088/1755-1307/6/3/432007.

Miller, K. G., Kopp, R. E., Horton, B. P., Browning, J. V. & Kemp, A. C. (2013). A geological perspective on sea-level rise and its impacts along the US mid-Atlantic coast. *Earth's Future*, 1(1), 3–18. doi:10.1002/2013EF000135.

Miller, K. G., Wright, J. D., Browning, J. V., Kulpecz, A., Kominz, M., Naish, T. R., . . . Sosdian, S. (2012). High tide of the warm Pliocene: Implications of global sea level for Antarctic deglaciation. *Geology*, 40(5), 407–410. doi:10.1130/G32869.1.

Mills, G., Buse, A., Gimeno, B., Bermejo, V., Holland, M., Emberson, L. & Pleijel, H. (2007). A synthesis of AOT40-based response functions and critical levels of ozone for agricultural and horticultural crops. *Atmospheric Environment*, 41(12), 2630–43. doi:10.1016/j.atmosenv.2006.11.016.

Mitrovica, J. X., Gomez, N., Morrow, E., Hay, C., Latychev, K. & Tamisiea, M. E. (2011). On the robustness of predictions of sea level fingerprints. *Geophysical Journal International*, 187(2), 729–42. doi:10.1111/j.1365-246X.2011.05090.x.

Molina, M., McCarthy, J., Wall, D., Alley, R., Cobb, K., Cole, J., . . . Shepherd, M. (2014). *What We Know: The Reality, Risks and Response to Climate Change*. Washington, D.C.

Mora, C., Frazier, A. G., Longman, R. J., Dacks, R. S., Walton, M. M., Tong, E. J., . . . Giambelluca, T. W. (2013). The projected timing of climate departure from recent variability. *Nature*, 502(7470), 183–87. doi:10.1038/nature12540.

Moretti, E. & Neidell, M. J. (2009). *Pollution, Health, and Avoidance Behavior: Evidence from the Ports of Los Angeles* (vol. 14939, no. 14939). National Bureau of Economic Research.

Moritz, M., Parisien, M., Batllori, E., Krawchuk, M., Van Dorn, J., Ganz, D. & Hayhoe, K. (2012). Climate change and disruptions to global fire activity. *Ecosphere*, 3(June), 1–22.

Moss, R. H., Edmonds, J. A., Hibbard, K. A., Manning, M. R., Rose, S. K., van Vuuren, D. P., . . . Wilbanks, T. J. (2010). The next generation of scenarios for climate change research and assessment. *Nature*, 463(7282), 747–56. doi:10.1038/nature08823.

Mote, P. W. (2006). Climate-driven variability and trends in mountain snowpack in western North America. *Journal of Climate*, 19, 6209–20. doi:10.1175/JCLI3971.1.

National Academy of Sciences. (2013). *Abrupt Impacts of Climate Change: Anticipating Surprises*. Retrieved from www.nap.edu/catalog.php?record_id=18373.

National Academy of Sciences & The Royal Society. (2014). *Climate Change: Evidence & Causes*. Washington, D.C.: National Academies Press.

National Association of Clean Water Agencies and Association of Metropolitan Water Agencies (2009). *Confronting Climate Change: An Early Analysis of Water and Wastewater Adaptation Costs.* CH2M Hill, Inc.

National Energy Technology Laboratory. (2010). *Estimating Freshwater Needs to Meet Future Thermoelectric Generation Requirements*.

National Research Council. (2010). *Advancing the Science of Climate Change*. Washington, D.C.: National Academies Press.

National Snow and Ice Data Center (2014). *Arctic Sea Ice Reaches Lowest Extent for the Year and the Satellite Record.* Retrieved from http://nsidc.org/news/press/2012_seaiceminimum.html.

Neumann, J., Hudgens, D., Herter, J. & Martinich, J. (2011). The economics of adaptation along developed coastlines. *Wiley Interdisciplinary Reviews: Climate Change*, 2(1), 89–98. doi:10.1002/wcc.90.

Neumann, J. E., Price, J., Chinowsky, P., Wright, L., Ludwig, L., Streeter, R., . . . Martinich, J. (2014). Climate change risks to US infrastructure: Impacts on roads, bridges, coastal development, and urban drainage. *Climatic Change*, 13. doi:10.1007/s10584-013-1037-4.

Nielsen, B., Strange, S., Christensen, N. J., Warberg, J. & Saltin, B. (1997). Acute and adaptive responses in humans to exercise in a warm, humid environment. *Pflugers Archive: European Journal of Physiology*, 434, 49–56. doi:10.1007/s004240050361.

NOAA. (2011). *State of the Climate: Drought for Annual 2011.* Retrieved from www.ncdc.noaa.gov/sotc/drought/2011/13.

NOAA. (2012). *NOAA National Climatic Data Center, State of the Climate: Drought for Annual 2012.*

NOAA. (2013a). *NOAA National Climatic Data Center, State of the Climate: Drought for Annual 2013.*

NOAA. (2013b). *United States Flood Loss Report—Water Year 2011.* Retrieved from www.nws.noaa.gov/hic/summaries/WY2011.pdf.

NOAA National Climatic Data Center. (2013). *Billion-Dollar US Weather/Climate Disasters 1980–2013*. Retrieved from www.ncdc.noaa.gov/billions.

Nordhaus, W. D. (1991). To slow or not to slow: The economics of the greenhouse effect. *Economic Journal*, 101, 920–37.

Nordhaus, W. D. (1994). *Managing the Global Commons: The Economics of Climate Change*. Cambridge, Mass.: MIT Press.

NPCC. (2013). *Climate Risk Information 2013: Observations, Climate Change Projections, and Maps* (pp. 1–38). New York: City of New York. Retrieved from www.nyc.gov/html/planyc2030/downloads/pdf/npcc_climate_risk_information_2013_report.pdf.

O'Loughlin, J., Witmer, F. D. W., Linke, A. M., Laing, A., Gettelman, A. & Dudhia, J. (2012). Climate variability and conflict risk in East Africa, 1990–2009. *Proceedings of the National Academy of Sciences USA*, 109(45), 18344–49. doi:10.1073/pnas.1205130109.

O'Neill, M. S. & Ebi, K. L. (2009). Temperature extremes and health: Impacts of climate variability and change in the United States. *Journal of Occupational and Environmental Medicine*, 51, 13–25. doi:10.1097/JOM.0b013e318173e122.

Olmstead, A. & Rhode, P. (1993). Induced innovation in American agriculture: A reconsideration. *Journal of Political Economy*, 101(1), 100–18.

Olmstead, A. L. & Rhode, P. W. (2011). Adapting North American wheat production to climatic challenges, 1839–2009. *Proceedings of the National Academy of Sciences USA*, 108(2), 480–85. doi:10.1073/pnas.1008279108.

Ortloff, C. & Kolata, A. (1993). Climate and collapse: Agro-ecological perspectives on the decline of the Tiwanaku state. *Journal of Archaeological Science*, 20, 195–221. doi:10.1006/jasc.1993.1014.

Osterkamp, W. R. & Hupp, C. R. (2010). Fluvial processes and vegetation—glimpses of the past, the present, and perhaps the future. *Geomorphology*, 116, 274–85. doi:10.1016/j.geomorph.2009.11.018.

Outdoor Industry Foundation. (2012). *The Outdoor Recreation Economy* (p. 18).

Park, J. & Heal, G. (2014). *Feeling the Heat: Temperature, Physiology and the Wealth of Nations*. NBER Working Paper No. 19725.

Patterson, W. P., Dietrich, K. A., Holmden, C. & Andrews, J. T. (2010). Two millennia of North Atlantic seasonality and implications for Norse colonies. *Proceedings of the National Academy of Sciences USA*, 107(12), 5306–10. doi:10.1073/pnas.0902522107.

Pechony, O. & Shindell, D. T. (2010). Driving forces of global wildfires over the past millennium and the forthcoming century. *Proceedings of the National Academy of Sciences USA*, 107(45), 19167–70. doi:10.1073/pnas.1003669107.

Peterson, D. L. & Littell, J. S. (2012). Risk assessment for wildfire in the western United States. In J. Vose, D. Peterson & T. Patel-Weynand (Eds.), *Effects of Climatic Variability and Change on Forest Ecosystems: A Comprehensive Science Synthesis for the US Forest Sector* (pp. 249–252). U.S. Forest Service.

Pfister, G. G., Walters, S., Lamarque, J. F., Fast, J., Barth, M. C., Wong, J., . . . Bruyère, C. L. (2014). Projections of future summertime ozone over the US. *Journal of Geophysical Research* 119, 5559–82. doi:10.1002/2013JD020932.

Phillips, K., Rodrigue, E. & Yücel, M. (2013). Water scarcity a potential drain on the Texas economy. *Southwest Economy*, 3–7.

Phillips, N. A. (1956). The general circulation of the atmosphere: A numerical experiment. *Quarterly Journal of the Royal Meteorological Society*, 82(352), 123–64.

Pielke, R. A., Gratz, J., Landsea, C. W., Collins, D., Saunders, M. A. & Musulin, R. (2008). Normalized hurricane damage in the United States: 1900–2005. *Natural Hazards*, 9, 29–42. doi:10.1061/(ASCE)1527-6988(2008)9:1(29).

Pierce, D. W., Barnett, T. P., Hidalgo, H. G., Das, T., Bonfils, C., Santer, B. D., . . . Nozawa, T. (2008). Attribution of declining Western US snowpack to human effects. *Journal of Climate*, 21, 6425–44. doi:10.1175/2008JCLI2405.1.

Pindyck, R. S. (2013). Climate change policy: What do the models tell us? (No. w19244). National Bureau of Economic Research.

Pollard, D. & DeConto, R. M. (2009). Modelling West Antarctic ice sheet growth and collapse through the past five million years. *Nature*, 458, 329–32. doi:10.1038/nature07809.

Porter, J. R., Xie, L., Challinor, A., Cochrane, K., Howden, M., Iqbal, M. M., . . . Travasso, M. I. (2014). Food security and food production systems. In C. B. Field, V. Barros, D. J. Dokken, K. J. Mach, M. D. Mastrandrea, T. E. Bilir, . . . L. L. White (Eds.), *Climate Change 2014: Impacts, Adaptation, and Vulnerability*. Contribution of Working Group II to the IPCC Fifth Assessment Report (pp. 485–533). Cambridge: Cambridge University Press.

Praskievicz, S. & Chang, H. (2011). Impacts of climate change and urban development on water resources in the Tualatin River Basin, Oregon. *Annals of the Association of American Geographers*, 101(2), 249–71. doi:10.1080/00045608.2010.544934.

Prudhomme, C., Giuntoli, I., Robinson, E. L., Clark, D. B., Arnell, N. W., Dankers, R., . . . Hannah, D. M. (2013). Hydrological droughts in the 21st century, hotspots and uncertainties from a global multimodel ensemble experiment. *Proceedings of the National Academy of Sciences USA*, 111(9), 3262–67. doi:10.1073/pnas.1222473110.

Pruski, F. & Nearing, M. (2002). Climate-induced changes in erosion during the 21st century for eight US locations. *Water Resources Research*, 38(12), 1298–1308. doi:10.10292001WR000493.

Radeloff, V. C., Hammer, R. B., Steward, S. I., Fried, J. S., Holcomb, S. S. and McKeffrey, J. F. (2005). The Wildland-Urban Interface in the United States. *Ecological Applications*, 15(3), 799–805. doi:10.1890/04-1413.

Rajagopalan, B., Nowak, K., Prairie, J., Hoerling, M., Harding, B., Barsugli, J., . . . Udall, B. (2009). Water supply risk on the Colorado River: Can management mitigate? *Water Resources Research*, 45, 1–7. doi:10.1029/2008WR007652.

Ramsey, J. D. & Morrissey, S. J. (1978). Isodecrement curves for task performance in hot environments. *Applied Ergonomics*, 9(2), 66–72.

Randall, D. A., Wood, R. A., Bony, S., Colman, R., Fichefet, T., Fyfe, J., . . . Taylor, K. E. (2007). Climate models and their evaluation. In Solomon, S., D. Qin, M. Manning, Z. Chen, M. Marquis, K.B. Averyt, M.Tignor and H.L. Miller (Eds.) *Climate Change 2007: The Physical Science Basis. Contribution of Working Group I to the Fourth Assessment Report of the Intergovernmental Panel on Climate Change* (pp. 589–662). Cambridge: Cambridge University Press.

Ranson, M. (2014). Crime, weather, and climate change. *Journal of Environmental Economics and Management*, 1999, 1–29. doi:10.1016/j.jeem.2013.11.008.

Ratnik Industries. (2010). Snowmaking 101. Retrieved from www.ratnik.com/snowmaking.html.

Rausch, S. & Rutherford, T. F. (2008). Tools for Building National Economic Models Using State-Level IMPLAN Social Accounts. Retrieved from www.mpsge.org/IMPLAN2006inGAMS/IMPLAN2006inGAMS.pdf.

Rehfeldt, G. E., Crookston, N. L., Warwell, M. V. & Evans, J. S. (2006). Empirical analyses of plant-climate relationships for the western United States. *International Journal of Plant Science*, 167(6), 1123–50. doi:10.1086/507711.

Reid, C. E., Neill, M. S. O., Gronlund, C. J., Brines, S. J., Brown, D. G. & Diez-roux, A. V. (2009). Mapping community determinants of heat vulnerability. *Environmental Health Perspectives*, 117(11), 1730–6. doi:10.1289/ehp.0900683.

Reilly, J., Paltsev, S., Strzepek, K., Selin, N. E., Cai, Y., Nam, K.-M., . . . Sokolov, A. (2013). Valuing climate impacts in integrated assessment models: The MIT IGSM. *Climatic Change*, 117(3), 561–73. doi:10.1007/s10584-012-0635-x.

Revesz, R. L., Howard, P. H., Arrow, K., Goulder, L. H., Kopp, R. E., Livermore, M. A., . . . Sterner, T. (2014). Global warming: Improve economic models of climate change. *Nature*, 508, 173–75. *doi:10.1038/508173a*.

Riahi, K. (2013). Preliminary IAM scenarios based on the RCP /SSP framework. In *Snowmass Conferences: Climate Change Impacts and Integrated Assessment (CCI/IA)*. Snowmass, Colo.: Energy Modeling Forum. Retrieved from http://web.stanford.edu/group/emf-research/docs/CCIIA/2013/7-31/Riahi IAM SSP & SSP-RCP replications.pdf.

Rignot, E., Mouginot, J., Morlighem, M., Seroussi, H. & Scheuchl, B. (2014). Widespread, rapid grounding line retreat of Pine Island, Thwaites, Smith and Kohler glaciers, West Antarctica from 1992 to 2011. *Geophysical Research Letters*, 41, 3502–9. doi:10.1002/2014GL060140

Rising, J., Hsiang, S. & Kopp, R. (2014). A Tool for Sharing Empirical Models of Climate Impacts. Retrieved from www.existencia.org/pro/files/poster_aggregator.pdf.

Roberts, M. & Schlenker, W. (2011). The evolution of heat tolerance of corn: Implications for climate change. In G. D. Libecap & R. H. Steckel (Eds.), *The Economics of Climate Change: Adaptations Past and Present* (pp. 225–251). Chicago: University of Chicago Press. Retrieved from www.nber.org/chapters/c11988.pdf.

Roberts, B. M. J. & Schlenker, W. (2013). Identifying supply and demand elasticities of agricultural commodities: Implications for the US ethanol mandate. *American Economic Review*, 103(6), 2265–95.

Rogelj, J., Meinshausen, M. & Knutti, R. (2012). Global warming under old and new scenarios using IPCC climate sensitivity range estimates. *Nature Climate Change*, 2(4), 248–53. doi:10.1038/nclimate1385.

Romero-Lankao, P., Smith, J. B., Davidson, D., Diffenbaugh, N., Kinney, P., Kirshen, P., . . . Ruiz, L. V. (2014). North America. In C. B. Field, V. Barros, D. J. Dokken, K. J. Mach, M. D. Mastrandrea, T. E. Bilir, . . . L. L. White (Eds.), *Climate Change 2014: Impacts, Adaptation, and Vulnerability. Contribution of Working Group II to the Fifth Assessment Report of the IPCC* (pp. 1439–98). Cambridge: Cambridge University Press.

Rotton, J. & Cohn, E. G. (2000). Violence is a curvilinear function of temperature in Dallas: A replication. *Journal of Personality and Social Psychology*, 78(6), 1074–81. doi:10.1037//0022-3514.78.6.1074.

Rotton, J. & Cohn, E. G. (2003). Global warming and US crime rates: An application of routine activity theory. *Environment & Behavior*, 35(6), 802–825. doi:10.1177/0013916503255565.

Rovere, A., Raymo, M. E., Mitrovica, J. X., Hearty, P. J., O'Leary, M. J. & Inglis, J. D. (2014). The Mid-Pliocene sea-level conundrum: Glacial isostasy, eustasy and dynamic topography. *Earth and Planetary Science Letters*, 387, 27–33. doi:10.1016/j.epsl.2013.10.030.

Roy, S. B., Chen, L., Girvetz, E. H., Maurer, E. P., Mills, W. B. & Grieb, T. M. (2012). Projecting water withdrawal and supply for future decades in the US under climate change scenarios. *Environmental Science & Technology*, 46, 2545–2556. doi:10.1021/es2030774.

RRC Associates. (2013). *Kottke National End of Season Survey 2012/13* (p. 1).

Sahoo, G. B. & Schladow, S. G. (2008). Impacts of climate change on lakes and reservoirs dynamics and restoration policies. *Sustainability Science*, 3, 189–99. doi:10.1007/s11625-008-0056-y.

Sahoo, G. & Schladow, S.G. (2011). Effects of climate change on thermal properties of lakes and reservoirs, and possible implications. *Stochastic Environmental Research and Risk Assessment*, 25, 445–56. doi:10.1007/s00477-010-0414-z.

Sakurai, G., Iizumi, T. & Yokozawa, M. (2011). Varying temporal and spatial effects of climate on maize and soybean affect yield prediction. *Climate Research*, 49(2), 143–54. doi:10.3354/cr01027.

San Francisco Bay Conservation and Development Commission (2011). *Living with a Rising Bay: Vulnerability and Adaptation in San Francisco Bay and on Its Shoreline*. Staff Report. San Francisco: San Francisco Bay Conservation and Development Commission.

Sathaye, J., Dale, L., Larsen, P., Fitts, G., Franco, G. & Spiegel, L. (2012). *Estimating Risk to California Energy Infrastructure from Projected Climate Change*. Retrieved from www.energy.ca.gov/2012publications/CEC-500-2012-057/CEC-500-2012-057.pdf.

Sathaye, J. A., Dale, L. L., Larsen, P. H., Fitts, G. A., Koy, K., Lewis, S. M. & de Lucena, A. F. P. (2013). Estimating impacts of warming temperatures on California's electricity system. *Global Environmental Change*, 23(2), 499–511. doi:10.1016/j.gloenvcha.2012.12.005.

Schlenker, W. & Roberts, M. J. (2009). Nonlinear temperature effects indicate severe damages to U.S. crop yields under climate change. APPENDIX. *Proceedings of the National Academy of Sciences USA*, 106(37), 15594–98. doi:10.1073/pnas.0906865106.

Schlenker, W., Roberts, M. J. & Lobell, D. B. (2013). Transformation is adaptation: US maize adaptability. *Nature Climate Change* 3(8), 690–91. doi:10.1038/nclimate1959.

Schneider, P. & Hook, S. J. (2010). Space observations of inland water bodies show rapid surface warming since 1985. *Geophysical Research Letters*, 37, 5. doi:10.1029/2010GL045059.

Scholes, R., Settele, J., Betts, R., Bunn, S., Leadley, P., Nepstad, D., . . . Taboada, M. A. (2014). Terrestrial and inland water systems. In C. B. Field, V. Barros, D. J. Dokken, K. J. Mach, M. D. Mastrandrea, T. E. Bilir, . . . L. L. White (Eds.), *Climate Change 2014: Impacts, Adaptation, and Vulnerability*. Contribution of Working Group II to the IPCC Fifth Assessment Report (p. 153). Cambridge: Cambridge University Press.

Scott, D., Amelung, B., Becken, S., Ceron, J.-P., Dubois, G., Gossling, S., . . . Simpson, M. C. (2008). *Change and Tourism: Responding to Global Challenges* (p. 256). Madrid: World Tourism Organization.

Scott, D., McBoyle, G. & Schwartzentruber, M. (2004). Climate change and the distribution of climatic resources for tourism in North America. 27, 105–117.

Scott, D., McBoyle, G., Minogue, A. & Mills, B. (2006). Climate change and the sustainability of ski-based tourism in eastern North America: A reassessment. *Journal of Sustainable Tourism*, 14(4), 376–98. doi:10.2167/jost550.0.

SDG&E. (2007). SDG&E Completes Service Restoration to All Fire-Impacted Customers. San Diego: San Diego Gas & Electric

Seager, R., Ting, M., Li, C., Naik, N., Cook, B., Nakamura, J. & Liu, H. (2013). Projections of declining surface-water availability for the southwestern United States. *Nature Climate Change*, 3, 482–86. doi:10.1038/nclimate1787.

Seki, O., Foster, G. L., Schmidt, D. N., Mackensen, A., Kawamura, K. & Pancost, R. D. (2010). Alkenone and boron-based Pliocene pCO2 records. *Earth and Planetary Science Letters*, 292(1–2), 201–11. doi:10.1016/j.epsl.2010.01.037.

Semenza, J. C., McCullough, J. E., Flanders, W. D., McGeehin, M. A. & Lumpkin, J. R. (1999). Excess hospital admissions during the July 1995 heat wave in Chicago. *American Journal of Preventive Medicine*, 16, 269–77. doi:10.1016/S0749-3797(99)00025-2.

Seneviratne, S. I., Nicholls, N., Easterling, D., Goodess, C. M., Kanae, S., Kossin, J., . . . M. Reichstein, A. Sorteberg, C. Vera, & X. Zhang (2012). Changes in climate extremes and their impacts on the natural physical environment. In C. B. Field, V. Barros, T. F. Stocker, D. Qin, D. J. Dokken, K. L. Ebi, . . . P. M. M. M. Tignor (Eds.), *Managing the Risks of Extreme Events and Disasters to Advance Climate Change Adaptation*. A Special Report of Working Groups I and II of the Intergovernmental Panel on Climate Change (IPCC) (pp. 109–230). Cambridge: Cambridge University Press.

Seppanen, O., Fisk, W. & Lei, Q. (2006). Effect of Temperature on Task Performance in Office Environment. Retrieved from http://escholarship.org/uc/item/45g4n3rv.pdf.

Shepherd, A., Ivins, E. R., Geruo, A., Barletta, V. R., Bentley, M. J., Bettadpur, S., . . . Zwally, H. J. (2012). A reconciled estimate of ice-sheet mass balance. *Science*, 338, 1183–89. doi:10.1126/science.1228102.

Sherwood, S. C. & Huber, M. (2010). An adaptability limit to climate change due to heat stress. *Proceedings of the National Academy of Sciences USA*, 107(21), 9552–55. doi:10.1073/pnas.0913352107.

Skaggs, R., Janetos, T., Hibbard, K. & Rice, J. (2012). *Climate and Energy-Water-Land System Interactions*. Alexandia, Va.: National Technical Information Service.

Spracklen, D. V., Mickley, L. J., Logan, J. A., Hudman, R. C., Yevich, R., Flannigan, M. D. & Westerling, A. L. (2009). Impacts of climate change from 2000 to 2050 on wildfire activity and carbonaceous aerosol concentrations in the western United States. *Journal of Geophysical Research*, 114(D20), D20301. doi:10.1029/2008JD010966.

Stabenau, B. E., Engel, V., Sadle, J. & Pearlstine, L. (2011). Sea-level rise: Observations, impacts, and proactive measures in Everglades National Park. *Park Science*, 28(2), 26–30.

Stern, N. (2013). The structure of economic modeling of the potential impacts of climate change: grafting gross underestimation of risk onto already narrow science models. *Journal of Economic Literature*, 51(3), 838–59.

St-Pierre, N., Cobanov, B. & Schnitkey, G. (2003). Economic losses from heat stress by US livestock industries. *Journal of Dairy Science*, 86(31), E52–E77. doi:10.3168/jds.S0022-0302(03)74040-5.

Strategic Environmental Research and Development Program. (2013). Assessing impacts of climate change on coastal military installations: Policy implications (p. 36). Retrieved from www.serdp.org/Featured-Initiatives/Climate-Change-and-Impacts-of-Sea-Level-Rise.

Strauss, B. H., Ziemlinski, R., Weiss, J. L. & Overpeck, J. T. (2012). Tidally adjusted estimates of topographic vulnerability to sea level rise and flooding for the contiguous United States. *Environmental Research Letters* 7, 014033. doi:10.1088/1748-9326/7/1/014033.

Street, I., Bin, O., Poulter, B. & Whitehead, J. (2007). *Measuring the Impacts of Climate Change on North Carolina*

Coastal Resources. Washington, D.C.: National Commission on Energy Policy.

Stroeve, J., Holland, M. M., Meier, W., Scambos, T. & Serreze, M. (2007). Arctic sea ice decline: Faster than forecast. *Geophysical Research Letters*, 34, L09501. doi:10.1029/2007GL029703.

Stroeve, J. C., Kattsov, V., Barrett, A., Serreze, M., Pavlova, T., Holland, M. & Meier, W. N. (2012). Trends in Arctic sea ice extent from CMIP5, CMIP3 and observations. *Geophysical Research Letters*, 39, L16502. doi:10.1029/2012GL052676.

Tagaris, E., Manomaiphiboon, K., Liao, K.-J., Leung, L. R., Woo, J.-H., He, S., . . . Russell, A. G. (2007). Impacts of global climate change and emissions on regional ozone and fine particulate matter concentrations over the United States. *Journal of Geophysical Research* 112, D14312. doi:10.1029/2006JD008262.

Tai, A. P. K., Mickley, L. J. & Jacob, D. J. (2012). Impact of 2000–2050 climate change on fine particulate matter (PM2.5) air quality inferred from a multi-model analysis of meteorological modes. *Atmospheric Chemistry and Physics*, 12(23), 11329–37. doi:10.5194/acp-12-11329-2012.

Taylor, K. E., Stouffer, R. J. & Meehl, G. A. (2012). An overview of CMIP5 and the experiment design. *Bulletin of the American Meteorological Society*, 93(4), 485–98. doi:10.1175/BAMS-D-11-00094.1.

Tebaldi, C. & Knutti, R. (2007). The use of the multi-model ensemble in probabilistic climate projections. *Philosophical Transactions of the Royal Society A: Mathematical, Physical and Engineering Sciences*, 365(1857), 2053–75. doi:10.1098/rsta.2007.2076.

Tebaldi, C., Strauss, B. H. & Zervas, C. E. (2012). Modelling sea level rise impacts on storm surges along US coasts. *Environmental Research Letters*, 7(1), 014032. doi:10.1088/1748-9326/7/1/014032.

Titus, J. & Richman, C. (2001). Maps of lands vulnerable to sea level rise: Modeled elevations along the US Atlantic and Gulf coasts. *Climate Research* 18(3), 205–28. doi:10.3354/cr018205.

Tol, R. S. J. (1995). The damage costs of climate change: Toward more comprehensive calculations. *Environmental and Resource Economics*, 5(4), 353–74. doi:10.1007/BF00691574.

Trenberth, K. E. & Fasullo, J. T. (2013). An apparent hiatus in global warming? *Earth's Future* 1, 19–32. doi:10.1002/2013EF000165.

Tu, J. (2009). Combined impact of climate and land use changes on streamflow and water quality in eastern Massachusetts, USA. *Journal of Hydrology*, 379(3–4), 268–83. doi:10.1016/j.jhydrol.2009.10.009.

Turney, C. S. M. & Jones, R. T. (2010). Does the Agulhas Current amplify global temperatures during super-interglacials? *Journal of Quaternary Science*, 25(6), 839–43. doi:10.1002/jqs.1423.

University of New Orleans & LSU. (2009). *Louisiana Tourism Forecast: 2009–2013*.

U.S. Bureau of Economic Statistics (2013). *US Travel and Tourism Satellite Accounts for 2009–2012*. Retrieved from http://www.bea.gov/scb/pdf/2013/06%20June/0613_travel_and_tourism_text.pdf.

U.S. Department of Agriculture. (2010). *2007 Census of Agriculture: Farm and Ranch Irrigation Survey* (2008) (vol. 3, p. 201).

U.S. Department of Agriculture. (2013a). *Crop Production 2012 Summary*. Retrieved from http://usda.mannlib.cornell.edu/usda/nass/CropProdSu//2010s/2013/CropProdSu-01-11-2013.pdf.

U.S. Department of Agriculture. (2013b). *US Forest Products*. Retrieved from www.fs.fed.us/research/forest-products.

U.S. Department of Defense. (2010). *Quadrennial Defense Review Report 2010*. Retrieved from www.defense.gov/qdr/qdr as of 29jan10 1600.PDF.

U.S. Department of Defense. (2012). *Base Structure Report Fiscal Year 2012 Baseline*. Washington, D.C.

U.S. Department of Defense. (2014). *Quadrennial Defense Review Report 2014*. Washington, D.C.

U.S. Department of Defense Science Board. (2011). *Trends and Implications of Climate Change for National and International Security*. Retrieved from www.acq.osd.mil/dsb/reports/ADA552760.pdf.

U.S. Department of Energy. (2012a). Hurricane Sandy situation report #5, October 30, 2012. *JAMA Neurology*, 71. doi:10.1001/jamaneurol.2013.4131.

U.S. Department of Energy. (2012b). *Petroleum Reserves*. Retrieved from http://energy.gov/fe/services/petroleum-reserves. Washington, D.C.

U.S. Department of Energy. (2013a). *Comparing the Impacts of Northeast Hurricanes on Energy Infrastructure*. Washington, D.C.

U.S. Department of Energy. (2013b). *US Energy Sector Vulnerabilities to Climate Change and Extreme Weather*. Washington, D.C.

U.S. Department of Homeland Security. (2009). *National Infrastructure Protection Plan: Partnering to Enhance Protection and Resiliency*. Washington, D.C.

U.S. Department of the Interior. (2012). *Colorado River Basin Water Supply and Demand Study*. Washington, D.C.

U.S. Energy Information Administration. (2013). *State Energy Data System 1960–2011*. Retrieved from www.eia.gov/state/seds/seds-data-complete.cfm.

U.S. Environmental Protection Agency. (2008a). *Clean Watersheds Needs Survey 2008 Report to Congress*, 22–23.

U.S. Environmental Protection Agency. (2008b). *Review of the Impacts of Climate Variability and Change on Aeroallergens and Their Associated Effects*. Washington, D.C.

U.S. Environmental Protection Agency. (2009). *Drinking Water Infrastructure Needs Survey and Assessment: Fourth Report to Congress*. Washington, D.C.

U.S. Environmental Protection Agency. (2010). *Guidelines for Preparing Economic Analyses* (vol. 2010). Washington, D.C.

U.S. Environmental Protection Agency. (2012). *Climate Change Indicators in the United States, 2012*. Washington, D.C.

U.S. Environmental Protection Agency. (2013). *Integrated Science Assessment for Ozone and Related Photochemical Oxidants*. Washington, D.C.

U.S. Global Change Research Program. (2009). *Global Climate Change Impacts in the United States*, ed. T. R. Karl, J. M. Melilo & T. C. Peterson. New York: Cambridge University Press.

U.S. Government Accountability Office. (2006). *Hurricane Katrina: Better Plans and Exercises Needed to Guide the Military's Response to Catastrophic Natural Disasters*. Washington, D.C.

U.S. National Research Council Committee on National Security Implications of Climate Change for Naval Forces. (2011). *National Security Implications of Climate Change for US Naval Forces Naval Forces*. National Research Council.

U.S. Navy. (2010). *Navy Climate Change Roadmap*. Retrieved from www.navy.mil/navydata/documents/CCR.pdf.

U.S. Nuclear Regulatory Commission. (2012). *Power Reactor Status Report for 2012*. Retrieved from www.nrc.gov/reading-rm/doc-collections/event-status/reactor-status/2012/#August.

Van Mantgem, P. J., Stephenson, N. L., Byrne, J. C., Daniels, L. D., Franklin, J. F., Fulé, P. Z., . . . Veblen, T. T. (2009). Widespread increase of tree mortality rates in the western United States. *Science*, 323(5913), 521–24. doi:10.1126/science.1165000.

Van Vliet, M. T. H., Ludwig, F., Zwolsman, J. J. G., Weedon, G. P. & Kabat, P. (2011). Global river temperatures and sensitivity to atmospheric warming and changes in river flow. *Water Resources Research*, 47, 19. doi:10.1029/2010WR009198.

Van Vliet, M. T. H., Yearsley, J. R., Ludwig, F., Vögele, S., Lettenmaier, D. P. & Kabat, P. (2012). Vulnerability of US and European electricity supply to climate change. *Nature Climate Change*, 2(9), 676–81. doi:10.1038/nclimate1546.

Van Vuuren, D. P., Edmonds, J., Kainuma, M., Riahi, K., Thomson, A., Hibbard, K., . . . & Rose, S. K. (2011). The representative concentration pathways: an overview. *Climatic Change*, 109, 5–31. doi:10.1007/s10584-011-0148-z.

Vecchi, G. & Knutson, T. (2008). On estimates of historical North Atlantic tropical cyclone activity. *Journal of Climate*, 21(14), 3580–3600. doi:10.1175/2008JCLI2178.1.

Vecchi, G. & Knutson, T. (2011). Estimating annual numbers of Atlantic hurricanes missing from the HURDAT database (1878–1965) using ship track density. *Journal of Climate*, 24(6), 1736–46. doi:10.1175/2010JCLI3810.1.

Vecchi, G. A. & Soden, B. J. (2007). Increased tropical Atlantic wind shear in model projections of global warming. *Geophysical Research Letters* 34, L08702. doi:10.1029/2006GL028905.

Villarini, G., Smith, J. A., Baeck, M. L., Vitolo, R., Stephenson, D. B. & Krajewski, W. F. (2011). On the frequency of heavy rainfall for the Midwest of the United States. *Journal of Hydrology*, 400, 103–20. doi:10.1016/j.jhydrol.2011.01.027.

Vine, E. (2008). Adaptation of California's Electricity Sector to Climate Change. San Francisco: Public Policy Institute of California

Viscusi, W. K. & Aldy, J. E. (2003). *The Value of a Statistical Life: A Critical Review of Market Estimates Throughout the World* (No. 9487). Cambridge, Mass.

Vissing-Jørgensen, A. & Attanasio, O. P. (2003). Stock-market participation, intertemporal substitution, and risk-aversion. *American Economic Review*, 93(2), 383–91.

Vose, J., Peterson, D. & Patel-Weynand, T. (2012). *Effects of Climatic Variability and Change on Forest Ecosystems: A Comprehensive Science Synthesis for the US Forest Sector* (vol. PNW-GTR-87, p. 282). Portland, Ore.

Vose, R. S., Applequist, S., Menne, M. J., Williams, C. N. & Thorne, P. (2012). An intercomparison of temperature trends in the US Historical Climatology Network and recent atmospheric reanalyses. *Geophysical Research Letters*, 39, 6. doi:10.1029/2012GL051387.

Vrij, A., Van der Steen, J. & Koppelaar, L. (1994). Aggression of police officers as a function of temperature: An experiment with the fire arms training system. *Journal of Community & Applied Social Psychology*, 4, 365–370.

Wada, Y., Wisser, D., Eisner, S., Flörke, M., Gerten, D., Haddeland, I., . . . Schewe, J. (2013). Multimodel projections and uncertainties of irrigation water demand under climate change. *Geophysical Research Letters*, 40, 1–7. doi:10.1002/grl.50686.

Walsh, J., Wuebbles, D., Hayhoe, K., Kossin, J., Kunkel, K., Stephens, G., . . . Somerville, R. (2014). Our changing climate. In J. M. Melillo, T. C. Richmond & G. W. Yohe (Eds.), *Climate Change Impacts in the United States: The Third National Climate Assessment* (pp. 19–67). Washington, D.C.: US Global Change Research Program. Retrieved from http://nca2014.globalchange.gov/report/our-changing-climate/introduction.

Walthall, C. L., Hatfield, J., Backlund, P., Lengnick, L., Marshall, E., Walsh, M.,... Ziska, L. H. (2012). *Climate Change and Agriculture in the United States: Effects and Adaptation*. Washington, D.C. Retrieved from www.ars.usda.gov/is/br/CCAgricultureReport02-04-2013.pdf.

Walthall, C. L., Hatfield, J., Backlund, P., Lengnick, L., Marshall, E., Walsh, M., . . . L. H. Z. (2013). *Climate Change and Agriculture in the United States: Effects and Adaptation*. USDA Technical Bulletin, 1935, 186.

Waring, K. M., Reboletti, D. M., Mork, L. A., Huang, C.-H., Hofstetter, R. W., Garcia, A. M., . . . Davis, T. S. (2009). Modeling the impacts of two bark beetle species under a warming climate in the southwestern USA: Ecological and economic consequences. *Environmental Management*, 44(4), 824–35. doi:10.1007/s00267-009-9342-4.

Warren, R. (2011). The role of interactions in a world implementing adaptation and mitigation solutions to climate change. *Philosophical Transactions of the Royal Society A: Mathematical, Physical and Engineering Sciences*, 369(1934), 217–41. doi:10.1098/rsta.2010.0271.

Wayne, P., Foster, S., Connolly, J., Bazzaz, F. & Epstein, P. (2002). Production of allergenic pollen by ragweed (*Ambrosia artemisiifolia L.*) is increased in CO2-enriched atmospheres. *Annals of Allergy, Asthma & Immunology*, 88(3), 279–82. doi:10.1016/S1081-1206(10)62009-1.

Wehner, M. F. (2012). Very extreme seasonal precipitation in the NARCCAP ensemble: Model performance and projections. *Climate Dynamics*, 40, 59–80. doi:10.1007/s00382-012-1393-1.

Wehner, M., Easterling, D. R., Lawrimore, J. H., Heim, R. R., Vose, R. S. & Santer, B. D. (2011). Projections of future drought in the continental United States and Mexico. *Journal of Hydrometeorology*, 12, 1359–77. doi:10.1175/2011JHM1351.1.

Weitzman, M. L. (2009). On modeling and interpreting the economics of catastrophic climate change. *Review of Economics and Statistics*, 91(1), 1–19.

Westerling, A. & Swetnam, T. W. (2003). Interannual to decadal drought and wildfire in the western United States. *Eos, Transactions American Geophysical Union*, 84(49), 545. doi:10.1029/2003EO490001.

Westerling, A. L. & Bryant, B. P. (2008). Climate change and wildfire in California. *Climatic Change*, 87(S1), 231–49. doi:10.1007/s10584-007-9363-z.

Westerling, A., Hidalgo, H. G., Cayan, D. R. & Swetnam, T. W. (2006). Warming and earlier spring increase western US forest wildfire activity. *Science*, 313(5789), 940–43. doi:10.1126/science.1128834.

Westerling, A., Turner, M. G., Smithwick, E. A. H., Romme, W. H. & Ryan, M. G. (2011). Continued warming could transform Greater Yellowstone fire regimes by mid-21st century. *Proceedings of the National Academy of Sciences USA*, 108(32), 13165–70. doi:10.1073/pnas.1110199108.

Weyant, J. & Kriegler, E. (2014). Preface and introduction to EMF 27. *Climatic Change*, 123(3–4).

White, M. C., Etzel, R. A., Wilcox, W. D. & Lloyd, C. (1994). Exacerbations of childhood asthma and ozone pollution in Atlanta. *Environmental Research*, 65, 56–68. doi:S0013-S9351(84)71021-8 [pii]\n10.1006/enrs.1994.1021.

Wilbanks, T., Bilello, D., Schmalzer, D. & Scott, M. (2012a). *Climate Change and Energy Supply and Use.* Washington, D.C.: U.S. Department of Energy.

Wilbanks, T., Fernandez, S., Backus, G., Garcia, P., Jonietz, K., Savonis, M., . . . Toole, L. (2012b). *Climate Change and Infrastructure, Urban Systems, and Vulnerabilities.* Washington, D.C.: U.S. Department of Energy.

Williams, A. P., Allen, C. D., Millar, C. I., Swetnam, T. W., Michaelsen, J., Still, C. J. & Leavitt, S. W. (2010). Forest responses to increasing aridity and warmth in the southwestern United States. *Proceedings of the National Academy of Sciences USA*, 107(50), 21289–94. doi:10.1073/pnas.0914211107.

Wilson, C. O. & Weng, Q. (2011). Simulating the impacts of future land use and climate changes on surface water quality in the Des Plaines River watershed, Chicago Metropolitan Statistical Area, Illinois. *Science of the Total Environment*, 409, 4387–4405. doi:10.1016/j.scitotenv.2011.07.001.

Wood, A. W., Maurer, E. P., Kumar, A. & Lettenmaier, D. P. (2002). Long-range experimental hydrologic forecasting for the eastern United States. *Journal of Geophysical Research*, 107(D20), 15. doi:10.1029/2001JD000659.

Wright, J. D. & Schaller, M. F. (2013). Evidence for a rapid release of carbon at the Paleocene-Eocene thermal maximum. *Proceedings of the National Academy of Sciences USA*, 110(40), 15908–13. doi:10.1073/pnas.1309188110.

Wu, X., Brady, J. E., Rosenberg, H. & Li, G. (2014). Emergency department visits for heat stroke in the United States, 2009 and 2010. *Injury Epidemiology*, 1(8), 1–8. doi:10.1186/2197-1714-1-8.

Wuebbles, D., Meehl, G., Hayhoe, K., Karl, T. R., Kunkel, K., Santer, B., . . . Sun, L. (2013). CMIP5 climate model analyses: Climate extremes in the United States. *Bulletin of the American Meteorological Society* 95, 571–83. doi:10.1175/BAMS-D-12-00172.1.

Wyon, D. (2000). Thermal effects on performance. In J. Spengler, J. M. Samet & J. F. McCarthy (Eds.), *Indoor Air Quality Handbook* (pp. 16.11–16.16). New York: McGraw-Hill.

Yancheva, G., Nowaczyk, N. R., Mingram, J., Dulski, P., Schettler, G., Negendank, J. F. W., . . . Haug, G. H. (2007). Influence of the intertropical convergence zone on the East Asian monsoon. *Nature*, 445(7123), 74–77. doi:10.1038/nature05431.

Yin, J. & Goddard, P. B. (2013). Oceanic control of sea level rise patterns along the East Coast of the United States. *Geophysical Research Letters*, 40, 5514–20. doi:10.1002/2013GL057992.

Yin, J., Schlesinger, M. E. & Stouffer, R. J. (2009). Model projections of rapid sea-level rise on the northeast coast of the United States. *Nature Geoscience*, 2, 262–66. doi:10.1038/ngeo462.

Yohe, G., Neumann, J., Marshall, P. & Ameden, H. (1996). The economic cost of greenhouse-induced sea-level rise for developed property in the United States. *Climatic Change* 32(4), 387–410. doi:10.1007/BF00140353

Yohe, G. & Tirpak, D. (2008). A research agenda to improve economic estimates of the benefits of climate change policies. *Integrated Assessment Journal*, 8(1), 1–17.

Yohe, G. W. (1990). The cost of not holding back the sea: Toward a national sample of economic vulnerability. *Coastal Management*, 18, 403–31.

Yohe, G. W. & Schlesinger, M. E. (1998). Sea-level change: The expected economic cost of protection or abandonment in the United States. *Climatic Change*, 38, 447–72. doi: 10.1023/A:1005338413531.

Youngblut, C. (2009). *Climate Change Effects: Issues for International and US National Security*. Retrieved from www.dtic.mil/cgi-bin/GetTRDoc?AD=ADA527880.

Zachos, J. C., Dickens, G. R. & Zeebe, R. E. (2008). An early Cenozoic perspective on greenhouse warming and carbon-cycle dynamics. *Nature*, 451(7176), 279–83. doi:10.1038/nature06588.

Zanobetti, A., O'Neill, M. S., Gronlund, C. J. & Schwartz, J. D. (2012). Summer temperature variability and long-term survival among elderly people with chronic disease. *Proceedings of the National Academy of Sciences USA*, 109(17), 6608–13. doi:10.1073/pnas.1113070109.

Zhang, D. D., Jim, C. Y., Lin, G. C.-S., He, Y.-Q., Wang, J. J. & Lee, H. F. (2006). Climatic change, wars and dynastic cycles in China over the last millennium. *Climatic Change*, 76(3–4), 459–77. doi:10.1007/s10584-005-9024-z.

Zhang, F. & Georgakakos, A. P. (2012). Joint variable spatial downscaling. *Climatic Change*, 111, 945–72. doi:10.1007/s10584-011-0167-9.

Ziska, L. H. (2008). Climate change, aerobiology, and public health in the Northeast United States. *Mitigation and Adaptation Strategies for Global Change*, 13, 607–13.

Ziska, L. H. (2010). Global climate change and carbon dioxide: Assessing weed biology and management. In C. and D. Rosenzweig & Hillel (Eds.), *Handbook of Climate Change and Agro-Ecosystems: Impacts, Adaptation and Mitigation* (pp. 191–208). Hackensack, N.J.: World Scientific Publishing.

Ziska, L. H., Teasdale, J. R. & Bunce, J. A. (1999). Future atmospheric carbon dioxide may increase tolerance to glyphosate. *Weed Science*, 608–615. doi:10.2307/4046118.

Ziska, L., Knowlton, K., Rogers, C., Dalan, D., Tierney, N., Elder M.A., Filley, W., . . .& Frenz, D. (2011). Recent warming by latitude associated with increased length of ragweed pollen season in central North America. *Proceedings of the National Academy of Sciences* 108(10), 4248–4251.

Zumbach, B., Misztal, I., Tsuruta, S., Holl, J., Herring, W., & Long, T. (2007). Genetic correlations between two strains of Durocs and crossbreds from differing production environments for slaughter traits. *Journal of Animal Science*, 85(4), 901–8.

ABOUT THE AUTHORS

PHYSICAL CLIMATE PROJECTIONS

DR. ROBERT KOPP is an associate professor of earth and planetary sciences and associate director of the Rutgers Energy Institute at Rutgers University. His research focuses on understanding uncertainty in past and future climate change.

D. J. RASMUSSEN is a scientific programmer at the Rhodium Group with expertise in weather, climate, and air pollution.

DR. MICHAEL MASTRANDREA is an assistant consulting professor at the Stanford University Woods Institute for the Environment and Co-Director of Science for the Intergovernmental Panel on Climate Change (IPCC) Working Group II Technical Support Unit.

DETAILED SECTORAL MODELS

DR. ROBERT MUIR-WOOD is the chief research officer of the catastrophe risk management firm Risk Management Solutions, Inc.

DR. PAUL WILSON is a senior director at Risk Management Solutions, Inc. He is in the Risk Management Solutions (RMS) model development team, leading the ongoing development of the RMS North Atlantic Hurricane and Storm Surge Models.

ECONOMETRIC RESEARCH

DR. SOLOMON HSIANG is an assistant professor of public policy at the University of California Berkeley's Goldman School of Public Policy and a faculty research fellow at the National Bureau of Economic Research. Dr. Hsiang is at the forefront of using econometrics to understand the social impact of climate change.

AMIR JINA is a postdoctoral scholar in Economics at the University of Chicago working on development and environmental economics.

JAMES RISING is a doctoral candidate in sustainable development and modeling at Columbia University.

INTEGRATED ECONOMIC ANALYSIS

MICHAEL DELGADO is a research analyst on the energy and natural resources team at the Rhodium Group.

SHASHANK MOHAN is a director at the Rhodium Group and leads the development and management of the company's suite of economic models and other quantitative tools.

PROJECT MANAGEMENT

KATE LARSEN is a director at the Rhodium Group and manages the firm's work on domestic and international climate-change issues.

TREVOR HOUSER is partner at the Rhodium Group, leading the firm's energy and natural resources work and is a visiting fellow at the Peterson Institute for International Economics.

COMMENTATORS

DR. KAREN FISHER-VANDEN is professor of environmental and resource economics at the Penn State College of Agricultural Sciences.

DR. MICHAEL GREENSTONE is the Milton Friedman Professor of Economics at the University of Chicago and director of the interdisciplinary Energy Policy Institute at Chicago (EPIC). His other current positions and affiliations include elected member of the American Academy of Arts and Sciences, editor of the *Journal of Political Economy*, faculty director of the E2e Project, head of the JPAL Environment and Energy Program, co-director of the International Growth Centre's Energy Research Programme, and nonresident senior fellow in economic studies at the Brookings Institution.

DR. GEOFFREY HEAL is the Donald C. Waite III Professor of Social Enterprise at Columbia Business School. He is a fellow of the Econometric Society, past president of the Association of Environmental and Resource Economists, recipient of its prize for publications of enduring quality and a life fellow, a director of the Union of Concerned Scientists, and a founder and director of the Coalition for Rainforest Nations.

DR. MICHAEL OPPENHEIMER is the Albert G. Milbank Professor of Geosciences and International Affairs in the Woodrow Wilson School and the Department of Geosciences at Princeton University. He is the director of the Program in Science, Technology, and Environmental Policy (STEP) at the Woodrow Wilson School and faculty associate of the Atmospheric and Ocean Sciences Program, Princeton Environmental Institute, and the Princeton Institute for International and Regional Studies. He is a member of the National Academies' Board on Energy and Environmental Studies.

LORD NICHOLAS STERN is the IG Patel Professor of Economics and Government and chair of the Grantham Institute for Climate Change and the Environment at the London School of Economics. He was knighted for services to economics in 2004 and became the 29th president of the British Academy in July 2013.

DR. BOB WARD is the policy and communications director at the Grantham Institute for Climate Change and the Environment at the London School of Economics. He is a fellow of the Geological Society.

FOREWORD AUTHORS

MICHAEL R. BLOOMBERG is the founder of the global financial-data services and media company Bloomberg LP. Between 2002 and 2013, he served as mayor of New York City and reduced the city's carbon footprint by 19 percent, revitalized the waterfront, implemented ambitious public-health and anti-poverty programs, expanded support for arts and culture, and increased graduation rates and private-sector job numbers to record highs. As a philanthropist, Bloomberg has given more than $3.3 billion in support of education, public health, government innovation, the arts, and the environment. In 2014, United Nations Secretary-General Ban Ki-Moon appointed Bloomberg special envoy for cities and climate change. In that role, Bloomberg works to highlight the climate work cities are doing and the critical role mayors can play in helping nations create ambitious carbon-reduction commitments. This builds on his role as president of the board of the C40 Climate Leadership Group, a network of megacities working to reduce global greenhouse-gas emissions. Bloomberg also serves as a cochair of Risky Business, an organization that is quantifying the economic risks American businesses face from climate change.

HENRY M. PAULSON JR. is a businessman, China expert, conservationist, and author. He is chairman of the Paulson Institute, which works to strengthen U.S.-China relations, and cochairman of the Risky Business Project and the Latin American Conservation Council of the Nature Conservancy. Paulson served as the 74th secretary of the Treasury under President George W. Bush and, previously, was chairman and chief executive officer of Goldman Sachs. He graduated from Harvard Business School and Dartmouth College.

THOMAS F. STEYER is an investor, philanthropist, and advanced-energy advocate. Before retiring from the private sector, Steyer founded and was the senior managing member of Farallon Capital Management. Steyer also founded several organizations aimed to accelerate the transition to an advanced-energy future, including Advanced Energy Economy, Center for the Next Generation, NextGen Climate, and Beneficial State Bank. Steyer serves on Stanford's board of trustees, where he and his wife founded the TomKat Center for Sustainable Energy and the Steyer-Taylor Center for Energy Policy and Finance.

INDEX